[illegible] ngs are cubs. (Source: (a) [illegible]

ENERGY EFFICIENCY IN PROCESS TECHNOLOGY

Proceedings of the International Conference on Energy Efficiency in Process Technology, held in Athens, Greece, 19–22 October 1992; organized by the Commission of the European Communities (DG XII) with the participation of the Centre for Renewable Energy Sources, the European Federation of Chemical Engineering, the European Chemical Industry Council, the Aluminium of Greece and Eurotherm.

ENERGY EFFICIENCY IN PROCESS TECHNOLOGY

Edited by

P. A. PILAVACHI

Commission of the European Communities, Brussels, Belgium

ELSEVIER APPLIED SCIENCE
LONDON and NEW YORK

ELSEVIER SCIENCE PUBLISHERS LTD
Crown House, Linton Road, Barking, Essex IG11 8JU, England

WITH 156 TABLES AND 709 ILLUSTRATIONS

British Library Cataloguing in Publication Data

Energy Efficiency in Process Technology
I Pilavachi, P A
660

ISBN 1-85861-019-2

Library of Congress CIP data applied for

Publication No EUR 14594 EN of the Commission of the European Communities Dissemination of Scientific and Technical Knowledge Unit, Directorate-General Information Industries and Innovation, and Telecommunications, Luxembourg

PREFACE

The main objective of the Community's energy policy consists in securing a sufficient energy supply to meet the present and future demand of its Member States, and in reducing the Community's dependence on imported energy. This is the reason why, since 1975, the Commission has been stimulating universities, industries and national laboratories of the EC Member States to perform R & D work aimed at energy saving. In spite of the temporary fall in oil prices, energy-saving measures will continue to remain imperative, as resources will continue to diminish rapidly. Research, development and demonstration effort in this area must therefore be maintained.

In this context, the Commission of the European Communities organized an International Conference on Energy Efficiency in Process Technology, with the participation of the Centre for Renewable Energy Sources, the European Federation of Chemical Engineering, the European Chemical Industry Council, the Aluminium of Greece and Eurotherm.

The Conference objective was to provide an international forum for the presentation and discussion of recent R & D relevant to energy efficiency, taking into account environmental aspects, in the energy-intensive process industries. A wide range of industrial sectors was covered, including new processes and equipment. Projects carried out within the present European Community JOULE Programme were included. The following is a partial list of topics addressed:

- Chemical Reactors
- Heat Exchangers
- Separation Processes
- Furnaces, Kilns and Ovens
- Combustion
- Process Integration
- Dynamic Simulation and Batch Processes
- Exergy Analysis
- Thermodynamic Cycles
- Efficient Production and Use of Electricity
- Sensors and Instrumentation

P. A. PILAVACHI

Conference Organization
Dr P. A. Pilavachi, DG XII, Commission of the European Communities

Scientific Committee
Prof. R. W. K. Allen, UKAEA—Harwell Laboratory, UK
Prof. D. Behrens†, Dechema, Germany
Prof. A. E. Bergles, Rensselaer Polytechnic Institute, USA
Dr E. N. Carabateas, General Secretariat for Research and Technology, Greece
Prof. M. da Graça Carvalho, Instituto Superior Técnico, Portugal
Prof. K. Cornwell, Heriot-Watt University, UK
Mr R. C. Dumon, CEC expert, France
Prof. G. Froment, Rijksuniversiteit Gent, Belgium
Prof. M. Groll, Universität Stuttgart, Germany
Prof. E. P. Gyftopoulos, Massachusetts Institute of Technology, USA
Prof. B. Kalitventzeff, Université de Liège, Belgium
Prof. N. Koumoutsos, National Technical University of Athens, Greece
Prof. N. N. Kulov, Russian Academy of Sciences, Russia
Prof. P. Le Goff, Institut National Polytechnique de Lorraine, France
Prof. Z. Leszczynski, Industrial Chemistry Research Institute, Poland
Prof. E. Macchi, Politecnico di Milano, Italy
Mr A. Mercer, ETSU, UK
Prof. H. L. J. Meunier, Faculté Polytechnique de Mons, Belgium
Prof. S. Pierucci, Politecnico di Milano, Italy
Dr P. A. Pilavachi, DG XII, Commission of the European Communities
Prof. K. E. Porter, University of Aston, UK
Prof. D. A. Reay, CEC expert, UK
Prof. S. S. Stecco, University of Florence, Italy
Mr B. P. ter Meulen, TNO, The Netherlands
Dr P. Trambouze, Institut Français du Pétrole, France
Dr P. Tzannetakis, Motor Oil, Greece
Prof. K. R. Westerterp, University of Twente, The Netherlands
Mr P. Zegers, DG XII, Commission of the European Communities

Guests of Honour
Mr Ioannis Paleokrassas
Minister of Industry, Energy, Technology and Commerce, Greece

Mr Georgios Contogeorgis
Former Minister for the National Economy, Greece
Former Member of the Commission of the European Communities

Administrative Issues
Mr G. Weidenbach, DG XII, Commission of the European Communities

From left to right: Mr G. Gorgias, Mr G. Contogeorgis, Dr P. A. Pilavachi, Mr I. Paleokrassas, Mr C. Nicolaóu, Mr N. Tavoularis. In the background: Dr E. Carabateas.

Conference Hall.

Plenary Session Dr P A Pilavachi

Plenary Session Mr P Zegers

Plenary Session: Mr I. Paleokrassas.

Plenary Session: Prof. N. Chrysochoides.

Plenary Session (front view)

Plenary Session (rear view)

CONTENTS

Session 3: Drying

Chairman: Prof. R. W. K. ALLEN (*UKAEA—Harwell Laboratory, UK*)

Session 4: Sensors and Instrumentation

Chairman: Prof. F. DURST (*Universität Erlangen-Nürnberg, Germany*)

Session 7: Heat Exchangers
Chairman: Prof. D. A. Reay (*CEC Expert, UK*)

Enhancement of Heat Transfer

Session 7: Heat Exchangers (continued)
Chairman: Prof. M. Groll (*Universität Stuttgart, Germany*)

Advanced Heat Exchangers

* Text in French.

Session 10: Process Integration
Chairman: Prof. S. PIERUCCI (*Politecnico di Milano, Italy*)

Session 10: Process Integration (continued)
Chairman: Mr J. KOSMADAKIS (*Motor Oil, Greece*)

Session 13: Chemical Reactors
Chairman: Prof. G. FROMENT (*Rijksuniversiteit Gent, Belgium*)

Session 13: Chemical Reactors (continued)
Chairman: Prof. K. R. WESTERTERP (*University of Twente, The Netherlands*)

Session 14: New Process Routes
Chairman: Prof. K. R. WESTERTERP (*University of Twente, The Netherlands*)

Session 15: Exergy Analysis
Chairman: Prof. E. P. GYFTOPOULOS (*Massachusetts Institute of Technology, USA*)

SESSION 8:

Furnaces, Kilns and Ovens

Chairman: Mr R. Dumon

ENERGY EFFICIENCY AND POLLUTION ABATEMENT IN GLASS FURNACES, BAKING OVENS AND CEMENT KILNS

M. G. CARVALHO and M. NOGUEIRA
Instituto Superior Técnico, Mechanical Engineering Department
Av . Rovisco Pais, 1096 Lisbon Codex, Portugal

ABSTRACT

The use of physically-based dedicated mathematical modelling for combustion chambers, load, preheaters and other associated equipment of ovens, kilns and furnaces, together with the implementation of advanced on-line sensors, may provide detailed knowledge of the very complex phenomena occurring in those industrial equipment. Application of this knowledge to the study of new design and operation concepts, turns possible optimization, integration and intensification of process furnaces. In the present paper this approach is discussed, together with research priorities, objectives and major achievements at present technological stage. Relevant developments under current joint European research project JOUE-0051-C "Energy Saving and Pollution Abatement in Glass-making Furnaces, Cement Kilns and Baking Ovens" are referred.

PARTICIPANT INSTITUTIONS

Participant Institutions in the JOULE Project "Energy Saving and Pollution Abatement in Glass Making Furnaces, Cement Kilns and Baking Ovens", Contract JOUE-0051-C:

Prof. M.G. Carvalho/Ing. M. Nogueira	Instituto Superior Técnico/Technical University of Lisbon — I.S.T. - Coordinator
Prof. Fred Lockwood	Imperial College of Science Technology & Medicine — ICST&M
Prof. Franz Durst/Dipl.-Ing. G. Dimaczek	University of Erlangen-Nuremberg — LSTM
Prof. Hector Meunier/Dr. P. Lybaert	Faculté Polytechnique de Mons — FPM
Prof. René Jottrand	Université Libre de Bruxelles — ULB
Dr. Christos Papadopoulos	Centre for Renewable Energy Sources — CRES
Dr. P. Verlaan/Dr. Ubo de Vries	TNO Nutrition and Food Research
Ir. Henk van Deventer	TNO Institute Environmental and Energy Research
Eng. Nelson Martins	Instituto da Energia — INTERG
Ir. Jean-François Bassine	Institut National Belge du Verre — INV
Dr. J. Richalet/Dr. Papon	ADERSA
Ing. Barros da Silva	Metal Portuguesa
Dr. Dimitri Hadjicostantis	TITAN Cement Company

INDUSTRIAL NEEDS FOR ENERGY SAVING

European industry is facing a sustained pressure to improve its productivity and product quality together with strong constraints about pollutant emissions and energy consumption. A positive response has to be based on optimization, intensification and integration of equipment and procedures. Improvements in energy efficiency, product quality, pollution abatement in ovens, kilns and furnaces are part of this efforts, but they cannot be achieved without an effective physically-based knowledge and understanding of the relevant thermal phenomena. This approach has to consider specific aspects of the industrial process itself, its interactions with the heating chamber (electrically heated or by combustion) and heat transfer efficiency. Improvements in the workability and quality of the products (reducing the level of rejections and increasing the process efficiency) and production intensification constitute an effective way for energy saving and pollution abatement. The efficiency of the heat recovering systems also cannot be neglected. Primary and secondary measures have to be studied in an integrated basis for whole system optimization.

The increasing awareness of the pollution constraints, expressed in severe legal restrictions, turns the emissions control a significant factor in the plant overall energy efficiency. On one hand the use of pollution control devices is, in general, damaging for the efficiency of the plant; On the other hand the improvement of the process and equipment efficiency will directly reduce the specific pollutant emission rate. Following this approach, pollution abatement may be obtained through:

- improving the performance of the thermal process (allowing the reduction of energy requirements);
- acting on the combusting process in order to decrease the emission of pollutants generated by combustion such as CO, NO_x and some classes of particles.
- using improved geometry arrangement and controlling strategies able to reduce chemical and physically generated emissions.

Low-cost pollution abatement is therefore attainable.

A POSSIBLE APPROACH

Mathematical modelling, based on computational fluid dynamics, and in particular 3-D modelling has to be explored as a powerful alternative to the empirical rules and global balance based steady-state modelling. 3-D modelling provides, at reasonable cost, a way to access the spatial distributions of main fields inside combustion chamber, electric heating chambers, heat recovering devices and load. These features allow an effective evaluation of the consequences for the process performance of modifications in operating conditions and design. A wide range for optimization is therefore open to explore primary and secondary effects. However, the application of computational fluid dynamics techniques to full scale industrial equipment remains a topic far from being closed since the physical description of the phenomena in the integrated system (heating chamber, heat recovering devices, load/furnace or kiln interaction) remains limited. Presently, general modelling procedures for fluid flow and heat transfer predictions in furnaces and kilns are well established. New developments should be aimed to explore advantages from a complete system representation, and control/operation and design integrated study.

Dedicated modelling allows an integrated representation of the furnaces, kilns and ovens systems with the consideration of physical interactions between the various sub-systems (load, heating chamber, heat recovery devices) and associated phenomena. Due to physically-based character of this approach, significant general know-how and experience may be transferred from and to several application fields and industrial sectors.

The continuous development of new sensors, the enlargement of on-line sensor capacities to traditionally non accessible variables and the application of recently developed techniques will allow **massive sensoring** of the analyzed equipment. An extensive characterization of the process will support and improve physical understanding of the phenomena, extended validation of modelling procedures. Possibilities for the application of new operation and design concepts, based on the control of traditionally non measurable critical variables, are therefore open.

The use of the above referred concepts, dedicated modelling and massive sensoring, for the control design and operation optimization in integrated design basis will turn possible production intensification, improvement of energy utilization efficiency, product workability, equipment maintainability reduction of energy costs of pollutant emissions abatement.

RESEARCH PRIORITIES AND EXPECTED ACHIEVEMENTS

A main goal of the research efforts involved in the optimization of process furnaces should be to develop engineering tools and knowledge able to assist the optimization of design, control and operation of kilns, ovens and furnaces. This goal will be achieved through the use of sophisticated computational modelling, advanced massive sensoring and its application to enhanced model based control strategies (integrating newly sensored variables and knowledge based concepts), as well as novel design solutions.

This approach has been followed in the current joint European research project *"Energy Saving and Pollution Abatement in Glass-Making Furnaces, Cement Kilns and Baking Ovens"* under the JOULE programme of DGXII, JOUE-0051-C (SMA). Modeling procedures are under development and validation enabling its use for a variety of furnaces, kilns and ovens.

The research objectives of this project may be stated as follows:

- Development and application of advanced sensors in harsh environments allowing the access to non conventional controllable variables and extensive characterization of the process.
- Development and validation of dedicated physically-based models of ovens, kilns and furnaces (comprising associated equipment and load) through the use of multidimensional and dynamically approaches. The modelling development may embody information obtained through the massive characterization of the equipment. Validation of the developed models have to be achieved through experimental characterization with advanced sensors considered in the previous objective..
- Study of more energy efficient design and operation concepts applying modelling facilities and embodying the new sensor capacities. Lower cost pollutant abatement possibilities can be concurrently considered.

BASIS FOR A DEDICATED MODELING APPROACH

Advances in computer power allied to advances in computational fluid dynamics science together with improved understanding of radiative heat transfer phenomena and chemical reactions have opened the possibility to obtain a detailed description of the process occurring in industrial furnaces. Several works reviewed below, had a critical importance in the development of well established procedures to simulate the main physical phenomena occurring in the thermal equipment.

Abou Ellail et. al. (1978) described a prediction method for three-dimensional reacting flows where a flux model for the thermal radiation was employed. The prediction procedure was applied to two industrial-type experimental furnaces from the French national gas corporation, "Gaz de France", in Toulouse and from the International Flame Research Foundation, IFRF, in The Netherlands. The fuel in both instances was natural gas.

A three-dimensional simulation of a combustion chamber was presented by Gosman et. al. (1980) in which a more accurate and efficient technique for the handling of the thermal radiation was used - the "Discrete Transfer" method of Lockwood and Shah (1980). The developed prediction procedure was applied to a cross-fired glass furnace burning U.K. North Sea natural gas.

Carvalho and Lockwood (1985) presented the computer simulation of an end-port regenerative furnace burning natural gas or oil. The completely three-dimensional character or the flow inside the combustion chamber was considered in the model. The continuity, momentum and energy, turbulent quantities, mixture fraction and it is variance transport equations were solved through the finite difference/finite volume method. The radiative heat fluxes were calculated through the discrete transfer method.

A similar procedure has applied by Semião (1986) to a ceramic glass smelting-kiln with oxygen-rich burning conditions. Carvalho et. al. (1987) have extended that work and used a two-dimensional axisymmetric model to simulate the burner region, providing with these results the inlet conditions for the three-dimensional calculations of the combustion chamber. The results were extensively validated with experimental data acquired in the furnace (Carvalho et. al., 1988). This model is based in the numerical solution of governing transport equations for momentum, mass, energy and species concentration besides a discrete transfer approach to solve the radiate heat transfer. The model was used to optimize the furnace operating conditions and, for example, the results have shown an 18% improvement in energy efficiency when oxygen enrichment is applied.

Carvalho, Lockwood, Papadopoulos and Semião (1990) presented a NOx formation / dissociation model based on the Zeldovich mechanism. This procedure was integrated in a three-dimensional flow and heat transfer model and applied to a side-port glass furnace.

Wieringa, PhElich and Hoogendoorn (1990) presented the modelling of a combustion chamber of a side-port glass furnace through a spectral solution of the radiative heat transfer in the combustion chamber enclosure. A zone model based procedure was used for the calculation of the radiative fluxes.

The application of modelling procedures in the design/control/operation optimization of industrial furnaces, namely on an integrated design basis, requires modelling procedures able to handle with the overall furnace system. Following this approach, modelling procedure for the calculation of 3-D fluid flow and heat transfer phenomena in an industrial furnace, considering through an integrated approach the combustion chamber (turbulent flow, thermal radiation, pollutants formation) and load (laminar buoyancy-driven glass-melt flow), has been recently developed (see for details Carvalho and Nogueira, 1992). The modelling procedure was applied to a medium-size end-port glass furnace.

A dynamically zero-one dimensional approach is proposed by Viskanta, Chapman and Ramadmyani (1990) to model gas fired batch and continuos recuperative furnaces. Particular attention was given to the radiative properties of the load and combustion products. A Hottel-zone-model based procedure was used to predict the radiative fluxes. The direct-flux-exchange-areas were calculated using a Monte-Carlo method.

A dynamical modelling procedure was presented by Meunier and Cambier (1991) for the simulation of the batch furnaces behaviour. The proposed procedure has been applied to help the designer in the conception of improved geometries and burner selection. Optimized heat transfer rates, and temperature schedules, meeting the constraints imposed at every point of the load, may be attained. The application of this simulator also allows the knowledge of temperature at the coldest points of the load where the temperature is not available by measurements.

A transient modelling procedure to solve a car-hearth metal heat treatment kiln was solved by Jicha and Vacenovsky (1992). Three heating zones were considered. Heat transfer conductive, radiative and convective heat transfer procedure was solved in basis of the SIMPLE algorithm.

Research in cement kilns process has been directed on flames and combustion phenomena as lower-grade coals and wastes will be used increasingly, and environmental regulations will require NO_x levels to be reduced. Mathematical models to simulate the heat transfer phenomena inside a cement kilns have been described in the literature namely by Ghoshdastidar and Unni 1989 and Kim, Lyon and Suryanarayana 1986. Work on coal combustion, burner aerodynamics and NOx generation to the cement industry have been extensively performed by Lockwood and Salooja, 1993 and Visser and Weber 1989.

Optimization of the energy efficiency in food industry has been in general limited by variety of complex geometries and configurations of the utilized thermal systems, by the small size of the majority of the plants and by the complexities of the involved phenomena together with the priority given to the product quality. Recently research attention has been directed to this subject as reviewed by Rask 1989. Numerical predictions of the baking process appear in the literature, DeVries, Sluimer and Bloksma (1988) and Carvalho and Martins (1991). This work considered baking oven heating processes through a detailed model of both oven and bread domains. The model incorporates a three-dimensional solution for turbulent flow, radiative and convective heat transfer in the baking chamber, and the heat transfer and "oven-rise" in the bread.

The flexibility of the above referred procedures can be clearly demonstrated by applications of similar three-dimensional codes to calculate the flow in industrial boilers. Robinson (1985) has presented a three-dimensional model for a large tangential-fired furnace of the type used in power-station boilers. In this work a six-flux model was used for the evaluation of radiate heat transfer. The model was strictly applicable only to gaseous- fueled furnaces. Abbas and Lockwood (1986) described the application of a fully three-dimensional mathematical model to the combustors of two large power station boilers: one front wall-fired and other corner-fired. Predictions of the aerodynamics of the flow were compared with experimental data obtained in cold models and combustion and radiation heat transfer characteristics were calculated for the corner-fired case using the "Discrete Transfer" method. The results were in fairly good agreement with the model data, suggesting the reasonable accuracy of the modelling procedures.

Other significant works in the boiler numerical simulation has been presented, namely the following in recent publications: Görner and Zinzer (1986), Sargianos, Agagnostopoulos and Bergeles (1990), De Michele, Pasini and Tozzi (1991) and Carvalho and Coelho (1991).

An extensive review of the numerous furnace modelling applications and type of models was presented by Meunier (1991).

However, in spite of the progress made during the last decade, mathematical models of furnace, kiln and ovens design have been far from sufficiently employed in the quest for improved furnace design. There would seem to be four principal reasons for this:

- in the early period of mathematical modelling development, the methods were insufficiently developed to represent a substantial improvement on existing design procedures.

- later on, when mathematical modelling emerged as a technique which could considerably assist the furnace design engineer, the involved sub-models are sophisticated and their use still demanding a level of commitment which industry, for the most part, was no prepared to make.
- lack of validation for the full scale furnace. Most models published in the literature have been validated at laboratory and semi-industrial scale. Such efforts are not without value but it is now recognized that its worth is very limited, because the several physical processes occurring in combustion chambers scale in different ways.
- the multi-dimensional prediction methods are highly computer time consuming. This is due to the geometrical scales disparity and the complexity of fluid mechanical, chemical and heat transfer processes.

Dedicated modelling approach aims to respond to the second and fourth reasons. A careful modelling representation of the actual furnace, oven or kiln system, and a correct exploitation of the specific aspects of each process will allow the set up of a faster modelling implementation and a significant reduction of the lead time needed to train the industrial user to have useful results. **Massive sensoring** will allow a correct validation of dedicated modelling codes against industrial data. Prediction codes capable of computing the three-dimensional characteristics of actual equipment to improved furnace and kiln design, control and operation optimization will therefore be available.

Application to the glass industry

The glass industry is significant of energy resources. Energy accounts for about 30-40% of production costs and at least 70% of all energy use is involved with melting, which is a just cause for making considerable efforts to improve the efficiency of melting furnaces and associated systems. Under this perspective, the use of 3-D physically-based models for turbulent combustion, radiation and soot and NO formation is an important tool.

Mathematical modelling of pollutants formation should be used to achieve the goals of energy-savings together with those of pollution-abatement. Detailed and validated predictions of nitric oxide emissions from a crossfire regenerative glass melting furnace were presented by Carvalho, Lockwood, Papadopoulos and Semião 1990. In this furnace the firing ports are located along the sides of the furnace. Figure 3 shows an application of the above referred modelling procedures to a container glass melting furnace. The furnace is of the end-port regenerative kind, in which the flame forms a loop within the combustion chamber. Some detailed predicted information is portrayed in figure 3 which shows respectively: the predicted flow pattern, temperature and NO fields within the combustion chamber, represented through an horizontal plane as well as the flow field and temperature distribution at the load surface.

The capacity of these modelling procedures for applications in energy efficiency optimization may de illustrated by the prediction of thermal efficiency variation for several actuating parameters. An application for the above referred end-port furnace in which the energy input has been varied between 6 and 8.4 MW is presented in table 1. In these table the calculated energy efficiency in the combustion chamber and pollutant emissions (SO_2, NO_x, soot and unburned fuel) are presented (for details see Carvalho, Coelho and Nogueira 1992).

One of the most important aspects requiring strategic research is the application of expert systems to the glass-making furnaces. Expert systems should incorporate information derived from the mathematical models to define optimum operating conditions.

TABLE 1
Effect of energy input variation on the thermal efficiency transferred to the glass and melting region

Fuel Energy Input	Thermal Efficiency	Total Heat Flux to the Glass	Maximum Flux Value	NO Outlet Flow Rate	SO2 Outlet Flow Rate	Soot Outlet Flow Rate	UHC Outlet Flow Rate
8.4 MW	0.457	3.933 MW	350.0 kW	2.08×10^{-3}kg	8.13×10^{-3}kg	8.26×10^{-4}kg	5.2×10^{-6}kg
7.6 MW	0.463	3.523 MW	314.9 kW	1.88×10^{-3}kg	7.52×10^{-3}kg	6.98×10^{-4}kg	1.25×10^{-5}kg
6.8 MW	0.470	3.278 MW	279.9 kW	1.19×10^{-3}kg	6.77×10^{-3}kg	5.70×10^{-4}kg	8.36×10^{-6}kg
6.0 MW	0.472	2.904 MW	230.5 kW	7.38×10^{-4}kg	6.10×10^{-3}kg	5.04×10^{-4}kg	8.11×10^{-6}kg

Application to Cement Industry

In cement industry energy is one of the most important operating cost. Reducing the energy consumption is of primary importance for cement industry. Environmental prediction will play an increasing part. All these benefits must be achieved within the constraints of maintaining high product quality.

Predictive studies directed to the development of expert systems and model based new design and operation concepts, have been carried out with the objective of to know the role and the dependence of several main variables. The considered operating conditions are referred in table 2. In table 3 a parametric study in which energy input variables have been changed is presented. This results were obtained with a mathematical model of the furnace system which considers the rotation effect on the temperature distribution at the walls internal surface. In figure 5 a) a sketch of the kiln geometry is presented. The temperature profile along the furnace length are plotted in figure 5 b). The effect of rotative motion of the kiln, introducing non-linearities in the temperature gradients, may be well observed in this figure, namely as far as the wall mean temperature profiles is concerned . In figure 5 c) the differences between both temperature profiles at furnace inlet and outlet also illustrates the above referred aspect. In this figure the temperature distribution at internal surface of the kiln refractory wall is plotted along the radial coordinate for two sections at load inlet and outlet regions (for details see Carvalho, Farias and Martins 1992).

The extension of current mathematical models to simulate unsteady-state conditions, such as those in cement plants, should also be attempted and linked with expert systems to control the whole plants. Research is needed on instrumentation for monitoring the process and, in particular, to measure mass flow rates, particle-size distributions and SOx and NOx concentrations.

TABLE 2
Dry process cement kiln considered working conditions.

$\omega = 1$ rad/s	Angular velocity
$\eta = 0.11$	Fraction of solid fill
$F_{int} = 3.54$ m	Kiln diameter
$\dot{m}_g = 14$ kg/s	Gas mass flow rate
$\dot{m}_g = 40$ kg/s	Charge mass flow rate
$\varepsilon_c = 0.75$	Charge emissivity
$\varepsilon_w = 0.7$	Inside wall emissivity
$\varepsilon_{w\ out} = 0.3$	Outside wall surface emissivity
$k_g = 0.145$ m^{-1}	Gas absorption coefficient
$k_{rf} = 0.23$ W/mK	Refractory wall thermal conductivity
$k_{st} = 29$ W/mK	Steel shell thermal conductivity

TABLE 3
Effect of the energy input variation on the temperature and heat fluxes inside a cement kiln.

Energy Input (MW)	Load Inlet Temperature (K)	Kiln Length (m)	Walls Thickness (m)	Outlet Load Temperature (K)	Gas Temperature (K)	Heat Flux to the Load (MW)	Heat Losses (MW)
61.6	**1123**	**65**	**0.23**	1663	1210	21.0	6.8
50.4	1123	65	0.23	1431	1194	12.0	5.1
61.6	**1023**	65	0.23	1615	1166	23.1	6.0
61.6	1123	**43**	0.23	1637	1313	20.0	4.9
61.6	1123	65	**0.18**	1654	1201	20.7	7.5

Application to the Baking Industry

The food industry is a significant user of energy and is a sector that differs from most of the other major energy users in that the size of processing unit can vary significantly. Here attention is devoted to baking ovens, where the main source of energy is natural gas and energy losses are due essentially to lack of optimization.

In contrast to the previously referred industries, this sector involves relatively low temperature environments, i.e. below 240°C, which are achieved mainly by indirect heating (direct gas firing is usually only used in biscuit ovens). Large energy losses have been explained by the process of evaporation and condensation. The fact that, baking time is mainly affected by the water vapor partial pressure and the crust formation depends upon the oven surface temperature and the consequent thermal radiation and the gas mixture circulation inside the oven, turns the process sensitive to a large number or operating parameters. Improved design of baking ovens should allow reduced oven temperatures, with radiation heat losses, and may eliminate the requirements of steam and ventilation: a reduction on energy consumption by 25% should be attainable. Only recently research attention has been directed towards this subject as reviewed by Rask 1989. Carvalho and Martins 1992 presented a detailed model of a baking oven. This model is based in the three-dimensional solution of the turbulent flow, radiative and convective heat transfer in the heating chamber and the heat transfer and "oven-rise" in the bread. The developed model has been applied to a baking oven of TNO Institute for Cereals, Flour and Bread (deVries 1991). The sketch of the baking oven and predictions are illustrated in figure 6. In the figure the predicted fluid flow pattern is presented through a vertical plane crossing the bread domain. The dynamical evolution of the temperature in the centre of the dough is also shown.

Strategic research is required to develop new oven concepts including multi-sensor systems to control the water vapour partial pressure, the crust colour and the level of forced air circulation. It is noted that research activities in this sector should be stimulated by the fact that Europe is a net exporter of baking ovens.

BASIS FOR A MASSIVE SENSORING APPROACH

In recent years, novel sensors have been developed for withstanding harsh environmental conditions occurring in process industry such as glass-melting furnaces and ceramic furnaces. For the application to high temperature environments new cooling devices allow the application of temperature sensitive sensors under these conditions. Research work in functional tests for different working conditions, taking into account high temperature, aggressive media and radiation as well as distortion caused by additional sources of heat production, has been carried out. These new developed sensors and additional sensors to be developed for direct access to non conventional controllable variables have shown new possibilities for process control in industry.

CONCLUSIONS

In the present paper the use of dedicated modelling and massive sensoring concepts for the control, design and operation optimization (in an integrated design basis) has been discussed. The expected consequences for the improvement of energy utilization efficiency, process intensification, reduction of energy costs of pollutant emissions abatement were reviewed. Applications to furnace, kilns and ovens systems were exemplified through ongoing developments. They are based on the use of dedicated modelling, through the numerical multi-dimensional solution of heat transfer fluid flow and chemical reactions, of combustion generated pollutant emissions, fuel specific consumption and load process.

The generality of the modelling and sensoring techniques discussed in the paper ensure a wide application range involving a large number of industrial situations.

New sensors able to operate under harsh environments are of unquestionable interest for the optimization of industrial furnaces, namely allowing control and monitorization of process critical variables.

The industrial relevance of the proposed approach is effectively attracting the industry to their implementation. Specialists in industrial automation systems, thermal systems design and process engineers have to be committed to the integration of those new concepts into the existing furnace kilns and ovens systems. So, a near future marketing of these achievements is an expectable consequence. This will certainly contribute for a continued development of the furnaces, kilns and ovens European technology and manufacturing industry competitiveness. Manufacturers, certainly, will be able to embody and promote such developments.

ACKNOWLEDGMENTS

Part of the work presented in this paper has been performed within the JOULE Project JOUE-0051-C entitled "Energy Saving and Pollution Abatement in Glass-making Furnaces, Cement Kilns and Baking Ovens." The authors are thankful to Dr. Pilavachi for the helpful suggestions and discussions.

REFERENCES

1. Abou Ellail, M.M.M., Gosman, A.D., Lockwood, F.C. and Megahead, I.E.A., Description and Validation of a Three-Dimensional Procedure for Combustion Chamber Flows. AIAA Journal of Energy, 1978, **2**, (2), 71-80.
2. Gosman, A.D., Lockwood, F.C., Megahead, I.E.A. and Shah, N.G., The Prediction of the Flow, Reaction and Heat Transfer in the Combustion Chamber of a Glass Furnace. AIAA 18th Aerospace Sciences Meeting, January, Pasadena, CA, USA, 1980, 14-16.
3. Lockwood, F.C. and Shah, N.G., A New Radiation Solution Method for Incoporation in General Combustion Prediction Procedures. 18th Symp. (Int.) on Combustion, The Combustion Institute, 1980.
4. Carvalho, M.G. and Lockwood, F.C., Mathematical Simulation of an End-Port Regenerative Glass Furnace, Proc. Inst. Mech. Engrs., 1985, **199** (C2), pp. 113-120.
5. Semião, V., Numerical Simulation of an Industrial Furnace. M.Sc. Thesis, University of Lisbon. *In Portuguese*, 1986.
6. Carvalho, M.G., Durão, D.F.G. and Pereira J.C.F., Prediction of the flow, reaction and heat transfer in an oxy-fuel glass furnace, Eng Comput., 1987, **4**, p. 23.
7. Carvalho, M. G., Oliveira, P. and Semião, V., Modelling of an Industrial Glass Furnace, Progress in Astronautics and Aeronautics, 1988, **113**, p. 363.
8. Carvalho, M.G., Lockwood, F.C., Papadopoulos, C., Semião, V., Predictions of Nitric Oxide Emissions from an Industrial Glass-Melting Furnace, Journal of the Institute of Energy, 1990, pp. 39-47, March.
9. Carvalho, M.G., Durão, D.F.G., Heitor, M.V., Moreira, A.L.N. and Pereira, J.C.F., The Flow and Heat Transfer Characteristics of an Oxy-Fuel Glass Furnace. Proc. 1st INFUB, Lisbon, Portugal, 21-24 March, 1988.
10. Wieringa,J.A., Phelich, J.J. and Hoogendoorn, C.J., Spectral Effects of Radiative Heat-Transfer in High Temperature Furnace Burning Natural Gas, J. Institute of Energy, september, pp.101.
11. Carvalho, M. G., and Nogueira, M., Analysis of Furnace Performance Through 3-D Physically-Based Modelling, XVI International Congress on Glass, Madrid, October, 1992.

12. Viskanta, R., Chapman, K.S. and Ramadmyani, S., Mathematical Modelling of Heat Transfer in Mich-Temperature Industrial Furnaces; Heat Transfer 90, Southampton, 1990.
13. Meunier, H. and Cambier, M., Simulation of a Batch Furnace for Metal Heat Treatment: Optimization of the Setpoint Schedule, Vol. 1, 2nd INFUB Conference, Vilamoura, Portugal, 1991.
14. Jicha, M. and Vacenovshi, P., Numerical Simulation of 3-D Fluid Flow and Heat Transfer in Chamber Car-hearth Furnace, ICHMT 2nd Int. Forum on Expert Systems and Computer Simualtion in Energy Engineering, Erlangen, Germany, 1992
15. Ghoshdastidar, P.S. and Unni, V.K.A., Heat Transfer in the Non-Reacting Zone of a Cement Rotary Klin", National Heat Transfer Conference, HTD-Vol. 106, Heat Transfer in Radiation, Combustion and Fires, 1989, pp. 113-122.
16. Kim, N. K. Lyon, J.E. and Suryanarayana, N.V., Heat Shield for High-Temperature Kiln, Ind Eng. Chem Process Des. Dev., 1986, **25**, pp. 843-849.
17. Lockwood, F.C., Salooja, A.P., The Prediction of Some Pulverised Bituminous Coal Flames in a Furnaces", Report on MMF 2 Investigation. IFRF doc. No. F 36/a/12, 1983.
18. Visser, B. M. and Weber, A. P., Computations of Near Burner Zone Properties of Swirling Pulverised Coal Flames, Report on MMF2 Investigation. IFRF doc. No. F 36/A/12, 1989.
19. Rask, C., Thermal Properties of Dough and Baking Products: a Review of Published Data, J. Food Eng, 1989, 9, pp 167-195
20. DeVries, G.A., Sluimer, P. and Bloksma, A. H., A Quantitative Model for Heat Transport in Dough and Crumb During Baking, Proc. of Int. Symp. "Cereal Science and Technology", Sweden, 1988.
21. Carvalho, M.G. and Martins, N., Heat and Mass Transfer in an Electrically Heated Natural Convection Baking Oven, 1st European and Third UK National Heat Transfer Conference, Birmingham, U.K., 1992.
22. Carvalho, M.G., Nogueira, M., Glass Furnace Efficiency Evaluation Through 3-D Modelling. Eur. Sem. on Improved Technologies for Rational Use of Energy in Glass Industry, Wiesbaden, Germany, 1992
23. Carvalho, M.G., Coelho, P. and Nogueira, M., Applicability of 3-D Mathematical Modelling of Thermal Equipment in the Development of Expert Systems. ICHMT 2nd Int. Forum on Expert Systems and Computer Simulation in Energy Engineering, Erlangen, Germany, 1992.
24. Robinson, G.F., A Three-Dimensional Analytical Model of a Large Tangentially-Fired Furnace. J. Institute of Fuel, Match, 1985, 116-150.
25. Abbas, A.S. and Lockwood, F.C., Prediction of Power Station Combustors. Imperial College of Science, Technology & Medicine, Mech. Eng., Report-Fluids Section, 1986.
26. Gorner K. and Zinser W., Prediction of Three-dimensional Flows in Utility Boiler Furnaces and Comparison With Expriments, ASME, 107th Winter Annual Meeting, U.S.A., 1986.
27. Sargianos, N., Anagnostopoulos, J., Bergeles, G., A Numerical Algorithm for Gas Combustion in Utility Boilers, Heat Transfer 90, Southampton, 1990.
28. DeMichele, G., Pasini, S., Tozzi, A., Simulation of Heat Transfer and Combustion in Gas- and Oil- Fired Furnaces, 2nd INFUB, Vilamoura, Portugal, 1991.
29. Carvalho, M. G. and Coelho P., Numerical Prediction of an Oil-Fired Water Tube Boiler, Eng Comput., 1990, **7**, p. 227.
30. Meunier, H., Modelling of Industrial Furnaces, 2nd INFUB, Vilamoura, Portugal, Vol. 1, 1991.
31. DeVries G. A., *private communication*, 1991.
32. Carvalho, M.G., Farias, T. and Martins, A., A Three-Dimensional Modelling of the Radiative Heat Transfer in a Cement Kilns, Proceedings of "Advanced Computational Methods in Heat Transfer II, 7-10th July 1992, Vol. I, pp. 141-160. Edited by L.C. Wrobel, C.A. Brebbia and A.J. Nowak.

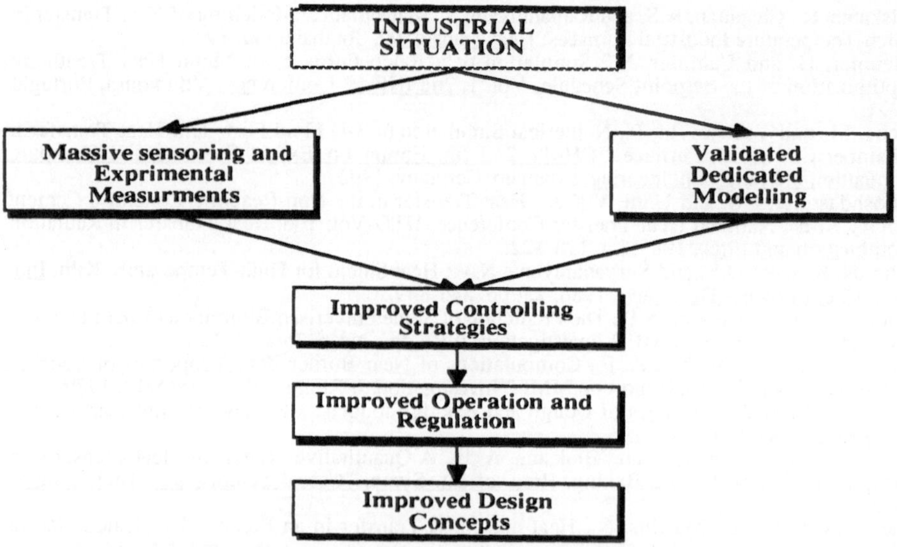

Figure 1. Considered strategy to respond to the industrial needs in energy saving and pollution abatement.

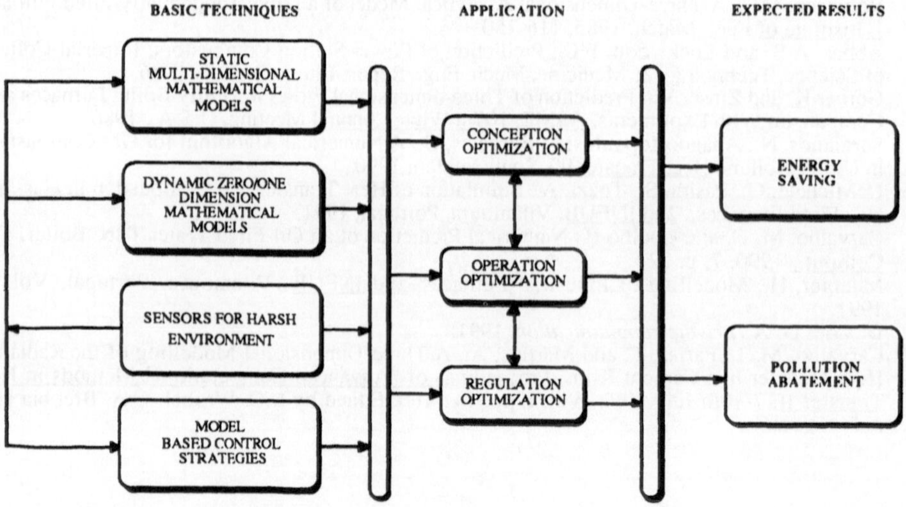

Figure 2. Approach for reduction of energy consumption and pollutant emissions in furnaces, kilns and ovens

a)

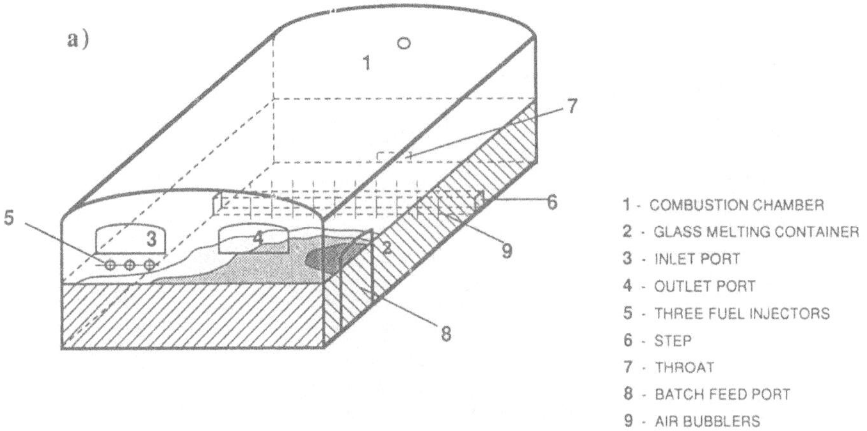
1
2
3
4
5
6
7
8
9
1 - COMBUSTION CHAMBER
2 - GLASS MELTING CONTAINER
3 - INLET PORT
4 - OUTLET PORT
5 - THREE FUEL INJECTORS
6 - STEP
7 - THROAT
8 - BATCH FEED PORT
9 - AIR BUBBLERS

b)

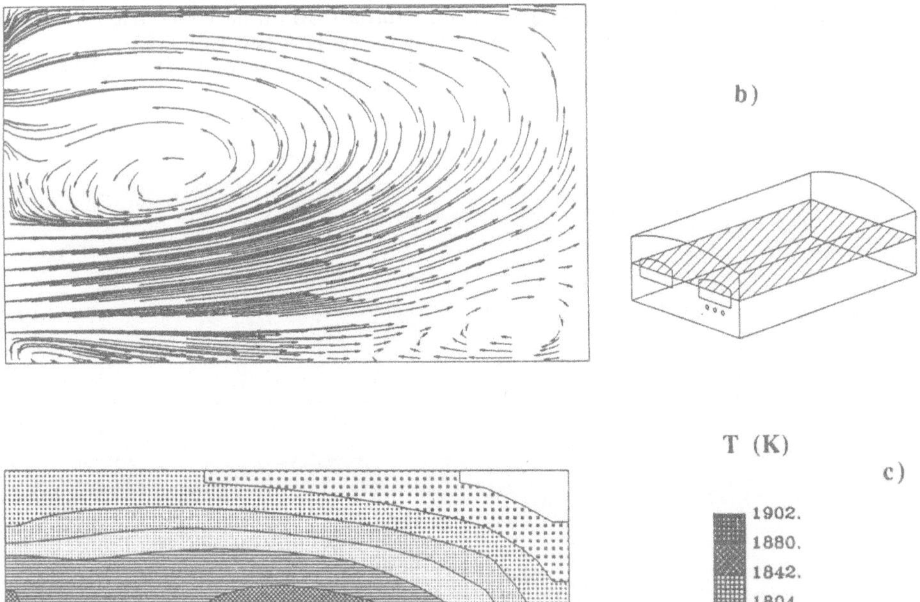

c)

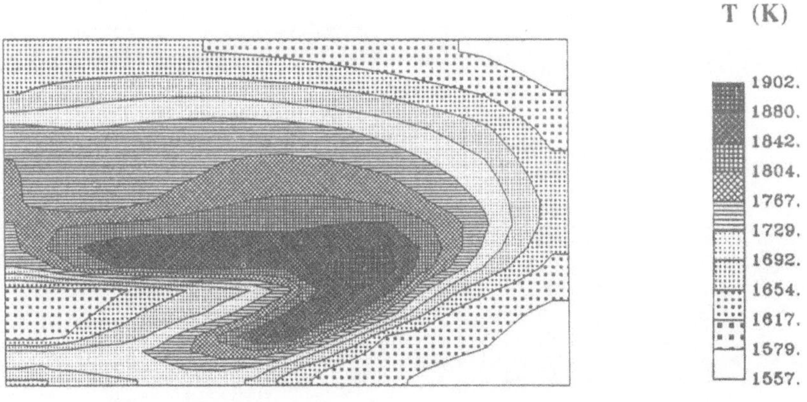
T (K)
1902.
1880.
1842.
1804.
1767.
1729.
1692.
1654.
1617.
1579.
1557.

d)

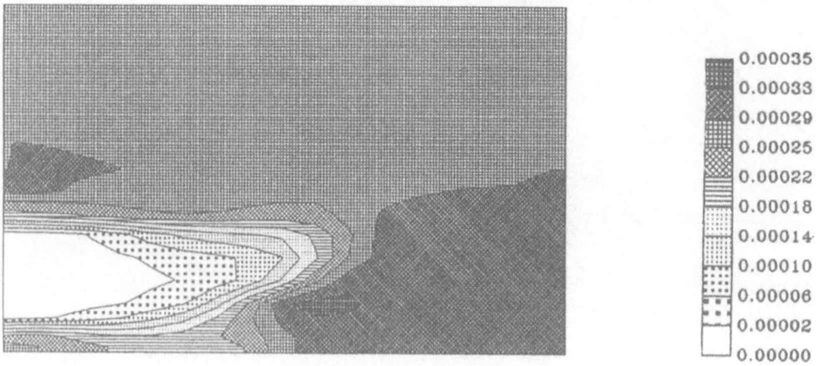

Figure 3 Prediction of the flow pattern, NO mass fraction and temperature fields inside the combustion chamber of an end-port glass melting furnace a) Furnace Geometry, b) Time averaged flow field in a horizontal plane crossing the burners row (time step for marker particules representation 0 25 s), c) temperature field at the same plane (K), d) NO mass fraction at the same plane

a)

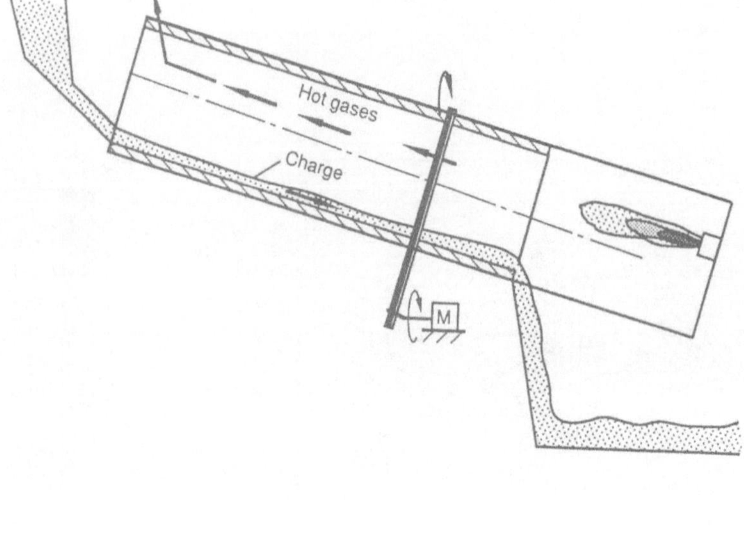

b)

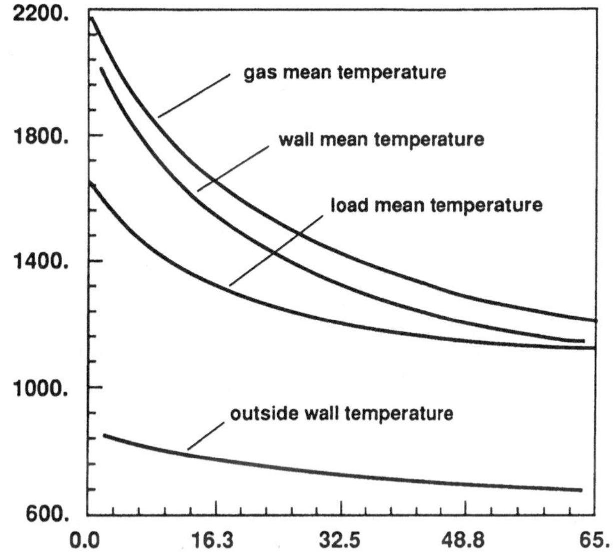

c)

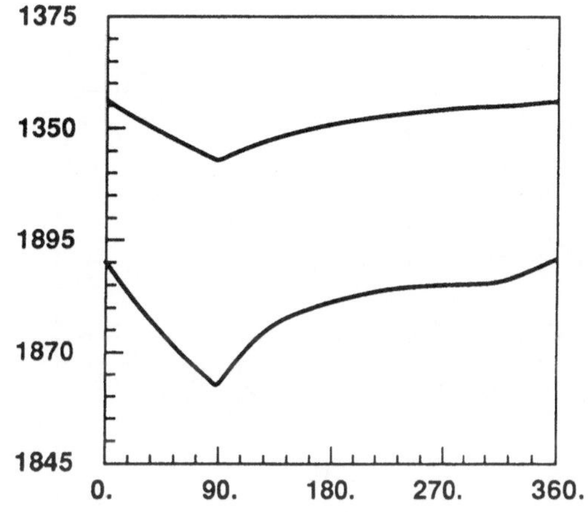

Figure 4. Prediction of the temperature distribution inside a dry process cement kiln. a) Kiln Geometry; b) Temperature profiles along the furnace length (K); c) Temperature profiles at refractory internal surface through a kiln section near the load inlet and outlet regions (K).

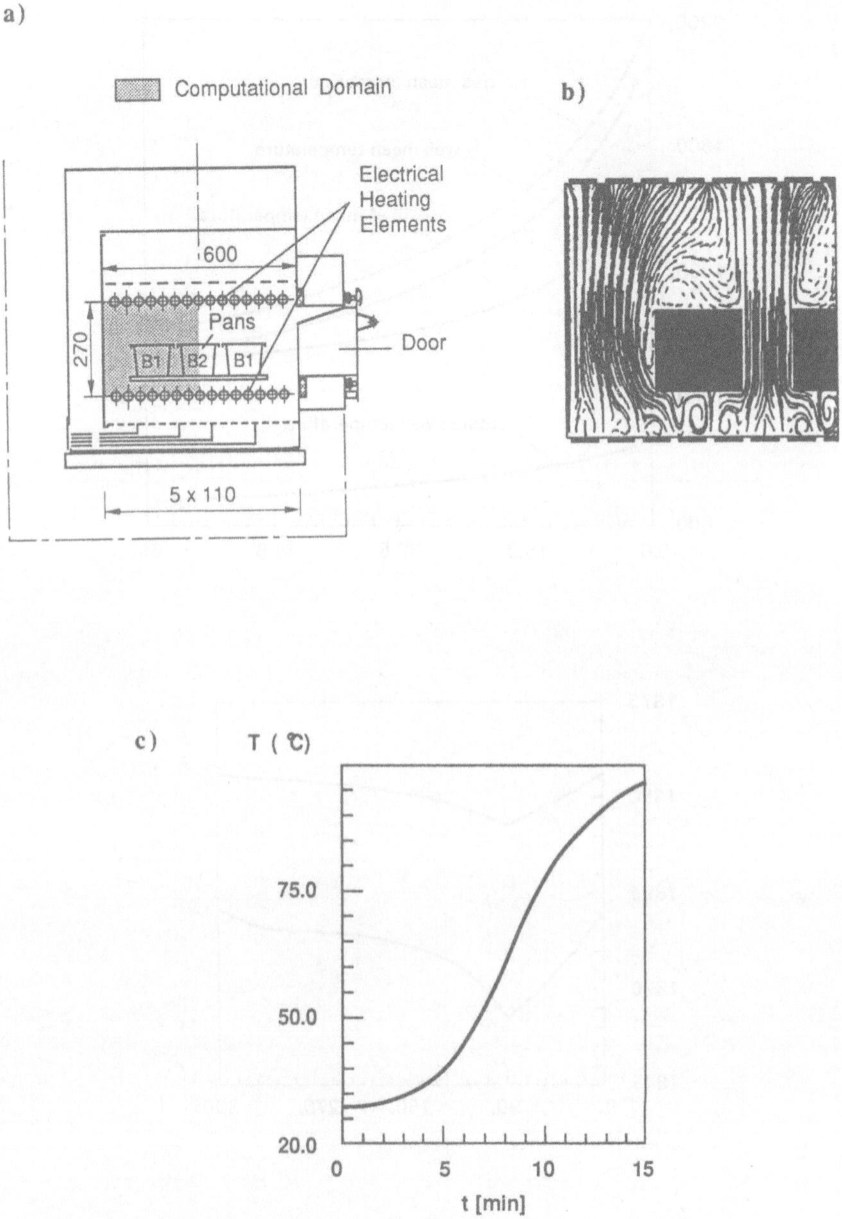

Figure 5. Prediction of the flow pattern and temperature distribution inside a baking oven and dough temperature evolution. a) Oven Geometry; b) Time Averaged Flow Pattern; c) Time dependent temperature evolution at the centre of the dough.

NUMERICAL SIMULATION OF THE COMBUSTION AERODYNAMICS INSIDE A ROTARY CEMENT KILN

T AVGEROPOULOS[1], J P GLEKAS[2], C PAPADOPOULOS[3]

1 TITAN Cement Company SA

2 3 Centre of Renewable Energy Sources /RUE Department

ABSTRACT

In this paper the utilization and validation of a computer simulation model which can be used for cement kilns, is presented The computational fluid dynamics code is used to explore kiln concepts and control strategies which can lead to energy savings and pollution abatement The CFD code used, utilizes a group of partial differential equations for the gas phase, the results compare satisfactorily with the available experimental data, ensuring that an inexpensive and easy to use mathematical tool is available for cement kiln heat balance calculations, data which is necessary for the optimum performance of cement kilns

INTRODUCTION

The reduction of pollutant emissions generated from cement plants, and the efficient use of energy can be realised if they are known in detail the flow, heat and mass transfer chatacteristics inside the rotary cement kiln

A widely used and well validated CFD code developed for flow and combustion applications was used to simulate the physical processes that prevail inside a full scale rotary cement kiln The results are validated against the data acquired during the operation of the kiln

The present paper is based on the work done by CRES and TITAN Cement Company SA involved in the project "*Energy Saving and Pollution Abatement in Glass-making furnaces, Cement kilns, and Baking ovens*", which is supported by JOULE Program This research project involves 13 research groups, namely

Instituto Superior Technico (I S T) Portugal, Imperial College of Science, Technology and Medicine (I C S T &M) - England, University of Erlangen/Nuremberg (L S T M - Erlangern) - Germany, Faculte Polytechnique de Mons (Mons) - Belgium, Universite Libre de Bruxelles (U L B) - Belgium, Centre for Renewable Energy Sources (C R E S) - Greece, TNO Cereals, Flour and Bread Institute (TNO Bread & Cereals) - Netherlands, TNO-Division of Technology for Society (TNO-Tech Society) - Netherlands, Instituto de Energia (INTERG) - Portugal, Institut National du Verre (INV) - Belgium, Adersa - France, Metal Portugessa - Portugal, and TITAN Cement Company SA - Greece

The main goal of this project is to develop "tools" to be used in the glass, cement and baking industry that can lead to energy savings and pollutant abatement

NUMERICAL METHOD

The CFD code used, predicts the flow/combustion behaviour of any axisymmetric combustion system Briefly the model is based on the time-mean equations for conservation of mass, momentum, species and energy for gas and particle phases The particle-phase equations are formulated in Lagrangian form and the coupling between phases is introduced through particle sources in the (Eulerian) gas-phase equations The standard k-ε turbulence model, a mixing-limited volatiles combustion model and the non-equilibrium diffusional radiation model are used

Phenomenological models are used to describe coal devolatilization and char combustion The model is formulated for axisymmetric flow in a cylindrical geometry The equations for the above mentioned physical processes are partial differential equations which are solved using the finite volume discretization technique The convective fluxes appearing in the finite-volume equations are approximated by the quadratic upstream weighted scheme (QUICK) of Leonard [2], while central differences are used for diffusion Because of the swirl intensity, the density variations encountered in the computations, the numerically non-diffusive QUICK scheme, the finesse of the meshes employed, the level of computational stability was slow

Therefore in order to prevent divergence several numerical experiments were performed to obtain the values of the underrelaxation factors which promote the stability of the numerical solution The pressure field is obtained using the PISO/SIMPLE algorithm [3]

APPLICATION OF THE METHOD

The code was applied to simulate the operation of a cement kiln, with the following characteristics

Length	80m
Diameter	5m
Slope	2 5 %
Speed	140 rph

Figure 1 presents schematically the cement kiln and the burner s geometry The simplified geometry that is actually simulated is defined by dotted lines, it is a tube a circular section of 5m, that is 67m long

Table I summarises the flow parameters that were used in the simulation

TABLE I

Fluid (Air)

Stream	Composition	Flowrate (m^3/h)	Temperature (°C)
1	Air	77,125	1000
2	Air (Swirl)	2,390	20
3	Coal+Air	5,560	60
4	Air	8,089	20

Density 1 18 Kg/m^3

Particle (Coal)

Material density	1200	Kg/m^3	C	44 7	%
Finess	90	µm	H_2	4 5	%
Paricle flow rate	3 06	Kg/sec	N_2	1 4	%
Net calorific value	6,200	Kcal/Kg	Volatiles	23 0	%
			Sulfur	0 9	%
			H_2O	3 2	%
			O_2	18 8	%
			Ash	13 5	%

In order to establish high mixing rates between the fuel (coal dust) and the air stream, and to enhance the flame stability, the special features of swirling flows are used in the operation of the kiln The swirl effect is introduced by stream No 2, with a swirl velocity component equal to w_2 = 48 m/s which corresponds to a swirl number S=1 27

First an isothermal run of the code was performed in order to establish an approximation of the flow filed inside the kiln, and then the combustion process was simulated. The combustion simulation was performed from the burner's exit to the endwall of the kiln (Figure 1).

Although the "tube" of the simulated kiln is not axisymmetric, some simplifications concernig the real geometry were made, which do not affect the characteristics of the physical processes we are interested in. Thus, the kiln was considered axisymmetric, allowing the calculations to be performed only in half of the kiln tube (from the central line to the wall).

Calculations were performed using meshes of 99x49, 197x49, and 244x59 in the terms of the grid points in the x and y directions respectively. The obtained results by the first two grids differ considerably especially at the exit of the kiln, where the discretization by the corse grid was poor enough. The results obtained by the fine grids show very little differencies, ensuring that the obtained results with the 197x49 mesh are grid independent.

The flowrates at the inlet, correspond to the following boundary conditions :

* For the velocities a turbulent boundary layer was considered, while the inlet conditions were :

  for stream No. 1 : u_1= 1.5 m/s

  No. 2 : u_2= 33 m/s

  No. 3 : u_3= 34 m/s

  No. 4 : u_4= 127 m/s

* The level of turbulence energy in the air stream at the inlet, was set to $k=0.05*U^2$, while the dissipation rate was described by the formula $\varepsilon = k^{1.5}/0.33R$ [4].

* For the temperature, the inlet conditions were :

for stream No. 1 : T_1= 1000°C

No. 2 : T_2= 20°C

No. 3 : T_3= 60°C

No. 4 : T_4= 20°C

The temperature distribution on the wall of the kiln, as it was measured by optical pyrometers during its operation, was used as boundary conditions for the outer wall. At the exit, Neumman boundary conditions were imposed.

RESULTS AND DISCUSSION

The computed geometry is shown in Figure 1 together with the two fine meshes used to cover the solution domain (197x49 and 244x59 meshes).

Figure 2 provides overall views of the predicted velocity and temperature contours, obtained by the 197x49 mesh.

Figures 3 to 6 show comparisons of predicted radial profiles by the three meshes for the velocity components (axial u, radial v, and swirl w), temperature and turbulence energy at different locations downstream the burner exit. The axial locations are non-dimensionalised by the diameter D of the burner.

The dependency of the results on the grid size is obvious when making comparisons between the results of the coarse and the fine meshes, with the greater differences observed far downstream from the burner's exit where the discretization by the coarse grid is not good enough.

During the operation of the kiln, the flame length was measured to be approximately 40 m. The calculations with the coarse mesh show that the flame penetrates into the kiln at a distance of nearly 52m from its entrance, while the fine meshes predictions are nearly 32m, which is close to the observed flame length.

Figure 6 present radial temperature profiles along the axis of the kiln. Due to the smoothing effect of the thermal radiation transfer these flatten rapidly with increasing distance. Near the burner and close to the axis of the kiln the predicted profiles show a temperature peak due to the intense mixing between the three air streams, and a quite uniform temperature region further away from the axis, at a temperature level of about 1000°C, which is the temperature of the secondary air at the inlet.

The above mentioned temperature peak (at y/R~0.4) shown in figure 6 (x/D=1) results in a corresponding density tough, and this can be assumed to cause a strong local forward acceleration counteracting the tendency towards the reduction of the flow speed, which is apparent from Figure 3 (x/D=1).

Figure 7 provide overall views of the predicted O_2, volatiles and products contours, obtained with the 197x49 mesh, for the first 20m of the kiln. Figures 8 to 10 show the predicted radial profiles of mean concentrations of O_2, products and volatiles along the axis of the kiln.

The volatile levels are high enough even at distances x/D=1 and 5 from the burner (Figure 10, x/D=1 and 5). The above pattern indicates that significant devolatization occurs very close to the burner. This is probably a result of the high inlet temperature of the secondary air stream and the well mixing of the air flow at the inlet. As a result the burning process starts very close to the burner (Figures 8 and 9). The small value of the volatiles level at a distance x/D=20 (Figure 10) suggests that this is the limit of the volatiles flame.

Further downstream, the burning process is more intensive and the products concentrations are high while the O_2 concentrations low, indicating highly reacted fluid The steep gradients of the O_2 products and volatiles concentrations show intense zone of devolatilization and combustion

As expected the O_2 and products profiles progressively flatten as the burner jet flow continues to spread and the unburnt hydrocarbon level diminish (Figures 8 and 9)

CONCLUSIONS

A CED code developed for flow and combustion applications was applied to simulate the combustion of coal inside a real rotary cement kiln

A step by step application of the code has been followed, ie first the isothermal case was calculated and then the combustion of the coal was simulated

Different grid dependency tests were performed, in order to obtain mesh independent solutions

The obtained results agree qualitatively with the physical patterns that prevail inside the kiln during its operation, as well as with data found in the literature for similar applications [1]

REFERENCES

[1] Hogg, S and Leschziner, M A, Second-moment computation of strongly swirling reacting flow in a model combustor, Lecture Notes in Physics, Numerical Combustion, 1989, 338-352

[2] Leonard B P, A stable and accurate convective procedure based on quadratic upstream interpolation, Comp Meths Appl Mech Eng, 1979, 19, 59-98

[3] Patankar, S V, and Spalding, D B, A calculation procedure for heat, mass and momentum transfer in three-dimensional parabolic flows, Int J Heat Mass Transfer, 1972 ,15, 1787-1806

[4] Dixon, T F, Truelove, J S, and Wall, T F, Aerodynamic studies on swirled coaxial jets from nozzles with diregent quarls, ASME, J Fluids Eng, 1983, 105, 197-203

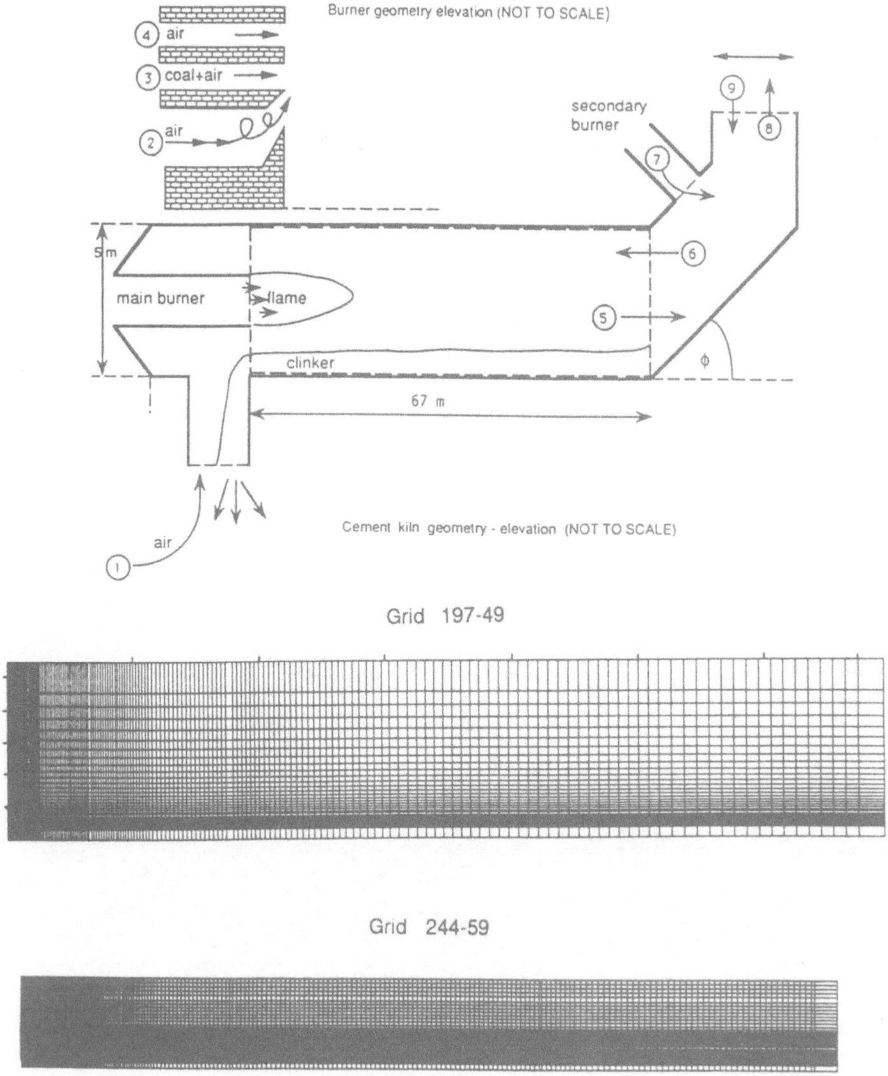

FIGURE 1 : The geometries of the cement kiln the main burner and the two fine meshes encountered in the calculations.

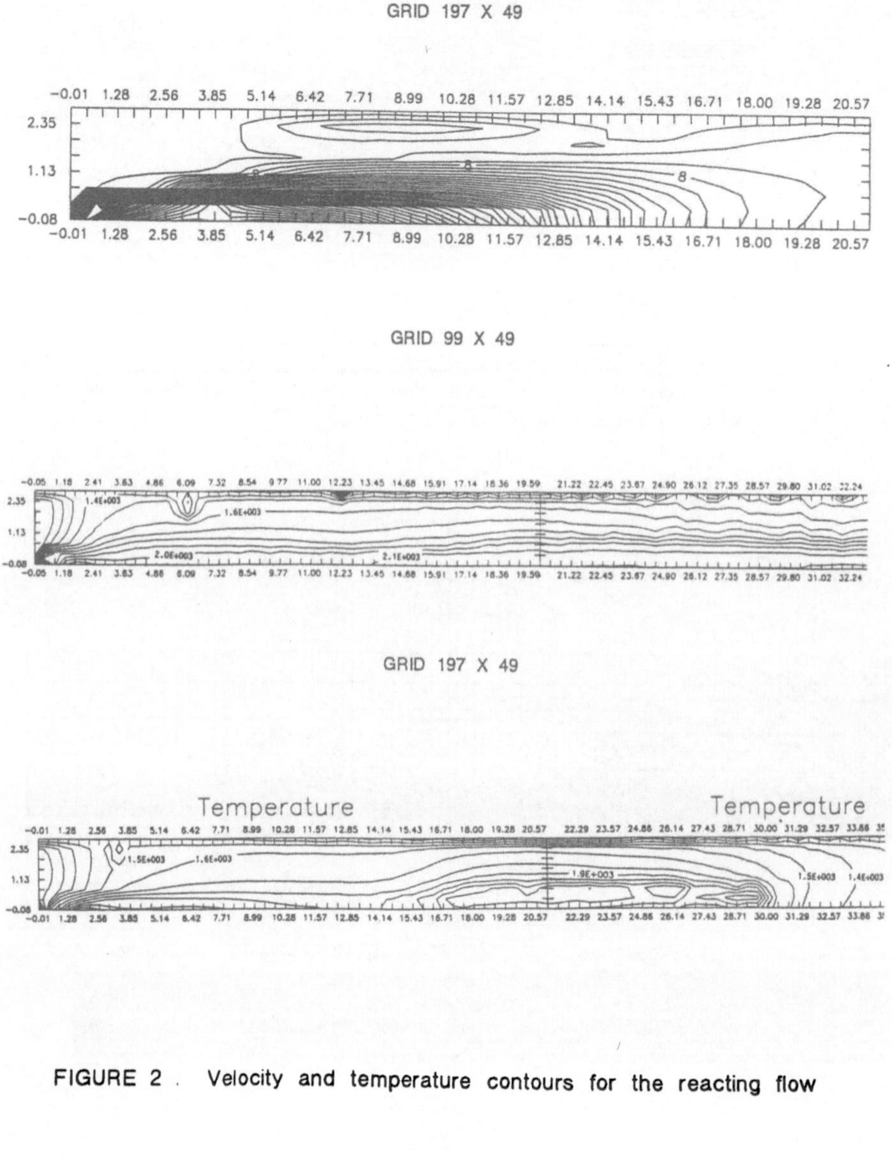

FIGURE 2 . Velocity and temperature contours for the reacting flow

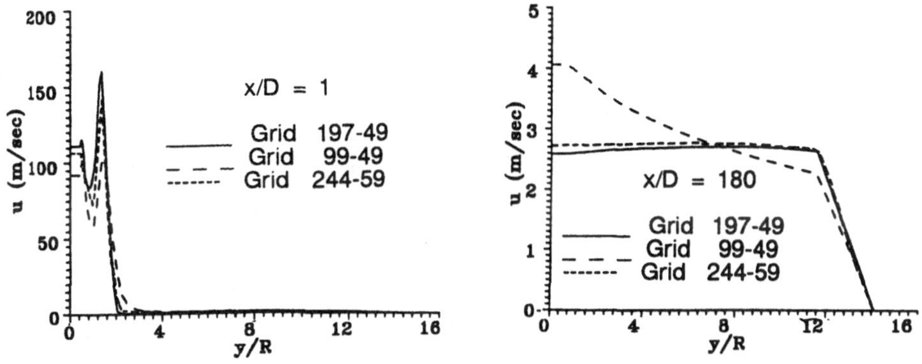

FIGURE 3 : Axial velocity profiles for reacting flow.

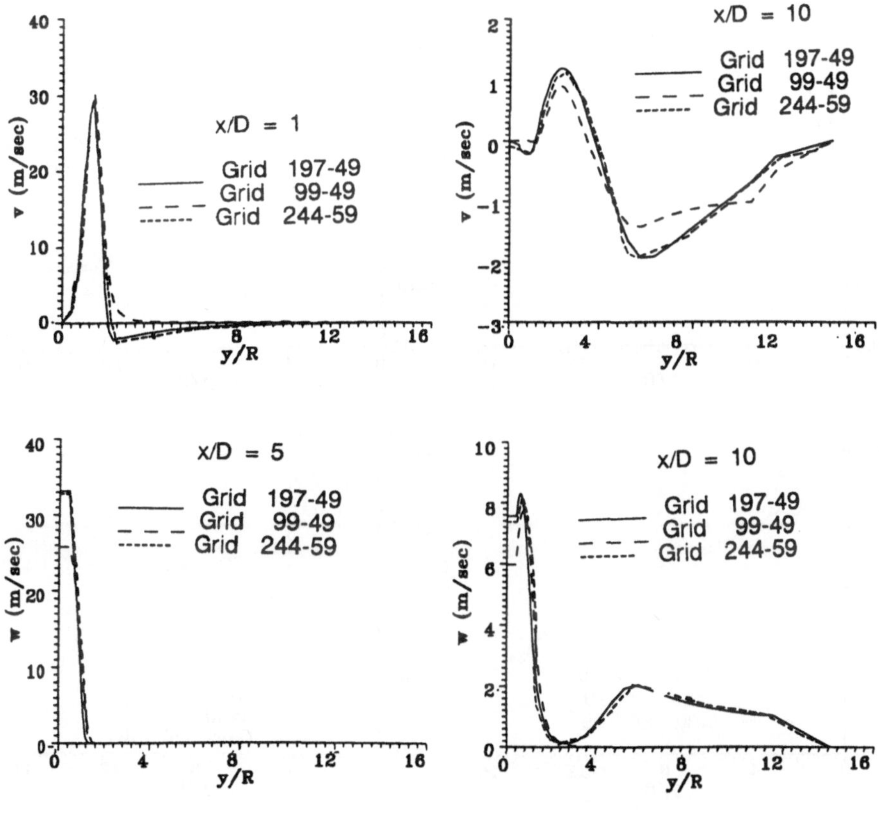

FIGURE 4 : Radial and swirl velocity profiles for reacting flow.

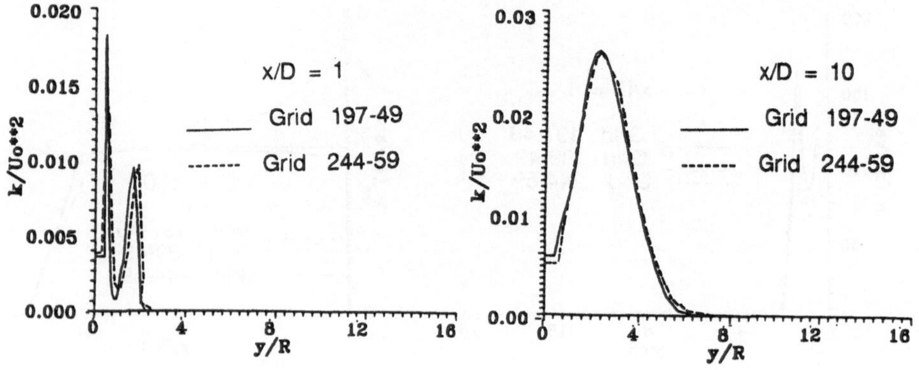

FIGURE 5 : Turbulence - energy profiles for reacting flow.

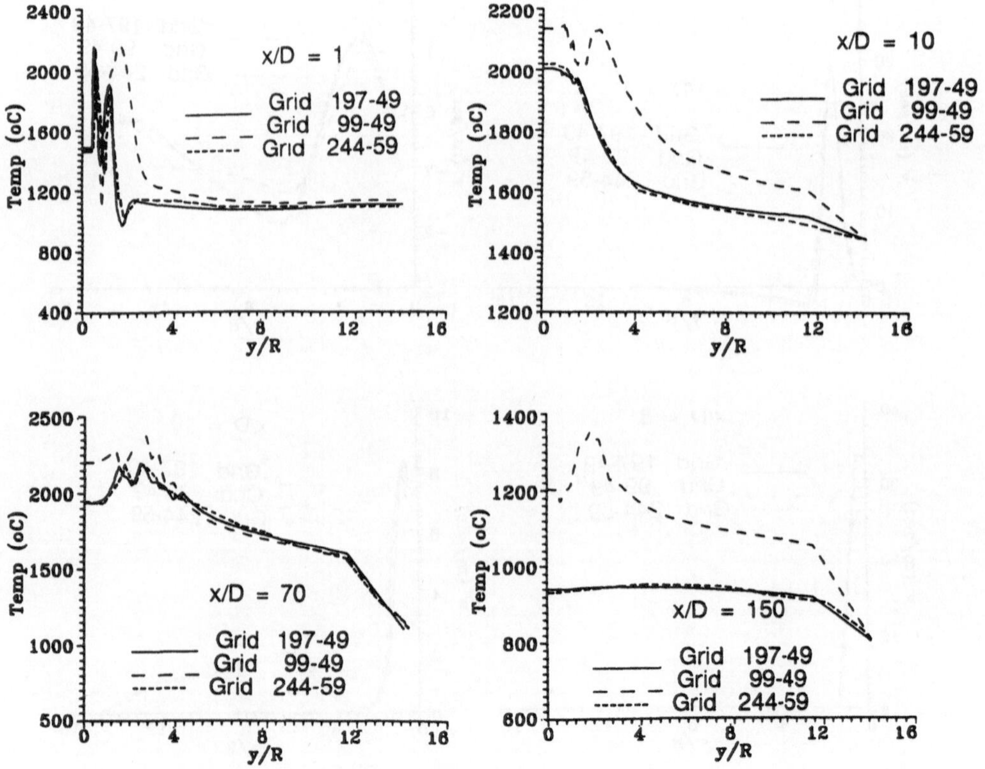

FIGURE 6 : Temperature profiles for reacting flow.

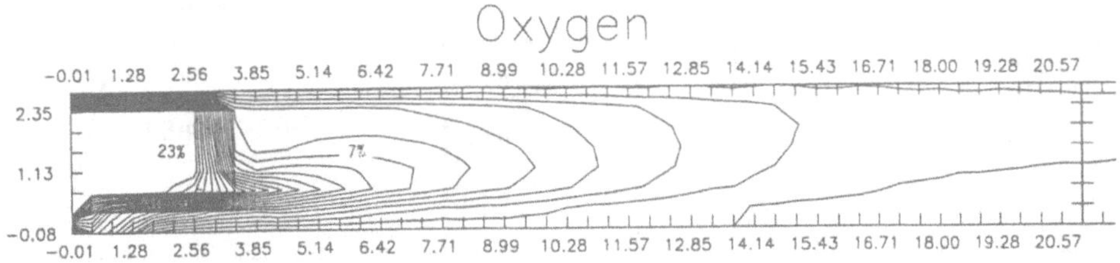

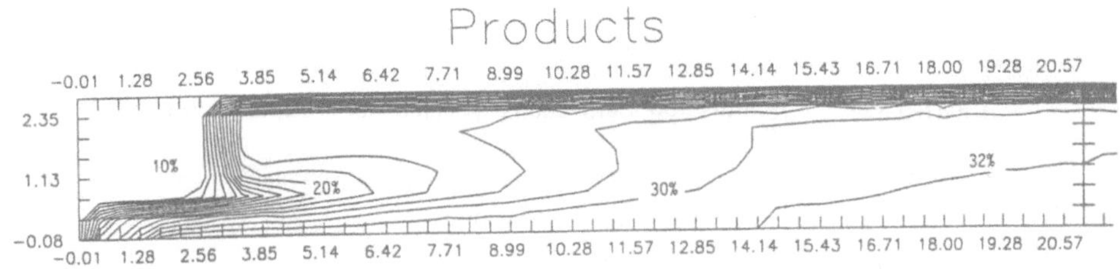

FIGURE 7 Oxygen and products contours for the reacting flow.

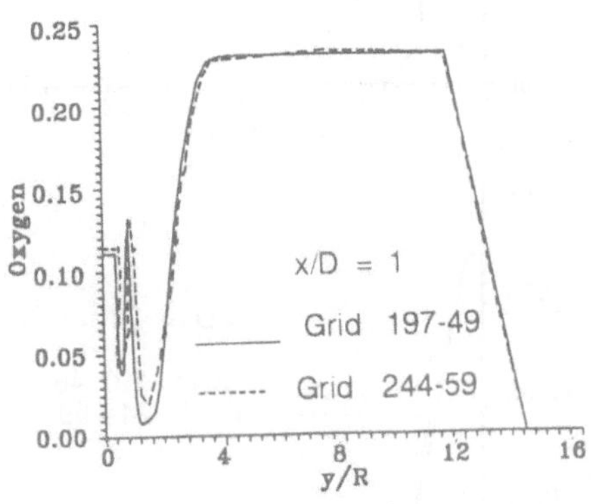

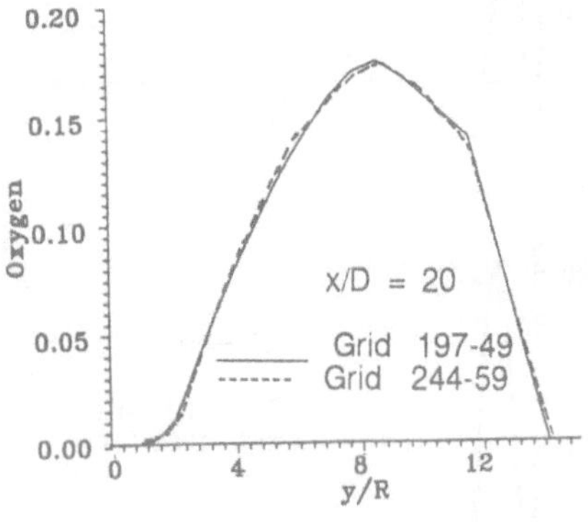

FIGURE 8 . Oxygen concentration profiles for reacting flow.

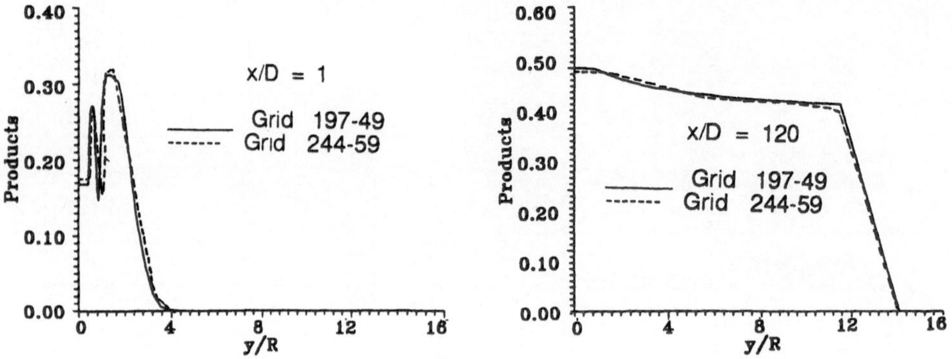

FIGURE 9 : Products concentrations profiles for reacting flow.

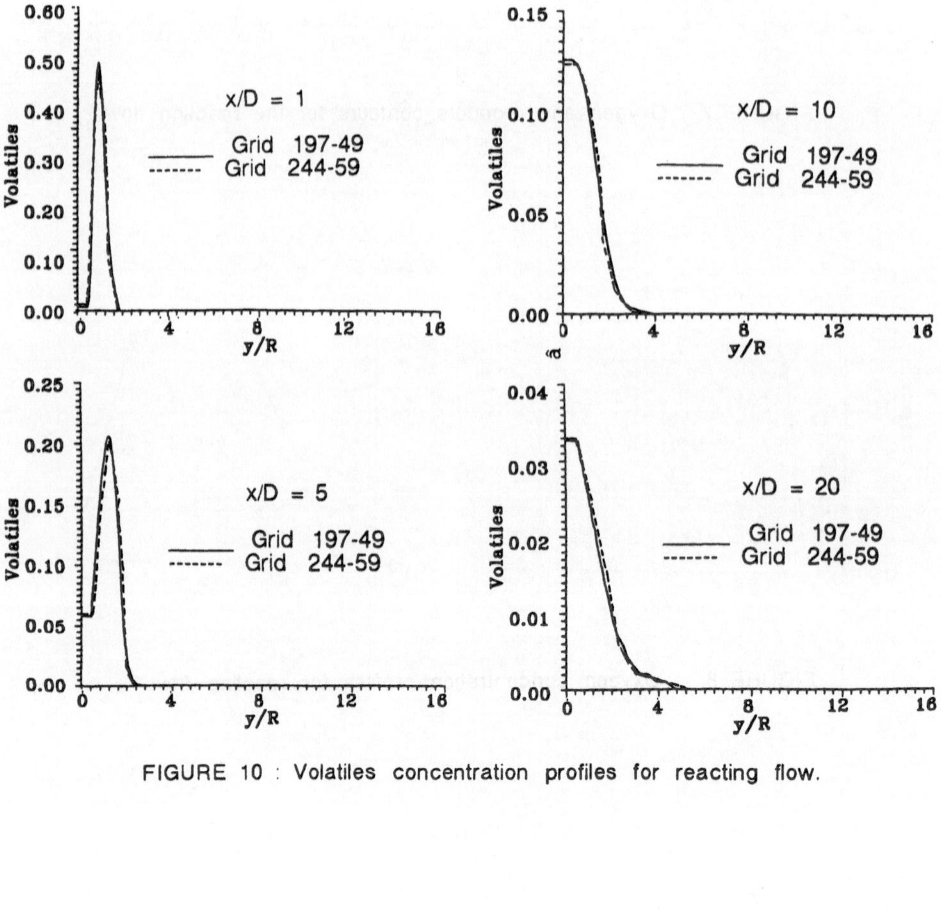

FIGURE 10 : Volatiles concentration profiles for reacting flow.

INDUSTRIAL LIME KILN COMPUTER DYNAMIC SIMULATION FOR IMPROVING EFFICIENCY

Bernardete Ribeiro and A. Dourado Correia
Laboratório de Informática e Sistemas
Departamento de Engenharia Electrotécnica
Universidade de Coimbra, P-3000 Coimbra, Portugal

ABSTRACT

The mathematical model of a pulp mill lime kiln and computer dynamic simulation are presented. The set of hyperbolic partial differential equations describing the chemical reactions, the mass balance, energy balance and axial solids transport is described. In order to obtain a model with an acceptable degree of complexity, the radial and angular variations for the solid and gas temperature equations are neglected. Since the wall represents an important thermodynamical link between kiln environment and surroundings the radial kiln wall variation was considered in the model. The objective is the optimization of the kiln operating conditions in industrial environment in order to increase product quality, decrease energy consumption and improve environmental conditions. The set of partial differential equations is solved by a finite difference method and by a finite element orthogonal collocation method with cubic piecewise polynomials as the approximating functions. The respective computational cost are compared. The results obtained by numerical methods are discussed in comparison with experimental data. Using the model, a study of energetic kiln behaviour is presented. The influence of various kinds of parameters on the energy process consumption is discussed. It is expected that this investigation will allow to enhance the knowledge of kiln behaviour and will be used for a further non conventional control approach.

INTRODUCTION

In this paper we compare results of computer simulation aiming at improvement of lime kiln efficiency in pulp and paper industry with those obtained in practice. Due to the highly complex kiln dynamics, a complete mathematical description would be forbiddingly complicated. The understanding of phenomena like heat transfer, mass transfer, and chemical reactions led us to a system of partial differential equations whose solution by a finite element orthogonal collocation method gives surprisingly precise results.

The remainder of the paper is organized as follows. In first part, a brief description of calcination process in chemical recovery cycle in pulp and paper industry is given. Next, the rotary lime kiln mathematical model is described and numerical methods to solve the set of partial differential equations are refered as well as computational costs. Following, main results concerning steady state kiln operation namely axial temperatures and composition profiles are given and discussed. A study of dust and lime solids content influence on kiln performance energetic index is presented. Next it is discussed the influence of step changes of control variables on principal measured variables in dynamic plant. Final conclusions are reported in the final paragraph of the paper.

CHEMICAL RECOVERY IN PAPER AND PULP INDUSTRY

In the Kraft process of paper and pulp industry the spent cooking liquor is recovered and regenerated. The final stage of the chemical recovery cycle is the conversion of green liquor to white liquor. In causticizing section, reburned lime (CaO) is used to regenerate the white liquor according to exothermic reactions described by Equations 1 and 2:

$$CaO + H_2O \longrightarrow Ca(OH)_2 + Heat \tag{1}$$

Calcium carbonate ($CaCO_3$) is obtained as a reaction subproduct and is separated from white liquor, washed and dewatered and finally calcined to regenerate reburned lime (CaO) which is ready to repeat the cycle.

$$Na_2CO_3 + Ca(OH)_2 \longrightarrow NaOH + CaCO_3 \tag{2}$$

The quality of reburned lime is very important parameter in the production of low cost, high quality white liquor. In the calcination process, the solids mainly calcium carbonate and inerts, are mixed with water originating the carbonate mud. Mud enters in the rotary

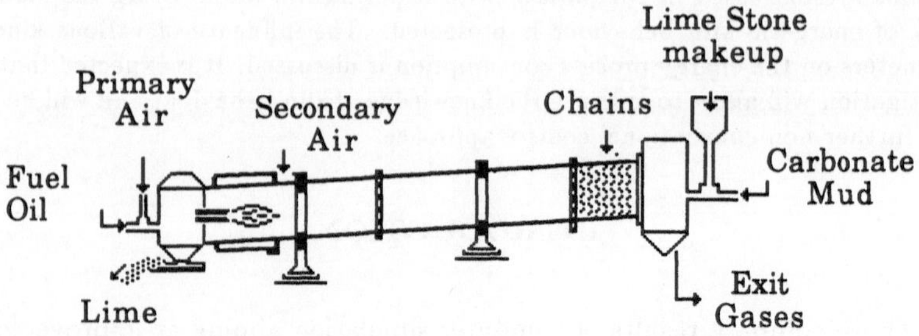

FIGURE 1: Rotary lime kiln calcination process

kiln highest extremity (see Figure 1) and turns down the kiln due to its inclination and rotation speed contacting the hot gases that flow countercurrently. During this process, lime mud suffers the drying, heating, calcination and burning processes until the final product (CaO) is obtained. The calcination reaction proceeds according to Equation 3:

$$CaCO_3 + Heat \longrightarrow CaO + CO_2 \tag{3}$$

The heat necessary for the above processes is obtained by fuel combustion which is injected at the lowest part of the kiln. Metalic chains are used in drying section to improve heat rate transfer from gas to the mud.

LIME KILN MATHEMATICAL MODEL

In Table 1 model dimensionless equations are presented with respective boundary and

Lime Kiln Model Dimensionless Equations				
		Solid Feed Bed		
Mass Balance	$\frac{\partial \overline{C}}{\partial t^*} + V_s^* \frac{\partial \overline{C}}{\partial z^*} = k^* \overline{r}$			(1)
Energy Balance	$(\frac{\partial T_s^*}{\partial t^*} + V_s^* \frac{\partial T_s^*}{\partial z^*}) = Q_s$			(2)
Material Transport	$\frac{\partial A_s^*}{\partial t^*} + \frac{\partial (V_s^* A_s^*)}{\partial z^*} + \frac{L}{V_N} \frac{D_s}{\rho_s} = 0$			(3)
Boundary Conditions	$z^* = 0$	$\overline{C}(z^*, t^*) = \overline{C}_B(t^*)$ $A_s^*(0, t^*) = A_{sB}$	$T_s^*(0, t^*) = T_{sB}$	(4)
Initial Conditions	$t^* = 0$	$\overline{C}(z^*, t^*) = \overline{C}_I(z^*)$ $A_s^*(z^*, 0) = A_{sI}^*$	$T_s^*(z^*, 0) = T_{sI}^*$	(5)
		Gas Phase		
Energy Balance	$\frac{\partial T_g^*}{\partial z} = Q_g$			(6)
Boundary Conditions	$z^* = 1$	$T_g^*(1, t^*) = T_{gB}^*$		(7)
		Kiln Wall		
Energy Balance	$\frac{\partial T_w^*}{\partial t^*} = F \frac{\partial^2 T_w^*}{\partial r^{*2}}$			(8)
Boundary Conditions	$r^* = 0$	$0 \leq z^* \leq 1$	$T_w^*(0, z^*, t^*) = T_g^*(z^*, 1)$	(9)
	$r^* = 1$	$0 \leq z^* \leq 1$	$\frac{\partial T_w^*}{\partial r^*} = Q_0$	(10)
Initial Conditions	$0 \leq r^* \leq 1$	$0 \leq z^* \leq 1$	$T_w^*(r^*, z^*, 0) = T_{wI}^*$	(11)

TABLE 1: Lime Kiln Mathematical Model

initial conditions. The equations are formulated [4][5][6] on the basis of several common assumptions: plug flow of material through the kiln, radiation and convection heat tranfer neglected in axial direction, no mixture of feed bed material in axial direction, feed bed specific heat and reaction heats constants, convection and emissivity coefficients constants, reaction rates determined by Arrhenius law and gas dynamics neglected due to its small residence time inside the kiln. Equations have been simplified with respect to the radial and angular coordinates, however, concerning the wall temperature equation, we also include radial variation. Since the wall represents an important thermodynamical link between kiln environment and its surroundings, the kiln performance may be improved if minimization of energy losses through the wall is achieved.

Right Hand Sides of Model Dimensionless Equations	
$r^* = A_i e^{-(\frac{E_i}{RT_s})}$	(12)
$Q_s = \frac{1}{\chi}\{\frac{R_{sg}}{A_s^*}(T_g^4 - T_s^4) + \frac{C_{sg}}{A_s^*}(T_g - T_s) + \sum_{j=1}^{m} Q_j\}$	(13)
$Q_g = \frac{1}{M_g^*}\{R_{gs}(T_g^{*4} - T_s^{*4}) + R_{gw}\frac{\partial T_w^*}{\partial r^*}\mid_{r^*=0} + C_{gs}(T_g^* - T_s^*) - M_B^* Q_F \frac{\partial C_F}{\partial z} + Q_{CO_2}^*(T_g^* - T_s^*) + Q_{H_2O}^*(T_g^* - T_s^*)\}$	(14)
$Q_0 = swa(T_a^{*4} - T_w^{*4}) + v_{wa}(T_a^* - T_w^*)$	(15)
$F = \frac{\alpha_w L}{V_N D}$	(16)
$D_s^* = D_{CO_2}^* + D_{H_2O}^*$	(17)

TABLE 2: Lime Kiln Mathematical Model

The dependent or state variables considered describe the distribution and transport of material inside the kiln, the chemical composition of the feed bed inside the kiln and the thermodynamical state of solid feed bed, gas and wall. Independent variables are axial coordinate, z^*, radial coordinate r^* and time coordinate t^*. The partial differential equations are first order hyperbolic except the last one which is parabolic of second order. Hyperbolic equations are defined in one dimension spacial domain, $T_z = \{z^* | z^* \in [0,1]\}$, whereas the parabolic equation is defined in two dimensions spacial domain, $\mathrm{T}_z \times \mathrm{T}_r$, $T_r = \{r^* | r^* \in [0,1]\}$.

NUMERICAL METHODS

Due to right hand sides nonlinearities presented in Table 2 of above equations an analytical solution is not workable and integration is obtained by selecting appropriate numerical methods. The methods of finite differences and orthogonal collocation on finite elements [2] were used for solution of partial differential equations. The finite difference scheme was selected according to the basic requisites: stability and consistency. The orthogonal collocation method belongs to the family of weighted residuals methods that are based on the approximation of the unknown function [8] by a linear combination of known functions. At any given instant of time t, each approximated solution of a component u_k is a piecewise polynomial of order previously choosen in each subinterval:

$$\tilde{u}_k = \sum_{i=1}^{N_{CPTS}} a_{ik}\Phi_i(x) \qquad k = 1..., N_{PDE} \tag{4}$$

where N_{CPTS} denotes the number of collocation points, N_{PDE} the number of partial differential equations, Φ the trial functions and the coefficients a_{ik} the unknowns to calculate. The trial functions were choosen under the condition to be linearly independent. In this work, cubic Hermite polynomial were selected since they assure the continuity of the function and its first derivative between consecutive elements. Equations 5 and 6 define the kiln wall temperature function and its first derivative approximations.

$$T_w(r^*, z^*, t^*) \approx \tilde{T}_w(u, z^*, t^*) = \sum_{i=1}^{4} a_{i+2k-2}(z^*, t^*)H_i \qquad k = 1, ...N_{Er} \tag{5}$$

$$\frac{\partial T_w(r^*, z^*, t^*)}{\partial r^*} \approx \frac{1}{h_k}\frac{\partial \tilde{T}_w(u, z^*, t^*)}{\partial u} = \sum_{i=1}^{4} a_{i+2k-2}(z^*, t^*)\frac{\partial H_i(u)}{\partial u} \; k = 1, ..., N_{Er} \tag{6}$$

where T_w is the kiln wall temperature and N_{Er} is the number of finite elements in which the radial coordinate was divided and H_i is the Hermite polynomial. Figure 2 shows that

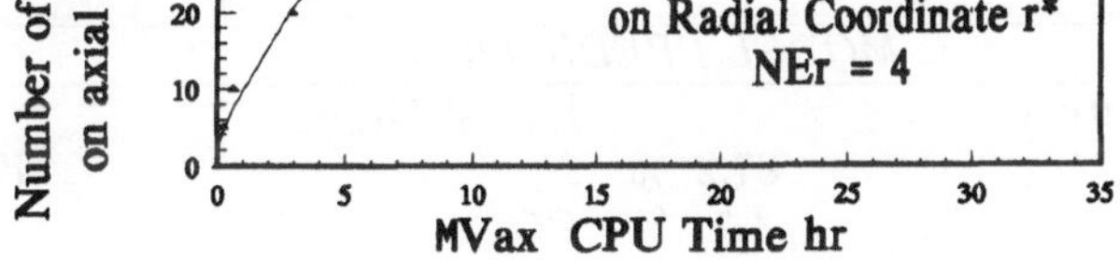

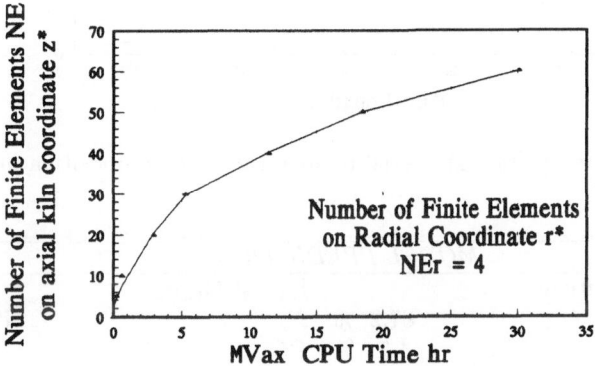

FIGURE 2: Computational cost in OCM

orthogonal collocation method on finite elements computational cost is higher than the finite difference method; however, the former presents better accuracy.

SIMULATION RESULTS

The model has been applied to an industrial unit with the operating and construction caracteristics shown in Table 3. Steady state kiln axial temperature and concentration profiles (Figure 3) obtained are compared with data from the real process and show a good agreement. Model predictions are the product composition at kiln outlet (hot end), combustion gas composition at kiln inlet (cold end), refractory peak temperature and performance index which are summarized in Table 4.

Kiln installation characteristics						
Kiln	Size (o.d× length in m)	Special Features	Process	Fuel	Production (tpd)	Energy (GJ/t)
	$3.0m \times 52.5m$	Satellite Coolers	kraft mill lime	oil	95-125	8-12

TABLE 3: Kiln features

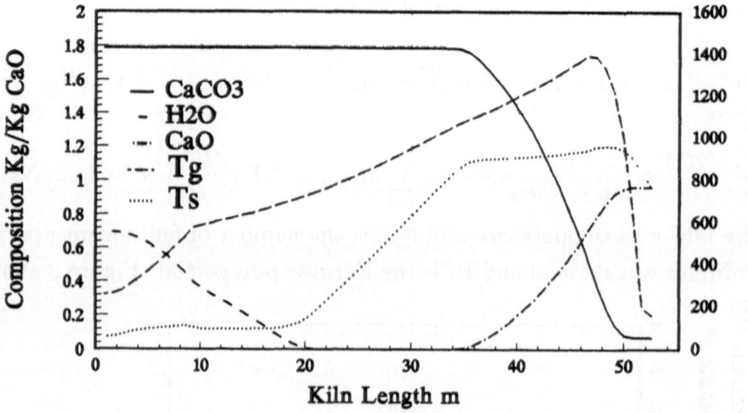

FIGURE 3: Steady state kiln axial temperature and composition profiles

MODEL PREDICTION				
Product Composition:		Exaust Gases		
CaO	87.2 %	N_2		72.56 %
$CaCO_3$	4.7 %	CO_2		25.67 %
Inertes	9.1 %	O_2		1.7 %
$T_{Product}$	757 C	$T_{exitgas}$		247 C
Refractory Peak Temperature		Performance		
$T_{refractory}$	1350 C	Performace Index	9.5	$GJ/t\ CaO$

TABLE 4: Steady State Simulation Results

Improving Kiln Efficiency

Influence of lime mud solids content, dust, burning conditions, feed impurities and rotation speed are analysed and discussed [1].

Lime Mud Solids Content and Feed impurities: Figure 4 shows that 5% reduction in water content of the mud fed to the kiln reduces the fuel requirement by about 500 MJ/t of kiln product. A reduction in water content of lime mud means that less heat is necessary

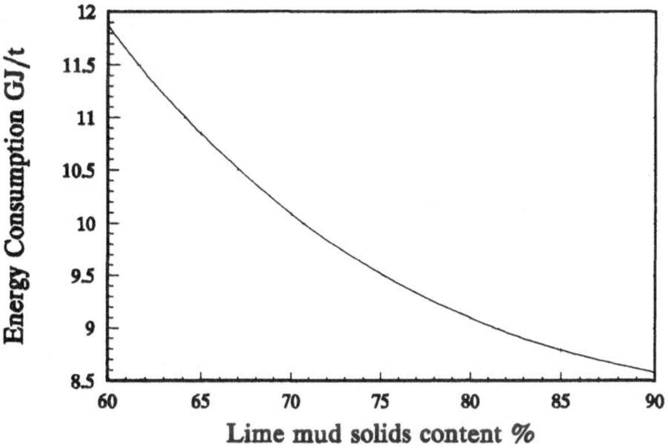

FIGURE 4: Effect of lime mud solids content on kiln performance

to warm up water in solids, vaporizate water and superheat the water vapor. Thus the fuel requirements are lower and energy savings higher. The reburning temperature (HET) at hot kiln end decreases with an increase in lime mud solids content reaching ≈ 980 C at the optimum solids content of 70%. Relatively to feed input impurities, an increase of 2% inerts results in 2.5% decrease in energy consumption because the calcination heat requirements are lower. However, the production rate is decreased; also minimization of feed impurities is required to avoid ring formation (or balls) which restrict production and kiln efficiency.

Combustion Air Control and Dust Minimization: Burning conditions and minimization of kiln dust circulation rate [7] influence the major kiln energy requirements. The amount of excess combustion air [3] reduces heat delivery from gas to the product. Also, incombustibles are increased and gases arrive at kiln feed end at higher temperature increasing energy losses. Simulation results show that an increase of 5% excess O_2 decreases % CaO in product by 6.25%, increase exit gas temperature by 5.7%, increase energy consumption by 2%. and increase dust by 6.7%. However, a certain amount of excess O_2 is necessary to control TRS. Relatively to dust circulation, computer results show that an increase of 2% in dust increase energy consumption by 1.1% and decreases the conversion in CaO by 2%. This is related with an optimum lime solids content of carbonate mud. If the moisture in lime mud increases or decreases with respect to a certain value, the production capacity also varies and therefore the energy consumption. An increase in solids content decreases the fuel consumption and increases dust losses (capacity and energy). Therefore an optimization of total costs requires an optimum solids content.

Dynamic Simulation

To evaluate the influence of kiln input parameters (induced draft rate, fuel rate, lime mud feed rate and speed of rotation) on output variables (reburning temperature, exit gas temperature, product quality and excess oxygen) step changes in kiln input parameters were achieved and dynamic kiln behavior observed. Figures 5 and 6 show a positive

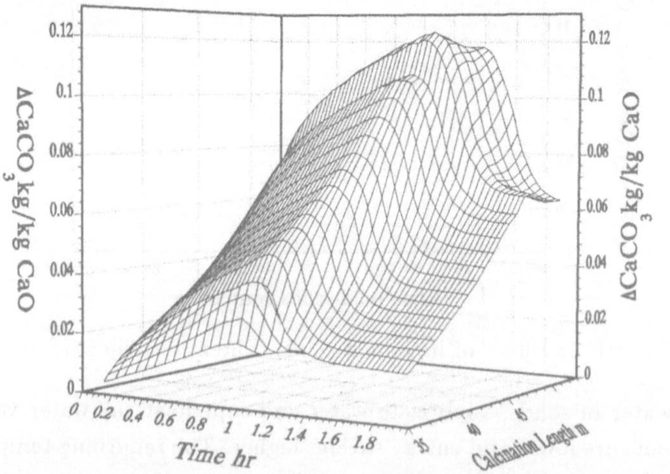

FIGURE 5: Histories of $CaCO_3$ variation

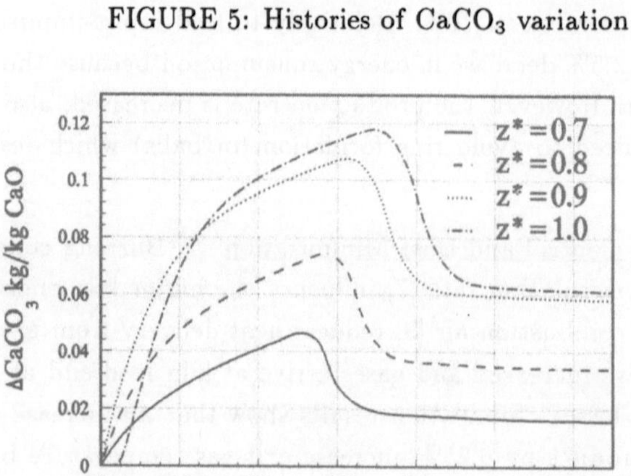

FIGURE 6: +5% Step change in kiln speed rotation

variation in $CACO_3$ composition to a +5% step change in kiln rotation speed. In this case, solids residence time decreases resulting in lower conversion in CaO. The maximum $CaCO_3$ variation occurs at the axial coordinate $z^* = 0.925$ in calcination zone and this wave of variation is propagated at solids velocity through kiln. Figure 7 shows time responses in Cold End Temperature (CET), Hot End Temperature (HET) and Burning

Zone Gas Temperature (BZT) when a +5% in draft flow rate is made. The heat is "pulled-

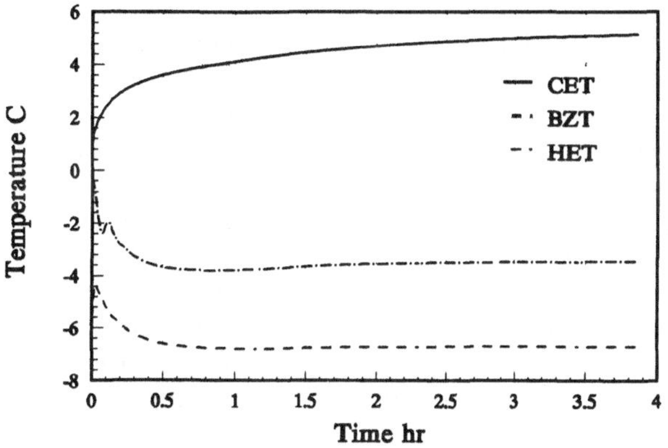

FIGURE 7: +5% Step change in draft flow rate

up" rising CET and lowering gas temperature (BZT) and solids temperature (HET) at hot kiln end.

CONCLUSIONS

A dynamic computer simulator of an industrial lime kiln was realized. A rigorous mathematical model incorporating heat and mass transfer mechanisms, kinetic of chemical reactions, material transport, and including the radial variation of temperature kiln wall and dust formation was used. The model was solved to simulate kiln behavior and has been tested against actual plant data obtained on a full size production kiln. Simulation results reproduce quite well the process in industrial environment and show that the kiln mathematical model is accurate enough and calibrated. Computer simulation was achieved for various input conditions at a given production rate and the desired operating conditions satisfying process limits at a minimum cost could be found. A parametric simulation study for improving kiln efficiency allowed to investigate the factors that influence kiln performance. Dynamic simulation results were presented for step changes on input variables and output variables expressing kiln transient behaviour were recorded. It is planned an advanced control system to drive kiln into stabilized operation which deals with non conventional approaches. It is considered that the lime kiln computer dynamic simulation presented will predict operational conditions for such industrial units and help in improving energy efficiency, product quality and equipment life. Further it is expected to use simulation curves for operator training and decision support.

ACKNOWLEDGMENT

Financial support from Instituto Nacional de Investigação Científica (INIC) and PORTUCEL is gratefully acknowledged

NOTATION

z^*	- kiln axial dimensionless coordinate	L	- kiln length
t^*	- time dimensionless coordinate	D	- kiln diameter
r^*	- kiln radial dimensionless coordinate	T_E	- scaling temperature
T^*	- dimensionless temperature	A_i	- frequency constant
T_s^*	- solids temperature	A_s^*	- cross section area feed bed
T_g^*	- gas temperature	D_s^*	- mass transfer coefficient
T_w^*	- wall temperature	R	- gas constant
Q_g	- heat supplied by gas	R_{ij}	- dimensionless radiation heat transfer coefficient
Q_s	- heat consummed or released by solids		
Q_0	- heat transfered between wall and environment	C_{ij}	- dimensionless convection heat transfer coefficient
Q_i	- heat consummed in chemical reactions	C_i	- composition of component i
r_j	- reaction rate j		

	Greek symbols		Subscripts				Superscripts
α_w	- thermal difusivity	g	- gas	s	- solids	$\tilde{u}$	- function approximation
Φ	- trial functions	w	- wall	B	- Boundary	$*$	- dimensionless variables
χ	- dust cycle	I	- Initial				

REFERENCES

[1] R. Bailey and T. Willison. Rotary lime kiln simulation - design technique for improving efficiency. *TAPPI*, 1:127–135, 1978.

[2] G. Carey and B. Finlayson. Orthogonal collocation on finite elements. *Chem. eng. Sci.*, 30:587–596, 1975.

[3] B. G. Jenkins and F. D. Moles. Modelling of heat transfer from a large enclosed flame in a rotary kiln. *Trans IChemE*, 59:17–25, 1981.

[4] B. Ribeiro and A. Dourado. Dynamic modelling and simulation of a pulp mill lime kiln. In *Modelling Identification and Control*, pages 330–333, IASTED, 1991.

[5] S. Saelid. *Modelling, Estimation and Control of a Rotary Cement Kiln.* PhD thesis, University of Trondhein, 1976.

[6] H. Spang. A dynamic model of a cement kiln. *Automatica*, 8:309–323, 1972.

[7] P. Uronen, H. Aurasma, and K. Leiviska. Static and dynamic modelling of a lime circulation loop. *Paperi ja puu*, 11:775–780, 1976.

[8] R. Vichnestsky. Use of functional approximation methods in the computer solution of initial value partial differential equations problems. *IEEE Transactions on Computers*, 22:1483–1501, 1969.

THE MODELLING OF THE GLASS PRODUCTION FURNACES

Mario Dente, Giulia Bozzano
Dipartimento di Chimica Industriale ed Ingegneria Chimica "Giulio Natta"
Politecnico di Milano, Italy

Ernesto Scorza, Vitaliano Torno, Gianfranco Bono
A.V.I.R. S.p.A., Italy

ABSTRACT

The glass production is one of the most energy intensive industries, so that an increasing attention has been provided in the last two decades towards a progressive rationalization of the process. The kernel of the process is constituted by the furnace, where the molten glass formation takes place. Significant savings on the energy consumption, quality of the product and general economy of the glass production have been obtained by improving its design and performance.

The rationalization of the glass production furnaces requires a better understanding of the fluid dynamic and heat transfer behavior both of the flames and flue gas and of the molten glass pool. Purpose of this work is to present a phenomenological and mathematical model that constitutes an essential tool for the improvements of the knowledge in this field. The type of furnaces examined is the so called "end-port" configuration. Attention has been specifically devoted to the description of the fate and rate of consumption of the composition blocks floating over the batch and conditioning, by means of the surface coverage, most of the heat transfer behavior of the furnace.

The effects induced on circulation movements of the molten glass by several mechanisms are taken into account: for example they are due to the presence of bubbling of air from the bottom of the glass bath, or natural convection promoted by electrical boosting or to superficial movements caused by the raw materials feeding devices, and so on.

These effects play a direct role in the heat transmission from the surface to the bottom of the bath. Several comparisons of the model prediction with data obtained from commercial furnaces of different geometry and operating characteristics are presented. The general agreement is satisfactory.

INTRODUCTION

The rising request of a better rationalization of the energy consumption, with the non secondary consequence of reducing the environmental effects produced by whatever fuel combustion, has put more interest in the last two decades, into a better understanding of the behaviour of kilns, ovens, furnaces. Sometimes, like in the case of glass industry, and particularly of the glass containers, the raw materials are cheap, the process (from a physico-chemical point of view) is simple, the source of energy and the way of giving it to the process (by combustion of a fuel with air) is well acquired. This is probably the most "mature" industry, in principle (since a lot of centuries). Relatively poor is the final product (the glass container), quite more expensive are the building materials of the furnaces, these ones are the real core of the process. The specific consumption of thermal energy for this product is relatively high, and therefore the environmental impact can be significant.

In the past two decades, but specifically in the very few last years, the interest for a better understanding of the mechanisms governing the glass furnaces behaviour has brought to a significant amount of research and mathematical modeling activities. After some pioneering activity, more finalized and very advanced works have been presented by Lockwood and oth. [2] and particularly by M.G. Carvalho and oth. [3,4,5,6]. This presentation will be concentrated essentially on some original aspects of our research, that is the composition blocks evolution [9], the estimation of the flames length [7], and some simplifications in solving the zoning method for the radiation problems [8]. The approach has been relatively elementary, however with a significant attention onto the phenomenological interpretation of the lot of experimental observations available from large commercial gas-fired end-port furnaces, producing glass containers. The three essential parts of the furnace that have been treated are: the combustion chamber, the glass batch and, more specifically, the interface between the two parts that is the composition blocks, their rate of consumption, their degree of surface occupation.

THE COMBUSTION CHAMBER AND THE LENGHT OF THE FLAMES

Due to their big size in the real cases the combustion chambers behaviour is dominated by the radiation (as could be easily demonstrated, and as it has been concluded by most of the researchers in the field), see for instance M.G. Carvalho and oth. [6]. As a matter of fact, the "jetting action" of the preheated air entering through the port (that is the inlet about every twenty minutes, when the symmetric other one is the outlet of the flue gas) is such to entrain an amount of gas over the length of the furnace, at least equivalent to 2.5÷3 times the total of the entering air. The furnace, in other words, is behaving as "well mixed". Therefore the "zoning method" proposed by Hottel and Sarofim [1], for evaluating the "direct exchange areas" and therefore the "total exchange areas" has been used for treating the global heat transfer problem. The presence of the gas, in terms of its absorbance and emissivity can be sufficiently described (as usually for combustion processes) by assuming the average properties of two gray gases and one transparent gas.

As it is well-known, the main difficulty in doing with heat radiation exchange (characterized by integral equation representing interactions surface-surface, surface-volume, volume-volume, and simplified to a sometimes very large set of algebraic equations by the zoning method), is the computational time needed for evaluating the direct heat exchange areas (together with the total number of volume and surface subdivision adopted). This fact can be dramatic for relatively small P.C. like those used often for the furnace on-line evaluation and control. In a recent work [8] original drastic simplifying rules, without a significant loss of accuracy, have been proposed and adopted for the time expensive evaluation of the "direct exchange areas". The net result has been to reduce the computing time of the combustion chamber walls, crown and gas temperatures, given the dispersion **heat transfer coefficients** and the **heat fluxes** towards the batch surface, to very reasonable values even with a large detail of the areas and the volumes.

The essential heat source (even in presence of electrical boosting under the batch surface) is given by the flames usually generated by three or four parallel burners under the ports. The evaluation of the length of the flames, their interference, and their volume is essential for the radiation heat fluxes to all the furnace, being them the primary source of heating. In a previous work [7] a method for predicting the multiple flames behaviour has been proposed, and verified on real commercial furnaces, with satisfactory results. With well designed burners the multiple flames have a global average length just reaching the position of the bubblers, or slightly more, without any U shape appearing. This is the case of the furnaces examined in this work.

As an example, in the case of Furnace 1 (3 parallel burners), the model is indicating a maximum length of the central flame of about 10 m.: the visual observation (see the picture enclosed) is indicating that the tip of the flame is just at the same level of the bubblers, about 10 m. from the end ports. The model proposed in that work evaluated the complete shape of the multiple flames: it shows that the ventilation of the preheated air makes the flames longer that the free single ones. For simplifying the calculations related to the radiation problems, the multiple flame has been assimilated to a rectangular parallelepiped with the average length, width and thickness of the real flame.

In the model simulation, another non-secondary aspect, is related to the alternance of firing: typically the time interval for the end-ports is 20 minutes. It has to be taken into account in evaluating the surface temperature of the internal refractories (crown and walls); as a matter of fact, the alternance has the effect (because of the heat capacity of the refractories) to smooth the maximum difference in the temperatures of the symmetrical opposite sides of the structure. The problem has been treated into a previous work [8] and the solution for that smoothing effect included into the present model.

FIGURE 1 Flames over the batch in a real furnace

THE GLASS BATCH MODEL

Most of the experimental informations usually available, regarding the batch behaviour are just related to the froth temperatures (sometimes), and to the floor temperature in some points along the central line. When more lateral temperature (e.g. close to the dog-houses) are available, they are indicating very modest temperature differences on the floor. It means that, practically most of the surface difference between the temperatures of the fired and non-fired sides are, practically, completely attenuated at the batch floor level, because of big heat transfer resistance of the bath. This fact suggests to use a two-dimensional model for the thermal balance equation into the batch, averaging over the surface transversal sections for the upper surface boundary conditions. As a local transversally averaged vertical velocity profile it is assumed a "law of similarity" that means:

$$v_x(y,x) \approx v_x(x)\, f(y)$$

where $v_x(x)$ is the mean average velocity of the melt, at the distance x from the end-ports. This simplifying assumption allows a simpler solution of the heat transfer equation into the batch, as presented in the work of G. Bozzano [9]. In the same just mentioned work, for taking into account the two main contributions to the vertical recirculating motions (helping the heat transport from the surface to the floor) an "effective vertical diffusivity" concept has been used. Together with the molecular and radiative diffusivity of the glass (depending on the glass type) the (eventual) contribution to the effective diffusivity given by the presence of boosters and bubblers, in the fusion region, has been estimated.

The resulting simplified heat transfer equation, better representing the experiments, is the following one:

$$v_x(x)\left(\frac{y}{h_{ve}}\right)^2 = \alpha_{eff}\frac{d^2T}{dy^2} - \frac{3\, q_{lat}}{\rho_{ve}\, Cp_{ve}}\left(\frac{y}{h_{ve}}\right)^2$$

(where the lateral heat dispersion have been lumped in a contribution per unit volume, varying parabolically with the distance from the floor, like the velocity) The vertical distribution of temperature resulting through the simplification, is in agreement with some available experimental data The surface ($y=h_{ve}$) boundary condition corresponds to consider that the net flux entering into the batch is that coming from the total radiation less that one devoted to the blocks heating, melting and heating of the produced froth The floor level ($y=0$) boundary condition corresponds to giving the heat dispersion through the floor (that is about a constant value)

The rate of consumption of the composition blocks

From the general behavior point of view, meaningful aspects of the furnaces producing glass containers, determining also the life of the (expensive) materials of their internals and the quality of the product, are
- the maximum temperature of the crown (that is conditioning the duration of the crown building materials)
- the surface temperature of the froth, over the melting batch, (that is strongly conditioning the corrosion rate of the refractory walls, around the flux line level)
- The temperature of the "floor of the melting batch", that is determining the "devitrification" of the produced glass incidentally, the strong dependence of the viscosity on the temperature, is such to reduce the effective residence time of the molten glass to a point where the solids and gases inclusions can hardly be removed from the batch (so creating defects into the product) **The rate of consumption of the composition piles or blocks is determining a lot of these aspects** The absorption of heat from the combustion chamber is essentially dominated by the "composition islands" as a matter of fact most of the energy has to be given for the glass forming reactions, for the melting and the heating at such a temperature to reach a viscosity sufficiently low to produce the glass and compensating the heat dispersion in the submerged areas

In a well-performed end-port furnace, for the production of "glass containers", boosted at the maximum load, about 70-80% of the total amount of the heat given to the bath are delivered to the composition piles for reacting and melting them the remaining part being devolved to further heating of the molten glass, and to the heat dispersions of the batch

Therefore the modeling of the rate of consumption of the "composition islands" is extremely useful for a realistic representation of the entire furnace behavior

It is important to point out that the presence of the composition piles induces buoyancy motions under the bottom-side of the blocks (because of the temperature difference existing between the surface of the molten glass batch and the adherent bottom-side of the composition blocks) These natural convection circulating motions, together with the ones induced by the eventual presence of electrical boosters (sometimes inserted into the melting batch), have such a distributed scale of development to make very unrealistic every pretension of detailed fluid-dynamics description into the molten glass batch

The proposed phenomenological model of the composition piles rate consumption is the following one

1) At the beginning, just after their feeding over the glass melting batch, the simple heating of the pile and the evaporation of the absorbed water is taking place Most of the heating to the powder, constituting the composition piles, is given by direct radiation to the upper side of the pile, natural convection is moderately contributing to the heating through the downside The temperature of the upper side surface is rising quickly, due to the internal poor heat transfer coefficient the bulk temperature of the composition islands is moderately modified The first phase of the evolution of the composition piles, after their exiting from the dog-houses over the melting batch is taking a few minutes Some reactions among solids powders begin to take place

2) When the upper surface temperature of the composition piles overcomes a critical temperature of about 1200°C (for lime-soda-silica glasses) the most important reactions start in producing the glass The cullet included into the composition also start to melt, helping the overall process Since this moment, the heat absorption arriving into the upper part of the blocks is about totally given to reaction, melting and heating of the glass-foam effluent from the surface of the pile The heating from the down-side is just contributing to rise the bulk temperature of the composition block As soon as the glass-forming reaction and melting start-up, the down flow of the just-formed glass foam takes place from the upper side of the composition A layer of glass foam is formed over the piles and flows over the batch-melt, the formed glass-foam moves its temperature rises over the batch, the froth is progressively liberating part of the gas bubbles content it releases still imperfect glass to the underlying batch, that little by little will give place to the refined glass Because of the large amount of bubbles inclusions and of the temperature level (150–250 °C more than the upper surface of the composition islands) the viscosity of the foam is quite lower than that of the corresponding glass, (so that it constitutes really a third phase over the bath, that moves quite fast towards the bubblers) Therefore the rate-determining step is essentially dominated by the internal heat transfer into the blocks

When the said critical surface temperature is reached, the thermal conductivity of the upper part of the block, where the molten glass is formed, progressively rises to a value close to that of the glass, so helping the internal heat diffusion The thickness of the block starts in being reduced From the downside of the block the heat transmission (through external natural convection and internal conduction) continues to contribute in increasing the average internal temperature of the unmelted part of the block [No penetration of melt into the pores of the unmelted material can take place]

A lateral consumption of the blocks contributes to reduce their horizontal section area, the linear rate of their lateral decrease in size, however, is only about one half of the vertical reduction in height so that when the final act of consumption of the block happens, it consists practically in a desegregation of the thin remaining block layer

The total consumption typically takes 20–30 minutes in a large end-port furnace (e g about 60–100 m^2 of batch surface) performed close to the maximum load This time is significantly depending on

a) the original thickness of the piles, as imposed by the composition throughput (i e by the furnace load), and by the characteristics of the batch feeders (particularly the stowing tool width, frequency and extension of the stow) A simplified expression giving the initial approximate average height of the piles for end-port furnaces with two symmetrical feeders is the following

$$H_{zo} = \frac{(1\ 25\ (1-f_r) + f_r)\ \mathrm{Load}}{v_{st}\ L_{st}}$$

b) on the thermal conductivity of the producing glass foam flowing over the piles (close to that of the molten glass at the same temperature) Therefore the instantaneous rate of consumption of the block, and its surface temperature, can be evaluated by solving the heat balance equation inside the block Accepting some simplifying hypotheses the blocks are assumed like infinite slabs, most of the heat diffusion into them can be considered unidimensional (in the vertical direction) The internal heat transfer flux is approximated by

$$q_{sup} = \{[(1-f_r)\ Cp_z + f_r\ Cp_{ve})]\ (T_{supz} - T_{z0}) + f_w\ Q_w + (1-f_r)\ Q_{reaz}\}\left(-\frac{dH_z}{dt}\right)$$

As a consequence the simplified expression for the local instantaneous thickness reduction of the block is the following

$$-\frac{dH_z}{dt} = \frac{U_{totsz}\ (T_{eqv} - T_z)\ (1\ 25\ (1-f_r) + f_r)}{\rho_z\left(Cp_{ve}\ \frac{U_{radz}\ (T_{eqv} - T_z)}{U_{radz} + U_{totsz}} + f_w\ Q_w + (1-f_r)\ Q_{reaz}\right)}$$

The heat flux given to the downside of the block (within the same approximation) is given by

$$q_{inf} = U_{totinfz}\ (T_{supb} - T_z)$$

3) The local batch surface temperature can be related to T_{supz}, ϕ_z etc, by the following balance, that takes into account the heating of the produced glass froth, the heating of the downside of the composition islands, the heat dispersion of the batch (through the side walls, the floor and the end sides) and the sensible heat given to the molten glass of the batch (the two last contributions are interactively given by the model of the glass bath)

$$T_{supb} = T_{supz} + \frac{(1-\phi_z) Q_{radve} - \phi_z U_{totinfz} (T_{supz} - T_z) - Q_{disp} + Q_{boost} - \rho_{ve} Cp_{ve} V_{ve} h_{ve} \frac{dT_{ve}}{dL_b}}{\phi_z U_{totinfz} + \frac{\phi_z\ Cp_{ve}\ cor\ U_{totsz}\ (T_{eqv} - T_z)}{(1\ 25\ (1-f_r) + f_r)\ (Cp_{ve}\ D + f_w\ Q_w + (1-f_r)\ Q_{reaz})} + (1-\phi_z)\ U_{radve}}$$

where

$$cor = 1 + \frac{h_z}{\sqrt{S_{z0}}}\sqrt{\frac{\phi_{z0}}{\phi_z}}$$

$$D = U_{radz} \frac{U_{radz}\,(T_{eqv}-T_Z)}{U_{radz}+U_{totsz}}$$

4) In the previous expression it appears the fraction ϕ_z of the surface blocks coverage (i.e. the total surface covered by the blocks divided by the total surface of the fusion part of the bath, till the bubblers, or the step position in absence of bubblers). The fraction ϕ_z can be deduced by comparing the rate of blocks thermal melting (dictated, as shown before, by the internal heat transfer into the blocks) with the material balance of the glass produced by the blocks. The resulting expression, by combination of the balances, if referred to the average of ϕ_z over all the bath, results as the following:

$$\phi_z = \phi_{z0}\left(1-\frac{h_{z0}-h_z}{2\sqrt{S_{z0}}}\right)^2$$

A similar expression can be used for the local instantaneous figure for ϕ_z, along the time of the evolution of the composition piles.

5) In the steady state, at least, and as a local average on the considered transversal section, the problem of the surface motions is playing a determining role. All the experimental evidence (through the video-cameras installed at the end of the furnaces) is showing that the composition islands move at such a velocity that they can hardly be just simply related to the reduction of the viscosity of the glass, at the highest temperatures of the surface. The interpretation of that phenomenon offered in this paper, following the experimental evidence, is the following one:

- the just generated glass froth, flowing down from the composition piles, is including gas and requires a certain amount of time to be freed from gaseous inclusions (in spite of the helping cullet fraction). Before of entering into the imperfect submerged formed "glass" (that is, still including the very small gas bubbles, to be freed through another significant amount of residence time) it forms a kind of superficial layer of glass froth, that is inclusive of relatively large bubbles, which liberation is not instantaneous. The layer is moving over the surface of the batch extremely faster than the submerged one: practically, its average movement is dominated by the composition blocks fusion rate. The froth layer we are referring to, has just a few centimeters of thickness (and it is conditioning the corrosion of the side walls at the neighboroughs of the flux-line, an experimentally very well known effect).

Most probably, the final releasing, even of the big bubbles (result of the coalescence of the smaller ones generated) through such a viscous medium, takes a relatively large amount of timeThe medium including large bubbles behaves like a distinguished mixed phase. In particularly it is different from the continuous glass phase (still including the smallest bubbles of gas) that is very slowly moving under the "fast" froth. The blocks of composition are, mainly, floating over and moving practically solidal with the superficial froth. Video-cameras, added (with control purposes) to some of the furnaces, show the movements of the composition blocks, and, therefore, the local velocity of the superficial froth layer. Moreover, the local and average amount of the batch surface coverage of blocks is made visible.

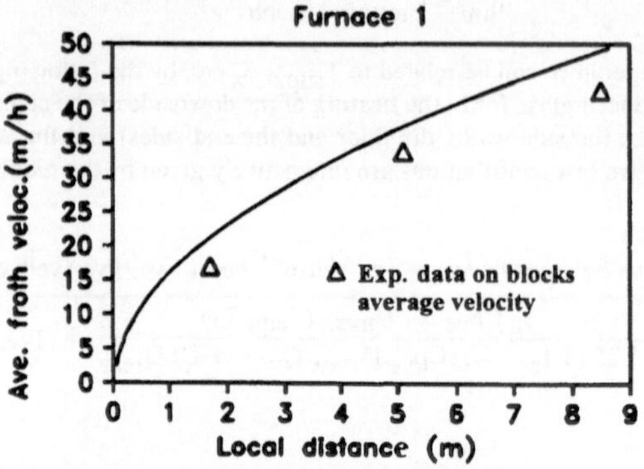

FIGURE 2 Superficial froth velocity averaged on the cross section: comparison with experimental data on average block velocity

The digital elaboration of the moving images have made possible (in some cases) the obtention of quantitative informations regarding the superficial velocities and the coverage degree of the batch (Of course, the elaboration of those data is not trivial for the P C connected to the video-cameras, particularly for deducing the surface velocity field)

Still most probably, from the experimental point of view, the local velocity of the froth has to be slightly larger than the observed velocity of the blocks, (the only one that can be digitally elaborated) The average velocity of the froth, has been modeled as a viscous layer continuously fed by the melting blocks (and continuously releasing glass to the underlying batch) The model estimates a few centimeters of thickness of the glass froth layer, and this range of figures is experimentally indirectly confirmed by the corrosion profiles of the side walls around the flux line

In Fig 2 a comparison of the calculated average traslational froth velocity at three different positions is presented, compared with the experimental figures deduced by the digital analysis of the video-cameras dynamic images, and together with the average coverage of the surface of the batch (till the bubblers line) Overposed onto the already mentioned froth surface motion (**traslational**) there is another significant **circulating** motion, induced by the "pumping action" of the batch feeders stowing tools (=the paddles)

The velocity in the twin vortex flow (for symmetrical feeding of the end-port furnaces) can be estimated by the paddles velocity, their width, and the width and length of the bath (till the barrier of the bubblers line)

The shape of the two circulating motion is about elliptical, (with a slight enlargement or restriction depending on the alternate presence or absence of the flames over the region) The result of the overposition is the presence, close to the central line, of a relatively fast stream directed towards the throat of froth (and blocks thereof) that can reach (at the maximum load) 70÷80 m/h (as detected in the case of Furnace 1 , Table 1) As a compensation, and close to the walls, a return stream towards the ports is appearing, characterized in the same case by velocities in the range of 20÷30 m/h at half the fusion region The net result is to offer to the composition blocks some recirculation opportunity, increasing their superficial residence time, till to their complete transformation into glass froth (However, if the load is exceeding a maximum, typical of glass type, furnace area and heating power compatible with the maximum allowable materials temperatures, part of the blocks can appear, after the bubblers and the step, over the refining region, worsening the behaviour of the furnace and the quality of the product A typical limitation to the specific load is in the range of 150 $Kg/m^2/h$)

Finally, the presence of the bubblers line induces further motions at the batch surface and particularly by generating return horizontal movements directed towards the ports (of course, at the expense of the circulation of the glass entrained by the rising big bubbles generated by the bubblers) The return velocity of this effect is strongly depending on the number of bubblers inserted on the line and on the flow of air for each bubbler The largest part of the horizontal extension of this effect can be approximately estimated equivalent to 2 0 times the depth of the batch

As a consequence, for such a distance, the froth and therefore the blocks are slowed down for the corresponding amount of the return velocity In order to give just an example, referring to a line of 15 bubblers, fed with a 0 5 Nm^3/h each one uniformly spaced on a width of 7 5 m, the resulting superficial return superficial return velocity is about 15 m/h at the bubblers line, declining to about 0 at 2 5 m behind A further aspect, that promotes an increasing of the froth velocity, is the presence of the step in the basin (However it can justify only differences between predicted and experimental figures of few m/h)

All these effects have been quantified in an approximated way in the model for the composition blocks consumption, obtaining satisfactory results, at least for the amount of available experimental cases

Some of the available experimental data regards the maximum (downward positive) velocity close to the center-line of the batch surface and the minimum (upward, negative) velocity close to the side walls The model offers predictions based on the overposition of the three main, previously mentioned, mechanisms Because of the existing uncertainties into the experimental data, we have preferred to refer to average figures, mediating between the under-flame and out-of the flame regions The Table 5 shows the comparison of the maximum (downward) and the minimum (upward) velocities at some points of the fusion region, in the case of Furnace 1

The Tables 1,2,3,4,5 and figures 3,4,5 show a very satisfactory agreement with real experimental data

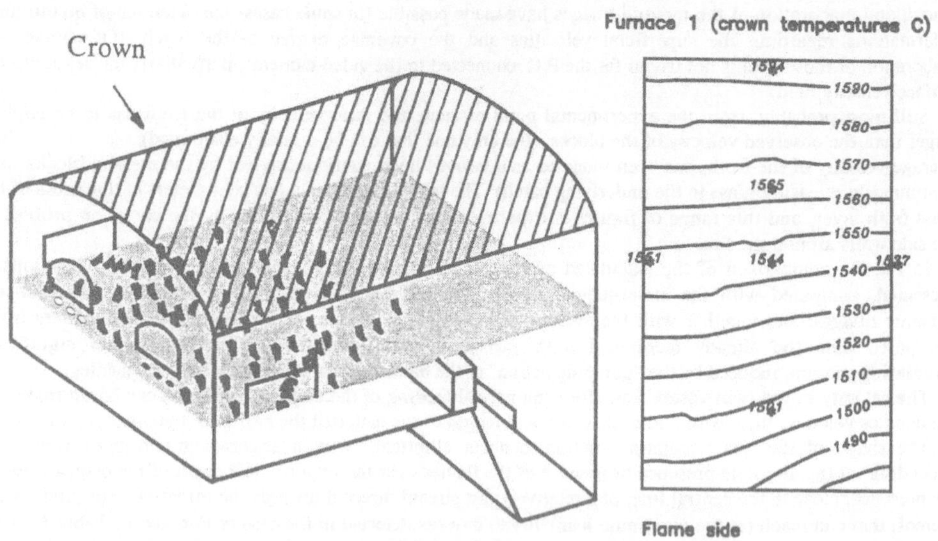

FIGURE 3 Sketch of a tipical modern end-port furnace and its crown temperature distribution (compared with the experimental data)

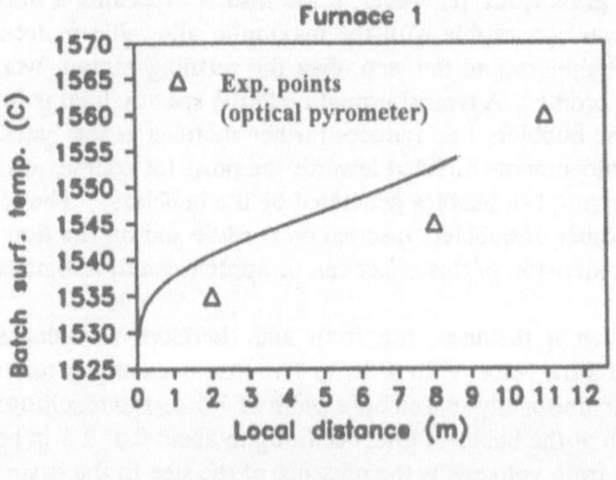

FIGURE 4 Froth surface temperature,average over the sections

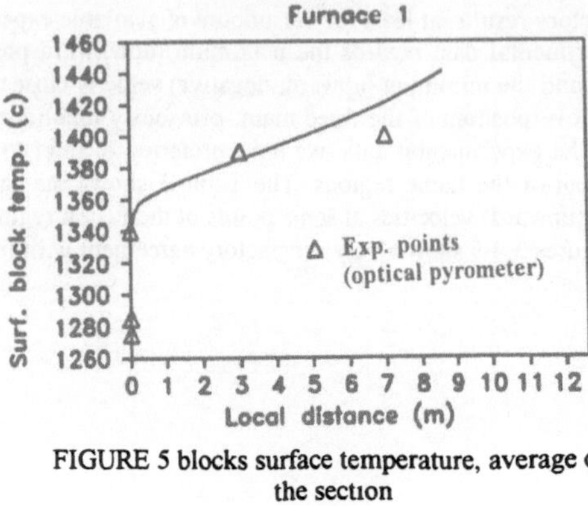

FIGURE 5 blocks surface temperature, average on the section

TABLE 1

Comparison between field data and model prediction

Furnace 1 — **Glass colour: green**

Geometrical data	Operating conditions	
Bath area 94 m^2	Load 298 ton/day	
Lenght 12 6 m	Cullet fraction 46 %	
N° of bubblers 15	Throughput/bubbler 0 3 m^3/h/bu	
Width of the stow pad 0 55 m/h	Electrical boosting power 1080 Kw	
Vel of the stowing paddle 134 4 m/h	Natural gas 1415 Stm^3/h	
	Exc air on st 1 11 Stm^3/Stm^3 St. air	
	Parasitic air 0 1 Stm^3/Stm^3Stoich air	
Variables	**Exp.**	**Sim.**
Tfloor(fusion region)	1358°C	1360°C
Tfloor(dog-houses region)	1332°C	1326°C
ϕz	32 %	31 %
Tsup froth(fusion region)	1545°C	1550°C
Tsup froth(dog-houses region)	1535°C	1534°C
Tcrown(dog-houses region)	1595°C	1500°C
Tcrown(fusion region)	1540°C	1540°C
Tcrown(refining region)	1597°C	1598°C
Tside wall(expose flame)	1586°C	1583°C
Tside wall(against flame)	1570°C	1573°C

TABLE 2

Comparison between field data and model prediction

Furnace 2 — **Glass colour: amber**

Geometrical data	Operating conditions	
Bath area 77 m^2	Load 235 ton/day	
Lenght 11 5 m	Cullet fraction 34 4 %	
N° of bubblers 15	Throughput/bubbler 0 4 m^3/h/bu	
Width of the stow pad 0 5 m/h	Electrical boosting power 948 Kw	
Vel of the stowing paddle 78 4 m/h	Natural gas 1093 Stm^3/h	
	Exc air on st 1 1 Stm^3/Stm^3 St air	
	Parasitic air 0 07 Stm^3/Stm^3Stoich air	
Variables	**Exp.**	**Sim.**
Tfloor(fusion region)	1350°C	1355°C
Tfloor(dog-houses region)	1320°C	1324°C
ϕz	n a	18 %
Tsup froth(fusion region)	1573°C	1577°C
Tsup froth(dog-houses region)	n a	1560°C
Tcrown(dog-houses region)	1475°C	1490°C
Tcrown(fusion region)	1565°C	1550°C
Tcrown(refining region)	1562°C	1562°C
Tside wall(expose flame)	n a	1540°C
Tside wall(against flame)	1530°C	1528°C

TABLE 3

Comparison between field data and model prediction

Furnace 3 — **Glass colour: white**

Geometrical data	Operating conditions	
Bath area 56 m^2	Load 170 ton/day	
Lenght 6 14 m	Cullet fraction 15 %	
N° of bubblers 0	Throughput/bubbler 0 m^3/h/bu	
Width of the stow pad 0 45 m/h	Electrical boosting power 450 Kw	
Vel of the stowing paddle 108 m/h	Natural gas 860 Stm^3/h	
	Exc air on st 1 1 Stm^3/Stm^3 St air	
	Parasitic air 0 07 Stm^3/Stm^3Stoich air	
Variables	**Exp.**	**Sim.**
Tfloor(fusion region)	1279°C	1286°C
Tfloor(dog-houses region)	1254°C	1245°C
ϕz	n a.	27 %
Tsup froth(fusion region)	1550°C	1552°C
Tsup froth(dog-houses region)	n a	1525°C
Tcrown(dog-houses region)	n a	1495°C
Tcrown(fusion region)	1548°C	1545°C
Tcrown(refining region)	1566°C	1560°C
Tside wall(expose flame)	1542°C	1539°C
Tside wall(against flame)	1530°C	1525°C

TABLE 4

Comparison between field data and model prediction

Furnace 4 — **Glass colour: semiwhite**

Geometrical data	Operating conditions	
Bath area 66 m^2	Load 192 ton/day	
Lenght 10 5 m	Cullet fraction 14 %	
N° of bubblers 0	Throughput/bubbler 0 m^3/h/bu	
Width of the stow pad 0 5 m/h	Electrical boosting power 0 Kw	
Vel of the stowing paddle 65 1 m/h	Natural gas 1062 Stm^3/h	
	Exc air on st 1 1 Stm^3/Stm^3 St air	
	Parasitic air 0 1 Stm^3/Stm^3Stoich air	
Variables	**Exp.**	**Sim.**
Tfloor(fusion region)	1279°C	1279°C
Tfloor(dog-houses region)	1240°C	1235°C
ϕz	n a	16 %
Tsup froth(fusion region)	1554°C	1556°C
Tsup froth(dog-houses region)	n a	1525°C
Tcrown(dog-houses region)	1480°C	1501°C
Tcrown(fusion region)	1547°C	1549°C
Tcrown(refining region)	1567°C	1571°C
Tside wall(expose flame)	1577°C	1568°C
Tside wall(against flame)	1565°C	1554°C

TABLE 5
Maximum (downward) and minimum (upward) velocities in the fusion region

Position (from end side port) (m)	Maximum velocity (m/h)		Minimum velocity (m/h)	
	Exp.	Sim.	Exp.	Sim.
2.0	+63.	+64.	-30.	-33.0
5.5	+70.	+60.	-24.	-20.
9.0	+80.	+85.	-20.	-28.

CONCLUSIONS

Some aspects, of the modeling of end-port glass furnaces have been presented. More specifically the attention has been concentrated on the mechanism of the composition blocks consumption and surface coverage. Some simplifying hypotheses for the behaviour of batch melt have allowed to easily solve the heat transfer equation and to deduce the values for the floor temperatures The interaction with the combustion chamber radiation equation gave back the determination of the refractory temperatures, in the most critical positions. The obtained results compare in a satisfactory way with several data of real industrial furnaces

NOMENCLATURE

h_z height of the pile
ϕ_z fraction of the batch covered by the piles
Q_{radve} heat flux towards the molten glass batch
Q_{disp} heat loss of the melting batch (per unit area)
Q_{boost} heat generated by the boosters (per unit area)
ρ_{ve} glass density
V_{ve} average molten glass velocity
L_b local distance
h_{ve} heigth of the melting glass batch
T_{ve} average melting glass temperature
U_{radve} radiative equivalent heat transfer coefficient to the melting glass
S_{z0} initial pile horizontal surface
ϕ_{z0} initial pile coverage of the batch
h_{z0} pile initial heigth
Q_{lat} side-walls heat dispersion, referred to the unit volume of the batch
T_{supb} batch surface temperature
v_{st} paddle velocity
L_{st} paddle width
T_{z0} composition initial temperature
f_w composition water content
Q_w heat of evaporation of water
Q_{reaz} heat of reaction
U_{totsz} pile surface globale heat exchange coefficient
T_{eqv} equivalent radiative temperature
T_z composition bulk temperature
ρ_z composition density
U_{radz} radiative equivalent heat transfer coefficient to the blocks
$U_{totinfz}$ global heat transfer coefficient at the downside of the blocks
f_r cullet fraction
Cp_z pile specific heat
Cp_{ve} glass specific heat
T_{supz} pile surface temperature

REFERENCES

1 Hottel, Sarofim, "Radiation Transfer", Mc Graw-Hill, (1967)
2 Lockwood, F C and Shah, N G (1980) "A new Radiation Solution Method for Incorporation in General Combustion Prediction Procedures" 18th Symp (Int) on Combustion, The combustion Institute

3 Carvalho, M G and Lockwood, F C (1985), "Mathematical Simulation of an End-Port Regenerative Glass Furnace", Proc Inst Mech Engrs , 199, (C2), pp 113-120

4 Carvalho, M G , Oliveira, P and Semião, V (1988), "A Three-Dimensional Modelling of an Industrial Glass Furnace", Journal of the Institute of Energy, September, 448, pp 143-156

5 Carvalho, M G and Nogueira, M (1990), "Mathematical Modelling of Heat Transfer in an Industrial Glass Furnace", Eurotherm Seminar No 17, Heat Transfer in radiative and Combusting Systems, October, Cascais, pp 8-16

6 Carvalho, M G , Nogueira, M (1991) "Glass Quality Evaluation via 3-D Mathematical Modelling of Glass Melting Furnaces", E S G Conference on the Fundamentals of the Glass Manifacturing Process, Sheffield, Sept

7 Borghini, A , (1989), "La fluidodinamica delle fiamme diffusive turbolente e lo scambio termico per irraggiamento in una fornace dell'industria vetraria" Thesis

8 Mai, S and Vailati Venturi, G , (1990), "Studio Teorico-Sperimentale dei fenomeni relativi alle fiamme turbolente, all'irraggiamento e scambi di calore nei forni dell'industria vetraria", Thesis

9 Bozzano, G (1991), "I forni per la produzione del vetro aspetti fluidodinamici e di scambio termico nel bacino ed analisi del comportamento non stazionario dell'intero forno", Thesis

NUMERICAL MODELLING OF CERAMICS FIRING KILNS APPLICATION TO THE IMPROVEMENT OF ENERGY EFFICIENCY AND PRODUCT QUALITY

P.LYBAERT, M.EL HAYEK, H.MEUNIER
Laboratoire de Thermique,
Faculté Polytechnique de Mons,
rue de l'Epargne, 56, B-7000 MONS, BELGIUM

ABSTRACT

In batch firing of ceramic products, product quality and furnace heat consumption are strongly dependent on a good temperature uniformity in the charge stackings. Hence, numerical simulators aiming at improving furnace operation and design have to be able to calculate the temperature field inside the charge. The purpose of this paper is to describe such a computational model, applicable to gas fired batch furnaces.
Model potentialities are illustrated by the application to the firing of refractory bricks. The influences of furnace geometry and burner control strategy on load temperature field and furnace efficiency are discussed.

INTRODUCTION

In batch firing of ceramic products, the load is placed inside a refractory-lined chamber of rectangular cross section (figure 1). The pieces of ware to be fired are arranged as one or several piles or stacks on the hearth. Heating is most often provided by gas burners. During the firing process, furnace atmosphere and composition are made to vary according to a given schedule.

In this process, final product quality is essentially dependent on a good temperature uniformity in the stackings: every piece of ware must have a similar thermal history. Hence, modern firing practice relies on different equipments and techniques which are claimed to promote uniformity, use of high velocity ("jet") burners and sequential control systems being the most used currently.

Although these techniques are now widely spread, their effectiveness has not yet received complete quantitative evidence and the ceramic industry still uses conservative firing cycles, i.e. with low heating rates and unuselessly long firing periods, in order to avoid unevenly fired ware.

Optimization of the firing cycle, which should lead to a better energy efficiency of the firing process while maintaining or even improving product quality, thus requires a better knowledge of the factors affecting temperature distribution inside the load. The purpose of the work is to use computational techniques to gain this knowledge.

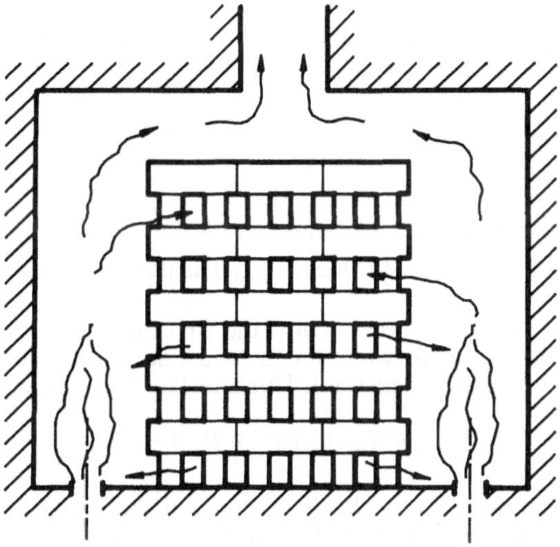

Figure 1. Furnace and charge geometry (not to scale).

The work is part of a cooperative research project entitled "Energy saving and pollution abatement in glass-making furnaces, cement kilns and baking ovens", supported by the EEC within the JOULE program. Eleven contractors are involved in the project: I.S.T. - Lisboa (P), project leader, Imperial College of Science - London (U.K.), C.R.E.S. - Athens (Gr), Instituto da Energia - Lisboa (P), University of Erlangen-Nurnberg (D), TNO - Environmental and Energy Research (N), Faculté Polytechnique de Mons (B), Université Libre de Bruxelles (B), Institut National du Verre (B), Adersa (F), TNO - Cereals, Flour and Bread Institute (N).

The simulation model will be first described. Model potentialities are then illustrated by the application to the firing of refractories. The influences on temperature uniformity of different factors - furnace geometry, location of burners and flue gas outlet, burner control strategy - are discussed.

DESCRIPTION OF THE MODEL

Load temperature non-uniformities in ceramics firing kilns result from different heat transfer phenomena: convection with the combustion gases, radiation through the gas and between combustion gases and ware surface, as well as piece to piece radiation inside the stacks, and conduction inside ware material. Hence, numerical models aiming at improving furnace operation and design have to be able to compute combustion gas flow and all these heat transfer components everywhere in the load.

Load arrangement is much too complicated to allow a detailed flow and heat tranfer calculation around every piece of ware. Therefore, the model relies on a simpler approach: the load is replaced by an "equivalent" anisotropic gas-solid porous medium, having macroscopic properties (volume porosity, directional friction factors,...). The equations describing the porous medium are then obtained by writing out mass, momentum and energy balances on a porous volume element. This modelling approach has proved to be fairly useful for designing heat exchangers in the nuclear industry [1,2].

Modelling Heat Transfer in the Solid Phase

The time-dependent ware temperature field is given by the energy conservation equation applied to the solid part of the load, which can be written

$$\frac{\partial}{\partial \tau}((1-\gamma_v)\rho_s c_s T_s) = \frac{\partial}{\partial x_i}\left(\lambda_{s,app}\frac{\partial T_s}{\partial x_i}\right) + (1-\gamma_v)Q_{s,r} + Q_{gs} \qquad (1)$$

In this equation, γ_v is the volumetric porosity of the porous medium, and $(1-\gamma_v)$ is the volume fraction of ware material in the load. $\lambda_{s,app}$ is the "apparent" thermal conductivity of the packing.

The equation includes two source terms: the first one, $Q_{s,r}$, is due to the interaction of the solid with the radiation field, while the second one, Q_{gs}, accounts for convective and radiative heat transfer between gas and solid. This interfacial volumetric heat rate can be written

$$Q_{gs} = h_{gs}A_{gs}(T_g - T_{si}) + Q_{gs,r} \qquad (2)$$

with h_{gs} being the gas-solid convective heat transfer coefficient and A_{gs} the interfacial area per unit load volume. The expressions used for the radiative components $Q_{s,r}$ and $Q_{gs,r}$ will be detailed later on.

Equation (2) takes into account that thermal conduction limits heat penetration into the solid, inducing temperature differences between the surface and the core of the ware. These temperature gradients are introduced in (2) by using a local interfacial solid temperature T_{si} instead of the local average solid temperature T_s. These temperatures are related to each other by expressing flux continuity at the interface, i.e.

$$h_{gs}A_{gs}(T_g - T_{si}) + Q_{gs,r} = h_{si}A_{gs}(T_{si} - T_s) \qquad (3)$$

In this relation, h_{si} is an "internal conductive heat transfer coefficient" that accounts for heat conduction inside the solid phase. h_{si} values can be found in the literature (see [3] for instance) for solids of simple shape submitted to a step change of temperature. Clearly, these values have to depend on time but a lower *constant* limit can be found for large times, when successive temperature profiles are similar to each other. This value is given by

$$h_{si} = C\lambda_s\left(\frac{S}{V}\right)_s = C\lambda_s\frac{A_{gs}}{(1-\gamma_v)} \qquad (4)$$

The constant C is a function of the shape of the solid. Orders of magnitude of C can be found by considering one dimensional bodies for which C values range from about 2.5 (slab) to 1.1 (sphere) [3].

Combustion Gas Flow and Convective Heat Transfer Modelling

The gas velocity and temperature fields are computed from the fluid mass, momentum and energy conservation equations. The following simplifying assumptions are made to derive the equations:

- combustion phenomena are neglected, i.e. the burners deliver combustion gases at theoretical (adiabatic) flame temperature;
- furnace atmosphere is incompressible;
- turbulent flow is assumed to prevail everywhere in the furnace.

Inside the stacks, according to Sha [1,2], the intersticial gas phase conservation equations have then the following general form, in tensor notation,

$$\frac{\partial}{\partial \tau}(\gamma_v \rho_g f) + \frac{\partial}{\partial x_i}(\gamma_i \rho_g u_i f) = \frac{\partial}{\partial x_i}\left(\gamma_i \Gamma_f \frac{\partial f^*}{\partial x_i}\right) + \gamma_v S_f - F_{gs,f} \tag{5}$$

with f being the conserved property (mass, momentum or energy) and f^* being the potential associated to the diffusive transport of f. In this equation, the γ_i's are the surface permeabilities of the porous medium. The expressions of the terms of the different transport equations are given in Table 1.

TABLE 1
Terms of the gas phase transport equations

Property	f	f^*	Γ_f	S_f	$F_{gs,f}$
Mass	1	0	0	0	0
Momentum	u_i	u_i	$\mu + \mu_t$	$-\frac{\partial p}{\partial x_i} + \rho_g g_i + S_{u,i}$	$R_{gs,i}$
Enthalpy	h_g	T_g	$\lambda + \lambda_t$	$Q_{g,r}$	Q_{gs}
Turbulence energy	k	k	$\mu + \mu_t/\sigma_k$	$P_k - \rho\epsilon$	$-P_{gs,k}$

with

$$S_{u,i} = \frac{\partial}{\partial x_j}\left(\mu_{eff}\frac{\partial u_j}{\partial x_i}\right) - \frac{2}{3}\frac{\partial}{\partial x_i}(\mu_{eff}\nabla\cdot\vec{u} + \rho_g k)$$

and

$$P_k = \mu_t\left(\frac{\partial u_i}{\partial x_j} + \frac{\partial u_j}{\partial x_i}\right)\frac{\partial u_i}{\partial x_j} + \beta g_i \frac{\mu_t}{Pr_t}\frac{\partial T_g}{\partial x_i}$$

The $F_{gs,f}$'s are distributed source terms accounting for the gas-solid interactions, i.e. gas-solid fluxes of f per unit volume of porous load. In the energy equation, it is the volumetric interfacial heat rate Q_{gs} given by equation (2). In the momentum equations, they are the three components $R_{gs,i}$ of the distributed flow resistance. They can be expressed from the velocity and the directional friction factors ξ_i as

$$R_{gs,i} = \gamma_v \xi_i \frac{1}{D_h}\frac{\rho_g u_i|\vec{u}|}{2} \tag{6}$$

The convective heat transfer coefficient and directional friction factors are dependent on the local velocity magnitude and direction, and are functions of the packing geometry. Empirical correlations have to be used to estimate their values. As heat transfer and pressure drop data on stackings of pieces are rather sparse in the open litterature, bench scale experiments have to be conducted to determine these correlations.

Turbulence is taken into account by using effective values of viscosity and thermal conductivity. Those turbulent values are calculated from the turbulence energy k and its dissipation rate ϵ by

$$\mu_t = C_\mu \rho_g \frac{k^2}{\epsilon} \quad \text{and} \quad \lambda_t = \frac{\mu_t c_g}{Pr_t} \tag{7}$$

Turbulence energy is given by its transport equation which is of the same form as equation (5) (see table 1). Gas-solid interactions result in an additional production term given by the relation

$$P_{gs,k} = C_k u_i R_{gs,i} \tag{8}$$

which simply states that part of the friction energy loss is converted into turbulence energy.

The dissipation ϵ is calculated from a mixing length l_m by

$$\epsilon = k^{3/2} / l_m \tag{9}$$

with the mixing length l_m being assumed to be proportional to the hydraulic diameter D_h of the packing, i.e.

$$l_m = C_l D_h = C_l \left(\frac{4\gamma_v}{A_{gs}} \right) \tag{10}$$

The free space, i.e. the space located between the furnace walls and the stacks as well as the empty space between adjacent stacks, is a homogeneous fluid volume that is described by the single phase fluid conservation equations. Those are identical to equations (5) with the porosity value and surface permeabilities being set equal to unity and the fluid-solid interaction terms dropped out. As to the turbulent viscosity and conductivity, they are calculated by the classical two-equation (k, ϵ) turbulence model [4].

Radiation Model

The radiation model is described in [6]. As for the computation of the fluid flow field, a detailed calculation of radiative heat transfer around every piece of ware is impossible. For radiation modelling, we will therefore follow the same approach as the one that is used to model the fluid flow: the charge is considered as an "equivalent" porous medium. The equations describing radiant heat transfer in the load are then obtained by a "cell model" of the porous medium: the porous structure is considered as a regular assembly of elementary cells (fig.2), on which radiative flux balances in the six directions are written out.

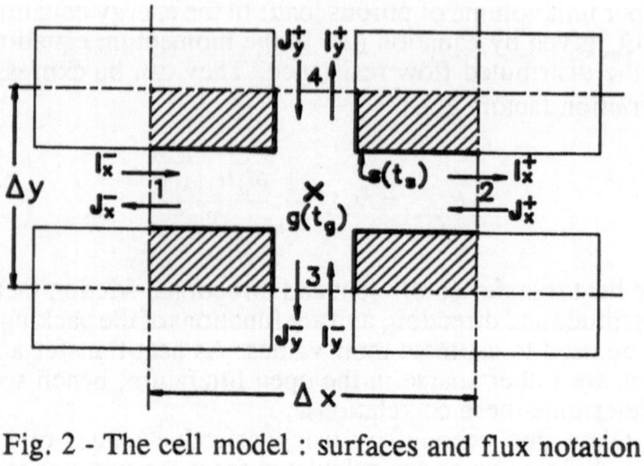

Fig. 2 - The cell model : surfaces and flux notation

These equations leads to a six flux approximation of the radiant transfer equation and are handled as in Gosman and Lockwood's work [7]. If I_i and J_i are the radiation fluxes in the positive and negative x_i directions, we define flux sums F_i by

$$F_i = (I_i + J_i)/2 \tag{11}$$

The F_i's are then given by a set of three differential equations i.e.

$$\gamma_i \frac{d}{dx_i}\left(\frac{2\gamma_i}{a_{si} + a_{gi} + 2s_{-ii} + 2s_{ij}} \frac{dF_i}{dx_i}\right)$$

$$+ 2a_{si}(\sigma T_{si}^4 - F_i) + 2a_{gi}(\sigma T_g^4 - F_i) + 4s_{ij}(F_j - F_i) = 0 \tag{12}$$

This model introduces the following volume-averaged radiative properties to account for the interaction of radiation with the porous medium: the directional solid and gas absorption coefficients, a_{si} and a_{gi}, account for absorption of radiant energy by the solid and gas phases respectively, the scattering coefficients s_{ij}'s are related to the fraction of the i-direction radiative flux that is reflected by the solid surface towards the j-direction. These coefficients are functions of the radiative properties of the fluid (absorption coefficient), of the surface properties of the solid (emissivity) and of the packing geometry (size and geometrical arrangement of the pieces). If the solid surface is perfectly diffuse and grey, they can be calculated by applying Hottel zone method [8] to the unit cell of the packing.

The absorption of radiant energy by the gas and by the solid leads to source terms in the gas and solid energy conservation equations which are evaluated by

$$Q_{g\ r} = 2a_{gi}(F_i - \sigma T_g^4) \tag{13}$$

and

$$Q_{s,r} = 2a_{si}(F_i - \sigma T_{si}^4) \tag{14}$$

while the interaction term associated to radiative heat transfer between the gas and the solid is given by

$$Q_{gs,r} = \left(\frac{GS_s}{V}\right)(\sigma T_g^4 - \sigma T_{si}^4) \tag{15}$$

with (GS_s/V) being the gas-solid total exchange area per unit porous volume.

This model is coupled with a classical six-flux radiation model in the free space [7]. The equations are identical to the above equations with the surface permeabilities set equal to one, scattering and solid absorption coefficients equal to zero and equal values for the gas absorption coefficients in the three directions.

Numerical Solution of the Equations

The model equations are discretised by a classical finite volume technique [9], using staggered grids for the different variables: velocity components are stored at the faces, while the other variables are located at the centre of the volume elements. A full implicit (Euler) method is used for the time integration. At every time step, the SIMPLER algorithm is used to solve the nodal equations.

APPLICATIONS

The model has been applied to the firing of refractory bricks. Two furnace geometries have been used. The first one is a furnace of rectangular section that can be approximated by a two-dimensional geometry. The second one is a typical industrial "cell furnace" that has to be simulated by a three-dimensional model.

2-D Furnace

Furnace and load geometry. The cross section of the 2-D furnace is represented on figure 1. The chamber is 1.9 m high and 2.3 m wide. The furnace is equipped with two rows of burners located in the hearth on both sides of the load. Each row of burners is simulated by a longitudinal slit. Installed firing power is about 300 kW per meter and per row. Flue gas exit is located in the middle of the roof or, alternatively, in the middle of the hearth.

The load is made out of alumina refractory bricks of standard dimensions (220 x 110 x 60 mm³) piled up in a single stack resting in the middle of the hearth. Packing arrangement is as represented on figure 1, with a brick spacing of 50 mm. This rather compact arrangement has a volume porosity of 0.46 and surface permeabilities of about 0.21 and 0.23 in the vertical and horizontal directions respectively. Load mass is 2700 kg per meter.

Operating conditions. Typical industrial firing conditions are applied to the furnace. Air/gas ratio is assumed to be constant, the excess air value being about 25 %. With Ekofisk natural gas, this gives a constant theoretical combustion gas temperature of about 1700°C and a maximum gas injection velocity of 80 m/s. Flue gas mass flow rate is controlled to vary according to the curve represented on figure 3. The total heating time is about 9.5 hours. A time step of 10 minutes is used in the simulation.

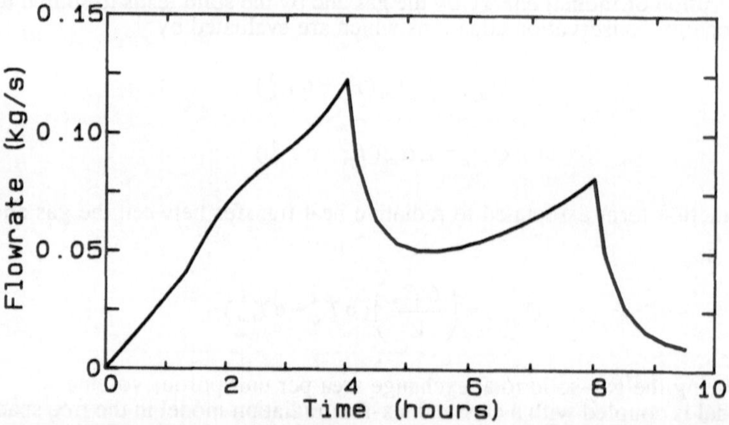

Figure 3. Firing of refractory bricks - Time evolution of flue gas mass flow rate

Properties of the load. The directional friction factors and convective heat transfer coefficients were measured on a small scale (about 1/4) packing for flow in the three directions []. The measured values are highly dependent on flow direction: vertical flow values are higher than values obtained for horizontal flow, by a factor of about four for the convection coefficients and by a factor of two for the friction factors.

The absorption and scattering coefficients have been computed by applying the zone method [8] to the unit cell of the packing. Values obtained at 800°C for a surface emissivity of 0.8 and CO_2 and H_2O concentrations of about 8 and 16 % (Natural Gas, 25 % excess air) are given in Table 2.

Discussion of the results. Three different furnace configurations have been simulated. In the base case (case 1), flue gas outlet port is located in the furnace roof and flue gas flow rate is modulated continuously according to the curve of figure 3. In the second case, flue exit port is in the middle of the hearth.

TABLE 2
Firing of refractory bricks - Volume averaged radiative properties
(T_f = 800 °C, Emissivities: $\epsilon_s = 0.8$, $\epsilon_g = 0.08$)

Scattering coef. (m^{-1})		Absorption coef. (m^{-1})	
s_{-xx}	0.045	a_{gx}	0.194
s_{-yy}	0.016	a_{gy}	0.174
s_{-zz}	0.045	a_{gz}	0.194
s_{xy}	0.112	a_{sx}	1.355
s_{xz}	0.000	a_{sy}	1.126
s_{yz}	0.112	a_{sz}	1.355
		GS_s/V	1.641

In the third case, furnace geometry is identical as in case 1 but a kind of "sequential control" is applied: turndown is achieved by switching on and off the burners and controlling the timing at which the burners are at high fire. This is simulated by dividing every time step in two parts. During the first part of the step, gas flow rate is equal to the maximum value, during the second part of the step, inlet gas flow rate is set equal to zero and heat transfer inside the furnace is due to natural convection and radiation.

In order to compare the different configurations, furnace performances are measured by four indicators: furnace efficiency is related to the average combustion gas and ware temperatures defined by

$$\overline{T}_g = \frac{1}{V}\int_V T_g dV \qquad \overline{T}_s = \frac{1}{V}\int_V T_s dV$$

while temperature uniformity can be evaluated by the standard deviations of gas and solid temperatures defined by

$$\sigma_{Tg} = \sqrt{\sigma^2_{Tg}} = \sqrt{\frac{1}{V}\int_V (T_g - \overline{T}_g)^2 dV} \qquad \sigma_{Ts} = \sqrt{\sigma^2_{Ts}} = \sqrt{\frac{1}{V}\int_V (T_s - \overline{T}_s)^2 dV}$$

Case 1 - Figure 4 shows the calculated time evolution of mean gas and solid temperatures. Mean temperatures at the end of the heating period are about 1430 °C. A maximum temperature difference of about 300 °C is reached after two hours firing. The rather high compacity of the load prevents gas from penetrating into the stacking and flue gas velocities through the load are rather small. This gives very low convective heat transfer rates between the gas and the pieces of charge and induces large gas-solid temperature differences.
Bad penetration of gas into the load also has a pronounced effect on gas and solid temperature uniformity (fig.5), inducing large temperature gradients during the first hours of the heating period. At about 700 °C, gas-gas and solid-solid radiative heat transfer starts to improve uniformity. At the end of the cycle, the furnace is quasi isothermal, with residual standard deviations of about 15-20 °C.

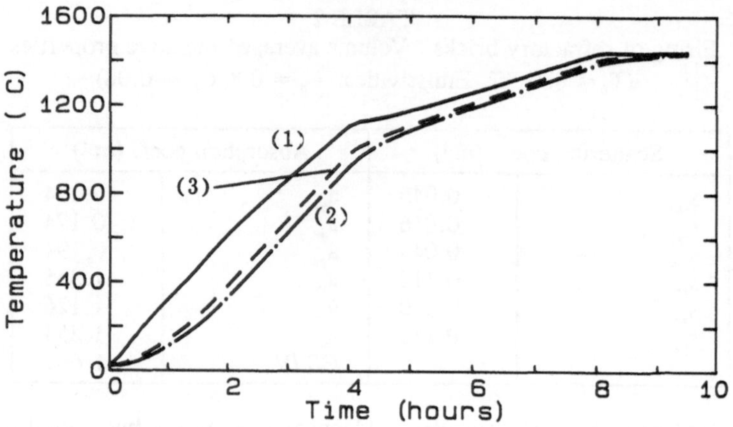

Figure 4. 2-D furnace - Time evolution of average temperatures
(1) combustion gases, (2) ceramic ware, (3) gas-solid interface

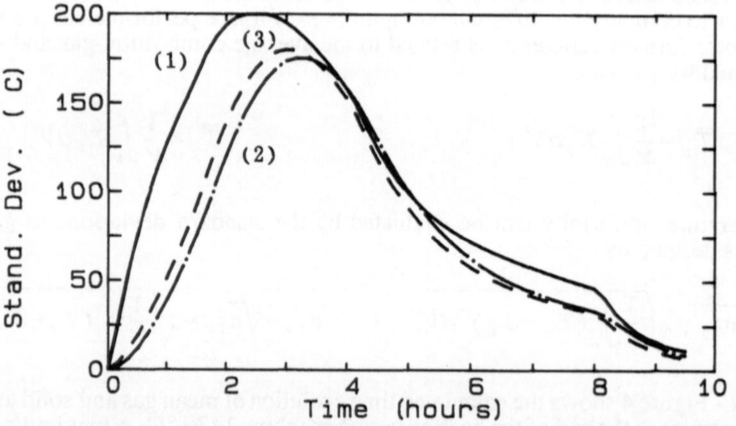

Figure 5. 2-D furnace - Time evolution of standard deviations of temperatures
(1) combustion gases, (2) ceramic ware, (3) gas-solid interface

Case 2 - When downdraught is used, the combustion gases are forced to flow through the load. This leads to an improved furnace efficiency: for the same total energy consumption, average ware temperature at the end of the heating period is about 1540 °C, i.e. 110 °C higher than for updraught firing. Solid temperature uniformity is also improved, the maximum standard deviation being about 15 °C lower than in case 1.

Case 3 - Sequential firing also improves both furnace efficiency and load temperature uniformity. This can be observed on figures 6 and 7, where the results obtained for the three configurations are compared. The improvement is rather small however. This should be due to the high compacity of the packing and the low convective heat rates achieved in firing of bricks. Sequential firing is expected to be more effective with less compact loads, as in sanitary ware or china firing. This will be investigated in the near future.

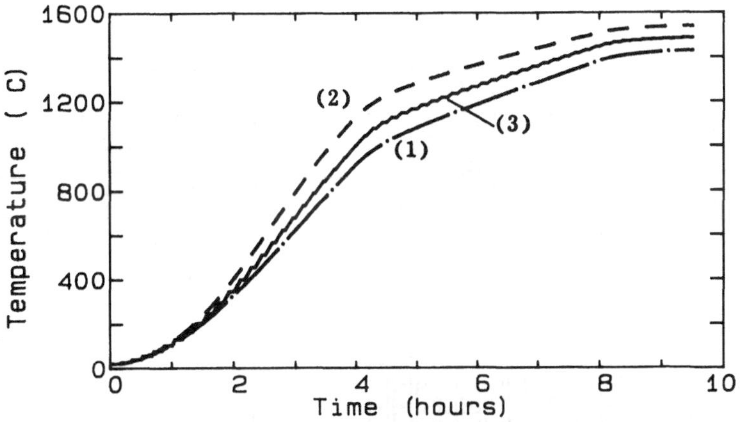

Figure 6. 2-D furnace - Time evolution of average ware temperature
(1) updraught, (2) downdraught, (3) sequential control

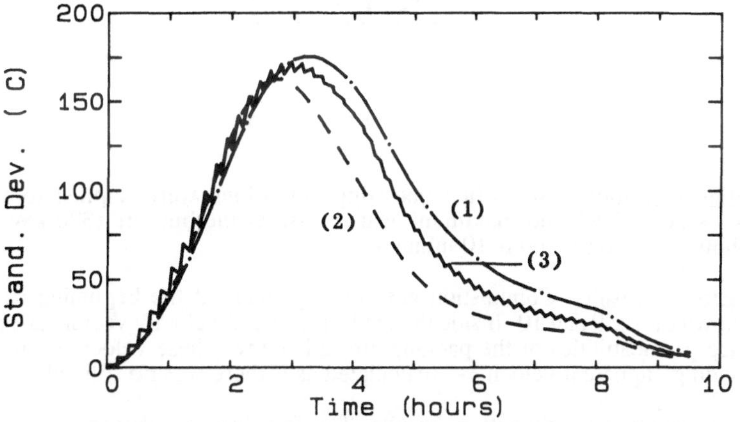

Figure 7. 2-D furnace - Time evolution of standard deviations of ware temperature
(1) updraught, (2) downdraught, (3) sequential control

3-D Furnace

Furnace and load geometry. The 3-D system geometry is represented on fig.8. The furnace is a cell type furnace of 11.3 m3 volume. It is equipped with four 300 kW high velocity jet burners, placed on the sidewalls of the furnace, at diagonally opposite corners. Flue gas flows out of the furnace chamber through a slit in the middle of the earth, between the two load stackings.

Every burner port has a 0.1 x 0.1 m2 square cross section. With natural gas burning with 25 % excess air, maximum gas velocity at burner port is about 80 m/s.

The load is made out of 5300 kg alumina refractory bricks piled up in two stackings on the hearth. Brick arrangement is the same as in the 2-D furnace.

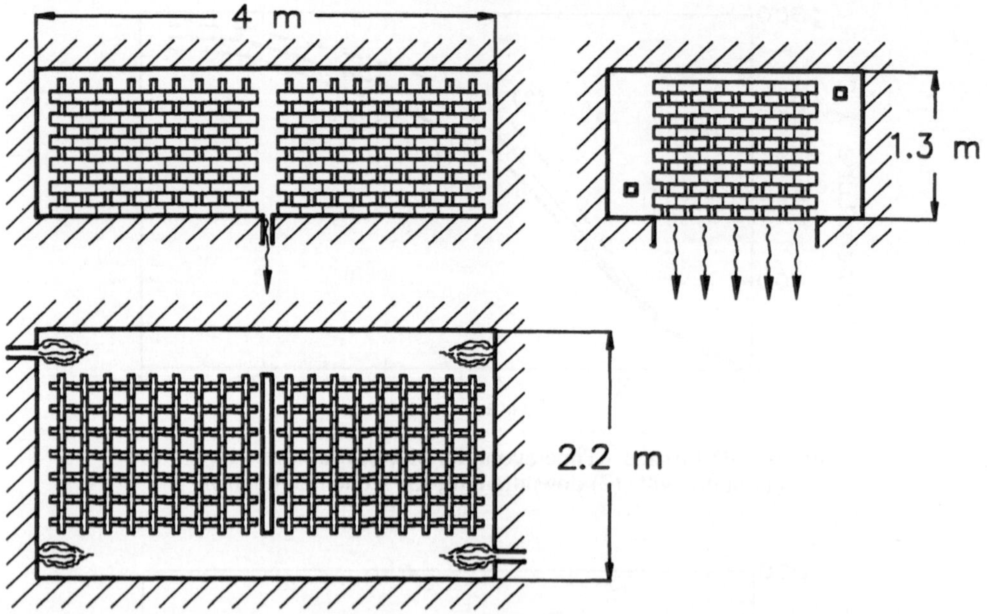

Figure 8. 3-D furnace geometry

Operating conditions. Simplified operating conditions were applied to the furnace: constant excess air (25 %) and maximum heat release at the burners (300 kW per burner) during two hours. The time step is 10 minutes.

Discussion of results. Combustion gas velocity fields at the beginning of the heating period are shown on fig.9 and 10. Inside the load, intersticial velocity vectors are represented. As the surface permeabilities of the packing are rather low, these velocities are higher than the corresponding "apparent velocities" (calculated as if there was no solid phase) by a factor of 4 to 5.

A very complex gas flow pattern results. The four high velocity gas jets induce in the free space around the load four contrarotating gas flows, two rotating in opposite directions about a vertical axis and two turning round a horizontal one. Gas penetrates the load mainly at both ends of the furnace and flows through the stackings towards the centre of the furnace and the flue gas outlet port. Intersticial gas velocities are rather low, i.e. less than 10 m/s.

This radial gas penetration into the load gives radial gas and solid temperature fields, with colder regions in the central part of the furnace.

The time evolution of the average temperatures and standard deviations are represented on figure 11 and 12. Although firing rate is higher than in the 2-D furnace, the standard deviations of solid and gas temperatures are slightly lower. This is due to a higher flue gas recirculation rate in the 3-D furnace. As for the 2-D furnace, radiant heat transfer start to improve ware temperature uniformity at about 600 °C.

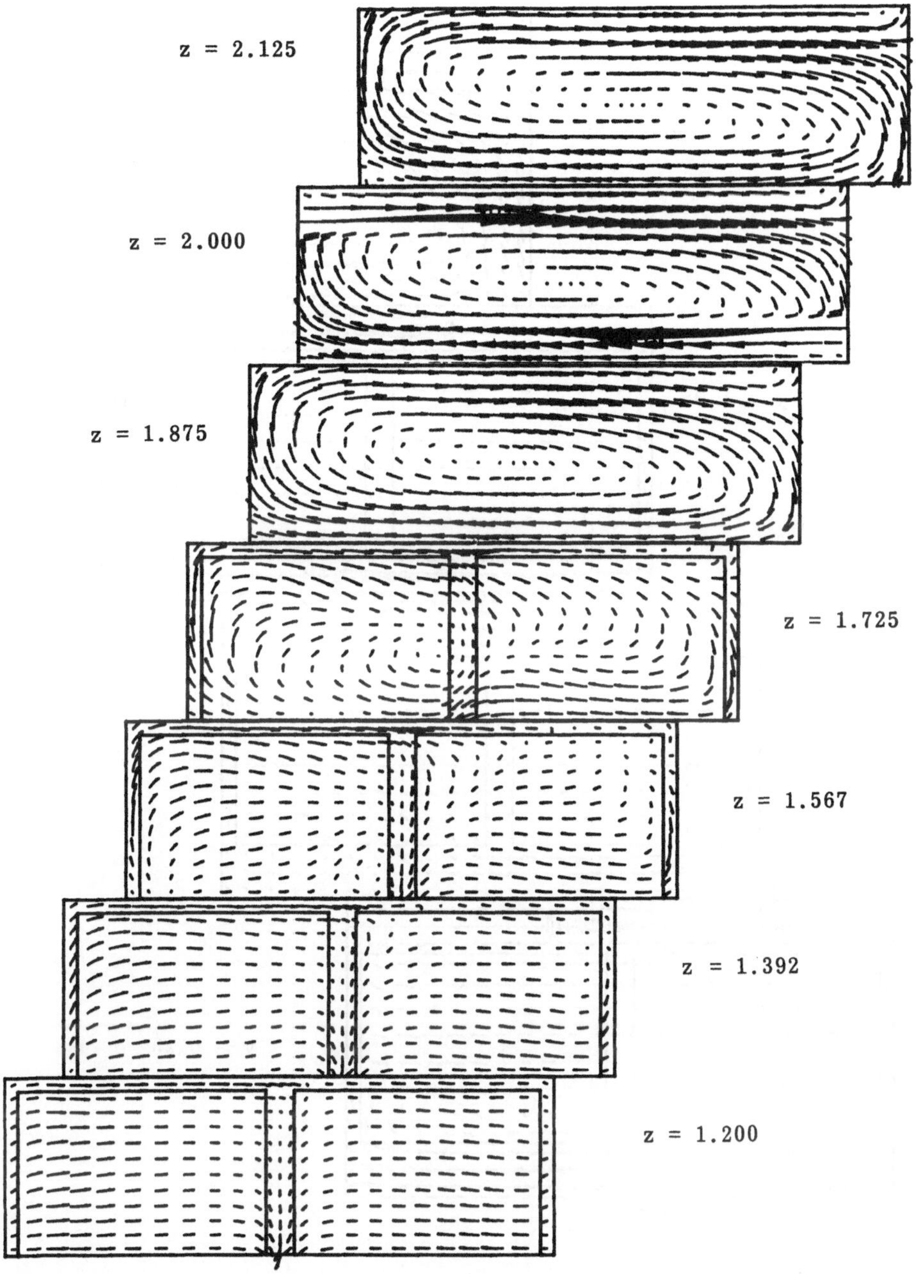

Figure 9.

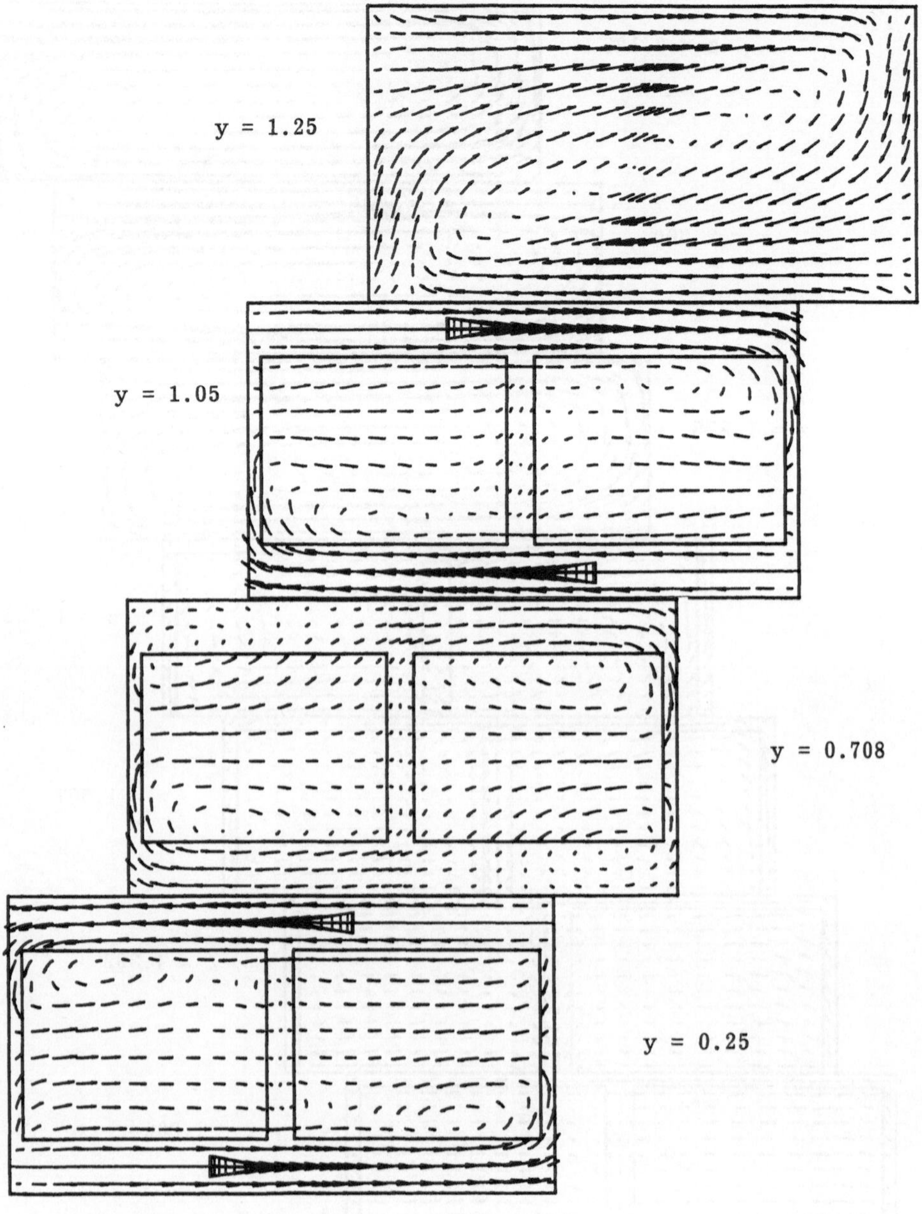

Figure 10.

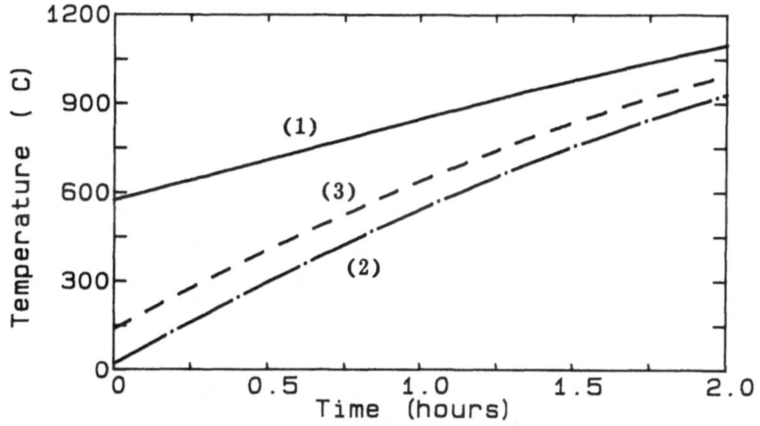

Figure 11. 3-D furnace - Time evolution of average ware temperature
(1) combustion gases, (2) ceramic ware, (3) gas-solid interface

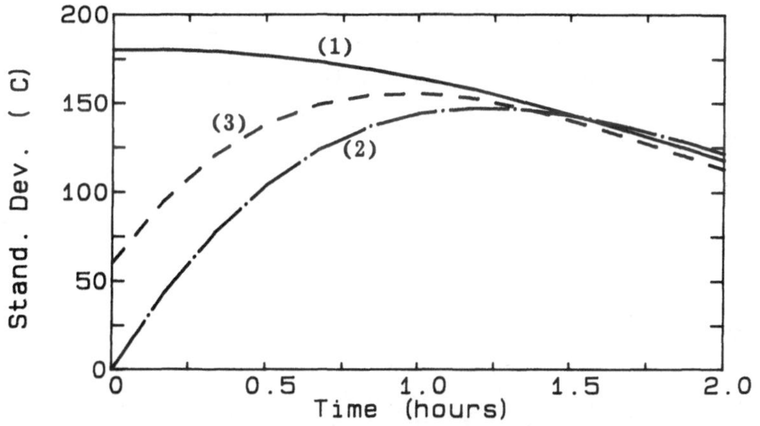

Figure 12. 3-D furnace - Time evolution of standard deviations of ware temperature
(1) combustion gases, (2) ceramic ware, (3) gas-solid interface

CONCLUSIONS

A numerical model able to compute convection, radiation and conduction heat transfer inside intermittent kilns has been developped. The model has been applied to the firing of refractories and has been used to evaluate the effect of different factors on ware temperature uniformity. In the near future, the model predictions will be compared with experimental results. Measurements will be obtained on a 3.75 m^3 pilot furnace which is constructed at the Faculty. We hope to also get some data from industrial furnaces.
After validation, the model will be used to provide design and operating guidelines aiming at improving energy efficiency of the firing process.

REFERENCES

1. Sha, W.T., Yang, C.I., Kao, T.T. and Cho, S.M., Multidimensionnal Numerical Modeling of Heat Exchangers, Journal of Heat Transfer, 1982, 104, pp. 417-425.

2. Sha, W.T., Numerical modeling of heat exchangers, in Handbook of Heat and Mass Transfer, ed. N.P. Cheremisinoff, Gulf Publishing Company, Houston, 1986, Vol.1, pp. 815-852.

3. Martin, H., Transient response to a step change of temperature, in Heat Exchanger Design Handbook, Hemisphere Publishing Corporation, 1986, Vol.2, pp. 2.4.3.1 - 2.4.3.12.

4. Launder, B.E., and Spalding, D.B., The numerical computation of turbulent flows, Computer Methods in Applied Mechanics and Engineering, 1974, 3, pp. 269-289.

5. Lybaert, P., Kharbouch, B., El Hayek, M., and Meunier, H., Modelling and simulation of combustion gas flow and heat transfer in charge stackings of ceramics firing kilns, in Proceedings of 2nd European Conference on Industrial Furnaces and Boilers (INFUB), Vilamoura, Algarve, Portugal, 2-5 April 1984.

6. Lybaert, P., El Hayek, M., and Meunier, H., Modelling of radiant heat transfer in charge stackings of ceramics firing kilns, in Proceedings of Eurotherm Seminar nr 21 - Heat Transfer in Semi-transparent Media, Lyon, France, 3-5 February 1992, pp. 217-227.

7. Gosman, A.D., and Lockwood, F.C., Incorporation of a flux model for radiation into a finite-difference procedure for furnace calculations, 14th International Symposium on Combustion, The Combustion Institute, 1972, pp. 661-671.

8. Hottel, H.C., and Sarofim, A.F., Radiative Transfer, McGraw-Hill, New York, 1967.

9. Patankar, S.V., Numerical heat transfer and fluid flow, Hemisphere Publishing Corporation, Washington DC, 1980.

USE OF FURNACE MODELLING TO IMPROVE ENERGY EFFICIENCY IN THE DEEPDRAWING STEEL SHEET INDUSTRY

H. MEUNIER, Professor; M. CAMBIER, Ir., Research Assistant
Faculté Polytechnique de Mons, Thermal Engineering Department,
rue de l'Epargne, 56, B-7000 Mons, Belgium.

PURPOSE OF THE WORK

Fabrication of steel sheet for deep drawing (car industry, electric household appliances, ...) requires progressively more severe specifications, with a drastic reduction in the accepted scattering of mechanical properties.
This imposes an accurate control of the final heat treatment, i.e. the annealing process. This is more easily achieved in batch furnaces, for which complicated temperature schedules may be followed, in particular for special steel grades.

This type of furnace is described hereafter and is depicted on fig. 1. A pile of sheet coils is heated, starting from room temperature, under an inert gas (generally, N_2 with a small amount of hydrogen) circulated by a fan inside a protective cover. This inner cover is then heated from the outside by tangential burners installed in the furnace wall (called "bell"). A finned tube recuperator, located in the flue gas outlet, preheats the combustion air up to a temperature essentially variable according to the flue temperature and flow rate.

At every point of the coils, the final steel sheet mechanical properties strongly depend on the temperature cycle at that point.
Metallurgical requirements appear on fig. 2. The main constraint is to reach a minimum given temperature and to maintain it during a given duration of time. Obviously, one has to check if this requirement is fulfilled at the coldest spot.
Furthermore, the temperature cannot go beyond a maximum imposed temperature. Obviously again, this concerns the hot spot of the load.

This is typically a "thick load problem". Indeed, due to the laminated structure, the radial heat conductivity of the coils is rather poor.
Some compensation is seeked by installing "convectors" between the coils, in order to heat the coil edges.
It is guessed - and measurements confirm it - that the cold spot is located about 1/3 thickness along the mid-height radius of the last bust one coil.

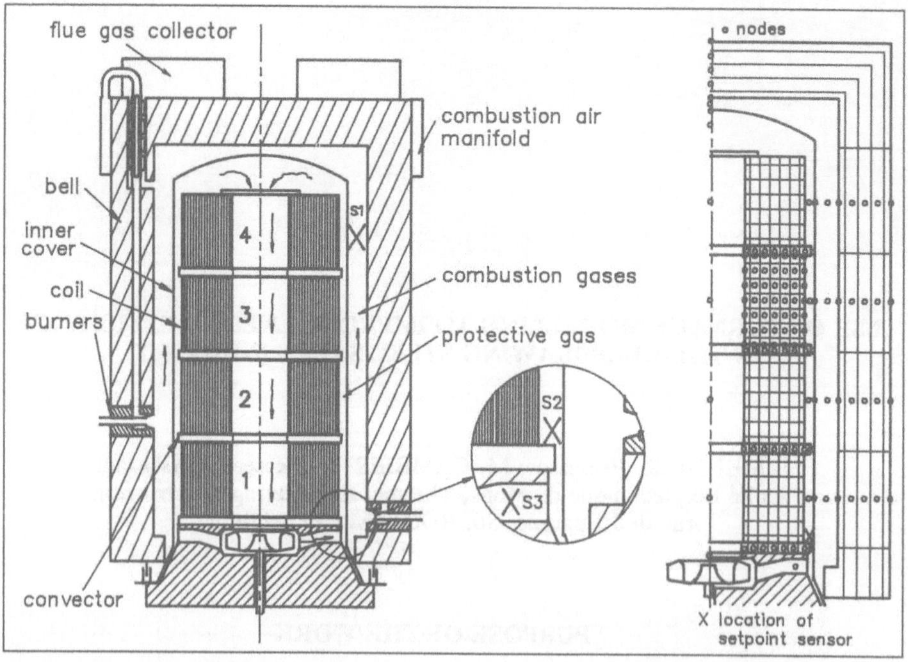

Figure 1. Batch coil annealing furnace : general arrangement, discretization.

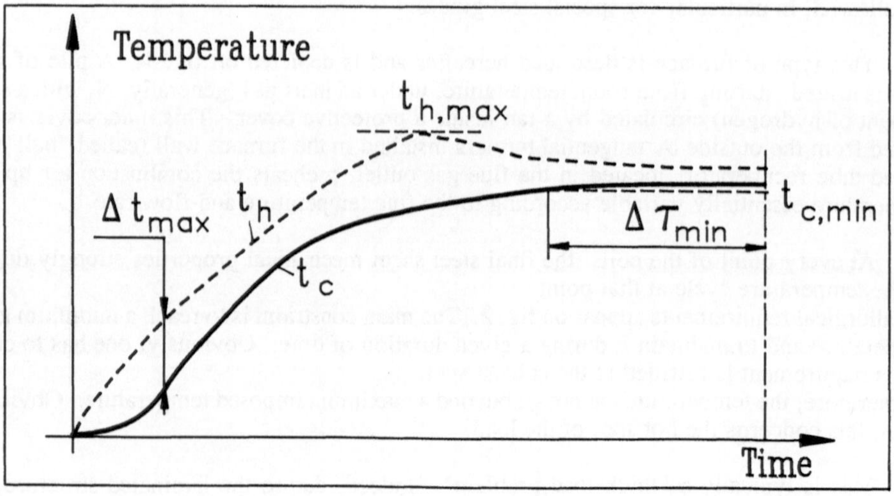

Figure 2. Metallurgical constraints.

Unfortunately, it is unfeasible in normal operation to embed thermocouples in the coils, and therefore to set up a closed control loop based on inner temperatures. Control is restricted to measurable temperatures (setpoints), generally in the protective gas (position S_2 on fig. 1 is frequently adopted, position S_3 also, but less often), in the furnace itself (position S_1), more seldom a coil surface temperature (generally, the lower edge).

Field measurements (fig. 3) show that these setpoint temperatures are far from representing, even approximately, the cold spot temperature. This situation always prevails when heating thick loads. The user, due to the uncertainty on the inner temperature field, is limited to an empirical approach consisting in testing a posteriori the effect on the product quality of slight empirical variations of the setpoint curve. The result of it is unusually long and energy consuming cycles. Furthermore, uncontrolled scattering of material properties leads to the rejection of products at this last stage of fabrication and again to higher energy consumption per unit of final product.

It should be emphasized that, at this last stage of the fabrication process, reduction of the rejects avoids the loss of the energy contained in the product itself, i.e. the energy consumed in all previous stages of the metallurgical process. A reduction on the reject proportion lead to a much larger overall energy saving and therefore, pollution abatement, that could be reached by installing energy saving equipment on the annealing furnace.
Anyway, both types of actions have been envisaged.

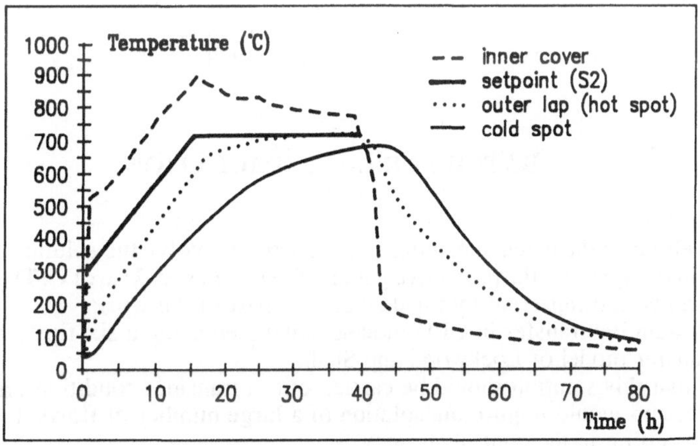

Figure 3. Measured coil temperature (bottom coil).

THE METHODS OF APPROACH

Two large steelmaking factories in Belgium started, in collaboration with the Faculté Polytechnique de Mons, a joint research programme aiming at improving the rational use of energy in the batch annealing process.

Two ways of approach have been tested :
- A better control of the inner temperature field in the coils by developing a simulator to predetermine the optimum setpoint temperature curve allowing to fulfil the metallurgical constraints at every point of the coils.
- Improving the process itself, by using a more conductive protective gas, such as hydrogen, to increase both the thermal conductivity of the coil, and the convection coefficients between the gas and the coils. The ultimate purpose is to improve temperature uniformity in the load.

The technology, related to the use of hydrogen as a protective gas, has been developed by several European furnace manufacturers. But, optimisation of the so-modified process has not been carried out as yet.
Furnace users are faced with new problems resulting from deep changes in the heating process and are still trying to define the new operating parameters, in particular, the setpoint curves. Use of furnace modelling and simulation has proved to be a powerful tool for solving the "teething troubles" of this new technology and for optimising the process.

It should also be mentioned that furnace manufacturers launched research programmes aiming at using regenerative burners on bell furnaces. Heat recovery on flue gases may be carried out more extensively than with conventional finned-tube heat recuperators and high energy saving may be expected. Again, simulation may help to check the validity of this approach prior to testing full-scale prototypes.

Furnace modelling is being promoted by the Commission of the European Communities through the Joule programme. A joint project on "Energy saving and pollution abatement in glass/ceramics furnaces, cement kilns and baking ovens" allows several research teams to work out modelling techniques, codes, ... to be applied to these industrial sectors. Obviously, the gained expertise could be very useful to other sectors, such as metallurgy.

BATCH FURNACE SIMULATION

Simplifications

Complete modelling of the batch annealing furnace would involve the solution of the two fluid flows existing in the system : the protective gas and the flue gases. Use of CFD (Computational Fluid Dynamics) would represent by far the heaviest part of the work.
Calculation of radiative transfer in CFD models is handled using a six-flux model, or better, the discrete transfer model of Lockwood and Shah.
It is reminded that this solution should be carried out in transient conditions, as prevailing in a batch furnace, and would require the solution of a large number of flows, the flue gas flow rate varying according to the control strategy.
Opportunities for the exploitation of CFD modelling is fast increasing owing to the rapid advances in computing power. As a result, it is now applied to many aspects of heating equipment design and to scientific research. However, running times are still prohibitive for plant exploitation.

Fortunately, for the problem on hand, the detailed solution of the fluid flow equations (Navier Stokes) may be avoided for extensive use in the plant.
The fluid flow is treated as a plug-flow, the repartition of the total flow rate between various circuits resulting from pressure drop considerations. The combustion pattern is oversimplified and is assumed instantaneous.
Convection coefficients result from the knowledge of the bulk flow and are computed by means of empirical correlations $Nu = F(Re, Pr)$.
The use of CFD is limited to the research stage, for the validation of the assumptions and for

providing empirical correlations.
Anyway, it is emphasized that a model should always be validated against measurements on full-scale furnaces.

The simulator

The knowledge-based model relies on three successive stages :

- modelling of heat conduction in the coils, involving anisotropy due to the presence of protective gas between the wraps;
- analysis of convective and radiative heat transfer in the space filled with protective gas between the coils and the inner cover;
- modelling of radiation and convection between the combustion gases and the outside wall of the inner cover, taking into account heat transfer and storage in the furnace walls.

Discretization of the system and the corresponding grid are shown at fig. 1 : a 2-D axisymmetric field is assumed for the coils and the bell walls.
A detailed description of the model may be found in another paper (Cambier and Meunier, 1990).

Conduction within the coils

Radial conductivity is described by a 1-D porous medium model taking into account conduction through the protective gas enclosed between the windings. A corrective term for conduction through contact points has been added. This term and porosity are related to the local temperature gradient.
This conduction model has been validated through industrial experiments, in which many thermocouples were imbedded into the coils or fastened on the inside or outside lap. The internal temperature field in the coil was computed from external measured values. Identification of computed values with the measured ones allowed to obtain the unknown parameters of the conduction model.
Tentative improvement of this model is still underway.

Heat exchanges in the space filled with the protective gas

The mean velocities and the repartition of the fan total flow rate between various hydraulic circuits (convector plates, central core) are computed from the pressure drop coefficients. These coefficients resulted from anemometric measurements for several types of convector plates and several coil stacking geometries.

At the outside and inside surfaces of the coil, the boundary condition involves, on one hand, convection with the protective gas and, on the other hand, radiative transfer with the inside surface of the inner cover (protective gas is assumed transparent).

Validation of this second stage of the model has been carried out thanks to the measurement of inner cover temperature at levels corresponding to each coil. Combined with the measured temperature of the coils, they allow to simulate only the space located under the inner cover, disregarding momentarily the combustion gases and the bell.

Heat transfer from combustion gases

The assumption of upward plug flow of the combustion gases did not allow to validate accurately the simulator. Furthermore, suction-pyrometer temperature measurements showed that a recirculation motion must be superimposed to the upward plug flow.
This effect is particularly important during the soaking period, when the net upward flow is low. An empirical correlation has been found between this recirculation factor and the net mass velocity through the gap between the bell and the inner cover.

Equations of the sensor and of the control system

Several locations for the setpoint thermocouple are usually adopted (see fig. 1) :
- in the protective gas : at the outlet of the fan diffuser or between the bottom coil and the inner cover;
- in the furnace;
- against the bottom edge of the first coil.

When the temperature setpoint sensor is located in the gas, its output should also be modelled, taking into account radiative and convective fluxes with the gas and the surrounding surfaces (in fact, the corresponding nodes).

A PID-controller adjusting the burner output may be represented by an additional differential equation. In fact, any control strategy leads to a set of equations linking outputs and inputs of the furnace simulator.

Validation of the simulator

Several experiments, aiming at validating the simulator, have been carried out in three different plants in Belgium. Many thermocouples were embedded into the coils. It was found out that the coldest spot is approximately located at 1/3 thickness from the inner lap.
A very large temperature difference is generated into the coil and the septoint is far from representing the temperature of the cold spot, especially when the protective gas is nitrogen. These field data are satisfactorily reproduced by the simulator, as attested by fig. 4. A similar conclusion applies to the validation of the fuel gas flow rate at the burners (fig. 5).

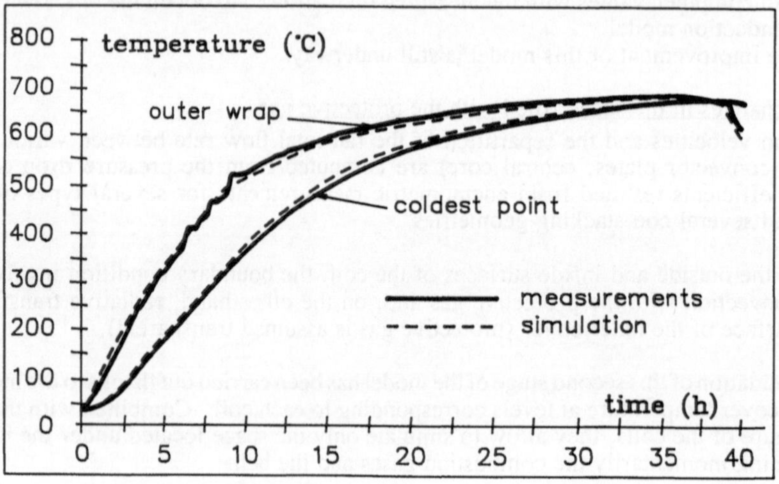

Figure 4. Validation : coil temperature.

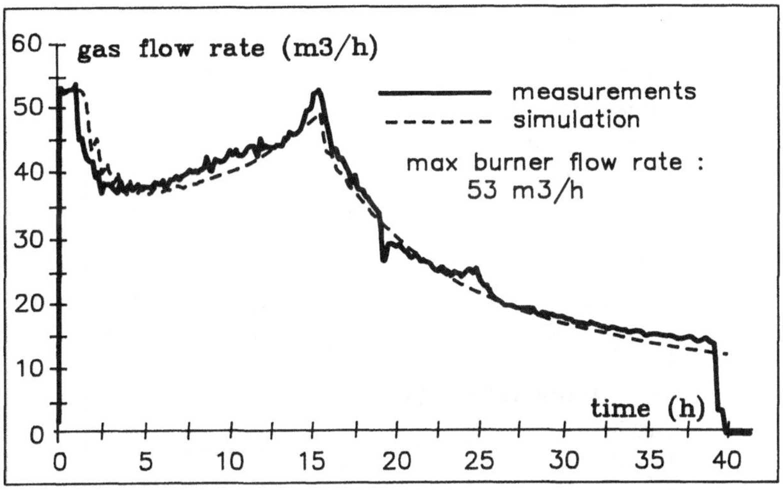

Figure 5. Validation : gas flow rate.

IMPROVEMENT OF THE CONTROL OF THE HEATING PROCESS : OFF-LINE OPTIMISATION OF THE SETPOINT TEMPERATURE SCHEDULE

The characteristics of the load vary very often : coil mass and dimensions, steel grade, sheet thickness, ... Furthermore, the metallurgical constraints, characterized by the parameters $t_{c,min}, t_{h,max}, \Delta\tau_{min}$, ... of fig. 2 may change with the desired properties of the final product. Off-line simulations with various setpoint curves for a given charge will lead to the "best" setpoint curve meeting all metallurgical constraints at every spot of the load, together with the shortest possible cycle and the lowest fuel consumption.

Various control strategies, using several setpoint sensors and shifting conditionally from one sensor to another, may also be tested and optimised. The strategy consisting in starting with a sensor located in the furnace and then shifting to sensors closer to the coils has proved to achieve a good compromise between conflicting requirements.

Temperature setpoint curves exhibiting a temporary overshoot (fig. 6) have been tested in a plant using a N_2-H_2 protective gas : the final temperature was more reliably attained at the cold spot. As predicted by the simulator, the cycle was drastically shortened and the furnace productivity was markedly improved. Consequently, the proportion of rejects fell. An average energy saving of 15 % is expected.

Another application was successfully achieved and led to a better recovery of lean residual gases.
Change of the heating gas are frequent in steelmaking factory : shift from the coke oven gas to blast furnace gas/natural gas mixtures- and inversely - may occur. The simulator is able to correctly predict the temperature staging between the various coils.

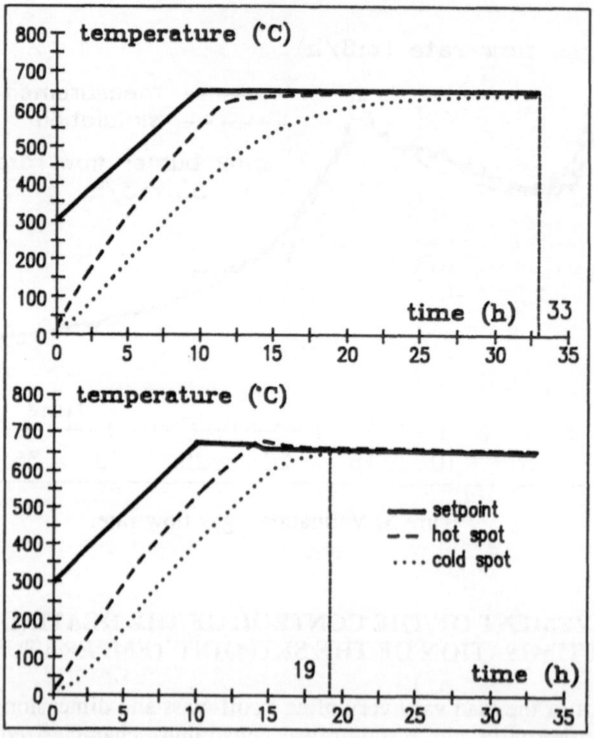

Figure 6. Optimum temperature schedule : effect of a setpoint overshoot.

THE IMPROVEMENT OF THE HEATING PROCESS BY CHANGING THE PROTECTIVE GAS

The use of pure hydrogen, instead of a mixture N_2 - 5 % H_2 as protective gas, has been studied. It has been experimentally shown that H_2 rapidly diffuses between the coil turns and increases the coil radial thermal conductivity. Better temperature uniformity is expected. Furthermore, convection coefficients between the gas and the coils are markedly improved.

Many existing furnaces has been revamped and converted to the use of hydrogen. The user is then faced with the definition of new setpoint curves.
Substitution of pure hydrogen to the nitrogen-rich protective gas may be easily predetermined with the help of the simulator.
The simulator has been tested and validated in two Belgian factories having shifted from HNX to H_2. Provided that appropriate data files have been entered for the thermophysical properties of the protective gas, validation was fairly good, with only a minor change of a parameter of the correlation giving the recirculation ratio of the flue gases.

The influence of the nature of the protective gas is globally shown on fig. 7, for a setpoint curve consisting in a ramp followed by a plateau.
Use of H_2 sharply decreases the temperature gradient through the coil, leading to a faster heating up of the coldest spot and accordingly an important shortening (almost 1/3) of the cycle.
Another result is a computed energy saving of about 10 %.

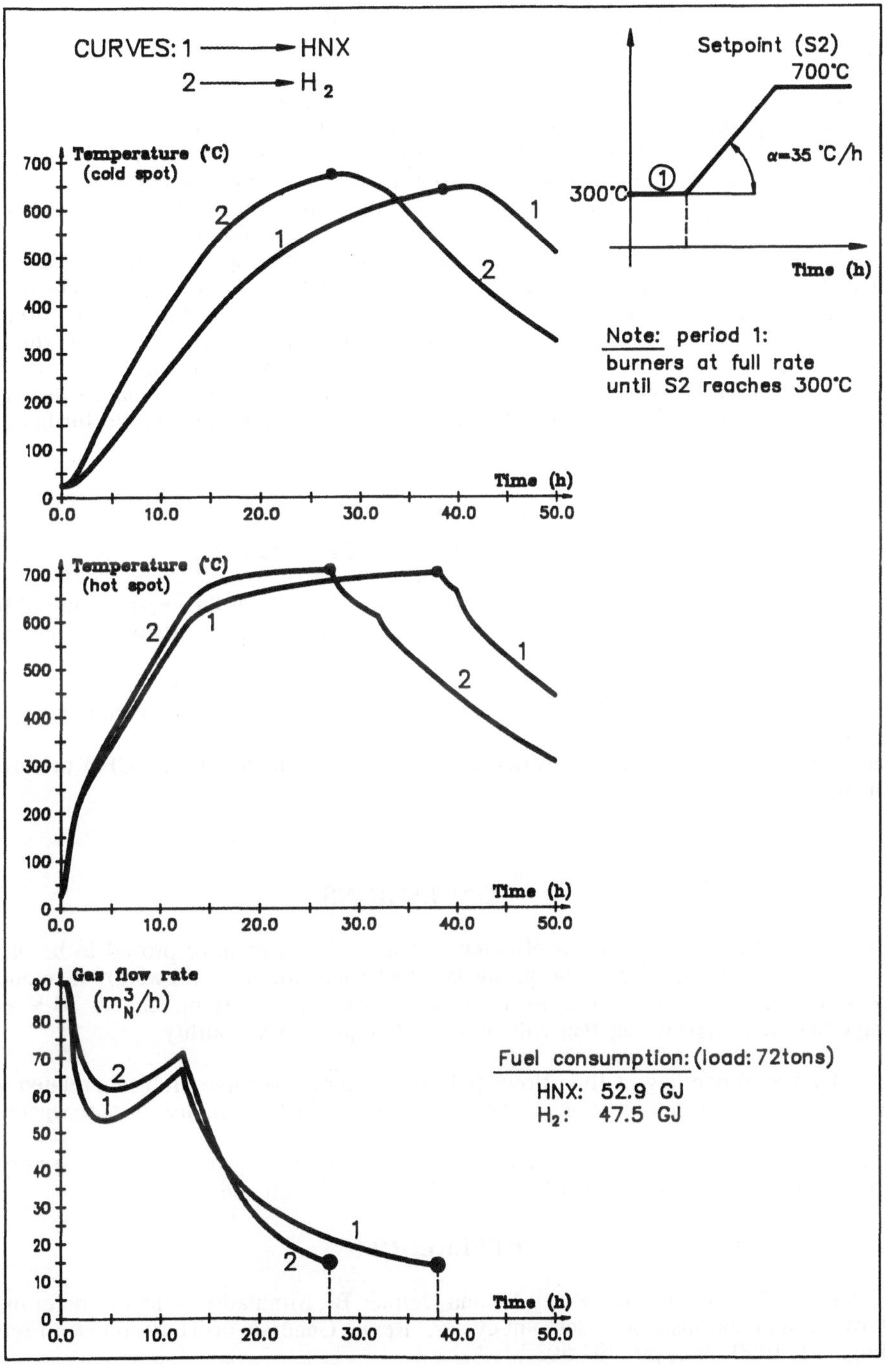

Figure 7. Improvement of the heating process by changing the protective gas.

But, obviously, temperature uniformity is improved, and an important reduction of the rejects is expected.

However, some drawbacks appeared during the first months of exploitation :

- temperature gradients building up inside the coils result in radial, circumferential and axial thermal stresses. A first study has shown that radial stresses appearing between successive coil turns account for sticking phenomena observed since the use of hydrogen.
- more rapid heating of the coil windings close to the core results in an expansion hindering the expulsion of pyrolysis gases. It resulted in "dirty" spots on the steel sheet.

Calculation of radial stresses requires the temperature field throughout the coil. This may be obtained from the model for various load configurations and temperature cycles. The influence of operating parameters, such as the slope of the ramp appearing in the setpoint curve, may be studied. On the same way, the influence of changing the protective gas fan (high-flow bases), of the burners installed power, may be predicted thanks to the model.
When these "teething troubles" are solved, it is obvious that less scattering in the material properties will be achieved, leading to a reduction of the rejects and hence to further energy saving for the whole process.

HEAT RECOVERY ON FLUE GASES

Running the simulator shows that a fuel saving as high as 40 % of the energy required for the annealing process itself may be achieved by regenerative burners instead of conventional finned-tube heat recuperator.
However, this result is obtained by assuming a simplified model for the flue gas flow.
Drastic reduction of flue gas flow rate could result in stratification and important temperature non uniformity of the coil pile.
In fact, the actual flow pattern is difficult to guess. It is planned to use CFD to solve this problem.

CONCLUSIONS

Physical modelling and simulation of batch annealing furnaces have proved to be valuable tools to improve the quality of the products (steel sheet for deep drawing), to reduce the proportion of rejects and hence to lower the overall energy consumption. A 20 % energy saving could be achieved, together with an improved process versatility.

Advanced modelling of fluid flows (CFD) is planned to solve problems related with a better selection of burner location, burning of residual fuel gases, use of autoregenerative burners.
The use of CFD, together with the rapid advances of computing power, will in a very near future help to solve many aspects of heating equipment design and operation.

REFERENCES

1. Cambier, M., Meunier, H., Remy M. and Delmée B., Simulation of heat transfer in batch furnaces and optimisation of setpoint cycles. Revue Générale de Thermique (in French + English), 1990, 346, pp. 494-503.

2. Lockwood, F.C., Shah N.G., A new radiation solution method for incorporation in general combustion prediction procedures. Proceedings 18th Int. Combustion Symp., Combustion Institute, Pittsburgh, 1980.

SESSION 8:

Furnaces, Kilns and Ovens
Continued

Chairman: Prof. H.L.J. Meunier

ON THE OPTIMISATION OF EXPANSION PROCESSES OF INDUSTRIAL MINERALS

T. BERDI, M. FOUNTI, EM. KAKARAS A. KLIPFEL and Z. NTOUROS
National Technical University of Athens
Mechanical Engineering Department, Thermal Engineering Section
Patission 42, Athèns-10682, Greece

ABSTRACT

The paper discusses the characteristics of expansion processes of industrial minerals and presents the thermodynamic analysis at a novel furnace layout proposed for the expansion process, combined with heat recovery measures aiming at achieving an integrated process.

Trajectories of expanded particles have been calculated in a conventional expansion furnace geometry and the presented results demonstrate the effects of particle size, density, initial velocity and angular velocity on the particle trajectories. The results have served as guidelines for the design of the novel furnace configuration. It is demonstrated that the use of the new type of furnace together with the high and low temepature heat recovery can contribute to considerable energy savings estimated at 45%.

INTRODUCTION: CHARACTERISTICS OF THE EXPANSION PROCESS

Industrial minerals, such as perlite -a generic term for naturally occuring siliceous volcanic rock-, have wide applications as construction (concrete and plaster aggregates, fillers), as high and low (cryogenic) temperature thermal insulation materials, as sound insulators, in the fire protection, in horticulture as well as in the field of filter aids [1]. Except for a few minor applications, perlite is used in its expanded form. When heated to a suitable point in its softening range, perlite expands four to twenty times its original volume. This expansion is due to the presence of 2% to 6% combined water in the crude rock. Employing various furnace configurations to heat-up the raw material, the combined water present in the mineral structure vaporises and reforms the softened particle shape. Steam escape creates a mineral "pop-corn" in which the original gas pockets within the rock are greatly expanded thus increasing the porosity of the structure and decreasing the original density. Temperatures of expansion vary between 760 °C. and 1,100 °C. Furnace residence time, particle size, feed rate, and cut off rate for particle collection are, among others, important parameters to be taken into consideration when producing an optimum low density product.

The horizontal stationary furnace, the co-current rotary furnace and the vertical furnace are commercially operated types of furnaces for the expansion of perlite [2,3]. The vertical furnace is the most widely employed type. Vertical furnaces are best suited for the expansion of dead perlites with a narrow size distribution at feed. Any of the furnace types may be operated with a pre-heater inserted between the feeder and the expanding furnace. Preheating allows a broader size distribution to be used in the feed, permits the use of livelier perlites, and reduces the amount of breakdown of the particles.

The amount of energy required to expand perlite varies widely according to the type of furnace, expansion process and crude perlite quality. According to [2], 400 MJ/m^3 can be regarded as an average through a typical plant, out of which 29% is consumed for the expansion process. These figures indicate that if an energy saving is to be achieved, this can be done in the expansion process, namely in the furnace. The introduction of new concepts in furnace design can lead to considerable energy savings. The success of the expansion mainly depends on achieving rapid heating up of the raw material to temperatures up to 1100 °C [3]. It is important to make efficient use of both raw material and fuel to produce an expanded aggregate conforming to close specifications of size distributions and density. Although the theoretical amount of energy required for the expansion process of raw perlite is about 942 kJ/kg, plants presently in operation consume approximately 3.35-5.024 MJ/kg.

Our theoretical calculations have indicated that 85% of the required amount of energy is consumed for heating up the raw material to the desired temperatures, 12% of the total energy is spent for the expansion process and the final 3% is consumed in the evaporation of the contained moisture. Additionally, existing expansion furnaces feature low volume fraction of particles and the gas-particle flow leaves the furnace at high temperatures.

The presently operating plants for the expansion of perlite do not make use of the heat contented in the exhaust gases and the hot expanded material, thus throwing away a significant amount of energy which can be used for preheating the combustion air and eventually the raw material. In order to exploit this energy a combination of heat recovery processes has to be applied in the design of an energy efficient plant.Also, the broad size distribution of raw material in combination to the abrupt changes in material density due to expansion render the operation of existing plants problematic.

FLOWFIELD CHARACTERISTICS AND PARTICLE TRAJECTORIES IN A CONVENTIONAL FURNACE

A typical vertical type expansion furnace used in Greece by Silver and Baryte Ores Minining Co., S.A. (purchased from the Perlite Corporation, USA) has been chosen to simulate the particle trajectories in order to quantify the required conditions for the fluidization of the particles. The burner (not modelled), is located at the bottom of the furnace and the hot gases escape from the top of the furnace. The raw perlite is fed into the side of the furnace at a point above the burner and all the expanded perlite is carried out at the top of the furnace. The flow is turbulent, upward flowing against the direction of gravity and laden with perlite particles, at constant but dilute concentrations.

The modelled furnace, based on the real conditions, as shown in Figure 1 is vertical with 0.640 m diameter and 5.92 m height. The entrance to the furnace has a diameter of 0.360 m, creating a diffuser type expansion geometry. At the exit of the chamber there is a sudden contraction to a diameter of 0.245 m. Table 1 shows the conditions of the flow and the properties of the gas and of the perlite particles with which the predictions have been performed. For prediction purposes the particles have been assumed to be spherical, having the properties of expanded perlite and with erosion properties similar to quartz-sand. The furnace walls were assumed to be made of Al-alloy. The predictions have been performed at 800^{0}C, simulating the exhaust gas flow rate, density and velocity as in the real furnace.

The particle relaxation time has been calculated using: $\tau_p = \rho_p d_p^2/18\mu f_D$, where ρ_p is the density of the particle, d_p its diameter and μ the viscocity of the gas, τ_w is the characteristic flow time-scale and c_D is the estimated drag coefficient.

Computational Approach

Single-phase flow: The predictions have been performed using a modified version of the TEACH code [4]. The modification of the standard version concerns the way which a geometry of a certain flow is introduced and handled [5].

For the present flow which is steady, incompressible, turbulent, axisymmetric, the standard time-averaged continuity and momentum equations were solved. The standard k-ϵ turbulence model was employed for the modelling of turbulent quantities. The resulting system was solved via a finite difference method based on a staggered grid arrangement, using the SIMPLE algorithm, the upwind differencing scheme and the Tri-Diagonal Matrix Algorithm (TDMA).

The diffuser type inlet and outlet were treated in a stepwise manner. The inlet U profile was set according to the calculated mean velocity from the flow rate, while k and ϵ were estimated from $k = 1.5(TuU)^2$ and $\epsilon = C_\mu^{3/4}k^{3/2}/l_m$, where Tu = 10 % and $l_m = 0.01 \times r_{inlet}$. The radial velocity was assumed to be zero at the inlet. The predictions have been performed with a non-uniform grid disrtibution of 140 x 43 nodes, as shown in Figure 2.

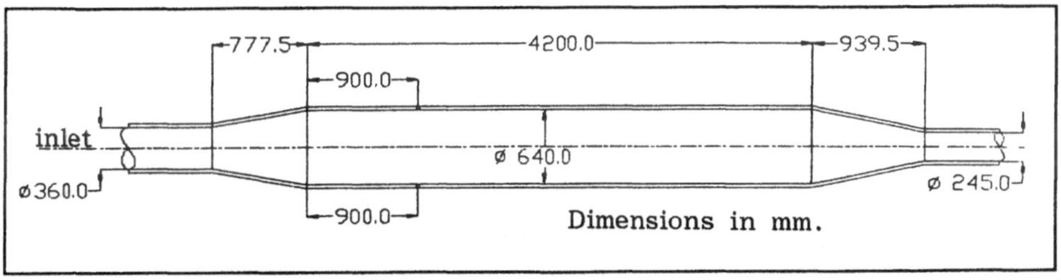

Figure 1: Vertical expansion furnace geometry used for the prediction of particle trajectories.

Particle trajectories and particle-wall collisions: Having predicted the flowfield for the single phase flow the particle trajectories were calculated by solving instantaneous particle motion (momentum) equations. The particle trajectories and the corresponding wall collisions have been calculated using a code developed at the NTUA for the prediction of erosion in furnace geometries [5]. The particle position at each location has been calculated using an iterative proceedure with a time step (equal to 1ms for this case) taking into account the calculated single phase flowfield. The fluid and particle velocity at every point has been calculated using Coons [6] polynomial which correlates the required velocity to four neighbouring points. Particle new positions and velocities have been iteratively calculated until they converged. The equations describing the particle motion have been analysed in the axial and radial furnace directions. Along the furnace walls and in the axial direction the particle motion is described taking into account drag, lift and Magnus, Saffman and gravity forces. In the radial direction the same forces are taken into account except gravity. The particle motion equations have been analytically solved.

When a particle approaches a wall, it is examined whether the new position lies within the boundaries of the flow or not. If the new particle position lies outside the boundaries, it is consired that the particle collides with the wall. The new radial position is considered on the wall and the axial position is calculated based on the new radial position and the tangent of the resulting angle, which is also the collision angle. The particle velocities and rebound angle have not been calculated using the empirical restitution coefficients given by Tabakoff et al [7] for different particle and wall materials. Viscous dissipation is modelled in the code but for the present flow conditions has not be taken into account. Similarly, the conditions for the Saffman force have not been fulfilled in the present case.

TABLE 1
Gas and particle flow properties

GAS FLOW (at 800°C)		Value of property				
Mass flow rate		$m = 1058{,}17$ kg/h				
Density		$\rho = 0.3243$ kg/m^3				
Viscocity		$\mu = 43{,}32 \times 10^{-6}$ kg/ms				
Maximum inlet velocity		$U_{max,in} = 10.8$ m/s				
		$Re_{max,in} = 2{,}9 \times 10^4$				
Maximum outlet velocity		$U_{max,out} = 20{,}87$ m/s				
		$Re_{max,out} = 3{,}83 \times 10^4$				
Bulk velocity		$U_{bulk} = 8.886$ m/s				
		$Re_{bulk} = 2{,}39 \times 10^4$				
PARTICLES (at 800 °C)		$\rho_p = 90$ kg/m^3				
d_p(mm)	U_p(m/s)	Re_p	C_D	τ_p(s)	τ_w(s)	τ_p/τ_w
3.5	2.14	63.1	1.32	0.40	0.042	33.9
2.5	2.14	40.0	1.70	0.25	0.042	17.2
1.0	2.14	16.0	3.06	0.05	0.042	2.76

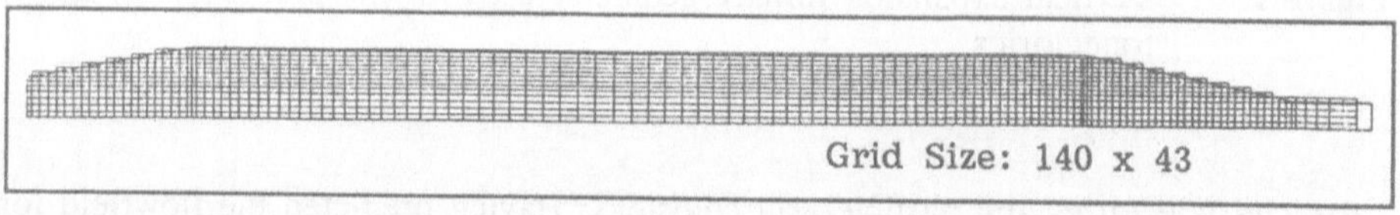

Figure 2: Calculation domain

Results

The single-phase flow results showed only local influence of the outlet geometry in the flowfield, also observed in [5]. For the two-phase flow predictions it can be generally stated that the particle trajectories do not follow the single-phase flow, presented in Figure 3.

Particles of 1 mm, 2.5 mm or 3.5 mm diameter have been injected in the flow field from an axial position of 1.67 m above the bottom of the furnace, with densities equal to 90 or 180 kg/m^3, with 45^0, 30^0 or 60^0 angle of incidence, with various initial velocities based on the theoretically calculated minimum fluidization velocity for the particles, and angular velocities equal to 0, +500 or -500 rad/s. The effects of varying the particle density, size, initial velocity and angular velocity in their trajectories in the furnace have been investigated. Only a few of the obtained results are shown here for reasons of space, in Figures 4,5,6 and 7. It can be observed that the particle size plays the most important role in the aerodynamics of the particle trajectories. Figures 4 and 5 show particle trajectories for the three initial particle sizes, angular velocity equal to -500 rad/s and initial velocities equal to 1.5 m/s and 4 m/s respectively. Only the 1.0 mm diameter particles are lifted up due to the low inertia values and they tend to follow the single phase flow velocity vectors. Larger particles, due to the effects of gravity and increased inertia, collide with the opposite furnace walls, rebounce several times and finally drop at the bottom of the furnace. If the particles are injected into the furnace without any angular velocity, as shown in Figures 6 and 7, with the same initial conditions as before, they may or may not collide with the furnace walls but are nevertheless lifted up and tend to follow the flow, independent of their initial size. Similar calculations, not presented here, have shown that the pattern of the particle trajectories is not significantly influenced by the particle density, which may suggest that heat transfer characteristics are not expected to after the particle aerodynamics significantly. Increasing the particle injection velocity does not considerably affect the particle trajectories, as shown by comparing Figures 4 to 5 and 6 to 7.

The predicted particle trajectories provide a first insight at the aerodynamics of expansion furnaces. Although several simplifications have been made, the general trends can be expected to simulate the real situation. The obtained results served as guidelines for the design of the proposed novel furnace. They have provided us with a first indication of the suitable particle size-range and particle initial velocities which would guarantee efficient fluidization of the particle inside the furnace. The development of a particle-expansion model taking into account heat transfer, particle-diameter increase and density changes due to the evaporation is required for a more detailed simulation of the expansion process.

THE PROPOSED FURNACE - THERMODYNAMIC ANALYSIS

Description of the proposed furnace layout

The proposed vertical furnace employs the concept of a bubbling and of a fast circulating fluidised bed. The concept allows the preheating of particles to take place under high volume fractions, whereas the expansion itself takes place at the upper part of the reactor. The tentative layout, is shown in Figure 8.

The proposed novel expansion furnace features a modular character that discriminates the preheating and heating up stages of the material from the expansion stage. Preheating of the raw material is achieved in a moving bed surrounding the main furnace. The material is further heated up in a bubbling bed at the bottom of the furnace. The rapid expansion takes place in a fast circulating bed. A cyclone is used to reassure the separation of the unexpanded material. The operation of the cyclone is based on the density differences between the expanded and unexpanded material. More than 85% of the total energy is used to preheat the particles and longer residence time are achieved at the lower part of the furnace. The expansion itself is a very fast process which takes place in the upper part of the furnace (i.e. the fast bed). The plant in Figure 8 comprises:

A. The raw material feed.

B. A moving bed, which surrounds the furnace and is used to cool the furnace walls and simultaneously pre-heat the raw perlite up to about 400 °C. It is estimated that upto 25% energy saving can be achieved at the moving bed.

C. A bubbling bed is positioned at the lower part of the furnace. The velocity of the air phase in the bubbling bed equals the minimum fluidization velocity of the larger raw perlite particles, as calculated in the first part of the paper. Temperatures in the bubbling bed

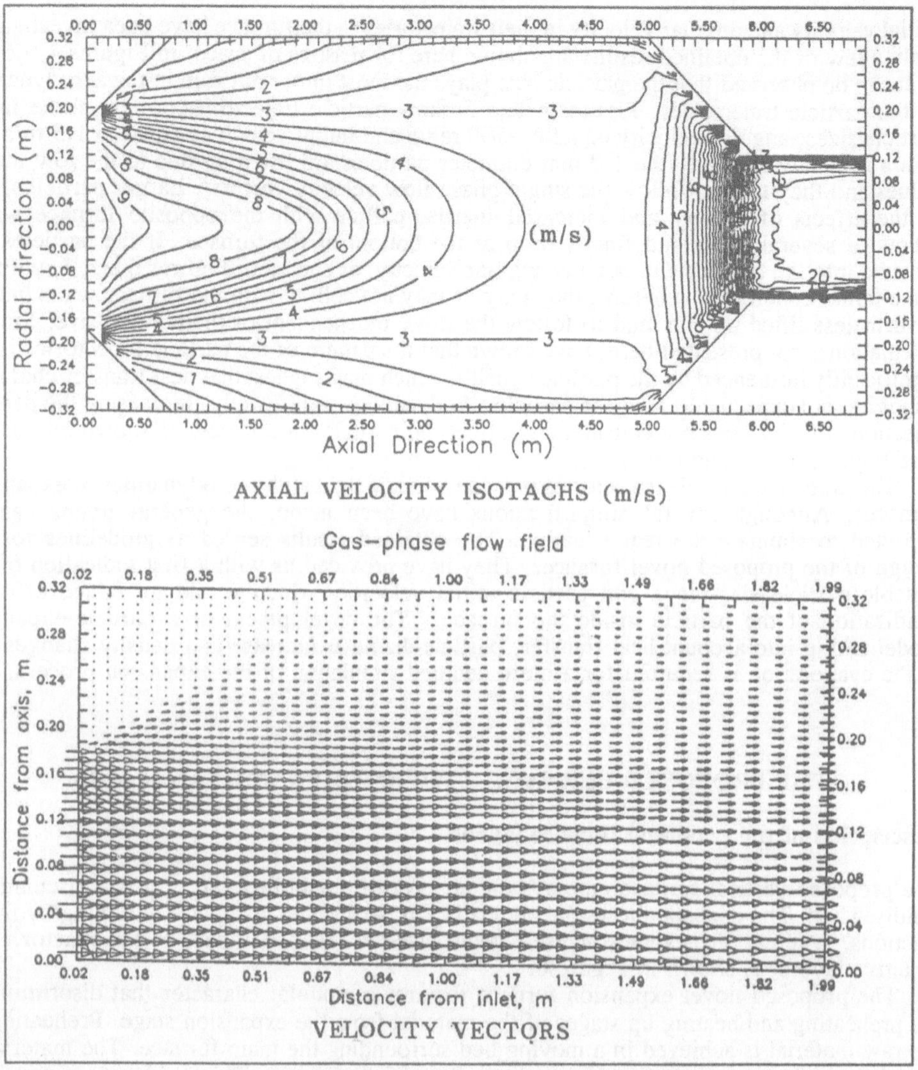

Figure 3: Predicted axial velocity isotachs and velocity vectors in vertical expansion furnace.

are about 550 °C. A first stage heating up of the material is achieved in the bubbling bed, whithout significant loss of the moisture content. The successful expansion of raw material depends mainly on its rapid heating at the expansion temperature (ca. 900 °C, depending on the raw material quality and size distribution).

D. <u>The expansion furnace</u>. (Fast circulating bed). Temperatures in the furnace vary between 900 -950 °C and the calculated volume loading of expanded material is 3%. The furnace wall temperatures are kept at about 750 °C to avoid melting of material on the walls. The calculations here been made for diesel fuel. The burner can be easily changed to other types, e.g. natural gas burner without major modifications to the furnace. The residence time

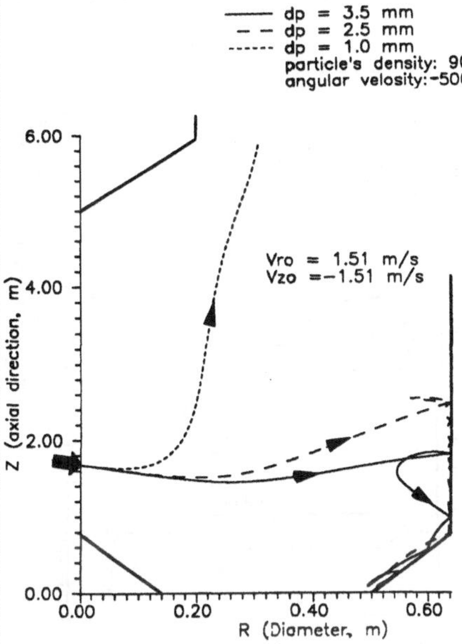

Figure 4: Effect of particle size

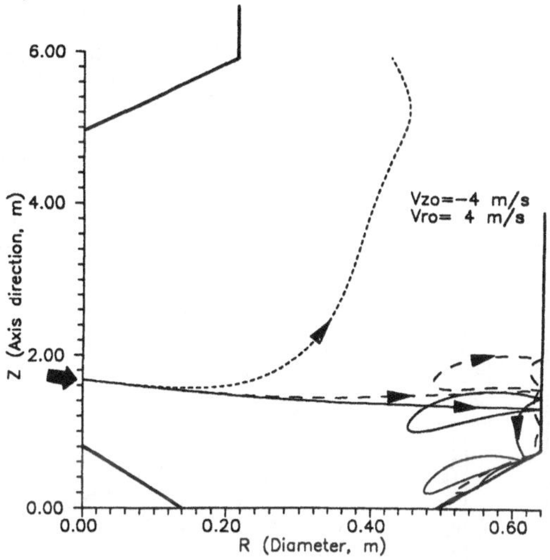

Figure 5: Effect of initial particle velocity

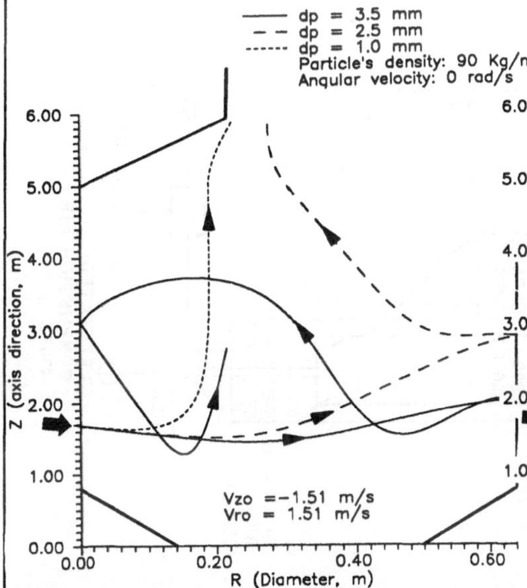

Figure 6: Effect of angular velocity

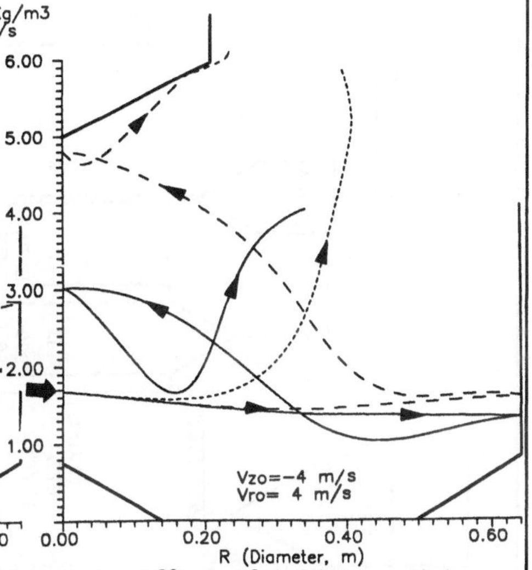

Figure 7: Effect of initial particle velocity

Figures 4, 5, 6 and 7: Effect of particle size, particle initial and angular velocity on the particle trajectories

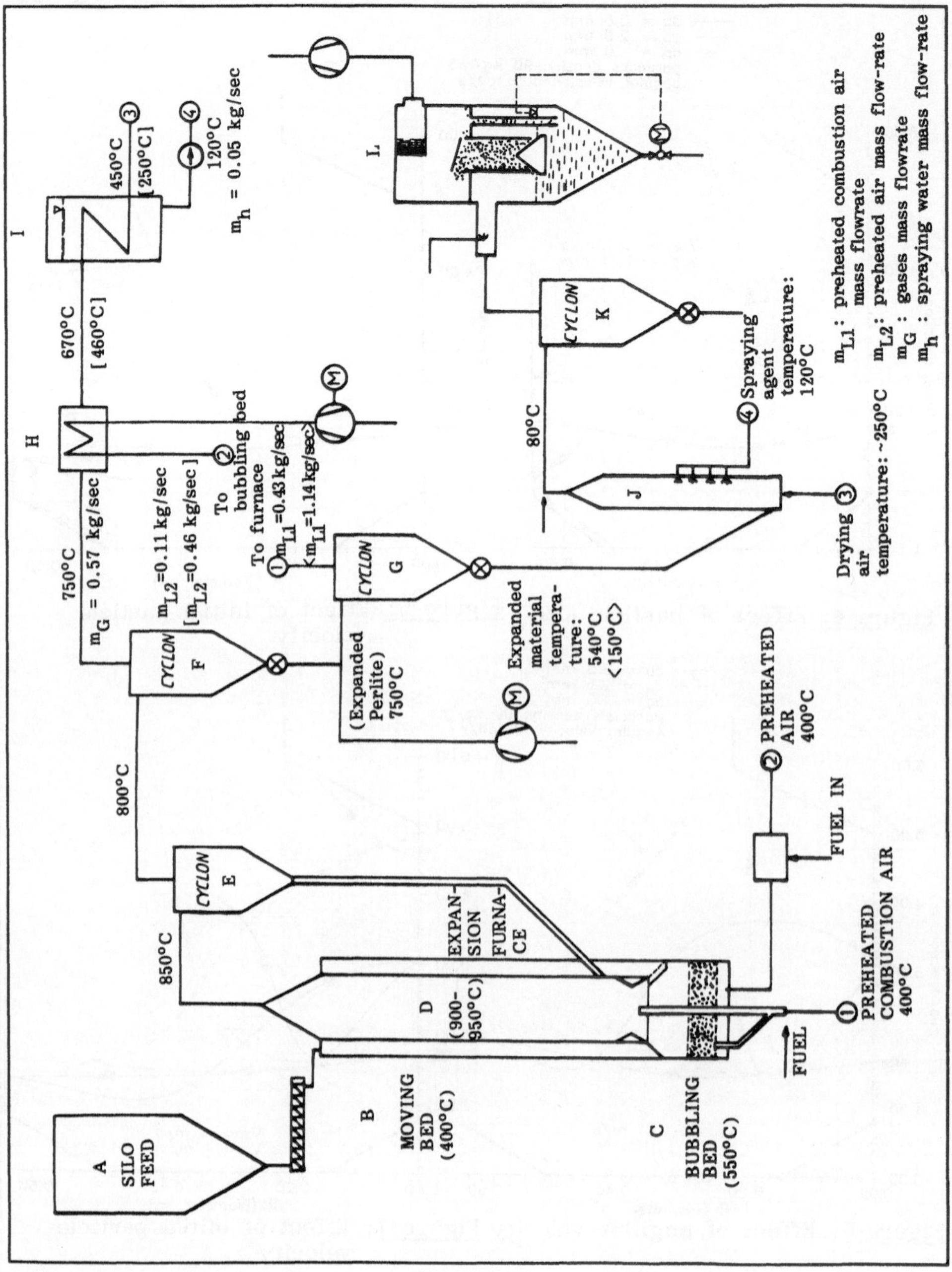

Figure 8: The proposed novel furnace design

of expanded material in the fast circulating bed is kept very low.

E. A <u>cyclone</u> is placed immediately after the expansion furnace. Expanded particles are separated from the unexpanded material which is externally either returned to the furnace or disposed. A operating plants, there is always a percentage of material that does not expand and has to be removed from inside the furnace and via the cyclone.

F. A second cyclone separates the expanded perlite from the hot gases.

G. A third cyclone is used as a cyclone preheater to preheat the combustion/transport air (1) which reenters the furnace with a temperature of about 400 °C. The expanded material after the cyclone contains an amount of energy at high temperature, about 540 °C. This energy can be saved increasing the air mass flow rate through the cyclone G. This case is in angles enclosed in Figure 8.

H. A surface heat exchanger is used with the exhaust gases from the first cyclone to preheat the air which is used in the bubbling bed (2).

I. A vessel is used to preheat the hydrophobization agent, e.g. silicone, utilising low temperature heat recovery, and to facilitate its pumping to the spraying area. The gases from the vessel at 250 °C are directed to the spraying area for drying up the moistened perlite and transport it. Gas temperature drops to 250°C either by increasing the mass flowrate through the heat exchanger H (case shown in brackets in Figure 8), or by doubling the spraying water mass flowrate. The case with no any additional temperature drop is also shown in Figure 8.

J. Hydrophobization Probes

K. The gases transporting the expanded and coated material exit the spraying area, are mixed with air to be cooled at about 80 °C and then enter the product classification cyclone(s).

L. The exhaust gases from the classification cyclone containing fine dust are passed through a "gas-washer" precipitator with a venturi scrubber which wets the dust and makes it easier to separate. The dust is retained and the exhaust gases are emitted to the environment. The plant also contains air and exhaust gas fans and fuel and spraying agent pumps.

Thermodynamic Analysis

The proposed furnace layout is based on the thermodynamic analysis of the expansion process and the aerodynamic behaviour of the expanded particles. The raw material features a broad spectrum of sizes from 0.2 mm to 3.5 mm, which results in a wide range of minimum fluidization velocities and terminal velocities of the expanded material calculated in the first part of the paper. The expanded material has to be quickly removed from the furnace (for example the minimum required terminal velocity of a 0.2 mm expanded particle is 0.4 m/s) and if low gas velocities are applied then no fluidization takes place for the larger particles (minimum required fluidization velocity for a 2 mm unexpanded perlite particle is 1.5 m/s).

The considered gas mass flow rate is sufficient for the proper aero-thermodynamic operation of the bubbling bed as it has been calculated from state diagrammes of the aerodynamics [8] and thermodynamics [9] of the bubbling bed. The heat transfer coefficient varies in the range of 350 - 400 $W/m^2 {}^\circ C$. The following parameters of the expansion process have been investigated:

- Residence time of particles in the bubbling bed and in the fast circulating bed.
- Temperatures required in the moving, bubbling and circulating beds which will exclude any damage on the prescribed physical properties of the expanded material.
- Required gas velocities.
- Effect of preheating and heating-up the raw material for various raw material qualities and size distributions.

From the heat and mass balance of the process shown in Figure 8 a specific energy consumption of 1.675 MJ/kg raw perlite is estimated. This represents a considerable energy saving in comparision to the 3.35 MJ/kg raw perlite required for the initial process. The higher energy efficiency is mainly due to the various preheating stages of the particulate phase which by no means affects the technical requirements of the expansion process.

The above theoretical analysis can not in any way substitute experimental investigation. Such an investigation determines more accurately the energy and fuel mass flowrates required for the expansion process, not only for the case analysed here, but also for alternate cases, i.e. if half of the preheated combustion air enters the bubbling bed. In this case, an increase in the heat transfer coefficient due to higher velocities is expected.

CONCLUSIONS

Particle trajectories have been calculated in a vertical, sudden expansion furnace. The predictions have demonstrated that particle initial size and angular velocity are the most important aerodynamic parameters influencing the transport of the expanded material.

A novel expansion furnace has been proposed for the expansion of industrial minerals and comprises a bubbling and a fast circulating fluidized bed. The concept allows the preheating of particles to take place under high volume fractions, whereas the expansion itself takes place at the upper part of the reactor. It has been calculated that energy savings upto 45% can be achieved in the novel furnace.

REFERENCES

1. Power, T., Perlite and Vermiculite. The market overlap, Industrial Minerals, 1986, also bul.of Perlite Institute Inc., 1987.

2. Team of authors, Perlite industrial exploration, UNIDO Czehoslovakia Joint Programme for international cooperation in the field of Ceramics, special consultants A. Lostak and Z.A. Engelthaler, 1984, Report No. JP/180/84.

3. Allen, M.J., The Perlite Institute Annual Meeting and Conference, Industrial Minerals, 1990, 69-73.

4. Gosman, A.D. and Ideriah, F.J.K., TEACH 2E : A general computer program for two dimensional turbulent, recirculating flows, Int. Report Imperial College, London, 1976.

5. Founti, M., Giannakopoulos, D., Kardamakis, S. and Klipfel, A., Experimental and computational investigation of a vertical particle-laden sudden expansion flow, 1st National Congress on Computational Mechanics, Athens, Greece, Sept. 1992.

6. Founti, M. and Berdi, T., Particle-wall collisions in confined gas-particle flows, the 1st National Congress on Computational Mechanics, Athens, Greece, Sept. 1992.

7. Tabakoff, W., Malak, M.F. and Hamed, A., Laser measurements of solid-particle rebound parameters impacting on 2024 Aluminium and Titanium alloys, AIAA Journal, 1987, **25**, 5, 721-726.

8. Molerus, O., Fluid-Feststoff-Stroemungen, Springer-Verlag, Berlin, 1982, p.64.

9. Schluender, E.-U. und Tsotsas, E., Waermeuebertragung in Festbetten, durchmischten Schuettguettern und Wirbelschichten, Georg Thieme Verlag, Stuttgart, 1988, p. 62.

QUALITY AND ENERGY CONTROL OF INDUSTRIAL BISCUIT BAKING

Ubo de Vries, Paul Verlaan and Maas van der Vliert
TNO Department of Cereal, Feed and Bakery Technology
Lawickse Allee 156701 AN, WAGENINGEN, The Netherlands

ABSTRACT

At present, industrial biscuit ovens are controlled by measuring the temperature in each oven section and keeping these temperatures at a predetermined value. A control system directly related to the quality of the biscuit where heat flux towards the product is controlled on basis of intrinsic product parameters (e.g. dimensions, moisture content, colour) is expected to reduce fluctuations in product quality and to result in a more efficient use of energy.
In the proposed control strategy a distinction is made between a heating and a drying period. During the heating period the product dimensions should be controlled. During the drying period control of moisture content and degree of browning are the objectives.
Width increase of dough pieces during baking is caused by cookie spread. Biscuit length is determined by the combined effect of cookie spread and elastic contraction. An additional increase in biscuit thickness occurs above a critical level of heat flux. During the drying period there is a practically linear relation between moisture loss and heat flux.

INTRODUCTION

It is estimated that during biscuit baking 10 % of the average energy input is used to raise product temperature and 20 % to evaporate moisture. The remaining 70 % is partly lost to surroundings and through exhaust gases. A significant portion of energy losses can also be attributed to variations in product dimensions. Biscuits are usually packed in tight wrapped roll-packs. Variations in the horizontal dimensions (length and width) result in a portion of biscuits that cannot be packed. These biscuits are usually ground to crumb and recycled back through the mixing process. Unfortunately, recycle percentages of 5 to 10 % are not uncommon.

It can be argued that variations in thickness also result in energy losses. Biscuit packets should contain a guaranteed minimum weight. Thickness variations result in a high average excess weight and thus in more energy needed to produce a roll-pack.

The cost of energy is a small percentage of the cost of ingredients. Recommendations to improve energy efficiency will therefore not be accepted if there is even a remote possibility that product quality will be affected.

Although product quality is claimed to be a top priority, biscuit oven control is limited to the machine level: A temperature is measured in each oven section and oven conditions are adjusted to keep this temperature at a predetermined value. This control loop can function properly and still lead to variations in product quality. Another disadvantage of this type of oven control is that it is not the temperature but the heat flux towards the product that should be controlled. In baking ovens convection and radiation have the same order of magnitude [1]. It is therefore unlikely that the measured temperature bears a constant relationship with total heat flux.

A control method directly related to the quality of the biscuit where heat flux towards the product is controlled on basis of intrinsic product parameters (e.g. dimensions, moisture content, colour) will result in a more efficient use of energy as well as reduce fluctuations in product quality. Therefore it is expected to find ready acceptance in biscuit industry. Detailed knowledge of the influence of oven conditions on product parameters and dedicated modelling of the baking process are essential steps in the development of this novel control strategy.

HORIZONTAL DIMENSIONS

Semi-sweet biscuits are manufactured by the use of steel rolls to form the dough into a continuous sheet. Dough pieces are cut from this sheet and baked. The height of the product increases considerably during the early stages of baking; this is called the oven spring. Horizontal dimensions also change after the dough pieces have been cut. The width of the baked biscuit (diameter perpendicular to the sheeting direction) is always larger than the width of the cutter. The length (diameter in the direction of sheeting) is usually, but not always, smaller.

Changes in dough piece dimensions after cutting were registered to detect the mechanisms that determine biscuit length and width (figure 1). The width of the biscuit appears to remain approximately constant after cutting. The diameter in the direction of sheeting shows an elastic contraction. After 30 minutes no more changes seemed to happen and the dough pieces were baked. During baking cookie spread occured in both directions to the same extent. Apparently the increase in width of a biscuit is caused by cookie spread. Changes in the direction of sheeting are caused by the combined effect of cookie spread and elastic contraction.

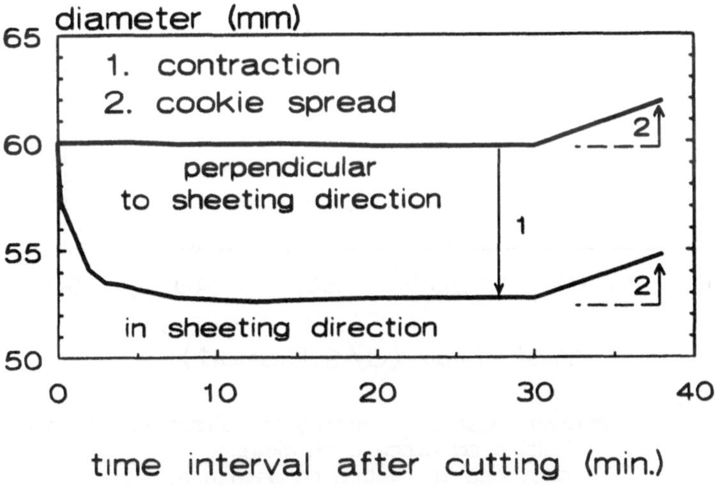

Figure 1. Dimensional changes after cutting.

In practice the time interval between cutting and baking is less than a minute. Contraction and cookie spread are expected to occur simultaneously during the early stages of baking.

THICKNESS

Biscuits were baked in a convection oven. Temperature set-points were varied between 180 and 300 °C. Superficial air velocities were chosen between 0 an 1.5 m/s. Heat flux towards the product was estimated using the air temperature measured with a suction pyrometer, temperatures of surfaces in the oven and of the biscuit. In figure 2 the final thickness is given as a function of the average heat flux towards the product during the first two minutes of baking. The initial thickness of the dough pieces was 1.5 mm. During baking with a relatively low heat flux (up till about 20 W/biscuit) the height increased to approximately 5.3 mm. In this range the biscuit thickness was not sensitive to minor changes in heat flux.

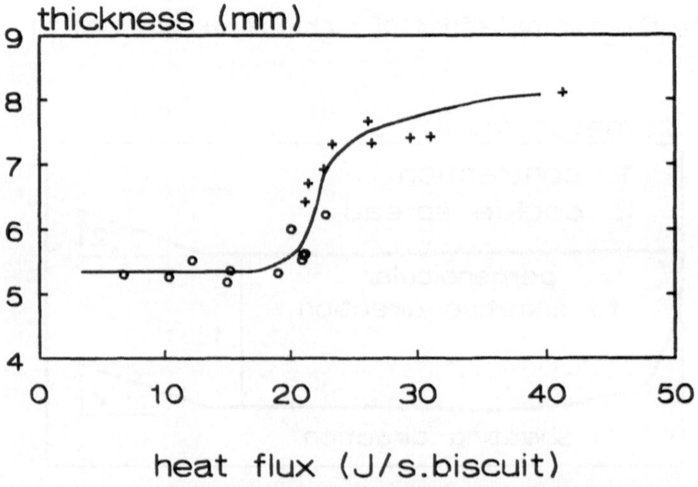

Figure 2. Effect of average heat flux during the first two minutes of baking on biscuit thickness.
(+ decrease in centre temperature)

An additional increase in thickness occurred when heat flux towards the product exceeded a critical level. This additional increase was accompanied by a decrease in centre temperature (figure 3). Both observations can possibly be explained by the development of a pressure difference between the biscuit centre and its surroundings during the early stages of baking. At a high heat flux more steam is generated in the centre than can diffuse towards the surface. The additional expansion will cause a pressure drop and a reduction of the water boiling-point.

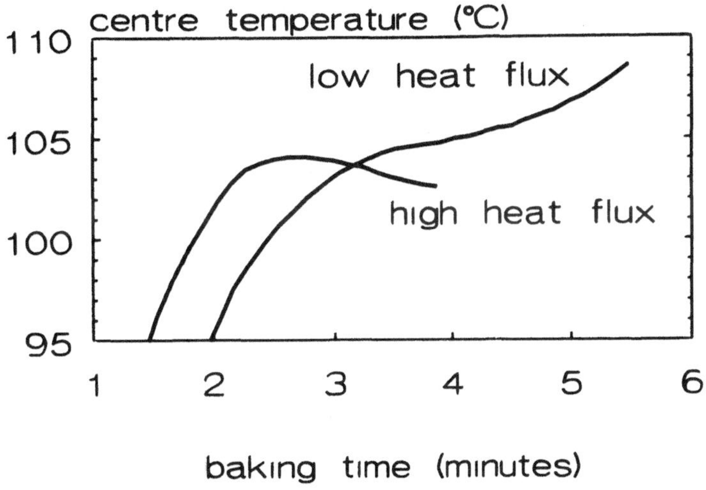

Figure 3. Centre temperature during baking with a high and a low heat flux

MOISTURE CONTENT

A model has been developed aiming at a simulation of simultaneous heat and mass transport within the product during baking, cooling and storage [2]. The heat transport mechanisms, conduction and evaporation-condensation are included in this model.

A distinction can be made between a heating and a drying period. In figure 4 the calculated partial water vapour pressure at a moment during the heating and a moment during the drying period are given. During the heating period an evaporation front with a relatively high partial pressure moves from the surface towards the centre of the product. Water vapour is transported from a high to a low partial pressure. As a consequence the moisture content in the centre of the product increases. The product loses moisture between front and surface. The drying period starts at the moment the evaporation fronts reach the product centre. From this point onward the centre will lose moisture as well.

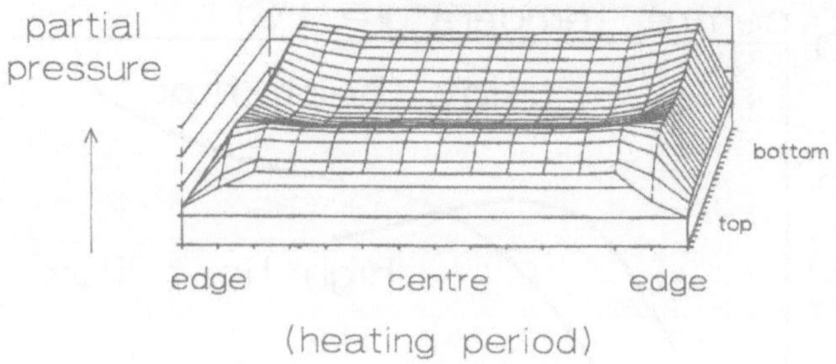

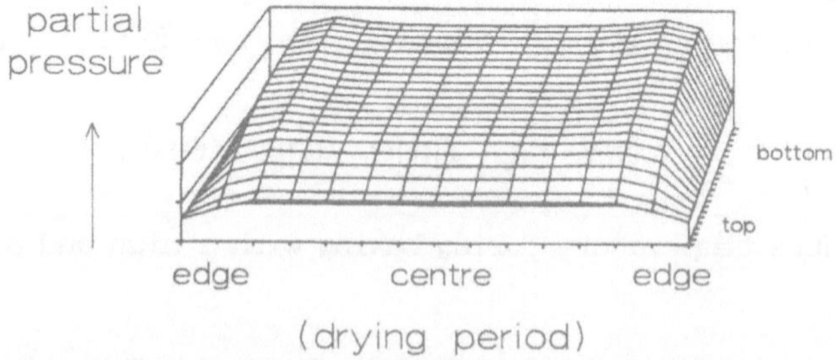

Figure 4. Partial water vapour distribution in a cross-section of a biscuit during heating and drying period.

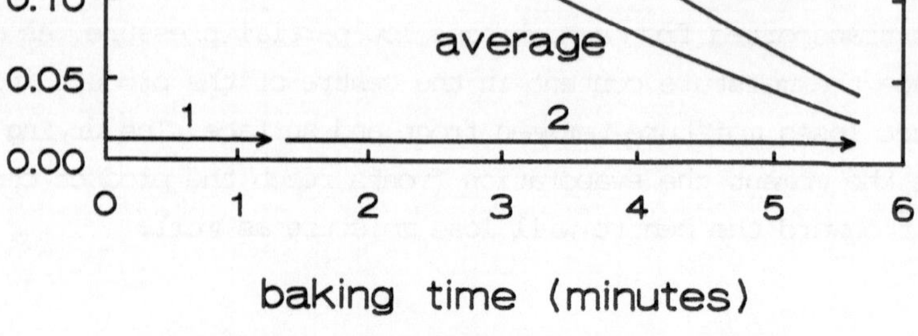

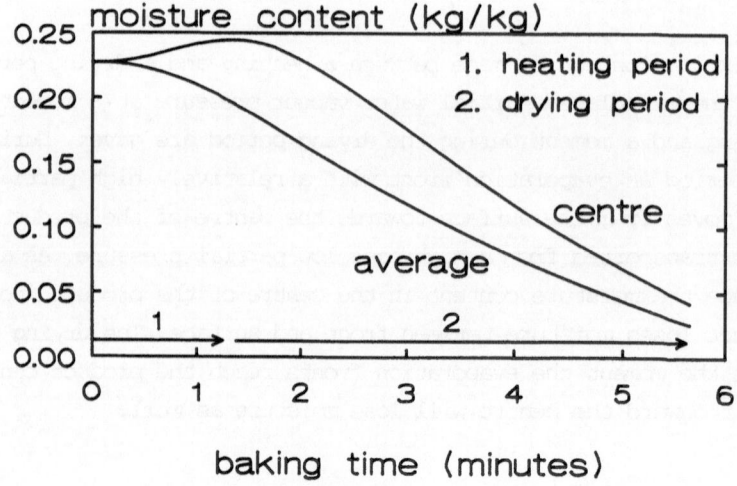

Figure 5. Calculated changes in moisture content during biscuit baking. (1=heating period, 2=drying period)

Calculated centre and average moisture contents are given in figure 5. The average moisture content decreases when the biscuit enters the oven. The evaporation-condensation mechanism causes an increase in centre moisture content during the heating period.

During the drying period the product temperature changes much less than during the heating period. Therefore the relationship between evaporation rate and heat flux is almost linear during the drying period (figure 6). The moisture content can be regulated by adjusting the heat flux in the last sections of the oven.

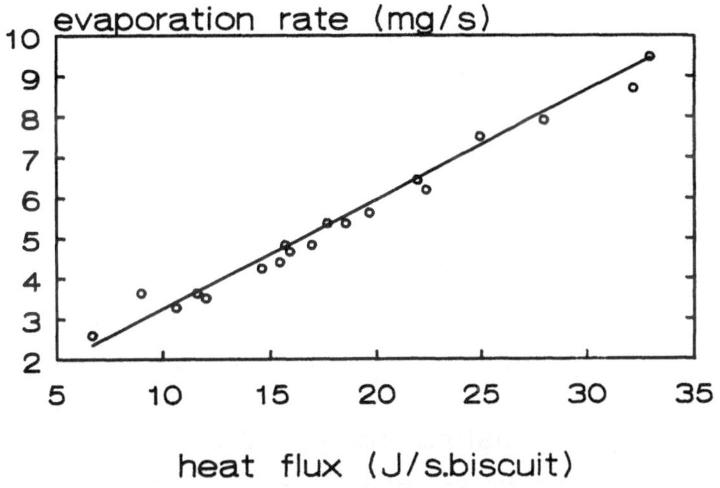

Figure 6. Relationship between evaporation rate and heat flux.

CHECKING

When a biscuit is removed from the oven there is an appreciable difference in moisture content of the inner, more moist, portion and the outer, drier, portion. Equilibration of the moisture content during storage can cause the feared phenomenon of checking [3]. If the centre becomes drier it will have a tendency to shrink. An increase in moisture content of the rim will cause a swelling of the material. Stresses that develop by this process can cause the formation of hair-line cracks days after baking.

In figure 7 the percentage of checked biscuits is given as a function of the average moisture content. Biscuits with a very high and with a very low moisture content don't develop hair-line cracks. A critical range of moisture contents is in between. The critical range is broader if forced convection is used.

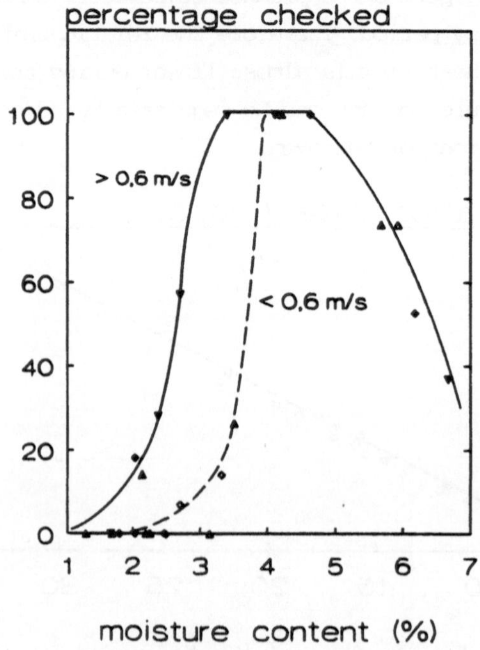

Figure 7. Checking versus average moisture content in biscuits baked with high and with low air velocity.

CONCLUSIONS AND RECOMMENDATIONS

The baking process can be divided into a heating and a drying period. During the heating period the evaporation-condensation mechanism causes migration of moisture towards the centre of the product. Final product dimensions are acquired during this period. The transition from heating to drying period is defined as the moment at which moisture content in the centre has reached a maximum and starts to decrease.
This distinction between heating and drying period can also be used in a control method. During the heating period dimensions should be controlled.

During the drying phase control of moisture content and degree of browning should be the objectives (figure 8).

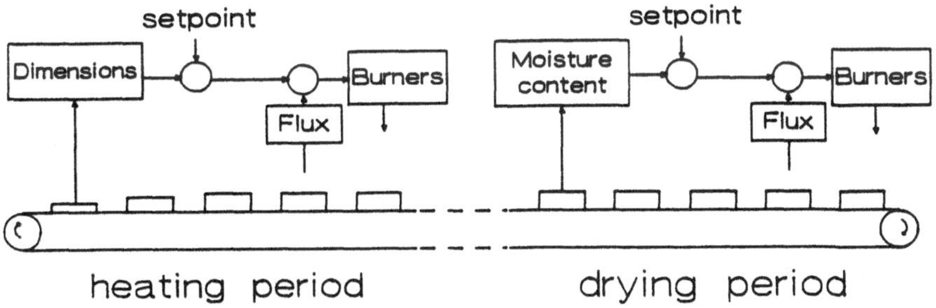

Figure 8. Block diagram of the proposed control strategy

The increased understanding of the baking process contributes to the development of the proposed control strategy. More knowledge is required about the kinetics of browning. Accurate sensors applicable in baking ovens are needed. Incorporation of oven spring in the mathematical model and combination of this model with a model of the baking chamber are other important milestones. At present efforts in this field are undertaken for bread baking [4,5].

REFERENCES

1. Standing, C.N.
   Individual heat transfer modes in band oven biscuit baking
   Journal of Food Science 39 (1974), 267-271.
2. De Vries, U.A., P. Sluimer and A.H. Bloksma
   A quantitative model for heat transport in dough and crumb during baking
   In: Cereal science and technology in Sweden, N.-G. Asp ed.
   STU Lund (1989), 174-188.
3. Dunn, J.A. and C.H. Bailey
   Factors influencing checking in biscuits
   Cereal Chemistry 5 (1928), 395-430.
4. M.G. Carvalho and N. Martins
   Mathematical modelling of heat and mass transfer phenomena in baking ovens
   Presented at CMEM 91 "Computational Methods and Experimental Measurements", 23-26 July 1991, Montreal, Canada.
5. Dalhuijsen, A.J.
   Baking better bread
   Applied Research 37 (1991) June/July, 6-7.

SESSION 9:

Combustion

Chairman: Prof. H.L.J. Meunier

CATALYTIC COMBUSTION: CURRENT STATUS AND IMPLICATIONS FOR ENERGY EFFICIENCY IN THE PROCESS INDUSTRIES

DAVID A. REAY
David Reay & Associates,
PO Box 25, Whitley Bay, Tyne & Wear, UK NE26 1QT.

ABSTRACT

A catalyst may be simply defined as a material which promotes a chemical reaction without taking part in it, and, in theory, without being affected by it. One complex reaction which can be promoted by a variety of catalysts is combustion.

Recently, with the growing interest in process intensification and the coincidental developments in compact heat exchanger technology and enhancement, new potential opportunities have arisen for the exploitation of catalyst technology in combustion systems across a range of applications, extending from power generation to compact reactors and energy recovery.

This paper discusses some of the activities in catalytic combustion currently being persued in Europe and the USA, and comments on the implications for energy efficiency.

It may be concluded at this time that while a number of major areas of development related to catalytic combustion have yet to be successfully concluded, catalytic combustion has considerable potential for low emission, energy efficient plant. This is particularly true in cases where environmental protection has to be given high priority.

INTRODUCTION

The use or improvement of catalytic processes, including catalytic combustion, have a number of implications with regard to the future design of utility and process plant. In addition to anticipated environmental pollution benefits, principally associated with the reduction in NOx afforded by the use of catalytic combustion, it is possible to foresee other advantages. There are a number of areas where catalyst technology can be used to intensify process plant, and in a significant proportion of these the perceived result will be a reduction in energy requirements and a lower plant capital cost. This is particularly true in situations where alternative, eg downstream, clean-up processes are currently the only alternative.

Some example areas include:

+ Radical redesign of process reactors.
+ Catalytic combustion in boilers, gas turbines and reciprocating engines.
+ Catalytic combustion on heat exchanger surfaces and in furnaces and process heaters.

The author's organisation, with the support of the UK Energy Efficiency Office and a number of major industrial sponsors, has recently investigated current international R&D activity in the area of catalytic combustion. The aim was to identify opportunities for cross-fertilization across existing and proposed applications of the technology, and to stimulate the more widespread use of, or further research into, catalytic combustion.

This paper draws on the data collected during the study, and the conclusions reached, in presenting a review of the technology and its application in a number of areas as listed above.

GAS TURBINES

Emissions from Gas Turbines.

The exhaust emissions from gas turbines are a function of the turbine design, fuel and operating conditions. The chemical species present arise from those in the inlet air and products of combustion. For gas turbines burning natural gas or distillate oil the major components are nitrogen, (the dominant component by volume), oxygen, carbon dioxide and water vapour. Minor components, which are those most likely to be subject to emission legislation, are typically:

Nitric oxide:	20-220 ppmv
Nitrogen dioxide:	2-20 ppmv
Carbon monoxide:	5-330 ppmv
Sulphur dioxide:	Trace-3 ppmv
Unburnt hydrocarbons:	5-300 ppmv
Particulates & smoke:	Trace-25 ppmv

The nitrogen oxides (NOx) are those most affected by emission legislation, hence the high activity directed at reducing NOx emissions. The NOx level from older gas turbines is around 200 ppm, and 100 ppm for modern machines. Turbine vendors are seaking to reduce emission levels to: NOx - 10 ppm; CO - 10 ppm; and hydrocarbons - 10 ppm.

Options for Emission Control on Gas Turbines.

The options available for the reduction of emissions from gas turbines are as follows;

Wet combustion control (steam or water injection).
Dry combustion control (dry low NOx or partially premixed flame).
Selective catalytic reduction.
Heat recovery steam reforming.
Partial autothermal reforming.
Partial catalytic oxidation.
Catalytic combustion.

The last four methods above involve what may be called upstream combustion or reaction. Heat recovery steam reforming (proposed by Jack Janes and the California Energy Commission), partial autothermal reforming (patented by ICI) and partial catalytic oxidation (patented by JARIX in Belgium) are based on external reforming of methane using steam, leading to hydrogen. The ICI system, illustrated in Fig. 1, is based on the use of a compact adiabatic reactor using the partial combustion of an air bleed from the air compressor to provide the heat for steam reforming. The amount of steam required is similar to that needed for convention steam-injected gas turbines (STIG), but will achieve much lower NOx levels.

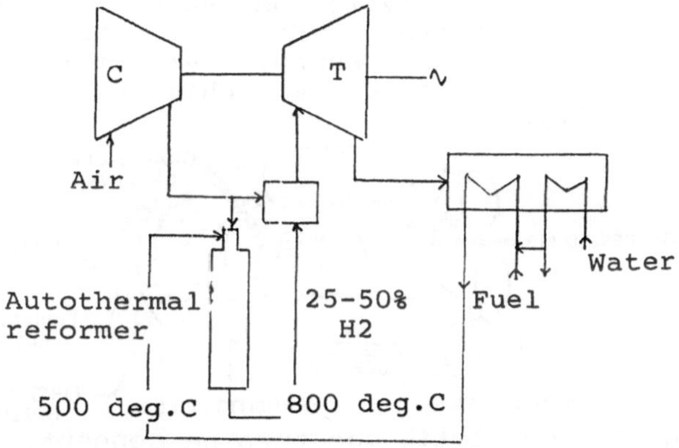

Fig. 1. The ICI Autothermal Reformer Concept.

The JARIX system is similar, and an economic analysis estimated that retrofitting a 1.27 MW gas turbine with a partial oxidation reactor, and using the combustible exhaust gases to help fire a lime kiln, would give a return on the investment of 1.4 years. However, much work remains to be done on reactor development.

Many of the major gas turbine vendors see catalytic combustion of methane in the gas turbine combustion chamber as the logical route to low NOx and realistic capital costs. Catalytic combustion enables stable combustion of lean mixtures of hydrocarbon fuels to take place without hydrogen addition.

However, two main technical barriers have to be overcome in order to ensure the success of catalytic combustion: The development of a high temperature, thermally shock resistant high voidage substrate; and the identification of a stable high temperature catalyst to operate at up to about 1250 deg.C.

Catalytic Combustion Developments.

Much work on gas turbine catalytic combustion was carried out in the 1970's, both in the UK and, predominantly, in the USA, spurred by new EPA regulations on emissions. A proportion of this research was on aircraft-type turbines, with fuels such as JP4. In addition to the 'routine' problems of catalytic combustion systems - those of reliability, maintainability and cost, catalyst poisoning by low quality fuels, the effects of engine vibration, start up and transient response, a particular area of concern was the introduction of premixed fuel to the catalyst. Good mixing and a uniform velocity profile at the entrance to the combustor were prerequisites for long catalyst support structure life, by avoiding hot spots. A conceptual catalytic combustion gas turbine of this period is illustrated in Fig. 2. Note that by grading the cells of the support structure, optimum combustion conditions could be achieved through the various stages, [1].

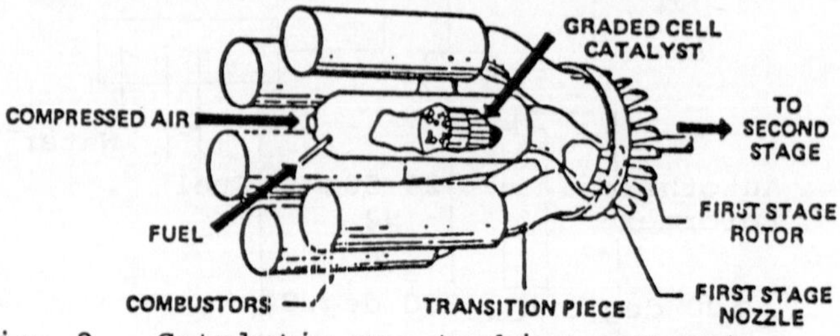

Fig. 2. Catalytic gas turbine concept.

Current activity in the field of gas turbine catalytic combustion is high. The Gas Research Institute in Chicago is working on the development of catalyst/substrates. Catalysts being considered are noble metals and non-noble metal oxides/chromates. The substrates are metal oxide ceramics, with technology transfer from the extensive effort put into automotive exhaust gas treatment. The overall approach involves the development of a catalytically stabilised combustor operating under very lean conditions. This is the 'roughing' stage where most of the methane is reacted. Once the gas temperature has risen, the residual methane is consumed in a combustion unit 'polisher' operating homogeneously. This dual approach avoids an excessively large catalytic unit which might otherwise be needed to deal with methane slip - the passage of unburnt gas.

The requirements of a catalytic combustion chamber for a current large industrial gas turbine operating on methane are as follows:

Space velocity: 2.7 x10E6/hr.
Preheat temperature: 350 deg.C.
Pressure: 10 bar.
Heat release rate: 980 MW/cu.m.
Catalyst exit temp.: 1080 deg.C.

To date the developments have been limited to 1 bar pressure (except for GE and ICI), while the best heat release rate has only approached 35% of that needed, with many of the attempts being significantly below even this. The GE and ICI collaboration [2] has achieved targets in terms of space velocity and catalyst exit temperature. Catalyst and catalyst support life testing has, in general, still to demonstrate that needed by the gas turbine vendors.

Companies active in the field include Rolls-Royce, GE, Toshiba, and Alzeta Corporation, who, with Johnson-Matthey, are developing catalytic combustion units for small cogeneration units of 50 kW plus.

There is interest from a number of companies in using catalytic combustion to exploit residual or waste fuels in gas turbines in an environmentally-friendly way. The burning of solvents is one possibility, and the use of catalysts to assist combustion of gas from landfill and other similar sources is proposed. Closed cycle (recuperative) gas turbines could also benefit from catalytic combustion, one idea being to coat the recuperator on one side with a catalyst to improve preheat temperatures.

CATALYTIC PLATE REACTORS

The common conventional reactors include adiabatic packed bed units and tubular reactors. Both can incorporate the catalyst on coated pellets, or, in the latter case, the tube wall may be coated. Both of these geometries are limited in performance by heat and mass transfer considerations, and the reactors are inherently large pieces of equipment.

As part of a research programme on process intensification commencing in 1975, laminar flow heat transfer was studied at ICI using closely spaced metal plates. Very high volumetric heat transfer coefficients were achieved which broadly confirmed the theoretical predictions [3].

It was soon recognised that such a heat transfer matrix could be the basis for a very intense catalytic reactor, provided that thin layers of highly active catalyst could be bonded to one or both sides of the plate. The inherent attraction of this approach is that it effectively short-circuits heat and mass transfer resistances between the reaction site and the heating or cooling medium. When the process reaction is endothermic, the heat needed to drive the reaction could, in principle, be provided by catalytic combustion on the other plate surface.

It is this last observation which links the catalytic plate reactor concept to catalytic combustion, and most of the applications of catalytic combustion described in this paper could benefit from the successful development of the plate reactor concept.

The intimate linking of the combustion heat source with the endothermic reaction process (and with a number of other potential heat and mass transfer situations), almost eliminates the overall heat transfer resistance. The long radiation paths needed for conventional furnaces are replaced by channel dimensions of one or two millimetres in plate matrices, with an obvious impact on the size of reactor needed. Such a comparison is made for a methane/steam reformer in Fig. 3.

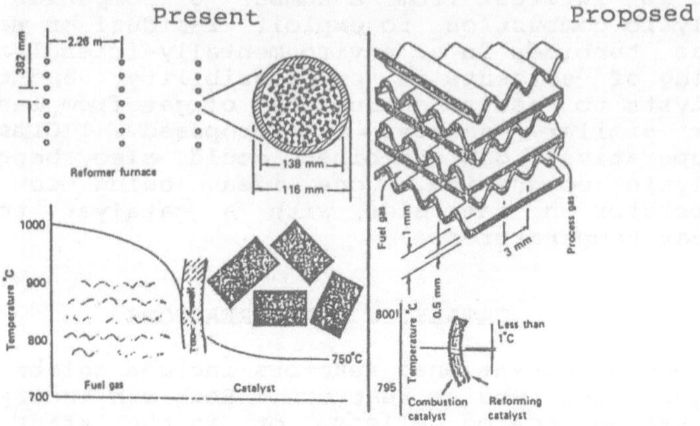

Fig. 3. Comparison of conventional and catalytic plate reformer.

It is beyond the scope of this paper to discuss the chemical process to which such a reactor might be applied. Suffice to say that concepts for constructing such a reactor, including the application of catalysts in small channels of an appropriate length, have been proposed, and are being persued by Professor Colin Ramshaw at Newcastle University. Fouling and catalyst regeneration are obvious areas to address, as is reaction stability and catalyst enhancement. The potential benefits in terms of energy and plant cost if a solution can be found are, however, considerable. One example quoted for the use of intensified reactors in an ethylene plant of 500,000 tpa capacity predicts energy savings of £5 million pa due to reduced steam generation. This is in addition to reduced capital costs.

CATALYTIC COMBUSTION IN BOILERS

As with the other combustion devices considered here, there are a number of options available for controlling the NOx emissions from boilers. Novel non-catalytic burners, as well as downstream treatment methods such as selective catalytic reduction are all being applied and persued in the quest for further performance improvements.

Catalytic combustion boilers were first mentioned in the literature in 1911, when Professor Bone of Leeds constructed a locomotive type boiler with tubes containing a catalyst in the form of porous fireclay mixed with felspar. The coke oven gas gave a boiler efficiency of 90%, compared to the 'norm' of 70% in conventionally fired boilers. Interestingly, one of the first applications of catalytic combustion was thus promoted in the context of energy efficiency, no mention being made of pollution control!

More recently, the emphasis has turned to environmental protection. One of the most comprehensive studies of catalytic combustion in boilers was at the initiation of the US Environmental Protection Agency in 1975, in conjunction with Acurex. Both watertube and firetube boilers were investigated. The concept in the case of the former was to locate tubes coated with the catalyst within the main tube bundle, as shown in Fig. 4, but water was excluded from these tubes (one reason being that this could inhibit light-off). A second, downstream, adiabatic catalytic combustor was located upstream of the convection section.

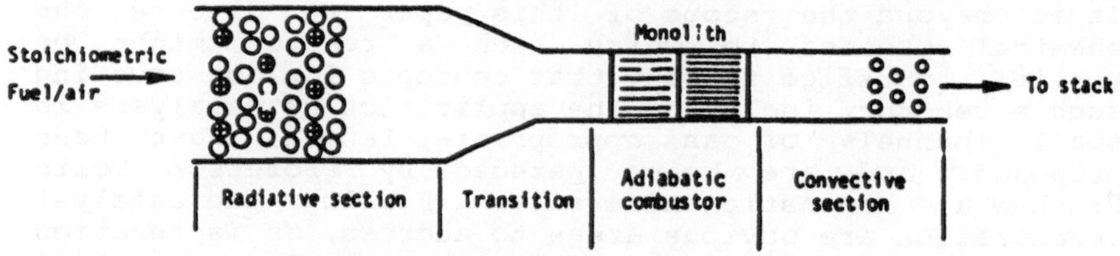

Fig. 4. Concept of radiative catalyst/watertube combustion system.

The catalysed tubes were alumina, with an alumina washcoat and platinum catalyst. While measured NOx levels were less than 2 ppmv, the heat release rate was in the radiative section was less than predicted, and the section was felt not to be fully suited to full scale development. Combustion efficiency in the first stage was less than 50%.

The workers also considered firetube designs employing catalytic combustion. One concept employed a graded cell catalyst as discussed in the context of gas turbines, with a preheat section to help initiate combustion. In a second design, a felt-like matrix material was considered as the catalyst element. Diffusion of the air/fuel mixture through the matrix led to combustion on the outer surface, which then radiates its energy to the water-cooled furnace wall.

More recently, within the last three years, patents have appeared describing work in Italy on both firetube and watertube boiler variants for domestic and industrial applications. In one such variant, the catalyst is applied in pellet-supported form, into which the tubes are inserted, as shown in Fig. 5.

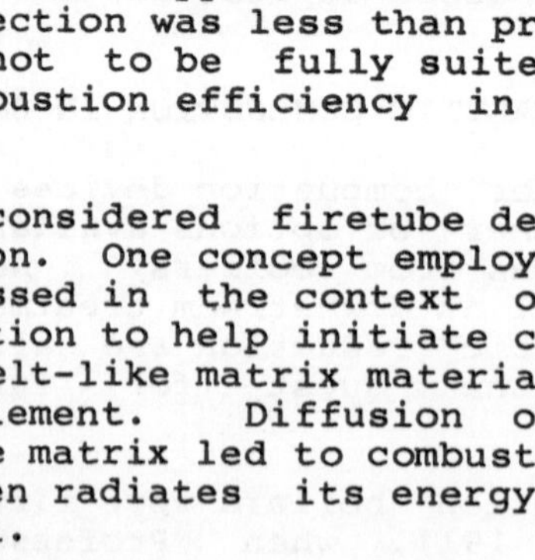

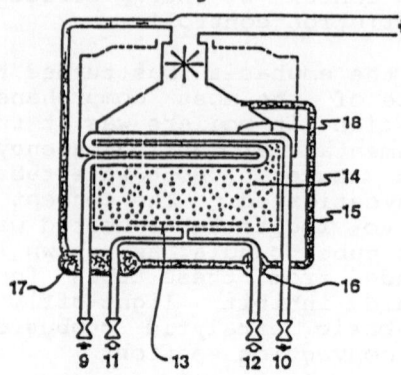

Fig. 5. Catalytic combustion boiler.

In the case of the firetube configuration, the pellets are inserted inside the tubes. In both cases the light-off is assisted by an electric heating element close to the methane/air mixture inlet (13 in Fig. 5).

The GRIV company in Italy [4] has approached the problems of catalytic combustion in boilers in a different way. Appreciating that a direct contact between the heat sink (ie heat exchanger tubes) and the catalyst would impair the reaction, the watertubes surround the catalyst bed in the form of a jacket. Complete methane combustion is claimed using pure or supported metal oxide catalysts. An example is a mixture of Cu and Cr oxides in various oxidation states, in particular copper chromite was used successfully over a period of 7 months, with little or no NOx emission.

The use of pellets in boilers has similar limitations to those described above in catalytic reactors. Ideally, a catalyst on the tube surface would be most effective, but in order to approach the performance of a conventional boiler two limitations have to be overcome. Firstly, the requirements which are currently necessary to sustain the combustion impose serious restrictions on the water side heat transfer rate. Secondly, the catalytic combustion conditions which would occur in any conventional practical boiler design lack the heat transfer potential of high driving temperatures which are typical of a conventional design. Hence a relatively greater surface area would be required.

These problems are not insoluble. For example, increasing the catalytically coated surface area relative to the water side area assists. It has been shown that the use of catalytically-coated tube inserts which radiate onto the boiler tube can cause a significant increase in the heat transfer rate and would not be directly affected by the water side heat transfer coefficient. A variety of other arrangements could be conceived which would rely on radiation or conduction to transfer heat from the extended surfaces to the boiler tube or to provide a hot gas source. However, it may necessitate a radical redesign of boilers in order to accommodate catalytic combustion, while retaining current volumes and efficiencies.

DIESEL AND GAS ENGINES

Several techniques are available or under development for the reduction of NOx emissions from Diesel and gas reciprocating engines; (note that some techniques are appropriate for gas engines only). From the point of view of NOx emissions it is important to distinguish between rich- and lean-burn engines. The former type operates with near-stoichiometric combustion, and there is thus little or no excess air in the exhaust. Lean-burn engines, on the other hand, have significant amounts of air in the exhaust.

There are two approaches to controlling NOx emissions, as with gas turbines. One involves modifications to the engine to limit NOx formation in the combustion process, while the other involves aftertreatment to reduce NOx concentrations in the exhaust. Those appropriate to reciprocating engines are principally catalytic reduction, engine modification, exhaust gas recirculation and pre-stratified charge.

What may be called 'primary' catalytic combustion in reciprocating engines is a topic which has received little attention in the literature. Perhaps the most publicised example was the Merritt engine developed to prototype form at Coventry Polytechnic. This uses a platinum catalyst coating in the combustion chamber located in the cylinder head, and the engine has run on low octane fuels. A glow-plug is used to assist catalyst light-off. Other companies have investigated catalysts on pistons and in the pre-combustion chamber (as in the Merritt engine), and catalysts have been supported here on wire mesh grids or on inserts of other forms with catalyst coatings. Small engines with catalytically-coated liners have also been studied.

One of the difficulties of catalytic combustion in engines, (and a problem also being addressed for petrol engine exhaust catalysts), is that of light-off. In the latter case, location of the catalyst near to the exhaust ports, or electric heating of the support structure, are possibilities, in particular to deal with emissions immediately the engine is started. For catalytic combustion in the chamber itself, the use of laser ignition may be a feasible alternative, and the method known as infra-red multi-photon dissociation, in which laser energy can 'target' specific components of natural gas, can lead to lower laser power requirements.

CATALYTIC HEAT EXCHANGERS AND RELATED EQUIPMENT.

The author's original view of a catalytic heat exchanger was a design based on finned tubes which would be coated with a catalyst. Surface combustion on the heat exchanger would give high heat transfer rates to the stream to be heated inside the tubes, offering, for example, the possibility of combining a catalytic fume incinerator and a process fluid waste heat boiler into a single package. The use of small catalytically-coated shell-and-tube heat exchangers for cabin conditioning in aircraft was an early application of another design, using oil mist as the heat source for the combustion.

The enhancement which can be achieved using surface combustion is a function of the reactants, catalyst and reaction rate. For a reaction rate of 10,000 W/sq.m, a surface heat transfer coefficient on a finned tube of 700 W/sq.m.K was predicted, an order of magnitude higher than a forced convection coefficient. No account was taken of the radiation component in this early analysis.

The concept has been applied more recently by L & C Steinmuller in a catalytically-coated tubular recuperator (US Patent 4903755, 1990), combining air preheat duties with NOx reduction. The manufacturer also proposes similar concepts based on coated rotating regenerator matrices.

A catalytic air preheater for a multi-stage down-hole high pressure combustor has also been proposed. In this instance, catalytic combustion of gases on one side of the heat exchanger provides heated air used to burn heavy oils at a later stage of combustion.

The advent of new heat exchanger technologies, such as the printed circuit heat exchanger, which could form the basis of compact catalytic plate reactors as discussed above, tube inserts and other enhancement devices which could be coated with catalysts, and the heat pipe/thermosyphon, open up the possibilitities considerably.

Catalytic Heat Pipes.

Heat pipes have already been integrated with catalytic combustion to a limited extent. In methanation plant, heat pipe heat exchangers have been used to isothermalise the reaction and to remove heat from adjacent cylindrical catalyst structures. More radical designs are possible, however, in which the heat pipe could have a catalyst directly applied to its outer surface, where appropriate. The features of heat pipes which lend support to such concepts are as follows:

(i) Isothermal operation. The heat pipe would be inherently safe and effective for the removal of reaction hot (and cold) spots.

(ii) Capability to deal with high heat fluxes, both radial and axial. For example liquid sodium heat pipes can handle 15 kW/sq.cm.

(iii) Heat pipes have a very low thermal inertia compared with solid thermal conductors. They can therefore be used to speed up the heating of a catalyst to assist light-off.

(iv) A range of heat pipe working fluids can be selected to cover all temperatures envisaged in catalytic combustion.

(v) Heat pipes can be used to maintain the surface temperature of a catalyst thereon essentially constant by using an inert gas buffer system. This could be employed to ensure accurate catalytic reaction control.

(vi) Heat pipe surfaces have been plasma-sprayed with materials such as alumina, and are therefore readily treatable as catalyst support members.

Catalytic Burners.

Catalytic radiant burners are applicable to a wide variety of industrial processes. The advantage of catalysing a burner plaque is that more heat is generated on it and radiated away, thus reducing the flame temperature and the NOx emissions. Alternatively, if the NOx level is kept constant, the output of the burner can be increased. Catalytic combustion will stabilise the flame and probably permit a higher turn-down ratio. Problems to be addressed include catalyst longevity, and loss of catalyst from some designs of plaque due to attrition of the outer surface. There are two ways in which catalytic combustion can be used:

(i) Catalytic radiant burners. These would give the same low NOx levels of 5-10 ppm as radiant plaque burners, but the heat flux per unit area of burner could be increased by a factor of up to 4. This would give a cost-effective replacement for electric radiant heaters in applications such as paper drying. The technology could also be used to reduce NOx levels on domestic gas boilers. The Gas Research Institute has already demonstrated that a platinum catalyst on conventional radiant burners will enable a doubling of the radiant heat output. Palladium is now being investigated.

(ii) Catalytic lean gas burners. Catalytic combustion would enable combustion of premixed lean gas mixtures and allow very low levels of NOx to be achieved on industrial furnaces. ICI has a catalyst which would enable flame temperatures of up to 1250 deg.C to be reached by either flue gas recycling or using excess combustion air.

Catalytic burners are already used in France for paint drying, and there is interest in extending the application to solvent drying/incineration.

CONCLUSIONS.

A number of applications, both existing and potential, for catalytic combustion have been reviewed. Of particular interest is the combination of reaction and combustion in novel plate reactors, but if this technology could be successfully developed, much wider process uses could be realised. Catalytic combustion is seen as one branch of process intensification, with implications for burners, heat exchangers and other unit operations.

REFERENCES.

1. Kesselring, J.P. et al. Design criteria for stationary source catalytic combustion systems. US EPA Report EPA-600/7.79.181, August 1979.

2. Orenstein, R.M. et al. Catalytic combustion for advanced gas turbines. Paper 108, Yokahama Int. Gas Turbine Congress, Yokahama, Oct./Nov., 1991.

3. Cross, W.T. and Ramshaw, C. Process intensification: Laminar flow heat transfer. Chem. Eng. Res. Dev., Vol. 64, pp 293-301, 1986.

4. US Patent 4953512. Methane catalytic combustion boiler for obtaining hot water for household and industrial uses. Assignee: GRIV Srl, Milan. Published 4 September 1990.

Modeling and Simulation of Hydrogen-Oxygen Combustion on Platinum Catalyst

Olaf Deutschmann and *Jürgen Warnatz*
Universität Stuttgart, 7000 Stuttgart 80, Germany

Mark D. Allendorf and *Robert J. Kee*
Sandia National Laboratories, Livermore, CA 94551, U.S.A.

Michael E. Coltrin
Sandia National Laboratories, Albuquerque, NM 8718, U.S.A.

ABSTRACT

Using computational methods, the catalyzed combustion of lean hydrogen-oxygen mixtures in a stagnation flow over a platinum surface and in a flat-plate boundary layer is considered. The analysis includes elementary chemistry in the gas phase as well as on the surface. The stagnation flow is modeled using a similarity transformation that leads to a one-dimensional boundary-value problem, whereas the flat-plate boundary layer is modeled by use of the boundary layer assumption. Results of the models are compared to two sets of experiments that determine (a) catalytic combustion and ignition limits in hydrogen-oxygen mixtures at low pressure (100 mTorr) and (b) OH concentration profiles in catalytically supported combustion at atmospheric pressure. The paper establishes the appropriate reaction mechanisms and interpretes the catalytic behavior in terms of the chemistry models.

INTRODUCTION

Catalytic combustion has a number of potentially important and practical applications like super-lean NO_x-free combustion and generation of low-temperature process heat. We expect that quantitative simulation capabilities can play an important role in accelerating the development in this field. Therefore, we have developed some new computational tools [1-4] that provide opportunities to analyze the elementary chemical processes that occur at gas-surface interfaces, and couple them to the surrounding fluid flow and to detailed gas phase reaction mechanisms.

Catalytic combustion has been known for more than 150 years [5], when *Davy* detected the ability of platinum surfaces to cause combustion of flammable mixtures "without flames," and catalytic combustion of hydrogen has been known since the famous work of *Langmuir* [6]. Nevertheless, there was inadequate quantitative knowledge of this process until recently, when studies of OH desorption from platinum catalysts were carried out using laser-induced fluorescence, LIF (see e. g. [7,8]), leading to some detailed insight into the surface oxidation mechanism of hydrogen which will be necessary in subsequent analysis of hydrocarbon catalytic combustion (experimental results presented e. g. in [9,10]).

The stagnation flow field and the flat-plate boundary layer are particularly amenable to analysis and to experimental investigation. The two-dimensional flow fields can be reduced to one-dimensional situations by a boundary layer or a similarity transformation. Furthermore, there are measurements available for these configurations, i. e. experiments of *Ljungström et al.* in a stagnation flow field [7] and of *Cattolica and Schefer* in a flat-plate boundary layer [11,12], which are accompanied by computations with a detailed gas-phase reaction mechanism (but global surface chemistry).

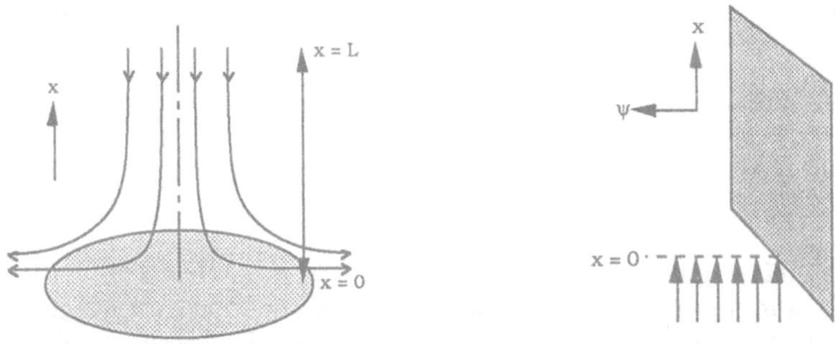

Figure 1. Sketch of stagnation flow field and flat plate boundary layer

THE STAGNATION FLOW FIELD

The Physical Problem

The stagnation-flow field that we consider is shown graphically in Fig. 1. At a distance of $x=L$ above a flat plate a uniform (independent of radius) downward velocity is imposed. In addition, the inlet-gas composition and temperature are also independent of radius and the radial velocity component is zero. By confining our attention to the center portions of the surface, we can neglect the edge effects, permitting use of the one-dimensional analysis.

We consider situations in which the surface is heated resistively. At low surface temperatures, the surface chemistry is sufficiently inactive. As the power is increased, however, the surface temperature becomes high enough to allow catalytic reaction that significantly raises the surface temperature ("ignition" point). Once the catalyst is "lighted," the imposed power to the surface can be reduced or eliminated and the surface will remain at the high combustion temperatures.

Mathematical Formulation

In a stagnation-flow field, scalar quantities (temperature and mass fractions) depend only on the distance from the surface and not on radial position. The boundary-value problem that we solve is stated as [13-15]

Mixture continuity:
$$\frac{1}{\rho}\frac{\partial \rho}{\partial t} = -\frac{\partial u}{\partial x} - 2V - \frac{u}{\rho}\frac{\partial \rho}{\partial x} = 0 \tag{1}$$

Radial momentum:
$$\rho\frac{\partial V}{\partial t} = \frac{\partial}{\partial x}\left(\mu\frac{\partial V}{\partial x}\right) - \rho u\frac{\partial V}{\partial x} - \rho V^2 - \frac{1}{r}\frac{\partial p}{\partial r} = 0 \tag{2}$$

Thermal energy:
$$\rho c_p\frac{\partial T}{\partial t} = \frac{\partial}{\partial x}\left(\lambda\frac{\partial T}{\partial x}\right) - \rho c_p u\frac{\partial T}{\partial x} - \sum_{k=1}^{K_g}\left(c_{pk}\rho Y_k V_k\frac{\partial T}{\partial x} + \dot{\omega}_k h_k\right) + S_q(x) = 0 \tag{3}$$

Species continuity:
$$\rho\frac{\partial Y_k}{\partial t} = -\frac{\partial \rho Y_k V_k}{\partial x} - \rho u\frac{\partial Y_k}{\partial x} + M_k\dot{\omega}_k = 0 \quad \left(k = 1,\ldots,K_g\right) \tag{4}$$

Equation of state:
$$p = \rho RT\sum_{k=1}^{K_g}\frac{Y_k}{M_k} \tag{5}$$

Surface species
$$\frac{dZ_k}{dt} = \frac{\dot{s}_k}{\Gamma} = 0 \tag{6}$$

In the governing equations the independent variables are x, the distance normal to the disk surface, and the time t. The dependent variables are the axial velocity u, the radial velocity υ which is scaled by the radius as $V=\upsilon/r$, the temperature T, the gas-phase species mass fractions Y_k, and the surface species site fractions Z_k. The meaning of the other symbols are: ρ= mass density, c_p= mixture specific heat capacity, M_k= molecular mass of species k, h_k= specific enthalpy, μ= viscosity, λ= thermal conductivity, p= thermodynamic pressure, R= universal gas constant, V_k= diffusion velocities (including thermal diffusion), $\dot{\omega}_k$ = chemical production rate by gas phase reaction, $\dot{s}_k$ = chemical production rate of species by surface reaction, K_g = number of gas phase species, K_s = number of surface species (not including bulk-phase species), Γ= surface site density. $(1/r)(dp/dr) \equiv \Lambda$ in the radial momentum equation is a eigenvalue of the problem [16,17]. The details of the chemical reaction rate formulation can be found in the user's manuals for the *CHEMKIN* [18] and *SURFACE CHEMKIN* [2] software. Details of the transport property formulation can be found in the user's manual for the *TRANSPORT* [19] software.

Equation (6) states simply the fact that in steady state the surface composition does not change. In some sense it could be considered a (possibly complex) boundary condition on the gas-phase system. However, because the surface composition is determined as part of the solution, Eq. (6) should be considered part of the system of governing equations.

The surface boundary condition becomes relatively complex in the presence of heterogeneous surface reactions. The gas-phase mass flux of each species to the surface j_k is balanced by the creation or depletion rate of that species by surface reactions, i. e.

$$j_k = \rho Y_k V_k = \dot{s}_k M_k \quad \left(k = 1,\ldots,K_g\right) \tag{7}$$

The radial surface velocities V are specified by a no-slip boundary condition as $V=0$.

The surface temperature is determined from an energy balance that considers conductive, convective and diffusive energy transport from the gas phase and radiation chemical heat release and resistive heating on the surface,

$$\lambda \frac{\partial T}{\partial x} - \sum_{k=1}^{K_g} \rho Y_k V_k h_k = \sigma\varepsilon\left(T^4 - T_w^4\right) + \sum_{k=K_g+1}^{K_g+K_s} \dot{s}_k M_k h_k + \dot{P}. \tag{8}$$

Here, σ= Stefan-Boltzmann constant, ε= surface emissivity, and T_w = wall temperature to which the surface radiates. The term $\dot{P}$ represents an energy source (here resistance heating in the surface).

At a height L above the surface, the boundary conditions are given as room-temperature mixture of hydrogen and oxygen at a specified velocity. Furthermore, we assume that the radial component of the inlet velocity is zero.

Numerical Solution Method

The computational solution of the stagnation-flow problem is accomplished with the program *SPIN* [20], which, in turn, uses the *TWOPNT* [21] software that implements a Newton/Time-Step algorithm [22].

THE FLAT-PLATE BOUNDARY LAYER

The Physical Problem

The flow field in the flat-plate boundary layer that we consider is shown schematically in Fig. 1. At a distance of $x=0$ at the bottom of a flat plate a uniform upward velocity is imposed (generated by a Mache-Hebra nozzle in the experiment). In addition, the inlet-gas composition and temperature are also independent of the distance y normal to the plate and the normal velocity component is zero [11,12]. We consider situations in which the surface has a constant temperature (as in the experiment).

Mathematical Formulation

The mathematical description of a boundary layer over a flat plate is given in detail elsewhere in the literature [23,24]. For the planar configuration given in Fig. 1, the conservation equations can be written as

Momentum: $$\rho u\frac{\partial u}{\partial x}+\frac{dp}{dx}=\rho u\frac{\partial}{\partial \psi}\left(\rho u\mu\frac{\partial u}{\partial \psi}\right)+\rho g \tag{9}$$

Thermal energy:

$$\rho u c_p\frac{\partial T}{\partial x}=\rho u\frac{\partial}{\partial \psi}\left(\rho u\lambda\frac{\partial T}{\partial \psi}\right)-\sum_{k=1}^{K_g}\dot{\omega}_k M_k h_k-\rho^2 u\sum_{k=1}^{K_g}c_{pk}Y_k V_k\frac{\partial T}{\partial \psi} \tag{10}$$

Species : $$\rho u\frac{\partial Y_k}{\partial x}=M_k\dot{\omega}_k-\rho u\frac{\partial}{\partial \psi}(\rho Y_k V_k)\qquad \left(k=1,\ldots,K_g-1\right) \tag{11}$$

where the symbols have the same meaning as in (1) - (8) and g = gravitational acceleration. The equation of state (5) and the surface species conservation (6) have to be added to this system. The independent variables are x, the distance along the plate surface, and $\psi=\int_0^y \rho u dy$, a density-weighted stream-function coordinate normal to the surface, where y is the physical space coordinate normal to the surface. The dependent varaibles in this parabolic differential equation system are p, ρ, u, T, Y_k and Z_k.

Numerical Solution Method

After discretization of the spatial derivatives (central differences on a fixed grid of ψ), the resulting system of differential/algebraic equations in x is solved with the code *DASSL* developed by *Petzold* [25].

PHYSICAL CHEMISTRY OF THE PROBLEM

Gas-Phase Reaction Mechanism

The gas-phase reaction mechanism that is shown in Table 1 is taken directly from modeling work on flame chemistry. Its validity has been established through numerous studies of flames, shock-tubes, flow reactors and stirred reactors. We simply apply it in this work without modification.

Surface Reaction Mechanism

We have developed a Pt surface reaction mechanism with associated rate expressions following the work of *Hellsing et al.* [26,27]. It consists of dissociative adsorption of both H_2 and O_2, formation of adsorbed H_2O via adsorbed OH, and desorption of H_2O. This mechanism is based on OH LIF measurements and is similar to reaction schemes postulated by *Lin et al.* [28] and by *Schmidt et al.* [29]. In our mechanism, these reactions were supplemented with additional reactions for the adsorption of H, OH, O, and H_2O. The complete surface reaction mechanism, the used pre-exponential factors A, the activation energies E_A and the sticking coefficients S are given in Table 2.

TABLE 1
Gas Phase Mechanism of Hydrogen Oxidation

Reaction		A(cm,mol,s)	b	E(kJ/mol)
1. H_2-O_2 Chain Reactions				
$O_2 + H = OH + O$		5.10E+16	-0.82	69.1
$H_2 + O = OH + H$		1.80E+10	1.00	37.0
$H_2 + OH = H_2O + H$		1.20E+09	1.30	15.2
$OH + OH = H_2O + O$		6.00E+08	1.30	0.0
$H_2 + O_2 = OH + OH$		1.70E+13	0.00	200.0
2. Dissociation/Recombination Reactions				
$H + OH + M = H_2O + M$	1)	7.50E+23	-2.60	0.0
$O_2 + M = O + O + M$		1.90E+11	0.50	400.1
$H_2 + M = H + H + M$	2)	2.20E+12	0.50	387.7
3. HO_2 Formation/Consumption				
$H + O_2 + M = HO_2 + M$	3)	2.10E+18	-1.00	0.0
$H + O_2 + O_2/N_2 = HO_2 + O_2/N_2$		6.70E+19	-1.42	0.0
$HO_2 + H = H_2 + O_2$		2.50E+13	0.00	2.9
$HO_2 + H = OH + OH$		2.50E+14	0.00	7.9
$HO_2 + O = OH + O_2$		4.80E+13	0.00	4.2
$HO_2 + OH = H_2O + O_2$		5.00E+13	0.00	4.2
4. H_2O_2 Formation/Consumption				
$HO_2 + HO_2 = H_2O_2 + O_2$		2.00E+12	0.00	0.0
$H_2O_2 + M = OH + OH + M$		1.20E+17	0.00	190.5
$H_2O_2 + H = H_2 + HO_2$		1.70E+12	0.00	15.7
$H_2O_2 + OH = H_2O + HO_2$		1.00E+13	0.00	7.5

1) enhancement: H_2O /20.0/
2) enhancement: H_2O /6.0/ , H /2.0/ , H_2 /3.0/
3) enhancement: H_2O /21.0/ , H_2 /3.3/ , O_2 /0.0/

TABLE 2
Surface Reaction Mechanism of Hydrogen Oxidation

Reaction	A(cm,mol,s),S	E_a(kJ/mol)	
1. H_2/O_2 Adsorption/Desorption			
$H_2 + Pt(s) = H_2(s)$	0.10		(stick. coeff.)
$H_2(s) + Pt(s) = H(s) + H(s)$	1.50E+23	12.5	
$O_2 + Pt(s) = O_2(s)$	0.046		(stick. coeff.)
$O_2(s) + Pt(s) = O(s) + O(s)$	5.00E+24	0.0	
2. Surface Reactions			
$H(s) + O(s) = OH(s) + Pt(s)$	3.70E+21	19.3	
$H(s) + OH(s) = H_2O(s) + Pt(s)$	3.70E+21	0.0	
$OH(s) + OH(s) = H_2O(s) + O(s)$	3.00E+24	100.5	
3. Product Adsorption/Desorption			
$H + Pt(s) = H(s)$	1.00		(stick. coeff.)
$O + Pt(s) = O(s)$	1.00		(stick. coeff.)
$H_2O + Pt(s) = H_2O(s)$	0.75		(stick. coeff.)
$OH + Pt(s) = H(s)$	1.00		(stick. coeff.)

RESULTS

Stagnation Flow Field

The first example is an (instationary) H_2-O_2 stagnation-point flow to a Pt surface with constant heating of the foil. The set of experiments [7] that we model involves a resistively heated platinum foil onto which flows a room-temperature mixture of 25% H_2 and 75% O_2 at a total pressure of 100 mTorr (13.2 Pa). Based on measured flow rates and the gas inlet diameter, we estimate the inlet velocity to be 3,000 cm/s. The ignition point is reported in terms of electrical current applied to the foil. By knowing the cross-sectional area of the foil and the resistance of platinum as a function of temperature, we convert the current to a power per unit surface area. Furthermore, the energy balance at the surface requires estimating the radiative losses from the foil which, in turn, requires estimating the emissivity of the platinum (~0.1 for Pt in the ignited case, ~0.3 for oxidized Pt in the unignited case [30]).

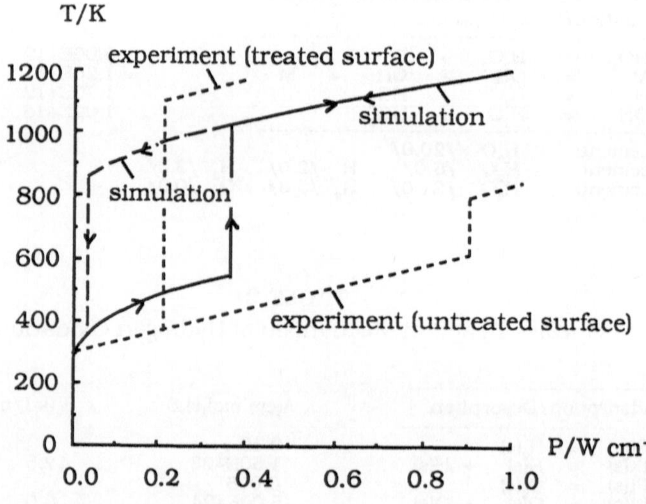

Figure 2. Ignition of 25% hydrogen-75% oxygen mixtures on Pt at p=100mTorr

Figure 2 gives the result of a simulation together with experimental curves for an un-cleaned platinum foil and a platinum foil which has been treated by a series of ignitions. The larger temperature jump in the experiment probably is due to chemical reaction and heat release on the backside of the foil.

Figure 3 shows the predicted temperature and species profiles in the gas-phase above the platinum surface for an imposed power of 0.8 W cm^{-2}. The fact that the temperature and the water concentration are highest at the surface are clear indicators that catalytic combustion is occuring (in contrast to gas-phase combustion induced by a hot surface). The OH radical is seen in low concentrations and is desorbed from the surface. The H-atoms and O-atoms are seen in very low concentrations (not shown in Fig. 3) and are adsorbed. A further indication of the catalytic behavior is seen by "turning off" the surface chemistry: In this case, the surface temperature is significantly lower, with its temperature supported only by the input power. The surface coverage consists mainly of O(s) before ignition and of free Pt(s) after ignition.

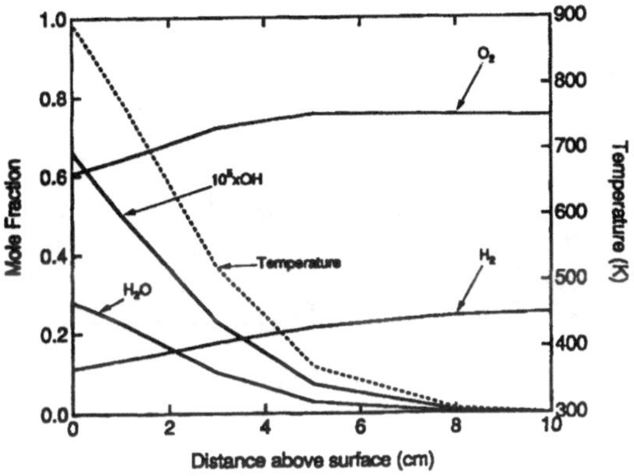

Figure 3. Temperature and mole fractions in the gas phase after ignition of a H_2-O_2 mixture on Pt (P = 0.8 W/cm²)

Flat-Plate Boundary Layer

Because our reaction mechanisms are based on elementary chemistry, they should be equally applicable to different flow geometries and pressure ranges. As a test of their validity, we simulated the *Cattolica and Schefer* [11,12] experiments for an atmospheric-pressure boundary-layer flow of H_2/air over a heated Pt surface. Our predicted OH profiles and the measurements are shown in Fig. 4.

It should first be observed that combustion under these conditions is largely a gas-phase process. Primary evidence for this observation is that the OH-concentration peaks in the gas phase boundary layer. Furthermore, *Cattolica and Schefer* found that combustion occurs in presence of a (presumably non-catalytic) quartz surface, which is also confirmed by our simulations. As observed in the experiments, we also find that gradients in the OH profiles to the surface appear when a catalytic surface is used, showing that the surface is a sink for OH radicals.

The mixture considered is close to the flammability limit and calculations are very sensitive to small changes in the gas-phase reaction rate coefficients. Using the reported experimental flow conditions, our calculation predicted OH-profiles qualitatively different from the measurements. However, increasing the surface temperature to 1220 K (the reported temperature was 1170 K) produced the good agreement shown in Fig.4, both in profile shape and peak magnitude. Given the high sensitivity to the surface temperature we believe that the observed OH profiles can be explained by inaccuracies in the measured surface temperature. Thus we conclude that our mechanisms satisfactorily reproduce the experimental results.

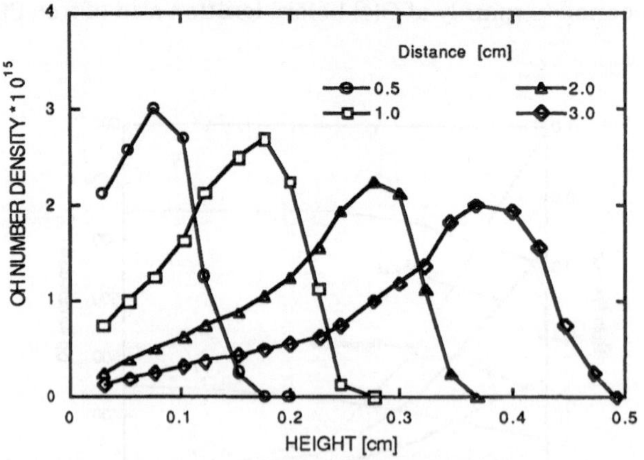

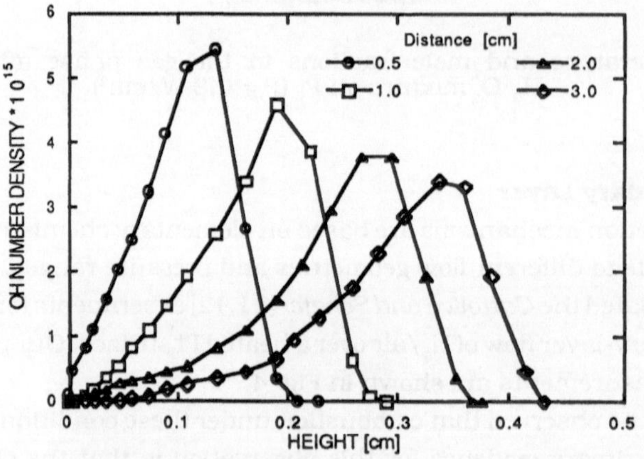

Figure 4. Stabilization of a hydrogen-air flame in a platinum flat plate boundary layer. Upper drawing: experiments; lower drawing: simulation

ACKNOWLEDGEMENTS

The *U. S. Department of Energy, Office of Basic Energy Sciences*, has provided support to this work. One of us (*J. Wa.*) has to thank *SandiaNational Laboratories*, Livermore, for their hospitality during performing this work.

REFERENCES

[1] M. E. Coltrin, R. J. Kee, and F. M. Rupley, "Surface Chemkin: A General Formalism and Software for Analyzing Heterogeneous Chemical Kinetics at a Gas-Surface Interface," Intl. J. Chem. Kin., 1991, to appear

[2] M. E. Coltrin, R. J. Kee, and F. M. Rupley, "Surface Chemkin (Version 3.7): A Fortran Package for Analyzing Heterogenous Chemical Kinetics at a Solid-Surface-Gas-Phase Interface," Sandia National Laboratories Report, *SAND90-8003*, 1990

[3] M. D. Allendorf and R. J. Kee, J. Electrochem. Soc. **139**, 1991, 841

[4] R. E. Mitchell, R. J. Kee, P. Glarborg, and M. E. Coltrin, Twenty-Third Symposium (International) on Combustion, The Combustion Institute, Pittsburgh, PA, 1990, 1169

[5] H. Davy, "Some New Experiments and Observations on the Combustion of Gaseous Mixtures", in: The Collected Works of Sir Humphrey Davy (J. Davy ed.), Vol. **6**, Smith, Elder, and Co., Cornhill, London, 1840

[6] I. Langmuir, Trans. Faraday Soc. **17**, 1922, 621

[7] S. Ljungström, B. Kasemo, A. Rosen, and T. Wahnström, E. Fridell, Surface Sci. **216**, 1989, 63

[8] D. S. Y. Hsu, M. A. Hoffbauer, M. C. Lin, *Surf. Sci.* **184**, 1987, 25

[9] L. D. Pfefferle, W. C. Pfefferle, Catal. Rev. - Sci. Eng. **29**, 1987, 219

[10] X. Song, W. R. Williams, L. D. Schmidt, R. Aris, Twenty-Third Symposium (International) on Combustion, The Combustion Institute, Pittsburgh, PA, 1991, 1129

[11] R. J. Cattolica, R. W. Schefer, Comb. Sci. Technol. **30**, 205-212 (1983)

[12] R. J. Cattolica, R. W. Schefer, Nineteenth Symposium (International) on Combustion, The Combustion Institute, Pittsburgh, PA, 1982, 318

[13] R. J. Kee, J. A. Miller, G. H. Evans, and G. Dixon-Lewis, Twenty-Second Symposium (International) on Combustion, The Combustion Institute, Pittsburgh, PA, 1988, 1479

[14] G. Evans and R. Greif, J. Heat Trans. ASME **109**, 1987, 928

[15] M. E. Coltrin, R. J. Kee, and G. H. Evans, J. Electrochem. Soc. **136**, 1989, 819

[16] G. Stahl, J. Warnatz, Combustion and Flame **85**, 1991, 285

[17] G. Dixon-Lewis, S. Fukutani, J. A. Miller, N. Peters, J. Warnatz et al., Twentieth Symposium (International) on Combustion, The Combustion Institute, Pittsburgh 1985, 1893

[18] R. J. Kee, F. M. Rupley, and J. A. Miller, "CHEMKIN-II: A Fortran Chemical Kinetics Package for the Analysis of Gas-Phase Chemical Kinetics," Sandia National Laboratories Report *SAND89-8009*, 1989

[19] R. J. Kee, G. Dixon-Lewis, J. Warnatz, M. E. Coltrin, and J. A. Miller, "A Fortran Computer Code Package for the Evaluation of Gas-Phase Multicomponent Transport Properties," Sandia National Laboratories Report *SAND86-8246*, 1986

[20] M. E. Coltrin, R. J. Kee, G. H. Evans, E. Meeks, F. M. Rupley, and J. F. Grcar, "SPIN: A Fortran Program for Modeling One-Dimensional Rotating Disk/ Stagnation-Flow Chemical Vapor Deposition Reactors," Sandia National Laboratories Report *SAND91-8003*, 1991

[21] J. F. Grcar, "The TWOPNT Program for Boundary Value Problems," Sandia National Laboratories Report *SAND91-8230*, 1991

[22] J. F. Grcar, R. J. Kee, M. D. Smooke, J. A. Miller, Twenty-First Symposium (International) on Combustion, The Combustion Institute, Pittsburgh, PA, 1986, 1778

[23] M. E. Coltrin, R. J. Kee, J. A. Miller, J. Electrochem. Soc. **131**, 1984, 425

[24] M. E. Coltrin, R. J. Kee, J. A. Miller, J. Electrochem. Soc. **133**, 1986, 1206

[25] L. R. Petzold, "A Description of DASSL: A Differential/Algebraic System Solver," Sandia National Laboratories Report *SAND82-8637*, 1982

[26] B. Hellsing, B. Kasemo, S. Ljungström, A. Rosen, and T. Wahnström, Surface Sci. **189/190**, 1987, 851

[27] B. Hellsing and B. Kasemo, Chemical Phys. Letters **148**, 1988, 465

[28] D. H. Hsu, M. A. Hoffbauer, M. C. Lin, Surf. Sci. **184**, 1987, 25

[29] W. R. Williams, C. M. Marks, L. D. Schmidt, Steps in the Reaction $H_2 + O_2 = H_2O$ on Pt: OH Desorption at High Temperature (to be submitted)

[30] Handbook of Chemistry and Physics, 63rd Ed., p. E-386, 1982

NO_X EMISSIONS FROM A FRONT WALL FIRED PULVERISED COAL BOILER

J.L.T. AZEVEDO*, A.J. BRANCO**, M.G. CARVALHO* AND C.F.M. COIMBRA*
* Instituto Superior Técnico - Mechanical Engineering Department
Av. Rovisco Pais - 1096 Lisboa Codex - Portugal
** Electricidade de Portugal
Av. Estados Unidos da América, 55 - 1700 Lisboa - Portugal

ABSTRACT

The present work is concerned with both axissymmetric and three dimensional pulverised coal combustors. The axissymmetric case is used to test and develop a numerical model which is confronted to experimental results including N related species distributions. NO emissions were measured for several operating conditions of a front wall fired pulverised coal boiler 720 MWt. These experimental results are compared to model predictions showing satisfactory overall agreement from the engineering point-of-view.

INTRODUCTION

The desirable insight about the behavior of flames into real furnaces improved and took benefits from the numerical tools developed to predict the combustion in different applications. The methods of analyzing pulverised coal combustion have been shifted from the empirical and global approaches to the phenomenological models since the early seventies [1]. The increase of computer power and a better understanding of turbulence and particle dispersion allowed the development of comprehensive numerical packages that could well predict main local properties in axissymmetric coal flames [2]. During the eighties, several groups of researchers developed their own numerical models which have been in some extend validated against experimental results obtained in laboratory semi-industrial furnaces with thermal power in the range from 0.1MWt to 10MWt [3-8]. Recently attention and investments on research of pulverised coal systems have moved the interest from improving efficiency to pollutants reduction and much of the investigation performed with numerical models has been centered on NO emissions because these can be easily reduced by changes in the aerodynamics and operating conditions of furnaces.

Despite the extensive number of studies in axissymmetrical geometries, the modelling of pulverised coal combustion in industrial-scale boilers are more limited mainly due to the lack of experimental results for model validation. The experimental analysis of single burners has been conducted in a number of laboratories (from which IFRF and Imperial College are examples) where a large number of measurements in different flames were obtained. The measurements performed nowadays include gas species concentrations with specific reactive-dominant species, temperatures and heat fluxes. Advances in instrumentation are being performed which will allow velocity measurements in the near future. The measurements made in large multi-burner boilers are restricted to the region close to the wall in utility boilers while more complete data sets are obtained in smaller scale combustors which are used for research purposes.

The objectives of the Joule Project (JOUE-0023) were the modelling of wall fired pulverised coal burners for pollutants emissions control. The work carried out to achieve such objective was a combination of experimental and numerical activities including basic research on CFD developments, experimental assessment and numerical predictions of single burners of different scales and of a full-scale industrial boiler. The partners involved on this project are NEI - International Combustion and Imperial College of Science Technology and Medicine from the United Kingdom, Instituto Superior Técnico and Electricidade de Portugal from Portugal and Centre for Renewable Energy Sources from Greece. Experimental information on single burner performance was obtained in the two laboratories in the U.K. (NEI and Imperial College). These results were used to develop numerical models to simulate single burners operation. The Portuguese task in this project is a combination of different activities. Many case tests were performed at IST to validate generally the models including the fuel NO model whose results were compared with experimental results for all different scales equipments. Measurements obtained by EDP in the flue gases of the industrial boiler allowed to identify the influences of operating conditions on NO emissions and to test the numerical code capabilities on the prediction of NO emissions with acceptable accuracy.

The results for the axissymmetric furnace are presented showing the limitations and performance of the model. The utility boiler under study is described simultaneously to the experimental campaign performed at the Sines Power Plant. The experimental and predicted results of NO emissions from the front wall fired boiler are compared in this section. In the last section some conclusions from the present work are outlined.

DESCRIPTION OF THE TEST FACILITIES

Semi-Industrial Furnaces

The semi-industrial furnaces considered in the JOULE project are those of Imperial College

(0.15MWt) and of NEI Combustion Limited (37 MWt). Both furnaces have a single burner being the first furnace downward fired and cylindrical (0.6 m in diameter) while the larger one is almost square and horizontal equipped with a low NO_X burner. These furnaces represent two extremes in size scales, the full scale found in an utility boiler and a small scale allowing detailed measurements of gas compositions and temperatures in a geometry simple to be modelled. The smaller burner considered in the present work consists on an annular secondary air inlet with controlled swirl intensity and a central tube with primary air transporting coal. Three swirl number (S_W=0.78,1.03 and 1.43) flames [9-11] were predicted using a gradual increase in flow rates keeping the excess air level at a constant value of 15%.

Utility Boiler

The boiler under consideration is a 300 MWe front wall fired pulverised coal boiler from Electricidade de Portugal (EDP - Electrical Company of Portugal) using bituminous coal, some times blended. The experimental investigation discussed here was carried out during the usual operation of the boiler.

Field tests were developed in the coal fired boiler n.3 of the Sines Power Plant (720 MWt) which furnace's main dimensions are indicated in figure 1. The burner arrangement in this

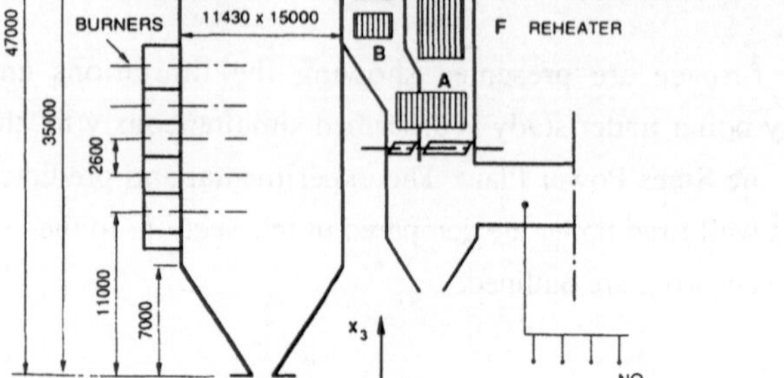

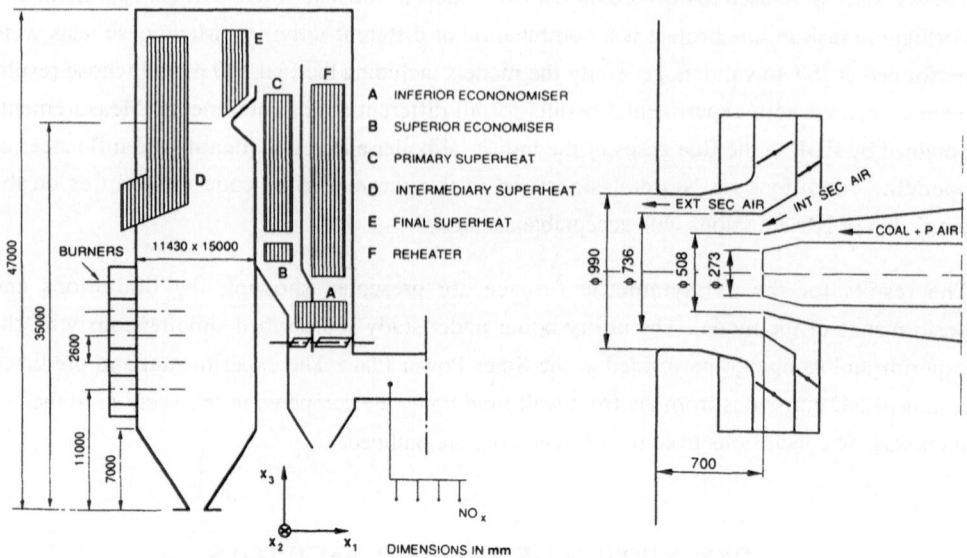

Figure 1 - Main dimensions of the industrial furnace

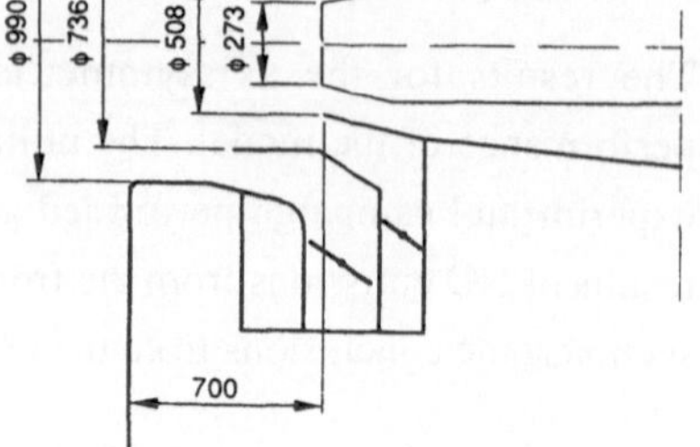

Figure 2 - Geometry of the burners

boiler is of 5 rows of burners installed in the front wall, being the 4 burners of each row feeded by a common wind box. Concerning the air staging capabilities, the burners are of dual controlled flow type as depicted in figure 2. The swirl number was estimated as 0.6 for the simulations.

Full-scale tests were performed after stabilization of the operating conditions and the recorded values are the average cross section values measured at the economizer outlet. The following chemical species were measured at this point: oxygen, carbon monoxide and NO_x ($NO + NO_2$ expressed in NO_2).

Sampling and Analyzing Equipment

The analyzing system used in the industrial tests is of extractive type comprising:

a) Gas sampling probe - Hartmann & Braun, with silicon carbide filters to retain the dust particles. Both internal filters and probe head are electrical heated.

b) Gas sampling line - Hartmann & Braun, electrically heated to prevent condensation. The gas line comprises the Teflon sample line, the heating element and a heat reflection and protection foil.

c) Gas cooler - To cool the sample gas down to the extent that its temperature does not drop below the dew point at any subsequent part of the analyzing system.

d) NOx analyzer - AC20M/Environment, for high concentration which works on the principle of chemiluminescence, generated by reaction of NO in the presence of Ozone.

Other analyzers were used in the experimental study (Hartmann & Braun, URAS-3G for CO and Magnos for Oxygen).

NUMERICAL MODEL

Gas Phase

The numerical model is based on transport equations for the mass, momentum, turbulent quantities, chemical species and enthalpy of the gas phase. These equations are of the general form.

$$U \nabla \rho \phi = \nabla \left(\Gamma_\phi \nabla \phi \right) + S_\phi + S_\phi^p \qquad (1)$$

where ϕ is the transported property, Γ_ϕ is the diffusivity of ϕ and S_ϕ and S_ϕ^p are respectively the source terms of ϕ from the continuum and the particulate phase respectively. [12-13]. The chemical species considered are volatiles, oxygen, and carbon dioxide.

The combustion rate of volatiles is considered using both a mixing rate based on the eddy-dissipation model [14] and a chemical kinetic rate for volatiles [15]. The consideration of the kinetic rate of volatiles combustion influenced the global volatiles burning rate only close to the burner [12] giving marginal improvements in the predictions.

The radiation in the models is handled by respectively a non-equilibrium diffusion model in the axissymmetric case while the three dimensional model of the boiler uses the discrete heat transfer model.

Particle Phase

The coal particles are simulated by a stochastic Lagrangian model. Momentum and energy equation are used to describe particle motion and temperature evolution.

$$\frac{\partial \vec{U}_p}{\partial t} = \frac{\vec{U} - \vec{U}_p}{\tau} + \vec{g} \tag{2}$$

The particle motion equation uses the instantaneous gas velocities simulated by superimposing to the mean flow field the turbulent motion simulated with the aid of the k-ε model. The importance of turbulence dispersion on pulverised coal combustion was revealed in a previous study [12]. An energy balance equation is applied to the coal particle to calculate temperature along the particle trajectory. The particle evolution is described by particle heating, drying, devolatilization and char burnout in sequence. This gives rise to terms in the particle energy equation which allow the calculation of temperature and particle evolution.

$$m_p\, c_p \frac{dT_p}{dt} = A_p\left[h\,(T - T_p) + \sigma\, \varepsilon_p\left(T^4 - T_p^4\right)\right] + K_{ch}\, A_p\, P_{O_2}\, H_c - H_{vol}\frac{dV}{dt} V_p - H_{H_2O}\frac{dm_{H_2O}}{dt} \tag{3}$$

For devolatilization the single first order reaction model with constants from Badziok and Hewksley [16] is used in the present work. The kinetic combustion rate of char is a combination of a kinetic rate from Field [17] combined with the diffusion rate. Volatiles stoichiometry are obtained from coal species balances and the char oxidation is considered to produce carbon dioxide.

Fuel NO model

The model for NO formation from the fuel nitrogen considers that nitrogen is released from coal in hydrogen cyanide at a similar rate as burnout. HCN is then subjected to two competitive reactions which are HCN oxidation producing NO and reduction of NO producing molecular nitrogen The kinetic rates of these reactions are according to De Soete [18]

$$R_{HCN \rightarrow NO} = 10^{10} \rho X_{HCN} X_{O_2}^{b} e^{33700/T} \quad kg/m^3 s \tag{4}$$

where the order of reaction with oxygen (b) is given by De Soete

$$R_{NO \rightarrow N_2} = 3 \times 10^{12} \rho X_{HCN} X_{NO} e^{30000/T} \quad kg/m^3 s \tag{5}$$

The NO formed is thus a balance of the NO produced by HCN oxidation and the NO reduced The results show that after the burning region HCN is totally consumed and the NO is just transported with the flue gases The rate of HCN oxidation has been adjusted by many authors [4, 19, 20] dealing with coal combustion within an order of magnitude higher than the value originally suggested by De Soete In the present work the best agreement for the industrial boiler was obtained with a value an order of magnitude higher while for the axissymmetrical case the constant indicated in equation 4 was retained

The reaction mechanism proposed by De Soete [18] although being one of the simplest, lumping some species represents the best compromise for modelling due to our limited knowledge on coal evolution during burnout A recent and extensive discussion and comparison between possible fuel NO models can be found in Boardman [20]

The NO reduction by carbon in the char particles in the axissymmetrical case was found to play a minor role in the overall balance of NO formation and thus it was disregarded [21] The thermal NO mechanism was also not considered and is expected to have a low contribution to NO emissions due to the relatively low temperatures in coal combustion and due to the large amount of NO produced from fuel bounded nitrogen

NUMERICAL RESULTS

Axissymmetrical Study

The two dimensional axissymmetrical study was carried out with the purpose of comparing the predictions with detailed measured results from the small scale downward fired cylindrical furnace This furnace installed at Imperial College of London has been used in the past years under the supervision of Prof Lockwood to generate a large amount of experimental results which are available to computer code validation and comparison

Detailed comparison with the present model results are presented elsewhere [12, 21] and here only the centreline evolutions are reported Figure 3 represents the centreline axial evolution of

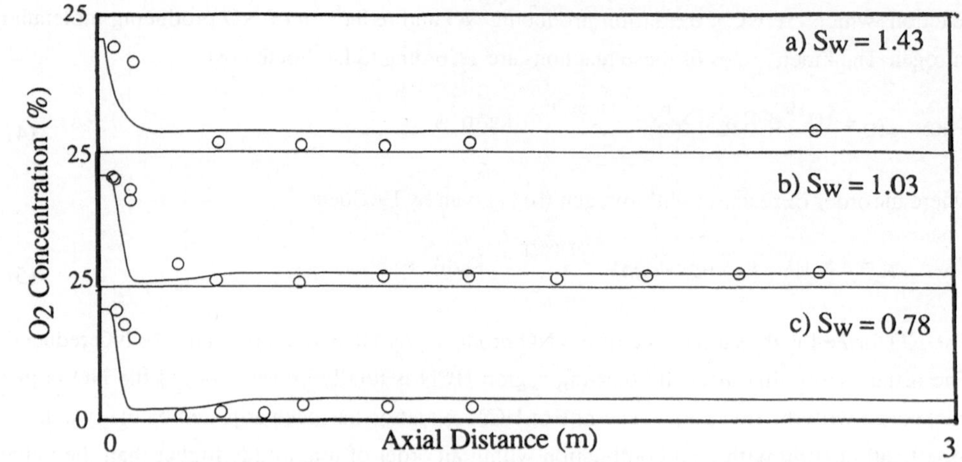

Figure 3 - Centreline oxygen concentration a) S_W=1.43, b) S_W=1.03, c) S_W=0.78.

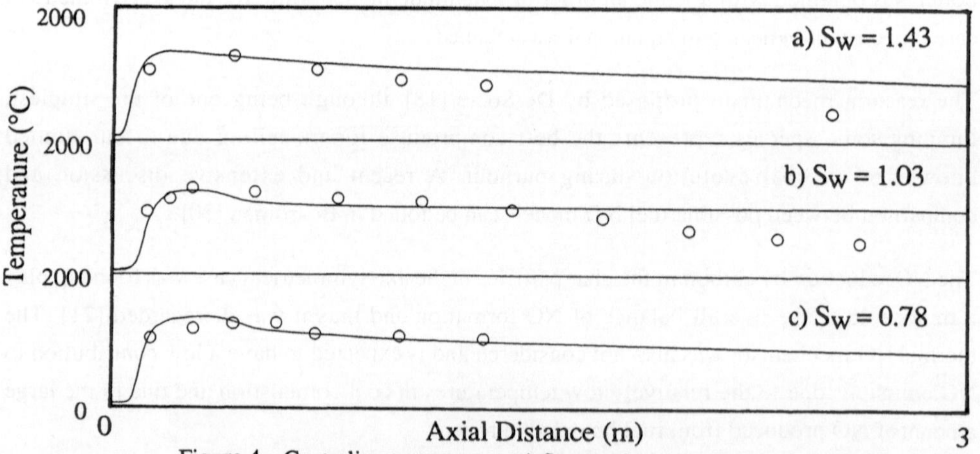

Figure 4 - Centreline temperature a) S_W=1.43, b) S_W=1.03, c) S_W=0.78.

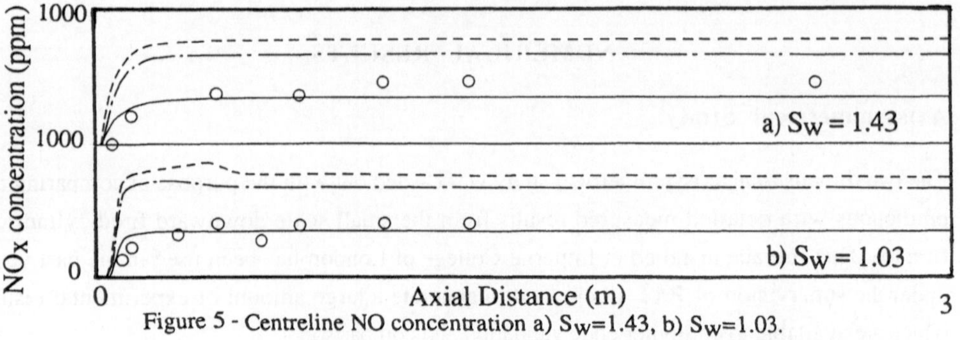

Figure 5 - Centreline NO concentration a) S_W=1.43, b) S_W=1.03.

(—— constant=10^{10}, —·— constant=3.1 10^{10}, - - - -Direct oxidation)

oxygen concentration showing that the central recirculation zone is predicted closer to the burner than suggested by the measurements. This deficiency is partially attributed to the difficulty of predicting accurately the swirling flow with the k-ε model although the model predicts the variation of the position of the central recirculation zone increasing swirl intensity. The overpredicted proximity of the central recirculating region promotes the reversal of particle trajectories which ignite and burn close to the quarl producing an initial high oxygen consumption. The temperature profiles on the centreline are shown in figure 4. Although less pronounced the temperature profiles are also influenced by the same model deficiency. At the end of the furnace the temperature is overpredicted possibly due to inappropriate boundary conditions.

Figure 5 shows the NO axial concentration profiles of the two higher swirl numbers, using the constant depicted in equation (4), this constant multiplied by 3.1 and considering direct oxidation of N from char to NO. The direct oxidation from N-char to NO overpredicts the NO emissions and furthermore the model sensitivity to operating conditions is almost lost which was also verified for the industrial scale study. The oxidation rate constant increased has been used by different authors in literature [19, 20, 22] but in the present model predictions the original constant performs better. This is not the case for the industrial scale study discussed below where a constant an order of magnitude larger gave the best results. The deficiencies in the predicted temperature towards the exit of the furnace should not have significant effects in the NO emissions as NO is almost constant after HCN consumption at the begining of the furnace. The model sensitivity to swirl is lower than measured. The predicted NO is a result of the prediction of all flow field, therefore suffering from the deficiencies of the model. The measurements considered show a small increase of NO emission with swirl although earlier results obtained in the same furnace showed the opposite behaviour [23]. In other studies [4] the swirl influence is small in the range of swirl numbers studied here.

Utility Boiler Study

The numerical study of the boiler was performed for several operating conditions including the use of the BOOS (Burners-Out-Of-Service) technique [13]. The calculation domain were restricted to the shadow region in figure 1, which permits comparison with the flue gases composition only with low reactive substances in the convection area. Since no HCN was found in the outlet of the calculation domain (top of the radiant section of the boiler) in the numerical studies it was assumed that NO concentration in this section is similar to the one verified in the economiser outlet.

Figure 6 shows the temperature, oxygen and nitric oxides distributions over a middle-plane section of the boiler. Interestingly, the predictions for the normal operating conditions corroborate the project assembling of the intermediary superheater (the radiation exchanger) since the higher temperatures are found in that neighbourhood The fuel lean regions of the combustor are characterized by oxygen concentrations above 5% As expected by applying the DeSoete's correlation, the NO mass fractions increase in the envelop of the flame, leading to values higher than 850 ppm in the fuel rich zones

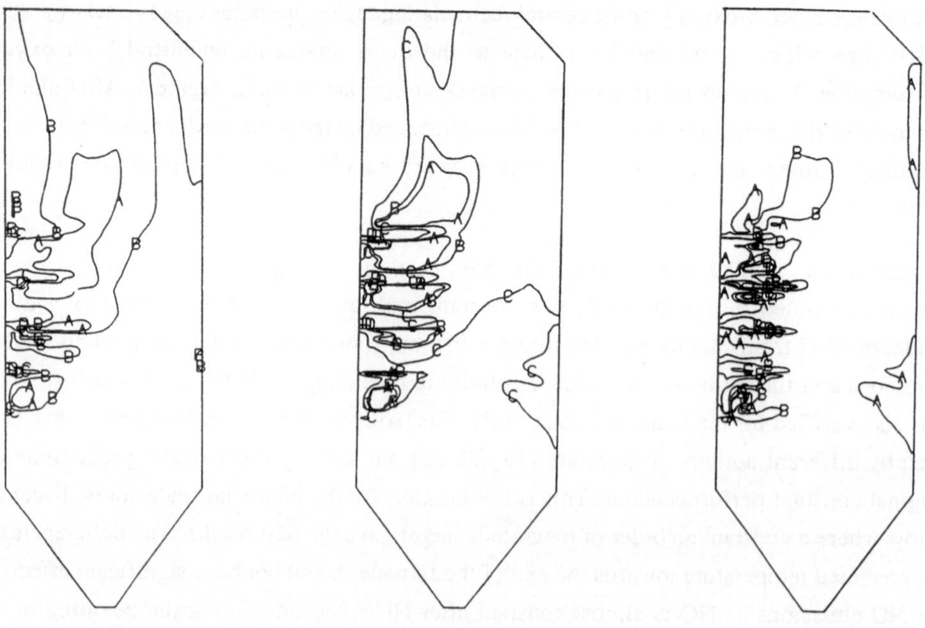

Figure 6 Results of the numerical predictions in the centre of the boiler (x_2 = 4 18m) *Left* Temperature profiles in K (A,2500, B,1750, C,1200, D,750) *Centre* Oxygen mass fractions in % (A,1, B,2, C,5 D 15) *Right* NO mass fractions in ppm vol (A,900, B,650, C,300, D,150)

For the standard case, which is addressed to 94 ton/h of coal consumption related to a primary air of 270 ton/h and a secondary air of 715 ton/h, the integrated value of predicted NO in the outlet of the calculation domain was found 723 ppm (6% O_2) which compares to the measured values for this operating condition in the range of 661 +/ 50 ppm (6% O_2) The slightly overpredicted value of the simulations is a consequence of the uncertain in the pre-exponential of the DeSoete's model, here adopted 10^{11} following the recommendation of Fiveland [19] to predict NO emissions in industrial equipments A good discussion about the fitting of measured values by adjusting these pre exponential values is included in Boardman [20]

Some parametric studies were also carried out to evaluate the sensitivity of the model to specific key variables The behaviour of the model is very satisfactory in relation to the mean particles diameter Increasing the averaged diameter by a factor of 2, the NO emissions decrease about 35% The measured values indicates roughly 38% of reduction for the same particles diameter variation (35µm to 70 µm) The response of the model to other specific variables as char oxidation rate and nitrogen concentration on coal was found similar to parametric studies in the literature [19]

CONCLUSION

A combination of validation of numerical models with measurements in both pilot plant and full scale boiler was conducted to study the NO emissions from a front wall fired pulverised coal boiler.

The detailed comparison of local values measured in the pilot plant showed some discrepancies attributed to the flow field prediction. For the industrial scale boiler the predicted NO emissions were found in agreement with the measured values and the sensitivity of the model is in agreement with the experiments. The rate constant for HCN oxidation which gives better results in the two cases considered, differs by an order of magnitude in the range of values reported in the literature. This value cannot be optimized unless all other flow and combustion results are fully compared with experimental values.

ACKNOWLEDGMENTS

This work is currently supported by the Commission of the European Communities under the contract JOUF-0023-C(EDB) entitled "The modelling of wall fired pulverised coal burners for pollutants emission control" of the JOULE subprogram of Solid Fuels. The partial financial support given by CAPES (Brazil) to one of the authors (C.F.M.C.) is gratefully appreciated.

REFERENCES

1. Gibson, M.M. and Morgan, M.A., Mathematical model of combustion of solid particles in a turbulent stream with recirculation, J.of the Institute of Fuel: 1970, 517-523.

2. Lockwood, F.C., Salooja, A.P. and Syed, S.A., A prediction method for coal-fired furnaces, Combustion and Flame 1980, **38**, 1-15.

3. Fiveland, W.A., Cornelius, D.K. and Oberjohn, W.J., A Numerical Model for Predicting Furnace Performance in Axisymmetric Geometries. *ASME* paper 84-HT-103, 1984.

4. Hill, S.C., Smoot, L.C. and Smith, P.J., Prediction of nitrogen oxide formation in turbulent coal flames, 20th Symp. (Int.) on Combustion, The Combustion Institute, 1984, pp 1392-1400.

5. Lockwood, F.C., Rizvi, S.M.A., Lee, G.K. and Whaley, H. , Coal combustion model validation using cylindrical furnace data, 20th Symp (Int.) on Combustion, The Combustion Institute, 1984, pp 513-522.

6. Zinser, N. and Schnell, U., Application of mathematical flame modelling to NOx emissions from coal flames, Fundamentals of Physical chemistry of Pulverised Coal Combustion, Ed. J. Lahaye, NATO ASI Series, Martins Nizhoff Publishers, 1987.

7. Truelove, J.S. and Williams, R.G., Coal combustion models for flame scaling, 22nd Symp. (Int.) on Combustion, The Combustion Institute, 1988, pp. 155-164.

8. Visser, B.M., Smart, J.P., Van den Kamp, W.L. and Weber, R., Predictions of Quarl Zone Properties of Swirling Pulverised Measurements and Coal Flames, 23rd Symp. (Int.) on Combustion, The Combustion Institute, 1990, pp. 949-955.

9. Abbas, T., Costa, M., Costen, P. and Lockwood, F.C., Nitrous oxide emission from an industry type pulverized coal burner, Combustion and Flame, 1991, **87**: 104-108.

10. Abbas, T., Costen, P. and Lockwood, F.C., The influence of near burner region aerodynamics on the formation and emission of nitrogen oxides in a pulverized coal-fired furnace. IC Report TF/91/16, to appear in Combustion and Flame, 1991.

11. Godoy, S., Hirji, K.A. and Lockwood, F.C., Combustion measurements in a Pulverised Coal-Fired Furnace, Comb. Sci. and Tech., 1988, **59**, 165-182.

12. Azevedo, J.L.T. and Carvalho, M.G.M.S., Modelling the near region of a swirling pulverized coal burner, 1st Int. Conference on Combustion Technologies for a Clean Environment, paper 15.4, Vilamoura, Portugal, 3-6 September, 1991.

13. Coimbra, C.F.M., Azevedo, J.L.T. and Carvalho, M.G., 3-D Numerical Model for Predicting NOx Emissions From a Pulverized Coal Industrial Combustor, Submitted to Combustion Science & Technology, 1992.

14. Magnussen, B.F. and Hjertager, B.H., On mathematical modelling of turbulent combustion with special emphasis on soot formation and combustion, 16th Symp. (Int.) on Combustion, pp. The Combustion Institute, 1977, 719-729.

15. Shaw, D.W., Zhu, X., Misra, M.K. and Essenhigh, R.H. , Determination of global kinetics of coal volatiles combustion, 23rd Symp. (Int.) on Combustion, The Combustion Institute, 1990, pp. 1152-1162.

16. Badzioch, S. and Hawkesley, P.G.W., Kinetics of thermal decomposition of pulverised coal Particles, Ind. Eng. Chem. Process Des. Dev., 1970, **9** (4): 521-530.

17. Field, M.A., Measurements of the Effect of Rank on Combustion Rates of Pulverised Coal, Combustion and Flame, 1970, 14, 237-248.

18. De'Soete, G.G., Overall reaction rate of NO and N2 formation from fuel nitrogen, 15th Symp. (Int.) on Combustion, The Combustion Institute, 1975, pp. 1093-1102.

19. Fiveland, W.A., Wessel, R.A. and Eskinazi, D., Pollutant model for predicting formation and reduction of nitric oxides in three dimensional, pulverized-fuel-fired furnaces, 24th Nat. Heat Transfer Conference, Pittsbusgh, 1987.

20. Boardman, R.D., Development and evaluation of a combined thermal and fuel nitric oxide predictive model, PhD. Thesis, Bringham Young University, U.S.A, 1990.

21. Azevedo, J.L.T. and Carvalho, M.G., Modelling Combustion and Fuel-NO_x in Pulverised Coal Flames, 6th Workshop on Two-Phase Flow Predictions, Erlangen 30/3-2/4, 1992.

22. Lockwood, F.C. and Milares, C.A., Mathematical modelling of fuel-NO emissions from PF burners. IC report, 1991.

23. Hirji, K.A., Lockwood, F.C., Rizvi, S.M.A., Offergeld, H. and Seitz, C.W., Measurements and prediction of NOx emissions in a coal fired combustor, Int. Specialists Meeting on Solid Fuel Utilization, The Portuguese section of the Combustion Institute, Lisbon, 6-9 July 1987.

A LOW NO_x GAS BURNER WITH A RADIANT FLAME

G.J. Nathan* R.E. Luxton+
* Research Fellow + Professor
Department of Mechanical Engineering, University of Adelaide, Australia.

ABSTRACT

A Precessing Jet (PJ) burner has been developed which enhances large scale turbulence by means of a naturally occurring fluid-mechanical phenomenon. A highly stable gas flame is produced without air swirl, quarl or a bluff body stabiliser. The resulting natural gas flame is lower in temperature and is more luminous than flames from many conventional burners so that it reduces NO_x emissions while achieving almost total burnout of combustibles. The high luminosity indicates potential for increased fuel utilisation efficiency in process industries where radiation heat transfer is important. For one configuration of the PJ burner firing natural gas in an 18MW industrial cement kiln, NO_x emissions are shown to be reduced by 75% with a simultaneous reduction in CO when compared with the original turbulent jet burner. Photographs of these flames show that the PJ burner produces a more rapidly spreading and more luminous flame. For another configuration, firing natural gas in a 25kW laboratory furnace, the radiation spectra of the flame produced by the PJ burner are compared with a high quality commercial swirl burner. The mean radiation intensity produced by the PJ burner is some twenty times that of the commercial burner in the visible spectrum and the normalised fluctuating intensity is about 50%, some five times that of the commercial burner in the visible range. Good potential for flame detectability is therefore apparent. Mean intensity is shown to increase by a factor of three with the addition of swirl to the air.

INTRODUCTION

Despite the increasing viability of non-traditional energy sources [2] the use of fossil fuels will continue for many years, particularly in high temperature processes which can be found for example in the manufacture of cement, steel and glass. Increased public awareness of the environmental impact of fossil fuel combustion is resulting in increasingly stringent emission standards, particularly with regard to the oxides of sulphur, carbon and nitrogen [1]. The first can be solved by the use of low sulphur fuels, by additives to capture sulphur during combustion or by exhaust gas scrubbing. The second requires that both the process and the combustion efficiency be maximised, while the third, though influenced by the quantity of nitrogen in the fuel, is strongly dependent on the combustion process. The programme of work described in the present paper addresses both the carbon and the nitrogen oxide emissions.

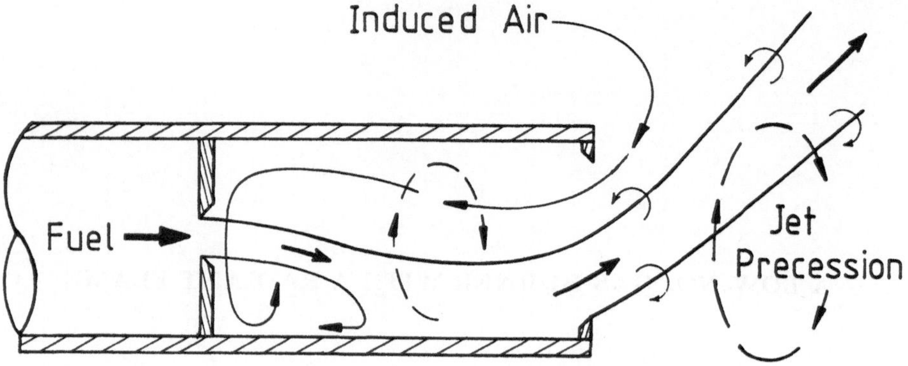

Figure 1: A diagramme of the Precessing Jet nozzle and a simplified schematic of the flow-field generated by it.

In minimising the specific fuel consumption, and so contributing to increased process efficiency, it is necessary to utilise available energy potentials, to maximise combustion efficiency and heat recovery, to minimise auxiliary power requirements (fans etc.) and to match the heat-flux characteristics of the flame (eg. its length and luminosity) to the requirements of the process. The latter is dominated by the characteristics of the mixing process between the air, fuel and combustion products which, where gas is available at high pressure, can be generated by the fuel jet, thus also saving auxiliary fan power. Likewise the emissions are strongly influenced by the mixing field through their dependence upon residence times at a given local temperature and mixture fraction. Intense, fine-scale mixing produces intense, high temperature combustion which results in high emissions of thermally generated NO_x.

The coincidence of high temperatures and lean combustion within the flame can be reduced by combustion staging [5] while flame temperatures can be reduced by flue gas recirculation [16]. However such systems are usually costly and are not readily applicable to some processes, notably those using rotary kilns. Instead, the emission of oxides of nitrogen can be reduced by modifying the spectrum of the turbulence field to produce an initial region of very large scale mixing between the fuel, air and combustion products, without a concomitant increase in fine scale mixedness. This appears to occur in the initial region of the Precessing Jet (PJ) flow which has been discovered at the University of Adelaide. Fuel-rich combustion then occurs at the interface of the fuel and air within the large structures resulting in a luminous, low temperature flame front. It is surmised that the energy cascade then extends the mixing progressively to smaller scales resulting in a naturally self-staged flame.

The PJ burner uses a naturally occurring fluid-mechanical phenomenon which causes a jet to be deflected through about sixty degrees from the nozzle axis and to precess [11,10]. The nozzle, shown in Figure 1, consists of an axi-symmetric chamber which has a large sudden expansion at its inlet and a small lip at its exit. The fuel jet which enters the chamber reattaches asymmetrically to the inside of the cavity wall. At the nozzle exit it is deflected at a large angle (typically 45^0) from the nozzle axis by strong local pressure gradients, as shown in simplified form in Figure 1. There are also strong azimuthal pressure gradients which cause the jet, and the entire flow-field within the chamber, to precess about the nozzle axis. The details of the experiments used to deduce this flow are described elsewhere [10,11].

The resulting initial flow field is dominated by turbulence of a scale many times that of

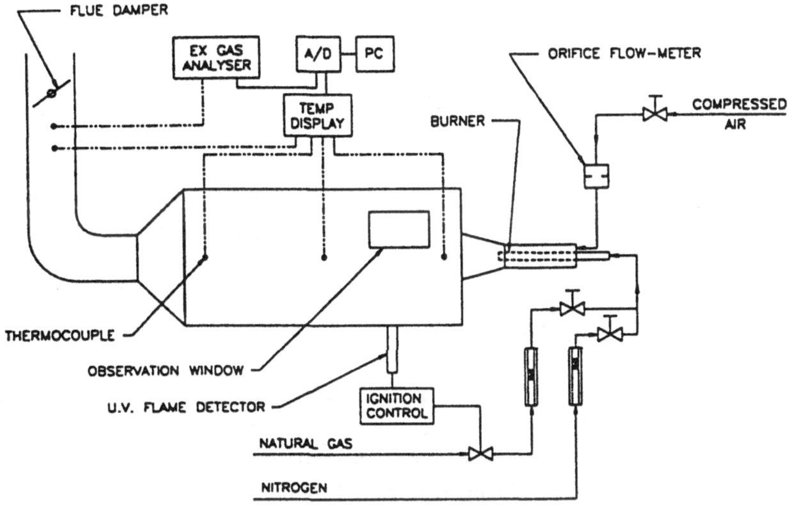

Figure 2: The laboratory furnace.

the nozzle, which has been shown to dominate the combustion process in a natural gas flame [13]. Those experiments, conducted in collaboration with the International Flame Research Foundation in IJmuiden, also showed that the PJ flame is more luminous, has lower peak temperatures and produces about 50% less NO_x than a comparable swirling flame. High flame stability has been demonstrated in earlier experiments with unconfined flames where the stand-off distance is reduced by an order of magnitude and the blow-off velocity increased by a factor of four relative to a simple turbulent jet. The flame has a very large turn-down capability [7,8]. Simulated blast-furnace gas (a $3MJ/m^3$ blend of CH_4 and N_2 with 1.5% H_2) has been burned stably with un-preheated air and without air swirl in a 25kW furnace [9]. The range of potential applications of the PJ burner is large because the mixing process is achieved directly by the fuel jet, obviating the need for a recirculation zone generated by either a bluff-body or air swirl in combination with an appropriately shaped quarl. In the present paper, the NO_x emissions and the visual flame characteristics in an 18MW cement kiln are presented along with laboratory experiments at 25kW which quantify the luminosity characteristics of the PJ flame. An account of the actual improvements in product quality and fuel efficiency in a cement kiln are documented elsewhere [4].

APPARATUS & METHOD

The 18MW dry Lepol rotary kiln used for the industrial trials is fired by natural gas and produces cement clinker. The kiln is a cylinder 3.1m in diameter and 34.2m in length inclined at 2^0 to the horizontal and rotated about its axis at about one rpm. After leaving the kiln, the clinker nodules drop on to a grate cooler where they pre-heat the combustion air to about 700 ^0C. A Lancom 3200 Flue Gas Analyser is used to measure the emissions.

The measurements of flame radiation were conducted in a small, uncooled, horizontally fired cylindrical furnace 1200mm long with a diameter of 600mm. The walls are lined with ceramic fibre. Quartz glass windows ensure complete transmission of radiation over a 300nm to 3500nm spectral band (Figure 2). Twelve adjustable radial swirl vanes are fitted to the air shroud to allow

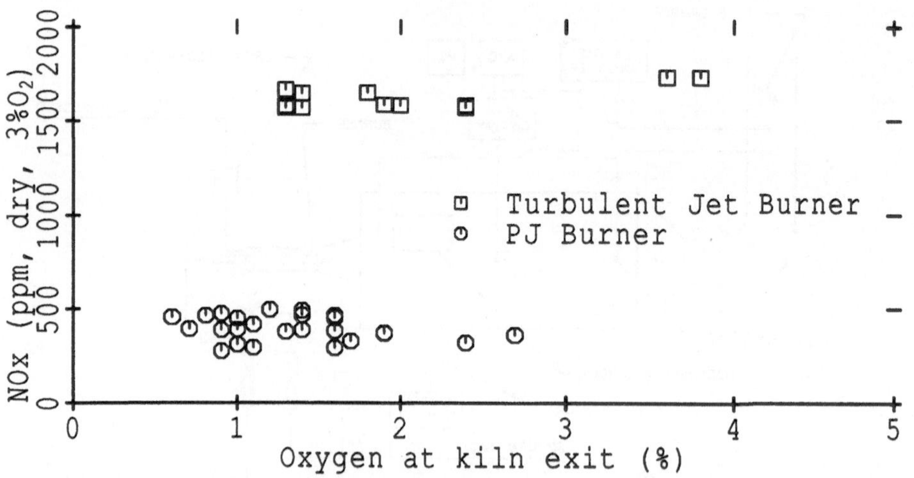

Figure 3: Emissions of Oxides of Nitrogen produced by both the Precessing Jet and the original turbulent jet burner.

variation of air swirl. The Lancom 3200 is used to measure the concentrations of O_2, CO, NO_x and flue gas temperature at regular intervals to determine when the furnace is operating under steady-state conditions. A butterfly baffle located in the flue about two diameters downstream from the gas sampling probe enables the furnace pressure to be kept positive, preventing the induction of air and corruption of the samples. Either the PJ burner assembly or the commercial burner assembly can be mounted to the furnace end-wall. Although separate air supplies and flame detection devices are used, the fuel through both burners is metered by the same Fisher and Porter rotameters with a stated accuracy of $\pm 2\%$ of the full scale reading. Both burners were fired at 25kW for the measurements reported.

Parallel radiation emitted from a 5mm diameter cross section of the flame is passed through two 5mm diameter holes separated by 2m before being transmitted through the mono-chromator and collected by the photometer. A high sensitivity photomultiplier tube used with a PAR Quantum Photometer, Model 1140, to measure radiation intensity. A standard Quartz-Iodine 200W Tungsten Filament Lamp was used for calibration [18]. Peak sensitivity is near 500nm and the response to wave-length is highly non-linear, the system being at least four orders of magnitude more sensitive to visible than to UV or IR radiation. Accuracy is therefore lower in the infra-red range. As data is collected from only a small cross-section of the flame it is here presented as a relative intensity.

The filter used for the measurement of the intensity spectrum is a Bausch and Lomb High Intensity Grating Monochromator with three alternative diffraction gratings: #33-86-25-02 for the visible (350nm$\leq \lambda \leq$800nm), #33-86-25-03 for the near infra red (700nm$\leq \lambda \leq$1600nm) and #33-86-25-04 for the mid infra red (1400nm$\leq \lambda \leq$3200nm). However, a diffraction grating transmits not only radiation of the specified wavelength but also that of higher orders (i.e. harmonic frequencies) which, while attenuated, can dominate the apparent reading if they occur near the peak sensitivity of the photometer. To avoid this problem low-pass optical filters are used when measuring infra-red radiation.

The output is logged on an IBM-XT personal computer using an RTI-800 analogue to digital board. Typically 5000 data points were sampled without filtering at 100 Hertz. This sample size is considered to be reasonable for the calculation of mean and rms statistics. In some cases 10 000 data points were collected to check for convergence of the calculation.

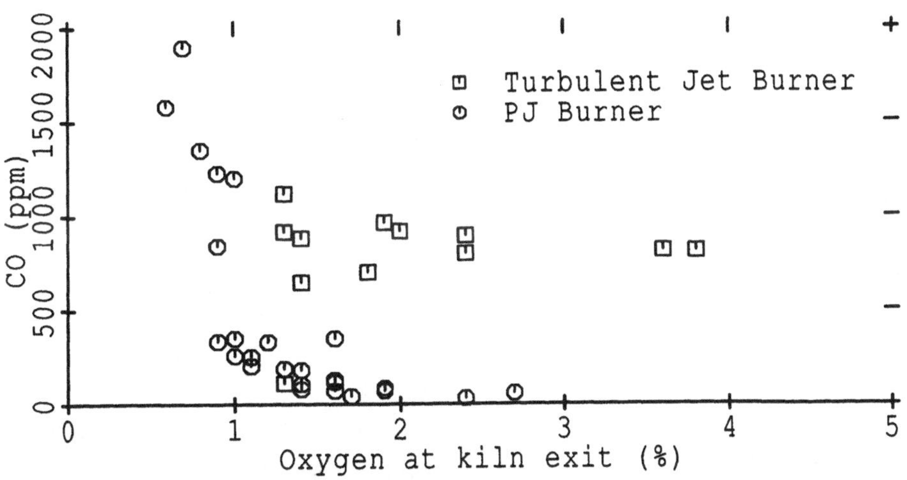

Figure 4: Emissions of Carbon Mon oxide produced by both the Precessing Jet burner and the original turbulent jet burner.

EXHAUST EMISSIONS IN THE CEMENT KILN

The NO_x emissions generated by the PJ nozzle during short term trials in an 18MW cement kiln are shown in Figure 3 and the CO emissions in Figure 4. They show that nitrogen oxides generated by the PJ burner are only about one quarter of those from the original turbulent jet burner while less carbon mon oxide remains unburned. This result is consistent with measurements elsewhere which also showed that peak flame temperatures and hence the generation of thermal NO_x in the flame are low [13,12].

RADIATION MEASUREMENTS

Fuel efficiency depends in part upon how well the heat flux profile of a flame matches that required by the process it serves. Within a cement kiln there are many heat transfer paths, both radiant and convective, which influence the temperature profile of the bed [17] but flame radiation is of major significance. A high rate of heat release close to the burner is desirable to ensure that the exothermic clinkering reaction at about 1400 ^{0}C occurs close to the kiln exit and to achieve maximum efficiency from the counter-flow heat exchange process. Flame impingement on the clinker must be avoided because a reducing atmosphere adversely affects clinker quality.

Photographs of the flames generated by the original turbulent jet burner and the PJ burner, respectively, are shown in Figure 5. They clearly show that the flame produced by the PJ burner is more bulbous, is established closer to the burner and is significantly more luminous. These observations are consistent with previous measurements of a small unconfined flame where the stand-off distance of the PJ flame was shown to be reduced by an order of magnitude relative to a that of a simple turbulent jet [7,8], and in the semi-industrial scale trials at the IFRF where

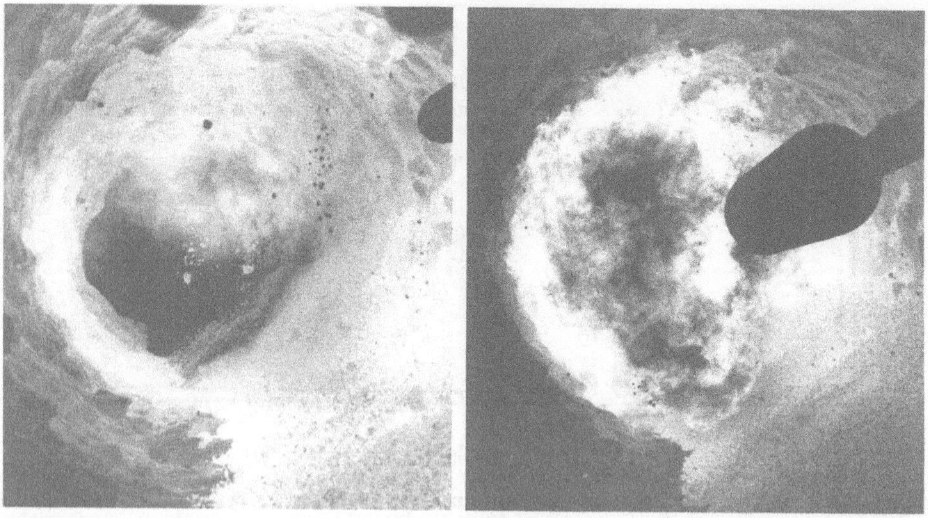

Figure 5: The flames produced by the original turbulent jet burner and the Precessing Jet burner, respectively, viewed from behind and below the nozzle, looking parallel to the kiln axis.

a highly luminous flame was observed [12,13]. More recent results obtained through an ongoing research and development programme with Adelaide Brighton Cement Ltd. has shown that, in comparison with a simple turbulent jet burner, the use of the PJ burner results in improvements in both product quality and fuel efficiency [4].

RADIATION MEASUREMENTS

When interpreting the laboratory scale measurements of flame radiation it is important to remember that flame radiation originates predominantly from two sources: (i) "black-body" radiation, usually in the infra-red range, from the high temperature sources within the flame, and (ii) chemi-luminescent and soot radiation, predominantly in the visible spectrum, which depend on the chemical composition of the fuel, the local stoichiometry and residence times.

Mean Radiation Spectra

The mean radiation spectra of the commercial and the PJ burners with both operated at a nominal 25kW and 5% flue O_2 are shown in Figure 6. The commercial burner produces a small recirculating core, in which the flame is moderately luminous, and a long blue tail. Consequently, radiation was measured both through the centre of the core and through the boundary between the core and the tail. Within the visible spectrum ($400 < \lambda < 700$) the mean intensity of the PJ burner flame is some 20 to 40 times greater than that of the commercial burner measured in the tail, and some 10 to 20 times greater than in the core. In the near infra-red wavelengths ($700 < \lambda < 1600$nm), the mean intensity of the PJ burner is typically a factor of 1.5 to 2 times larger. The maximum intensity of both burners occurs at a wavelength of about 1000nm. This implies that the PJ burner produces more sooting within the flame than does the commercial burner which could imply either that more of the combustion occurs under fuel-rich conditions, or that the residence times at sub-stoichiometric conditions are increased.

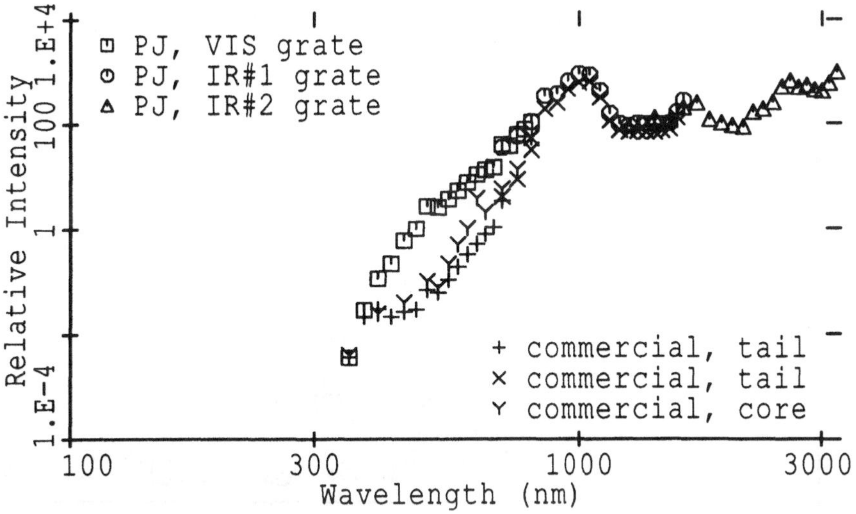

Figure 6: Comparison of the mean radiation intensity (relative scale) from the flames of the commercial and PJ burners in the visible and near infra-red spectra.

The mean temperatures of the flame and the walls are known approximately from thermocouple measurements. Hence their respective theoretical (black-body) wavelengths of maximum radiation intensity can be estimated using Wein's displacement law, $\lambda_{max}T = 2896.6(\mu\text{mK})$. Assuming $T_w \approx 900^0$C and $T_{flm} \approx 1300^0$C, the blackbody wavelengths of peak intensity are $\lambda_{m-w} \approx 2500$nm and $\lambda_{m-flm} \approx 1800$nm. Thus it can be seen that the longer wavelengths measured are contaminated by the radiation from the wall, while those in the visible and near infra-red range are a reasonable representation of flame radiation characteristics.

Fluctuating Radiation Spectra

The spectral distributions of the fluctuating components of the flame radiation, represented by the normalised intensity, $I'/\bar{I}$, are shown for both burners in Figure 7. The most significant result is that, within the visible spectrum, $I'/\bar{I}$ is approximately 5 to 10 times larger for the PJ burner than the commercial burner at the edge of the bright flame core and is some 3 to 10 times larger in the brightest part of the core. The magnitude of the fluctuating signal for the commercial burner is largest at the blue end of the visible spectrum while it is large throughout the visible spectrum for the PJ burner. The absolute value of $I'/\bar{I}$ for the PJ burner is about 50%. For $\lambda \geq 800$nm both burners produce a flame with $I'/\bar{I} \approx 6\%$.

The above result suggests that a substantial portion of the combustion within the PJ flame occurs in turbulent structures (flamelets) which are large compared with the 5mm diameter measuring area an interpretation consistent both with results from the laser sheet visualisation and pdf's measured at the International Flame Research Foundation (IFRF) [13] and observations of high speed cine schlieren photography of an unconfined flame [6]. By contrast the mixing and combustion within the swirling flame from the commercial burner is at a smaller scale than the 5mm observation port, giving smaller fluctuations. The large fluctuations in radiation intensity have proved to be a significant advantage in providing reliable flame detection in an 800MW (thermal) boiler [15].

The Effect of Air Swirl

Figure 8 shows that, at 25kW and with 4% flue O_2, the mean radiation intensity at $\lambda = 650$nm

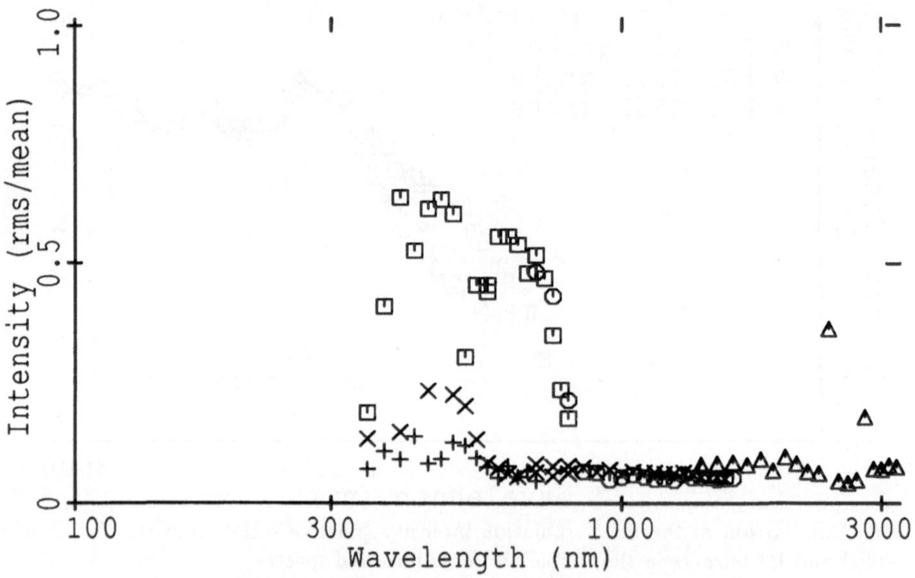

Figure 7: Comparison of the normalised rms radiation intensity ($I'/\bar{I}$) from the flames of the commercial and PJ burners in the visible and near infra-red spectra (symbols as per Figure 6).

increases by a factor of about three as the Swirl Number is increased from 0 to 1.14, a trend consistent with visual observations in the 2MW trials at the IFRF [12]. However the normalised fluctuating intensity is almost independent of swirl intensity, falling only slightly with increased swirl.

CONCLUSIONS

In comparison with a simple turbulent jet burner in an 18MW cement kiln, the Precessing Jet (PJ) burner produces a flame which is established much closer to the nozzle, spreads more rapidly and is more luminous. The NO_x emissions were measured to be 25% of the turbulent jet flame with CO emissions reduced simultaneously. The radiation characteristics of the PJ burner have been quantified in a 25 kW furnace with un-preheated air. Measurements show that in comparison with a commercial burner, which stabilises the flame with air swirl, the PJ burner produces:

1. a mean radiation intensity which is ten to forty times greater in the visible spectrum and about 70% greater in the near infra-red spectrum;
2. the normalised radiation intensity in the visible spectrum is typically 50%, an increase by a factor of five. At the near IR wavelengths the rms radiation is comparable;
3. a mean radiation intensity (at 650nm) which increases strongly with air swirl without significant change in the rms;
4. a flame whose radiation signature is easier to detect with industrial flame detectors.

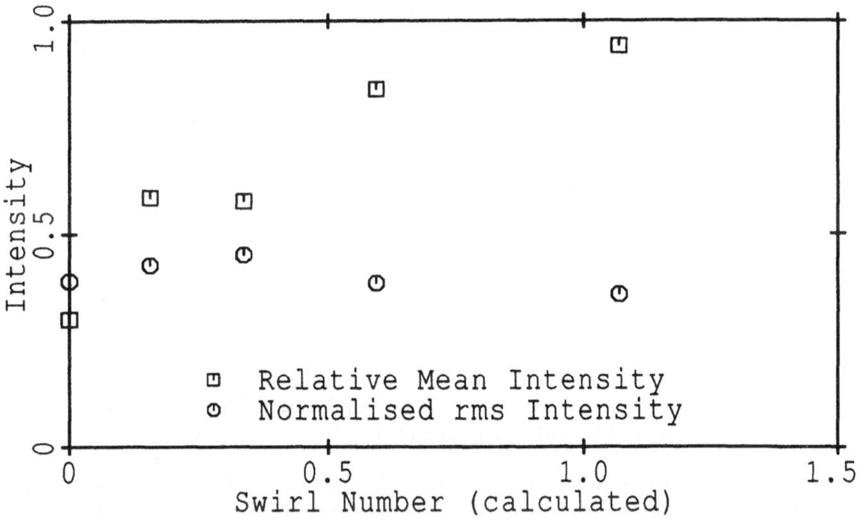

Figure 8: The effect of air swirl on the mean intensity ($\bar{I}$ and the normalised fluctuating intensity ($I'/\bar{I}$) of radiation from the PJ burner flame at 650nm.

The above results suggest that the PJ burner could make valuable contributions to improving energy efficiency in, and reducing NO_x emissions from, process industries which burn natural gas, particularly those requiring significant radiation. Because it does not depend upon air swirl, the PJ burner has potential to benefit a wide range of process industries.

ACKNOWLEDGEMENTS

The measurements in the cement kiln would not have been possible without the support and cooperation of Adelaide Brighton Cement Limited, its Chief Process Engineer Mr. C.G. Manias and the staff of the Angaston Works, in particular Mr. S.A. Balendra. The laboratory trials were partially funded by a contract from the Gladstone Power Station of the Queensland Electricity Commission. Radiation measurements were conducted using equipment lent by the Defence Science and Technology Organisation, Salisbury, and the Department of Physics, University of Adelaide. The support of the Australian Research Council and the S.A. State Energy Research Advisory Committee are also gratefully acknowledged.

References

[1] Bowman,C.T. (1992) "Control of Combustion Generated Nitrogen Oxide Emissions: Technology Driven by Regulation", Twenty-Fourth Symposium (International) on Combustion, The Combustion Institute, Sydney, Australia.

[2] Kaneff,S. (1990) "A Benign Sustainable Future based on Renewable Energy", Australian Institute of Energy Conference, 15–17 July, Adelaide, Australia.

[3] Luxton,R.E., Nathan,G.J. and Luminis Pty. Ltd. (1988), "Mixing of Fluids", Patent Application No. 16235/88, Australian Patent Office, April 1988, International Patent Application No. PCT/AU88/00114, April 1988.

[4] Manias,C.G. and Nathan,G.J. (1992) "The Precessing Jet Gas Burner - A Low NO_x Burner Providing Process Efficiency and Product Quality Improvements", International Kiln Association Conference, Oct. 26–28, Toronto, Canada.

[5] Mulholland,J.A. and Lanier,W.S. (1985) "Application of Reburning for NO_x Control to a Firetube Package Boiler", Journal of Engineering for Gas Turbines and Power, **107**, 739.

[6] Nathan,G.J. (1988) "The Enhanced Mixing Burner", PhD Thesis, Dept. Mech. Eng., The University of Adelaide.

[7] Nathan,G.J. and Luxton,R.E. (1988) "A Stable, Un-Premixed Gas Burner with Infinite Turn-Down Ratio", First European Conference on Industrial Furnaces and Boilers, March 1988, Lisbon, Portugal.

[8] Nathan,G.J. and Luxton,R.E. (1991) "The Entrainment and Combustion Characteristics of an Axi-symmetric, Self Exciting, Enhanced Mixing Nozzle", Third ASME-JSME Thermal Engineering Joint Conference, March 17–22, 1991, Reno, Nevada.

[9] Nathan,G.J. and Luxton,R.E. (1991) "Flame Stability and Emission Characteristics of the Enhanced Mixing Burner", Second European Conference on Industrial Furnaces and Boilers, April 2–5, Algarve, Portugal.

[10] Nathan,G.J. and Luxton,R.E. (1991) "The Flow field Within an Axi-symmetric Nozzle Utilising a Large Abrupt Expansion", Recent Advances in Experimental Fluid Mechanics, ed. F.G. Zhuang, International Academic Pub., 527-532.

[11] Nathan,G.J. and Luxton,R.E. (1991) "Mixing Enhancement by a Self-Exciting, Asymmetric Precessing Flow-Field", ed. J.A. Reizes, Fourth International Symposium of Transport Phenomena, **4**, 1511–1521, Sydney, Australia, July 14–18, 1991.

[12] Nathan,G.J. and Smart,J.P. (1991) "An Investigation of the Combustion Characteristics of the Enhanced Mixing Burner firing Natural Gas at 2MW", IFRF Doc. No. F90/y/9, International Flame Research Foundation, Ijmuiden, The Netherlands.

[13] Nathan,G.J., Luxton,R.E. and Smart,J.P. (1992) "Reduced NO_x Emissions and Enhanced Large Scale Turbulence from a Precessing Jet Burner", Twenty-Fourth Symposium (International) on Combustion, The Combustion Institute, Sydney, Aust.

[14] Nathan,G.J. (1992) "The Radiation Characteristics of the Precessing Jet Burner firing Natural Gas in a 50kW Furnace", Internal Report to Gladstone Power Station, Queensland Electricity Commission, Dept. Mech. Eng., The University of Adelaide.

[15] Nathan,G.J. and Luxton,R.E. (1992) "Results from Full-Scale Trials firing the Precessing Jet burner with Natural Gas in Boiler No.6", Internal Report to Gladstone Power Station, Queensland Electricity Commission, Dept. Mech. Eng., The University of Adelaide.

[16] Offen,G.R., Eskinazi,D., McElroy,M.W. and Maulbetsch,J.S. (1987) "Stationary Combustion NO_x Control", Journal of the Air Pollution Control Association, **37**, (7), 864–871.

[17] Pearce,K.W. (1972) "A heat transfer model for rotary kilns", Fourth Symposium on Flames and Industry, Imperial College, London, 19–20 Sept., pp 67-75.

[18] Stair,R., Schneider,W.E. and Jackson,J.K. (1963) "A New Standard of Spectral Irradiance", Applied Optics, **2**, (11), 1151-1154.

EXPERIMENTS IN LARGE SCALE COMBUSTION SYSTEMS AND THE CHARACTERIZATION OF THE BURNING EQUIPMENTS

JOÃO CASSIANO, MANUEL HEITOR, ANTÓNIO MOREIRA AND TITO SILVA
Instituto Superior Técnico, Technical University of Lisbon, Department of Mechanical Engineering, Av. Rovisco Pais 1096 Lisboa, Codex, Portugal

ABSTRACT

Measurements of the incident wall heat fluxes and local flame temperatures and major species concentrations were obtained in a large industrial 250MWe utility boiler and are presented and discussed together with results obtained in a 350kW model swirling burner. The work is aimed to improve knowledge of the thermal processes involved in large-scale oil-fired combustion systems and quantify the extent to which combustion is limited by chemical kinetics, as well as the existence of zones characterized by temperature gradients or those with excess CO. The results in the model burner include the rate at which liquid droplets spread in a turbulent swirling flame and are of interest to improve knowledge of the mixing process of the fuel with air in practical burners.

INTRODUCTION

Current requirements of reduced emissions and improved efficiency of industrial burning equipments, together with the increased use of low-quality residual fuels, has emphasized during the last decade the need for accurate methods of analysing furnace performance and heat transfer. It is known that plant efficiency gains obtainable by improving combustion heat release efficiency may be negligible in many cases, but a large potential exists for increasing overall efficiency by, for example, lowering excess air levels, while at the same time decreasing pollutant emissions [1-2]. Detailed knowledge of the combustion characteristics of the burning systems is therefore required to attempt to optimize current operating conditions and, also, to allow to validate and refine calculation methods which can extrapolate and interpolate experimentally acquired information and be used as valuable design tools [3-5]. This paper is aimed to help to achieve these objectives and considers experiments conducted in a 250MWe residual fuel-oil front fired utility boiler, which are complemented by detailed measurements in a 350kW model burner in order to increase knowledge of the aerodynamics of near burner zones.

It should be noted that measurements of heat transfer characteristics of power station boilers are very expensive and limited by the geometry, time and number of instruments and skills required and the few works reported in the literature are concerned essentially with plants typically below 80MWe [6-11]. The results are subject to inaccuracy [12-13], but provide useful information to improve and predict the performance of utility boilers and are extended here to a 250MWe oil-fired boiler. It is obvious that in general the results lack the detail required to interpret the combustion processes in the near burner zone, and here we attempt to solve this problem by using a model burner in a laboratory environment. The flow arrangement is typical of burners used under "dual" burning of liquid and gaseous fuels and comprises axial injection of fuel into co-axial streams of swirling gas and air.

Recent use of advanced instrumentation in comparatively simple liquid-fuelled burners [e.g., 14-16], have shown that an important effect of the swirl flow is to centrifuge the fuel droplets to large radii and, therefore, to cause local regions of high concentration and turbulence suppression. This has important consequences for practical burners since the fuel tends to be removed from the flame zone and turbulent fluctuations assist mixing controlled combustion. A compromise is required between the need for swirl to stabilize the flame and to increase the residence time through the formation of a recirculation zone, and the need to avoid centrifuging of fuel away from the flame, [e.g., 17]. It is important (e.g., for modelling purposes) that the physical understanding is available individually for each process involved, and here we present a detailed analysis of the combustion characteristics and mixing of swirl-stabilized liquid-fuelled flows with and without combustion.

The following section describes the utility boiler and the experimental method used, together with sample results obtained under real operating conditions. The study of the model burner is reported in Section 3 and the last section summarizes the main conclusions of the work. The assessment of accuracy of the experimental methods used are based on previous experiments and are presented in condensed form.

COMBUSTION MEASUREMENTS IN THE UTILITY BOILER

Boiler Geometry and Operating Conditions

The utility boiler used throughout this work is a subcritical, natural circulation, front wall oil-fired boiler, with a pressurized combustion chamber and a divided convection chamber where steam reheating and primary steam superheating takes place. The furnace is top-supported and the membrane walls were constructed by welding rows of vertically oriented water cooled tubes, forming the side panels. The furnace dimensions are approximately 19.9 m height and 8.5 m depth. It is fired with fuel oil by 12 burners arranged in an array of 4 burners in 3 different levels. The burners consist of a central tube through which the fuel-oil is pulverized into the furnace at the burner throat. Secondary air (i.e., the main combustion air) flows from the windbox through air registers and enters the furnace through an annulus formed between the oil nozzle and the burner throat (0.9 m diameter). At maximum continuous superheated steam flow rate (i.e., 771 ton/h at 162 bar and 540°C, which is necessary for driving a 250 MWe alternator) the residual fuel oil is supplied at a rate of 56 ton/h with 5% excess air.

The fuel oil, which is described in detail in Reference [18], includes 11% of asphaltenes and 220ppm of vanadium (mass basis), being out of the design specifications for the boiler. The extent to which this affects the heat transfer characteristics of the furnace is an important objective of this work and is discussed in the section below. In addition, the experiments reported here were carried out for either clean or dirty surface conditions, corresponding to water injection rates between superheaters of 76 ton/h and 110 ton/h, respectively. The boiler has inspection ports on 7 levels, which have provided the necessary physical access to the probes used through the work described here. Each port is referred here by two digits, the first concerned with its level and the second with its position as shown in Figure 1.

Experimental Method and Results

Measurements of total incident wall heat flux, together with those of mean gas temperature and major gas species, were obtained through the inspection ports and sample results are briefly analysed here.

A total heat flux meter was designed and used to measure the convective and radiative heat flux incident on the walls from a solid angle of 2π steradians up to a maximum of 500 kW/m^2, in the way suggested by [19]. It consists of a water-cooled probe which houses a cylindrical block, the front surface of which is serrated and blackened to give high radiation absorptivity. The probe was calibrated making use of a purpose built black body furnace, which was periodically used between consecutive measurements. The response time of the meter is of the order of three minutes and, based on the repeatability of the results, an accuracy of +/- 5% can be claimed for the measured values of heat flux.

Figure 1 shows the distribution of total incident wall heat flux along the side and rear walls of the boiler, for nominal net electrical generation (i. e., 250 MWe). The values varied between 200 kW/m^2 at port 1.6 and 316 kW/m^2 at port 3.4 with values measured on the rear wall greater than those on the side walls by 10 %. The distribution is asymmetric with the values measured on the right side greater than those on the left side of the boiler up to 22%. These results are important to evaluate the combustion characteristics of the boiler for a number of reasons [e.g., 20]. First, they allow to identify problems in the plant which are commonly caused by excessive local rates of heat transfer and may give rise to dry-out within a tube, tube corrosion or heavy ash and slag formation. Second, they provide information to be used during the commissioning of new plants and allow to refine and validate design methods.

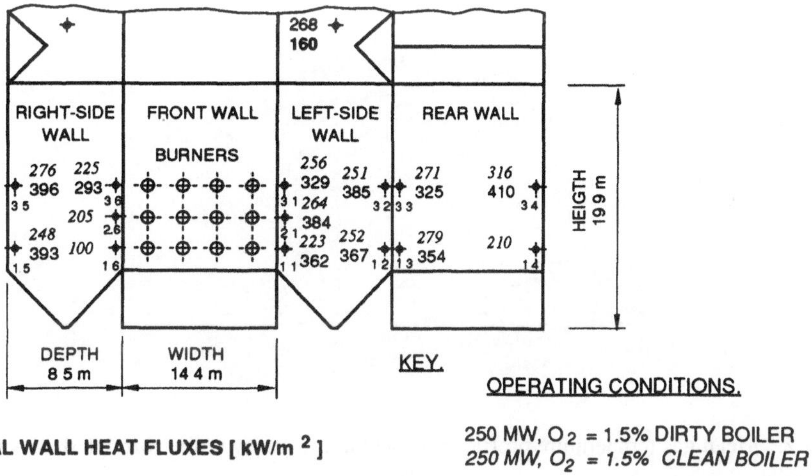

Figure 1. Distribution of total incident wall heat fluxes for two different wall conditions.

As an example, the results of Figure 1 include the analysis of the effect of the wall conditions on the heat transfer characteristics of the boiler. Typical incident heat fluxes for dirty walls (i.e., after several weeks of running the boiler at full power) are in general 20% higher than those obtained just after the boiler has been cleaned. An important overall consequence is the alteration of the wall emissivity and the consequent increase of about 100°C in the exhaust gases temperature, as measured at the boiler nose.

Gas temperatures were measured using a 3.5 m long, double-shielded, water-cooled suction pyrometer, which consists of a Pt/Pt-13%Rh thermocouple made from 350 μm diameter wires, which are protected by a system of radiation shields from heat losses to the wall chamber. The flow rate through the pyrometer could be varied so that the measurements presented here could be obtained independently of the suction velocity [12].

Figure 2 presents mean temperature profiles obtained through different ports in the boiler and shows that, with the exception of the near burner zone, the temperature of the combustion gases is uniform in each level with values in the range 1500-1600°C near the rear-wall and close to the ash-tray. At the boiler nose the temperature of the combustion gases decreases to 1050°C.

Major gas species concentrations were obtained by a gas sampling probe consisting of a sampling tube of 2 mm diameter mounted in a water cooled jacket of 50 mm diameter and 3.5 m length. The probe is connected to dedicated gas analyzers by a sampling system, which comprises a water cooled condenser, a diaphragm pump a calcium-chloride drier and a cotton wool filter. Dry analysis of $O2$ were made with a paramagnetic analyzer, while those of CO_2, CO and NOx were made with infrared analyzers. The analysis are precise within 0.5 % of full scale, but larger errors are likely to result from the sampling process. Gas sampling flow rates could be varied by means of a bypass valve installed in the sampling system, and the measured species were found to be independent of the suction velocity in the range of 1 l/min to 5 l/min.

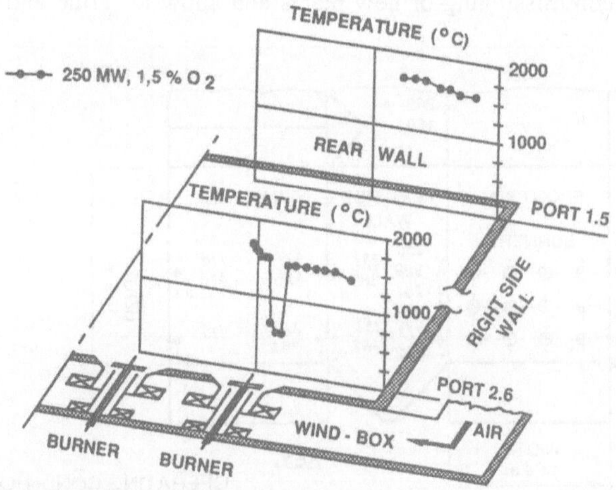

Figure 2. Mean gas temperature profiles obtained in the utility boiler through the inspection ports.

The results, Figure 3, show near uniform values across the boiler with the expected concentration of CO_2 and O_2 at the boiler nose, but with excess CO around the rear wall and close to the ash-tray. This is likely to be related with the inadequacy of the present furnace geometry for the fuel and operating conditions being used, which promotes the deterioration of the boiler tubes in this zone.

The results presented in the paragraph above were used to run an overall energy balance in order to compare the actual operating conditions with the design values of the boiler and to estimate the current wall temperatures. This is important to assess the consequences of using low-quality residual fuels, as far as tube corrosion and heavy ash slag formation are concerned. The calculations were conducted assuming radiation as the dominant mode of heat transfer to the walls, as analysed previously [11, 21], and considering an uniform wall temperature.

Based on the actual value of 245MW for the overall vaporization heat load in the combustion chamber, the estimated gas emissivity is 0.47 and, therefore, higher than the design value of 0.39. The wall temperatures were estimated to lye around 450°C, that is about 15 to 20°C higher than the typical design values. Although the calculations are subject to a number of assumptions, this explains observations made during boiler maintenance and motivates the need to improve and tune the burning equipments to the fuels being burned.

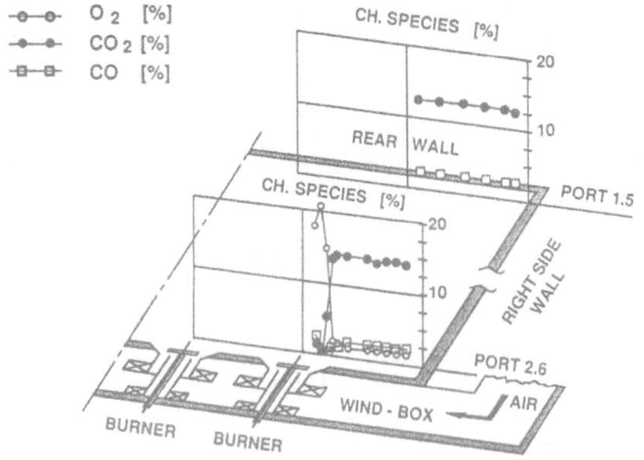

Figure 3 Profiles of mean concentrations of major species measured in the boiler through the inspection ports.

DETAILED ANALYSIS OF A TYPICAL BURNER

The previous sections focused on the analysis of a full scale utility boiler burning a residual fuel. We now turn to the study of a model burner in order to provide the necessary details of the near burner zone found in combustion equipments. The flow geometry and the experimental method used throughout the work are briefly described and sample results are presented. Further details can be found in References [22, 23]. It is evident that the extrapolation of the results to the analysis of full scale burning systems must be carried out carefully and, for example, requires consideration of the effects of thermal radiation. The results presented below provide, however, the necessary detail to analyse the aerodynamics of near burner zones and to validate and refine calculation methods which should be used to extrapolate the experimentally acquired information. To help to achieve these objectives, results are presented for reacting and non-reacting conditions in order to assess the extent to which current physical models used to simulate isothermal turbulent flows can be used under combusting situations.

The Burner Geometry and the Experimental Method

The flow configuration used throughout this work comprises a commercial fuel atomizer with an external diameter of 23mm, assembled in a low velocity co-flow of propane gas (54mm O.D.), which is externally surrounded by a high velocity co-flow of air (84mm O.D.). A diverging quarl typical of those found in full-scale burners, was located at the burner exit and could be removed to permit the measurement of boundary conditions. Swirl can be imparted to both streams by means of fixed blades at 45° with resulting swirl numbers, estimated from the geometry of the blades, equal to S_o=0.77 and S_i=0.85 for the outer and inner streams, respectively. Most of the measurements presented here were obtained for reacting and non-reacting conditions in the absence of liquid fuel. The reacting flow corresponds to bulk velocities, defined as the ratio between the flow rate and the cross sectional area, equal to U_o=30m/s (Re_o=49500) in the air stream and U_{gas}= 1.8ms (Re_i=3000) in the gas stream, corresponding to a flame with an air to fuel volumetric ratio (AFR) equal to 27.6 and a heat load of about 350kW. The non-reacting flow analysed throughout this work was obtained by replacing the propane gas by air with the same momentum flux, which results in a Reynolds number equal to 4000 in the inner flow.

The origin of the axial axis, x, is taken at the center of the exit plane of the model burner and the tangential velocity is taken positive in the anticlockwise direction, as facing the burner. The burner is located vertically directed upwards and the symmetry of the flow was verified by measuring several complete radial profiles in the horizontal plane.

Velocity measurements have been obtained with a dual-beam laser-Doppler anemometer [22], based on an argon-ion laser light source at 514.5mm (1W nominal) with sensivity to the flow direction provided by light-frequency shifting from acoustic-optic modulation (double Bragg cells) with a resulting shift of the Doppler signal in the range 0-10MHz. The half-angle between the beams was 4.92° and the calculated dimensions of the measuring volume at the e^{-2} intensity were 1.528 and 0.132mm.

Detailed mean temperature measurements were obtained with bare-wire thermocouples fabricated from 0.080mm diameter Pt/Pt: 13%Rh wires, which were supported on 0.500mm diameter wires of the same material an placed on a long "L"-shaped probe. Mean concentrations of major species were measured on a wet basis with a flame ionization detector for unburned hydrocarbons and on a dry basis with infrared analysers for CO and CO_2 and a paramagnetic analyser for O_2. A "L"-shaped, water-cooled stainless-steel probe of 7mm O.D. was used for sampling.

Results and Discussions

Figure 4 shows the measured streamline distributions and indicate the most salient features of the mean flow in the vicinity of the burner head for reacting and non-reacting conditions. The two flows have similar patterns with qualitatively similar distributions of mean velocity (not shown here due to lack of space), which are typical of those observed in highly rotating flows and include a central swirl driven recirculation zone surrounded by an annular forward flow region where the maximum tangential velocities occur with absolute values about 80% of the annular bulk velocity. Despite the qualitative similarities between the two flows, combustion induces significant quantitative differences: the mean axial and tangential velocities increase because the density is lowered and the axial and angular momentum must be conserved; the recirculated mass flow rate decreases from 67% in the non-reacting flow to 14% in the reacting flow as a result of the reduction in density; the length of the recirculation zone decreases by 32.5% and its maximum width increases by about 12%.

The reverse flow zone of the reacting conditions is characterized by uniform concentrations of the major species and high mean temperatures, Figure 5, as in other recirculating flames [24]. The stoichiometric line (i.e. $\tilde{f}$ = 0.060) occurs for radii larger than those characterized by zero velocity and, consequently, the recirculation zone is fuel rich.

The reverse flow itself consists primarily of burnt exhaust gases in such a way that fuel chemical equilibrium is unlikely to provide a realistic description of its thermochemistry. The maximum values of CO and unburnt hydrocarbons occur along the reacting shear layer for zones of near local stoichiometry and, therefore, suggest that chemical limitations imposed on the fuel and CO burn-up rates may be small, at least far from flame extinction. However, the performance of the flame stabilization process depends upon the turbulence exchange between the recirculation flow and the main flow.

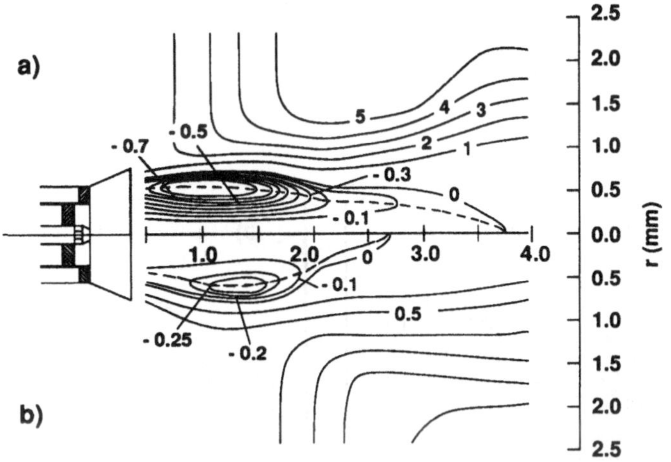

Figure 4. Measured streamline distributions for: a) non-reacting; and b) reacting (AFR=27) flows, Re=49500

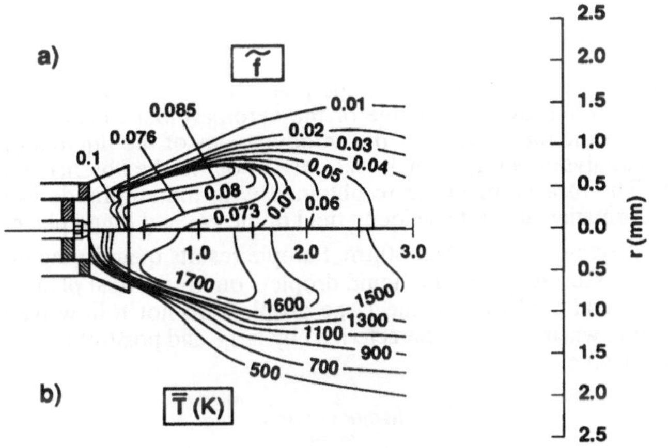

Figure 5. Distribution of measured scalar characteristics
a) Mixture Fraction; b) Gas Temperature

Inspection of figure 6 shows that the general levels of velocity fluctuations for both reacting and non-reacting conditions are small inside the recirculation zone and large in the highly strained annular shear layer. In this region turbulence is strongly anisotropic with $\widetilde{u''^2}\,max=1.33\widetilde{v''^2}\,max=2.0\widetilde{w''^2}\,max$ in the reacting flow, with $\widetilde{v''^2}$ and $\widetilde{w''^2}$ increasing considerably towards the stagnation point, in a way similar to that observed in highly recirculating flows downstream of baffles. The iso-contours of turbulent kinetic energy for the reacting flow show that maximum values occur in the vicinity of the rear stagnation point and analysis of the probability-density distributions suggests the presence of some flow periodicity.

The evidence of periodic oscillations in the frequency spectra of the velocity fluctuations could only be observed in the vicinity of the free stagnation point where the spectrum of radial velocity fluctuations have shown a peak around 30Hz. Analysis has shown that when this frequency is scaled with the flow parameters in the shear layer adjacent to reverse flow zone, the local Strouhal number is of the order of those characterizing the "bursting" phenomena in turbulent shear layers [22]. Although the energy contained in the frequency peak is only about 4% of the total spectral energy, the observation suggests the likely importance of intermittence in the mixing process between the hot fluid inside the recirculation zone and the heterogeneous density field, in a way similar to that observed in other comparatively simple flows with density gradients.

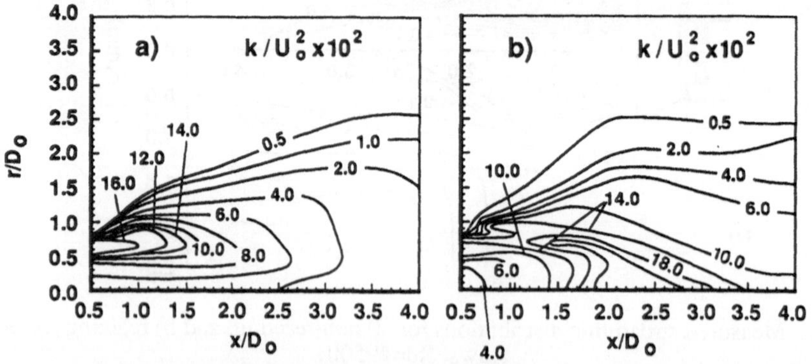

Figure 6. Distribution of measured turbulent kinetic energy
a) Non-reacting Flow; b) Reacting Flow, AFR 0 27.6

The foregoing analysis considered the single-phase flow in the vicinity of the burner head, which is important to improve knowledge of the aerothermochemistry of typical burning systems. We now turn to the analysis of the aerodynamics of the liquid droplets within the reverse flow zone analysed before, which was carried out in the absence of combustion for ease of analysis. The measurements were obtained with the amplitude technique of [23], which provides information about the velocity field of the gas and liquid phases, with the latter including droplet diameters larger than 40µm. Sample results quantifying the projections of the velocity vectors associated with the liquid droplets on the vertical plane of symmetry are shown in Figure 7, and confirm that the large droplets do not follow the mean gas flow around the centerline, where they are characterized by large and positive axial velocities within the gas recirculation zone.

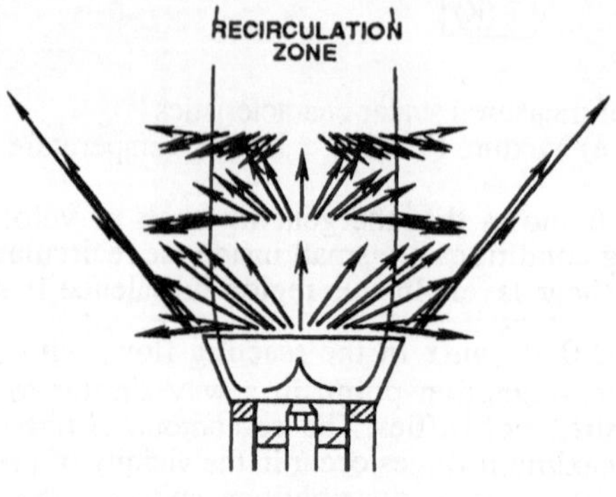

Figure 7. Time-averaged velocity vectors of the liquid phase along the vertical plane of symmetry at several stations downstream of the burner exit.

As the distance to the nozzle increases, the droplets decelerate due to mean aerodynamic drag and, at least some of them, recirculate through the edges of the recirculation zone, being centrifuged away to the high velocity regions established around the gas recirculation zone. This qualitative picture of the flow was shown to be independent of the flow rates of air and liquid, which stems from the dominance of the swirl recirculating air flow on droplets dispersion in burner arrangements. The implication for practical burning equipment is that the fuel droplets tend to be removed from the flame zone, giving rise to zones of high concentration of unburnt fuel. The consequent effects on the release of unburnt pollutants is a major concern, and its solution is dependent upon the knowledge of particle dynamics in turbulent flows.

CONCLUSIONS

Measurements of the incident wall heat fluxes and local flame temperatures and major species concentrations were obtained in a large industrial 250MWe utility boiler and are presented and discussed together with results obtained in a 350kW model swirling burner. The work is aimed to improve knowledge of the thermal processes involved in large-scale oil-fired combustion systems and show that the local gas temperature reach maximum values around 1650° close to the burners, but decay up to 1350°C far from them, where the measured profiles show near uniform distributions. The results quantify the extent to which combustion is limited by chemical kinetics effects, as well as the existence of zones characterized by increased pipe wear due to excess CO levels. The total incident wall heat fluxes are in the range 250-400KW/m^2 and determine the performance of the burner operating conditions.

The aerodynamics of the near burner flows are analysed making use of laboratory flows with a heat load up to 350kW. Single- and two-phase flows established in the vicinity of a swirl model burner are discussed and the results used to improve knowledge of the processes of turbulent combustion and particle dispersion in strongly recirculating flows, such as those typical of industrial burning systems. The mean flow includes a central swirl driven recirculating zone surrounded by an annular forward flow region where the maximum tangential velocities occur with absolute values about 80% of the annular bulk velocity. The total recirculated mass flow rate is 14% of the income air flow and this gives rise to a reverse flow zone characterized by uniform concentrations of the major species and high mean temperatures, as in other recirculating flames. The stoichiometric line (i.e. $\tilde{f}$ = 0.600) occurs for radii larger than those characterized by zero velocity and, consequently, the recirculation zone is fuel rich. With exception of a narrow zone close to the fuelling device the concentration of carbon monoxide and unburned hydrocarbons vanishes within the recirculation zone and the former exhibits maximum values along the flame front equal to 3.5%. This results from a balance between the initial CO formation due to the C_3H_8 oxidation, and the subsequent oxidation of CO to CO_2 within the recirculation zone. Results obtained for non-reacting conditions show that the upstream part of the recirculation zone is dominated by the presence of large droplets, which may affect the evaporation rate in the practical flows. The droplets are centrifuged away to the high velocity region established around the recirculation zone and lead to a loss of fuel from the ignition zone.

ACKNOWLEDGMENTS

The support and technical assistance of the staff of the Portuguese Power Company, EDP is gratefully acknowledged. Thanks are due to Messrs. Carlos Carvalho, Jorge Coelho and Janós Rohaly, for help in the preparation of the manuscript. The large scale experiments have been supported by the Commission of the European Communities under the ESPRIT Project N. 2192, AIMBURN, while the laboratory work has been performed under the scope of the SCIENCE Project SC1-0459.

REFERENCES

1 Breen, B.P. and Sotter, J.G. (1978). Reducing Inefficiency and Emissions of Large Stream Generator in the United States. Prog. Energy Combust. Sci., **4**, pp. 201-220.
2 Lawn, C.J. (1987). Principles of Combustion Engineering for Boilers, Academic Press.
3 Carvalho, M.G.M.S. and Coelho, P. (1990). Numerical Prediction of an Oil-Fired Water Tube Boiler. Intl. J. Engn. Computations, **7**, pp. 227-234.
4 Görner, K. and Zinser, W. (1986). Investigation into Turbulent Particle Transport in Utility Boiler Furnaces. Eighth Members' Conference of the International Flame Research Foundation.
5 Lockwood, F.C., Papadopoulos, C. and Abbas, A.S. (1988). Prediction of a Corner-Fired Power Station Combustor, Combust. Sci. and Tech., **58**, pp. 5-23.
6 Wall, T.F. and Stewart, I. McC. (1975). Tests on the Spectral Radiation from a Large Pulverized-Coal Flame, J. Inst. Fuel, May 1971, pp. 235-240.
7 Abraaham, K.U. and Rajan, S. (1983). Measurements of Furnace Heat Transfer on a Corner-Fired, Pulverized Coal Boiler, Journal of the Institute of Energy, December 1983, pp. 217-224.
8 Robinson, G.F. (1985). A Three-Dimensional Analytical Model of a Large Tangentially Furnace, Journal of the Institute of Energy, September 1985, pp. 116-150.
9 Boyd, R.K. and Kent, J.H. (1986). Three-Dimensional Furnace Computer Modelling. Twenty-First Symp. (Intl.) on Combustion, The Combustion Institute, pp. 265-274.
10 Bonin, M.P. and Queiroz, M. (1991). Local Particle Velocity, Size and Concentration Measurements in an Industrial Scale Pulverized Coal Fired Boiler. Comb. and Flame, to appear.
11 Butler, B.W. and Webb, B.W. (1991). Local Temperatures and Wall Radiant Heat Flux Measurements in an Industrial Scale Coal Fired Boiler, FUEL, **70**, pp. 1457-1464.
12 Heitor, M.V. and Moreira, A.L.N. (1992). Probe Measurements for Scalar Properties in Reacting Flows. In: "Combusting Flow Diagnostics", eds D.F.G. Durão et al., Kluwer Academic Publishers, pp. 79-136.
13 Hottel (1986). Background and Perspectives on Temperature Measurement in Furnaces. AIChE Symp. Series, **82**, pp. 1-12.
14 Mao, C.P., Wang, G. and Chigier, N.A. (1986). An Experimental Study of Air-Blast Atomizer Spray Flames. Twenty First Symp. (Intl.) on Combustion, The Combustion Institute, pp. 665-673.
15 McDonnel, V.G. and Samuelsen, G.S. (1988). Evolution of the Two-Phase Flow in the Near Field of an Air-Blast Atomizer under Reacting and Non-Reacting Conditions. Proc. 4th Int. Symp. on Appl. of Laser Anemometry to Fluid Mechanics, paper 15.1, Lisbon.
16 Hardalupas, Y, Taylor, A.M.K.P. and Whitelaw, J.H. (1990). Velocity and Size Characteristics of Liquid Fuelled Flames Stabilized by a Swirl Burner. Proc. R. Soc. London A, **428**, pp. 129-155.
17 Liu, CH., Nouri, J.M., Whitelaw, J.H. and Tse, D.G.N. (1989). Particle Velocities in a Swirling, Confined Flow. Combust. Sci. and Tech., **68**, pp. 131-145.
18 Cassiano, J., Heitor, M.V. and Silva, T. (1991). Temperature, Species and Heat Transfer Characteristics of a 250MWe Utility Boiler. Submitted for publication in Comb. Sci. and Tech.
19 Chedaille, J. and Braud, Y. (1972). Measurements in Flames, E. Arnold Publ.
20 Neal, S.B.H.C., Northover, E.W. and Hitchcock, J.A. (1982). Some New Devices for the Measurement of Heat Flux in Power Station Boiler Furnaces. J. Inst. Energy, March, pp. 8-11.
21 Viskanta, R. and Menguç, P. (1987). Radiation Heat Transfer. Prog. Energy Combust. Sci, **13**, pp. 97-160.
22 Durão, D.F.G. , Heitor, M.V. and Moreira, A.L.N. (1991). Turbulent Transport Processes in Swirling Recirculating Non-premixed Flames. Proc. 8th Symp. on Shear Flows, Munich, September 9-11.
23 Durão, D.F.G., Heitor, M.V. and Moreira, A.L.N. (1992). Flow Measurements in a Liquid Fuelled Burner. In:"Applications of Laser Techniques to Fluid Mechanics", eds R.J. Adrian et al., Springer-Verlag, pp. 163-182.
24 Heitor, M.V. (1992). On the Analysis of Turbulent Heat Transfer in Recirculating Flames. Intl. J. Heat and Fluid Flow, to appear.

SESSION 10:

Process Integration

Chairman: Prof. S. Pierucci

MAKE USE OF THE UTILITY SYSTEMS TO OBTAIN FLEXIBLE HEAT EXCHANGER NETWORK SATISFYING THE MINIMUM ENERGY REQUIREMENT

KALITVENTZEFF B.
Professor at the University of Liège, Sart Tilman, Bât B6, B- 4000 Liège.
General manager of Belsim s.a., Allée des Noisetiers, B-4031 Liège (Angleur)

ABSTRACT

The technology for process energy integration ("Pinch Technology" and subsequent developments of it) is often regarded as lacking of flexibility. Theoretical MER (Minimum Energy Requirement) targets are too often not feasible because of match restrictions between process streams. Operability of integrated processes appears sometimes to be poor when modifying the production rate; start-up procedure is not enough taken into account in the process energy integration.

When integrating chemical process and steam utility network altogether, these drawbacks can be removed advantageously. In this communication, we will illustrate the application of innovative procedure for utility system integration by discussing an industrial case study. The example shows the benefits that can be obtained when integrating the chemical process with the utility network. Problems of restricted matches, pressure level selection of the utilities and overall integration of the production unit into the production site can easily be dealt with. The example also stress the importance of using integrated computer tools and especially the sequential use of targeting, simulation, synthesis and optimization tools.

The methodology used to solve the problem is a three steps procedure that has been named AGE : Analyse, Generate, Evaluate. We will show that the problem definition is not unique and that new goals and constraints will normally appear during the problem solving procedure. We will stress the importance of having adapted methods but also corresponding computer tools to solve specific problems.

INTRODUCTION

When we look for energy savings, we often face a paradox. On one side, most of chemical processes are divided into several different sections that have to be kept as much as possible independent for different technological reasons, and on the other side, the energy integration tends to match the sections together to save energy. The interconnection of different sections may be very effective in terms of energy, but a global analysis will show that most of the benefit obtained has to be reconsidered for technological reasons like process flexibility, safety, maintenance, start-up, reliability or layout. In the industrial application used as illustration, the process is divided into five sections. Table 1 shows the Minimum Energy Requirement (MER) of each of the sections considered separately. The minimum energy requirement of the five sections integrated together corresponds to 38% of the sum of the MER of the separated sections. For flexibility reasons the process engineers ask for keeping the sections independent. They do not accept process-process exchange between sections, except for the sections S2 and S3. The integration of S2 and S3 is the only one that can be taken into account. As the results are confidential, all the heat loads and flowrates of this communication have been normalized with respect to the heat load input of the reference case - last row of table 1 - MER with connections allowed between streams of S2 and S3. The integrated solution without restricted matches ("Whole process integrated") corresponds to 41,2% of the reference. The energy penalty due to forbidden connections is thus very important .

TABLE 1 : Results of the Minimum Energy Requirements targeting.

Sections	Heat In (kW/100kWref)	Heat out (kW/100kWref)
S1	32,1	35,7
S2	56,0	23,3
S3	20,7	17,4
S4	00,0	70,0
S5	00,0	42,2
Sum of independent sections	108,9	188,5
Whole process integrated	41,2	120,8
S2+S3 integrated	67,9	31,8
With restricted connection	100,0	179,6

A better target is reached by integrating the utilities because they make exchange between sections possible. Steam will be produced in one section and condensed in another, giving mechanical power production as a subsequent benefit.

INTEGRATION OF THE UTILITIES

A chemical process is always imbedded in a larger industrial system : figure 1. Raw material is transformed into products, by-products and waste by means of energy. The energy is supplied to the chemical process by the utilities that we classify into two categories : the raw utilities and the intermediate utilities.

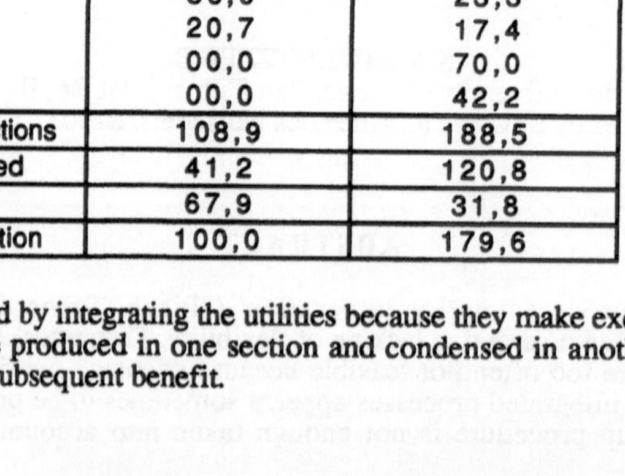

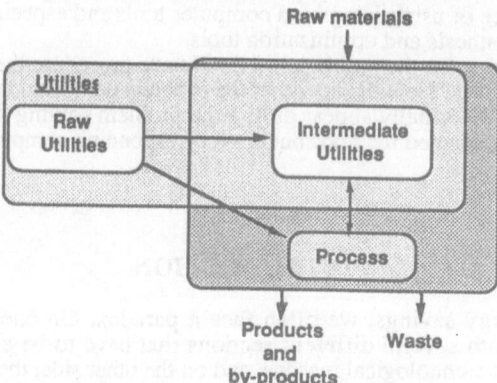

Figure 1 : two categories of utilities.

The Raw utilities

The raw utilities are introduced in the system where they produce the required thermal or mechanical effect and then they are rejected from the system as waste. Raw utilities are fuel, water, air and electricity. Processing such utilities delivers for instance fumes of a furnace or

exhaust of a gas turbine as hot utilities. This may require specific unit that leads to high investments. The multiplication of such units will be avoided.
In our example, the hot raw utilities is natural gas that is burned to produce the thermal requirements at high temperature in section S2 (figure 2a). The cold raw utilities will be river and refrigerated water and air.
The characteristics of the raw utilities is determined by external conditions : fuel composition, ambiant temperature, river temperature, etc... That may introduce problems of flexibility that have to be taken into account.
We have experienced that to solve every energy integration subproblem it is interesting to apply a three steps procedure that we name AGE for Analyse, Generate, Evaluate. Let us illustrate the AGE procedure for integrating the hot raw utility (figure 2).

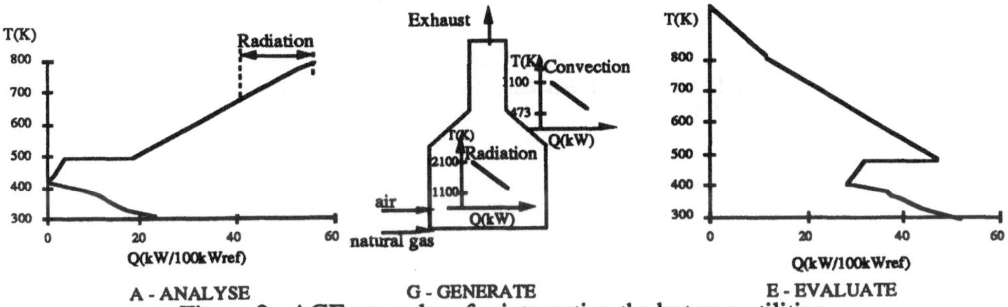

Figure 2 : AGE procedure for integrating the hot raw utilities.

a) Analyse
The shape of the composite curve gives the MER and shows a high temperature level for the hot utility. This indicates that the hot utility must be delivered by burning natural gas (chosen by the process engineer). For technological reasons, the high temperature thermal requirement must take place in the radiation section of the furnace. This additionnal constraint has to be satisfied and will determine the minimum flowrate of natural gas.
b) Generate
We calculate the simulation of the furnace for a given natural gas composition and an air excess. That gives us the inlet temperature and the composition of the fumes. The target temperature of the fumes is chosen so that it respects technological and environmental constraints.
c) Evaluate
The optimal flowrate of the natural gas is determined by integrating the fumes to the process and by taking into account the constraints identified in the Analyse step. The "Grand composite" curve of the section S2 obtained for the remaining problem resulting from the hot raw utility integration is given figure 2c. The heat transfered in the radiation section is eliminated from the problem.

In the example, the integration of the hot utility is defined by technological considerations rather than by rational use of energy. The natural gas flowrate calculated is 8,1 $\frac{Nm^3/h}{100kWref}$. The "Grand composite" curve is highly modified after this step and a new integration problem is defined. It leads to a new energy target and increases the gap between non integrated and integrated process as indicated table 2.

TABLE 2 : Results of the MER targeting with fumes integration

Sections	Heat In (kW/100kWref)	Heat out (kW/100kWref)
S1	32,1	35,7
S2	00,0	27,9
S3	20,7	17,4
S4	00,0	69,9
S5	00,0	42,2
Sum of independent sections	52,8	193,1
Whole process integrated	00,0	140,3
S2+S3 integrated	07,2	31,8
With restricted connection	39,3	179,6

If the sections are kept independent, the MER corresponds to both hot and cold utility requirements. For the whole integrated process, only cold utilities are required as shown on the "Grand composite" curve : figure 3.

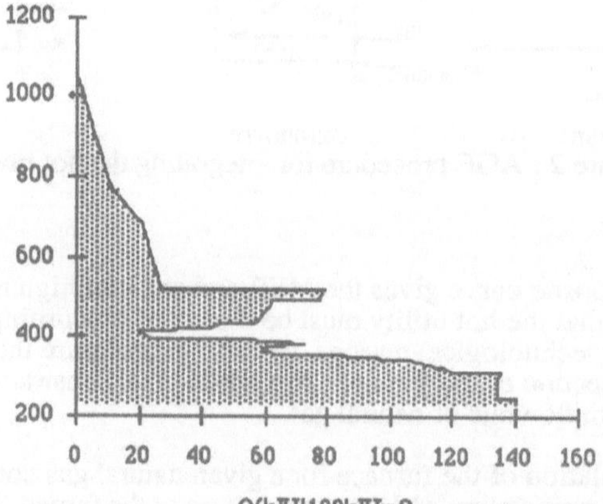

Figure 3 : "Grand composite" of the whole process.

It can be seen that a lot of heat is transfered accross the pinch; this could have been easily avoided by rejecting fumes at the atmosphere at a higher temperature, but anyway without changing the utility cost. We keep the fumes into the general remaining problem because its heat potential can maybe used elsewhere.

The intermediate utilities.

The intermediate utilities are produced by exchange with other utilities or with the process itself. Most of the intermediate utilities include interlinked hot and cold streams. They are usually cycles like refrigeration or Rankyne cycles or steam networks. Such utilities offer high opportunity for the integration and high degree of flexibility since their operating conditions can be manipulated. Steam network is a typical hot intermediate utility. It presents a lot of advantages : steam can easily be transported from one place to another. Steam heat content is latent heat, that provides good heat transfer coefficient, small heat exchange area and less expensive heat exchangers using well known and specialized technology. The distribution of the steam through the network, makes steam available all over the process. That improves flexibility and reliability. If one steam production unit fails, for example if a process section is shut down, the steam will be supplied by increasing steam furnaces heat load and the installation may often still continue to operate. Unit

start-up will be performed using similar procedure. For mechanical power production the same flexibility is obtained. Producing mechanical requirements by the steam turbines of the network will prevent failure in the electricity supply.
Refrigeration cycles are intermediate cold utilities. They can be integrated by exchanging with process streams. Flexibility is obtained by manipulating pressure levels and flows of the cycles within prescribed ranges. Integrating the mechanical power production may increase the profit of using refrigeration cycle or mechanical vapour recompression.
A similar AGE procedure is used to integrate the intermediate utilities. Both steam network and refrigeration cycles have to be integrated. We will only examine the case of the steam network.

The steam network integration.

a) Analyse

The Analyse step is based on the analysis of the shape of the "Grand composite" curve, but as explained here above, it implies also the identification of the technological constraints of the process and its environment.
The first technological constraint is the restricted matches between sections to keep them independent. This may be overcome by the steam network. The "Grand composite" curve of the integrated process (figure 3) is a threshold problem and it features self-sufficient zones. By analysing the shape of the "Grand composite" curves of each zone separately, one can remark that steam may be produced in section S4 (figure 4). In this case, the steam production is a cold utility. The steam produced can be used as hot utility in section S3. If the flowrate and the pressure of the steam produced in section S4 is high enough to satisfy the steam requirements of the section S3 the two sections will be interconnected through the utility network but will remain independent in term of process-process exchange. By using the steam network and by choosing the appropriate pressure, the MER of the integrated process can be reached. Futhermore, as the integrated process is a heat producer, the analysis of the "Grand composite" curve shows that steam may be exported by the process.
The pressure levels proposed were chosen according to the existing steam network available on the process site. They seems to be appropriate with regard to the shape of the composite curves of each sections.

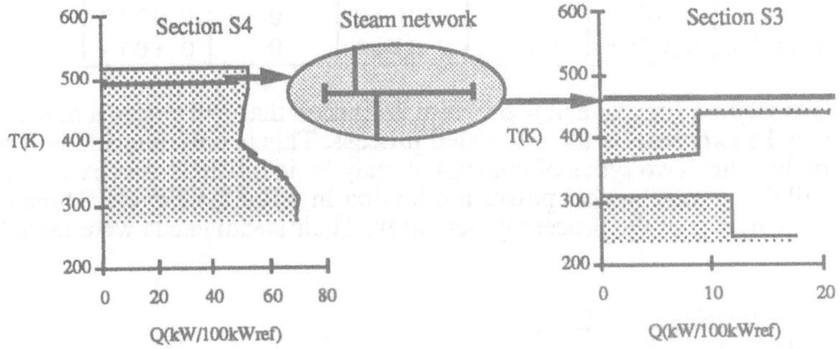

Figure 4 : Avoid restricted matches penalty using steam network.

b) Generate

Steam production and steam consumption are simulated according to the chosen pressure levels. The steam production implies heating from the deaeration drum (saturated liquid at 1,5 bar) to the saturated steam at the given pressure level. Steam consumption has been computed based on condensation heat only at the given pressure level. Steam at different pressure level have different costs. Steam production has a negative cost since it corresponds to a benefit; steam consumption has a positive cost.

c) Evaluate

We have developed a mathematical formulation that allows to calculate the optimal integration of the utilities. This allows to select among a list of given utilities the one that will provide the minimum energy requirement at minimum cost and determine their optimal flowrates. The mathematical formulation is a MILP (mixed integer linear programing) problem presented in Maréchal (ref 3.). It is possible to add constraints to the MILP problem in order to represent the restricted matches between sections. This mathematical tool has been used to calculate the optimal flowrate of the utilities in the example.

The utilities proposed are cooling water, refrigerated water, refrigeration cycle and obviously the steam production and consumption at different pressure levels as given in table 3 (30, 11 and 1,5 bars). We have calculated the flows of the utilities with and without connection restriction. The calculation without restriction shows only steam export; when restricted matches are considered, the steam export is increased but part of it is directly recycled from the steam network to the process : scheme of figure 4. As HP steam is exported and MP steam imported, mechanical power may be produced by expansion. A technological contraint limits this production because steam must be superheated.

TABLE 3 : Steam characteristics.

	LPin	MPin	HPin	LPout	MPout	HPout
Pressure (bar)	1,5	11	30	1,5	11	30
Tin(°C)	111 *v	184 *v	234 *v	111 *l	111	111
Tsaturation (°C)	111	184	234	111	184	234
Tout(°C)	111 *l	184 *l	234 *l	111 *v	184 *v	234 *v

*v saturated vapour, *l saturated liquid

TABLE 4 : Steam flowrates for optimal integration (given for 100kWref)

	LPin	MPin	HPin	LPout	MPout	HPout
Pressure (bar)	1,5	11	30	1,5	11	30
Flow (t/h)						
Without restriction	0	0	0	0,0613	0	0,0315
With restriction	0,0224	0,0808	0	0,0701	0	0,1011

The results of the AGE procedure suggest a new problem definition that will define a new target. We found that steam may be exported by the integrated process. This is only interesting if there are steam consumers on the site. Two types of consumers may be identified : process units that will use steam as hot utilities or mechanical power production in a Rankyne cycle. Three other processes were present on the site of the process under study. Their steam needs were identified, they are given table 5.

TABLE 5 : Other process unit requirements

Process unit	Pressure (bar)	Flow (t/h)
U1	16	0,026
U1	4	0,026
U2	11	0,077
U3	4	0,046

(flows given for 100 kWref)

It was decided not to calculate the MER of the other processes but only to consider their demands as constant temperature streams. Doing so, the integration is extended to the whole site. A new AGE procedure is performed to target the minimum energy requirement of the process site. The "Grand composite" curve with the new demands is given in figure 5 (curve a).

One can remark that a pinch point appears due to the new requirements; energy has to be introduced in the system. This pinch point will be taken into account to performe the next AGE procedure. As there is nó need for producing 30 bar steam, it was decided that the unit under study will supply medium pressure steam instead of high pressure steam. This leads to a small loss in mechanical power production but helps in the process control. The pressure of the medium pressure steam can be manipulated to control the operation of the reactor that produces most of the steam.

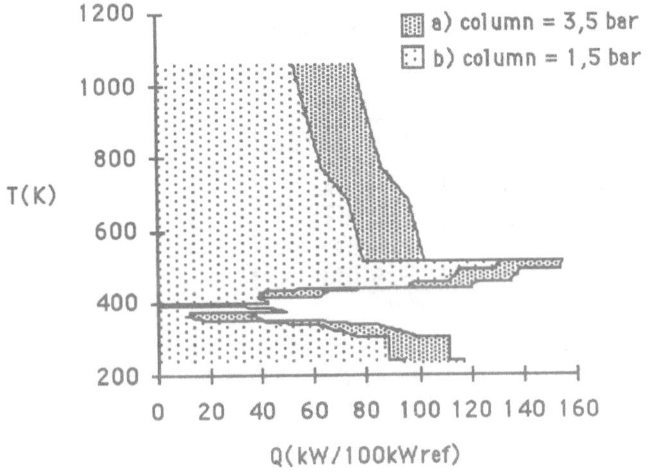

Figure 5 : "Grand composite" curve of the whole site.

Process modifications

The analysis of the shape of the "Grand composite" of the whole system indicates the opportunity for mechanical vapour recompression (from below to above the pinch point) or for distillation column pressure modification. Among the streams present in the neighbourhood of the pinch point, we found a column boiler above the pinch point (Analyse step). By decreasing the pressure of the column, it was possible to place the boiler below the pinch point (Generate step by simulation). That leads to a 23,2 kW/100kWref savings (31% of the MER) after verification of the operability of the new operating point of the distillation column (Evaluate step). The resulting "Grand composite" curve is given figure 5 (curve b). Table 6 gives the MER results before and after process modification. It is important to undertand that this modification is only interesting if the integration is extended to the whole site. Table 6 indicates that no benefit is obtained if the section S1 is considered alone. This new energy target corresponds to the whole site integration. It can only be reached when using the utility network.

TABLE 6 : MER for the whole process site.

MER	Heat In (kW/100kWref)	Heat out (kW/100kWref)	Curve figure 5
Process site			
Before modification	75,5	116,8	a) Column= 3,5 bar
After pressure modification	52,3	93,6	b)Column=1,5 bar
S1 section			
Before modification	33,0	36,6	
After pressure modification	33,0	36,6	

Steam network integration.

The utilities have to be chosen for the new problem by analysing the composite curve figure 5 (curve b). It was decided to use a steam network, inlet steam being supplied by a single furnace for the whole site. This steam will be produced at 30 bars and superheated to 410°C. As the pressure is higher than the maximum pressure required (16 bars), the steam network will include a steam expander with multiple draw off. The mechanical power production will be a subsequent benefit of the integration. Several other steam network technologically feasible structures have been tested using the same AGE procedure. The best one has been chosen. It is given in figure 6. The table 7 summarizes the results that define the target for the synthesis of the heat exchanger network. This concludes the targeting step of the study.

TABLE 7 : Steam balance for the unit under study.

	Header	Flow (T/h)
Exportation	MP 11bar	0,094
	LP 1,5 bar	0,08
Importation	HP 30 bar	0,0018
	MP 11 bar	0,028
	LP 1,5 bar	0,082

(The flowrates are expressed for 100 kW ref)

HEAT EXCHANGER NETWORK AND STEAM NETWORK SYNTHESIS AND OPTIMIZATION.

The synthesis of the heat exchanger network is out of the scope of this communication. Let us only mention that the steam network has been used to reduce the complexity of the MER heat exchanger network by allowing heat to cross the pinch point. The final structures are obtained after a simultaneous non linear optimization of the steam network and the heat exchanger network to minimize the costs including investment and operating cost. Table 8 summarizes the steam balance of the unit under study and table 9 gives some operating cost results.

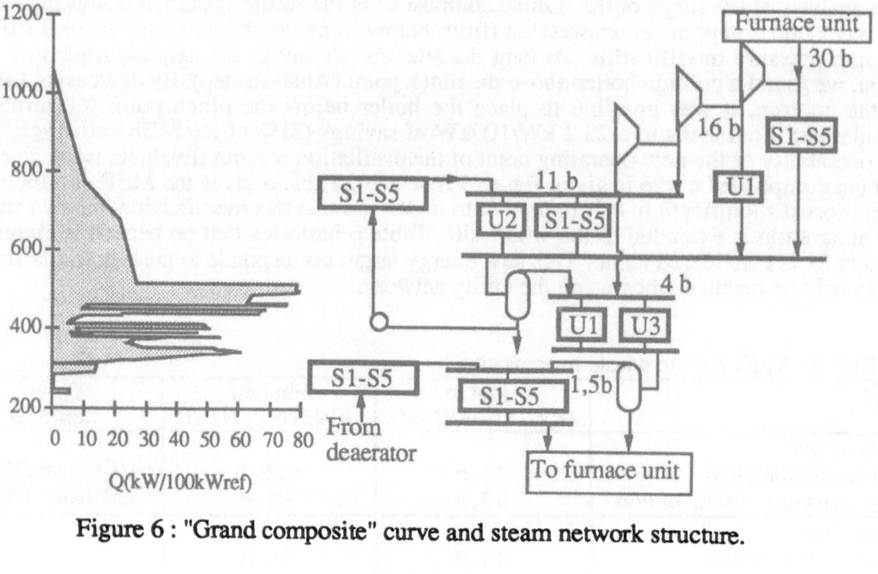

Figure 6 : "Grand composite" curve and steam network structure.

TABLE 8 : Optimized steam balance of the unit under study.

	Header	Flow (T/h)
Exportation	MP 11bar	0,095
	LP 1,5 bar	0,077
Importation	HP 30 bar	0,0019
	MP 11 bar	0,029
	LP 1,5 bar	0,082

(The flowrates are expressed for 100 kW ref)

TABLE 9 : some final results.

Type	Value	Unit	Cost (MU/year)
Natural gas	8,08	Nm3/h	0,66
HP 30 bar	0,09	T/h	0,72
Cold utilities			0,29
Total operation			1,66
Mechanical power	6,87	kW	-0,16
Total			1,51

(The values are given for 100 kW ref and arbitrary monetary units (MU))

The solution takes all the technological constraints into account that have been identified during the study by the process engineers. For example, the control of reactors using the steam pressure level is possible, the sections may work at lower levels independently, start-up procedures are easy even if the process is highly integrated. Each modification of the operating condition will have an influence on the steam network efficiency, but not on the operability of the installation.

CONCLUSIONS

Mathematical and computer tools, concepts and know-how

To perform energy integration, we combine mathematical and computer tools, concepts and process know-how. The computer tools are of different types. They can be used
- for collecting data : control computer, process data bases, data reconciliation;
- for generating data : simulation;
- for optimizing data : simulation and optimization.

The iterative aspect of the problem resolution requires integrated computer tools.
Several concepts have to be understood to propose optimal solutions, among these the concepts of the composite curves analysis and the utility integration are of first importance. The consistency of the methodology is also crucial.
The engineer know-how defines constraints that have to be added in order to propose feasible solution. The process knowledge will help to select among the solutions those that are feasible, that present enough advantages and acceptable cost.
We have identified a three steps procedure that we name AGE for Analyse-Generate-Evaluate. This procedure has been applied iteratively during the study. The solution of each AGE gives results and leads often to the problem redefinition, in term of battery limits, new constraints or more precise calculations.

Utility network integration

The integration of the utilities in a process is an important step of an energy integration study. It takes place in a general methodology for process energy integration and optimal heat exchanger network synthesis presented in Maréchal (ref 1.) and implemented into a software named SYNEP (ref. 4). The utility network allows to provide feasible integrated solutions that takes the restricted matches into account, it allows also to extend the battery limits of the study without increasing the complexity of the solution proposed. The integration of and through the utilities allows to overcome problems of flexibility, layout, operability and unit start-up that are often disadvantages of integrated design. The importance of integrating the utilities is illustrated by an industrial example where technological constraints were very restritive. The solution features a lot of advantages and overcome most of the disadvantages mentioned by the process engineers.

ACKNOWLEDGMENTS

SYNEP research is supported by the Commision of the European Communities (contracts JOUE-0009-B and EN3E-0136-B). The authors wish to express their thanks for this support.

REFERENCES

1. Maréchal F. and Kalitventzeff B., SYNEP1 : a methodology for energy integration and optimal heat exchanger network synthesis, Computers and Chemical engineering Vol 13, No 4/5, 1989.
2. Linnhoff B., User guide on process integration for the efficient use of energy, The institution of chemical engineers 1982.
3. Maréchal F., Kalitventzeff B., Heat and power integration, a MILP approach for optimal integration of utility systems. 22nd Symposium of the Working Party on Use of Computers in Chemical Engineering, COPE-91, Barcelona 1991.
4. SYNEP user's manual, Belsim s.a., Avenue Pré-Aily, Centre Socran, 4900 Angleur (Belgium).

A RETROFIT DESIGN MODEL FOR IMPROVING THE OPERABILITY OF HEAT EXCHANGER NETWORKS

Katerina P. Papalexandri and Efstratios N. Pistikopoulos*
Centre for Process Systems Engineering
Department of Chemical Engineering
Imperial College of Science, Technology and Medicine
London SW7 2BY

ABSTRACT

A systematic framework for simultaneously improving the flexibility and structural controllability of heat exchanger networks is presented in this paper. Based on a multiperiod hyperstructure representation where all retrofit alternatives are accounted for, the proposed retrofit strategy involves an iterative scheme between a multiperiod mixed-integer nonlinear programming (MINLP) model, which includes explicit controllability requirements, and a flexibility analysis subproblem. The resulted retrofitted network structure features minimum annualized cost, while being (i) flexible to operate within a specified range of uncertain stream flowrates, inlet temperatures and/or heat transfer coefficients, and (ii) structurally controllable for a given set of disturbance and control variables.

INTRODUCTION

Heat integration of chemical processes has been one of the most active research fields, with the main focus on synthesis of heat exchanger networks, as the energy efficiency of a plant greatly depends on properly integrating heat sources and cold sinks. Maximum energy recovery at low investment cost has been so far the main objective of heat exchanger network (HEN) synthesis and retrofit design problems. However, due to the changing environment a chemical plant is bound to operate, operability considerations, such as flexibility and controllability, must be taken into account, since economical optimality of a heat exchanger network under fixed operating conditions does not necessarily guarantee even feasible operation under varying and uncertain conditions.

The problem of systematically incorporating operability aspects in the synthesis and retrofit design of heat exchanger networks has recently begun to receive an increasing attention in the literature (Gundersen, 1992). Floudas and Grossmann (1985, 1987) developed a

*Author to whom all correspondence should be addressed

synthesis procedure for obtaining heat exchanger networks that are flexible to operate over a specified range of uncertain stream flowrates and inlet temperatures. Saboo and Morari (1984), and later Colberg et al. (1986), provided useful insights into the problem, introduced a "resilience" index for measuring the flexibility of HEN, and proposed a flexibility index "target" approach for obtaining flexible heat exchanger networks. Kotjabasakis and Linnhoff (1986) employed an evolutionary design approach based on the concept of "downstream path" and information obtained from "Sensitivity Tables". Cerda and coworkers (1989, 1990) proposed a synthesis methodology for obtaining flexible heat exchanger networks introducing the concept of "transient" and "permanent" process streams to describe the uncertainty involved.

The problem of including control objectives in the process synthesis level for heat exchanger networks has been considered by Georgiou and Floudas (1989). They coupled the network superstructure concept and a disturbance rejection criterion to obtain network structures with minimum disturbance propagation. A comprehensive multiobjective optimization approach for analyzing the interactions of design and control was recently developed by Luyben and Floudas (1992). Also, recently, Skogestad and coworkers (1992) considered flexibility and controllability aspects for the problem of "optimal" bypass placements in the synthesis of operable heat exchanger networks.

The approaches presented above are mostly applicable for the synthesis and grassroot design of heat exchanger networks. The retrofit design problem, however, is different and more complex than the grassroot design (Grossmann et al. 1987, Gundersen, 1989). Yee and Grossmann (1987) addressed the issue of reassigning exchangers, and proposed a systematic two-stage retrofit approach. Floudas and Ciric (1989) further categorized the possible network structural modifications and extended the generalized match network hyperstructure representation to the retrofit case. While both approaches are restricted to the case of fixed operating conditions, Papalexandri and Pistikopoulos (1992) recently proposed a multiperiod hyperstructure network representation with which retrofitted heat exchanger network designs can be obtained, featuring minimum total annualized cost and being able to operate over a number of operating conditions.

This paper addresses the problem of redesigning heat exchanger networks in order to improve their operability. Flexibility is introduced via an iterative scheme between a retrofit model and a flexibility analysis step, whereas structural controllability aspects are explicitly incorporated within a multiperiod retrofit model through the use of the disturbance propagation criterion.

PROBLEM DEFINITION

The problem to be addressed in this paper can be stated as follows:

Given is an existing heat exchanger network consisting of **X** heat exchangers of known area **EA**. The network involves a set **HP** of hot process streams, a set **CP** of cold process streams and a set of hot and cold utilities **HU** and **CU** respectively. The flowrate heat capacities **F**, inlet temperatures T^I and/or heat transfer coefficients **U** are not fixed, but vary within a range of values, $\{\mathbf{F}^{LO}, \mathbf{F}^{UP}\}$, $\{\mathbf{T}^I_{LO}, \mathbf{T}^I_{UP}\}$, $\{\mathbf{U}^{LO}, \mathbf{U}^{UP}\}$, respectively. The outlet temperatures T^O are fixed or constrained through some specification. A minimum temperature approach ΔT_{min} is also specified to determine feasible heat exchange in each exchanger (not the actual utility consumption, HRAT).

Given are also (i) a flexibility target, that is a range for the uncertain parameters where feasible operation of the network is desired, and (ii) a controllability target, that is to prevent a given set of "control" variables to be affected by a given set of "disturbance" parameters (total disturbance rejection).

The objective is then to obtain a retrofitted heat exchanger network design at minimum total annualized operating and structural modification cost, which simultaneously meets the flexibility and controllability targets.

The following assumptions are made: (i) The heat exchangers are of the countercurrent type. (ii) No phase changes are allowed. (iii) Constant heat capacities are considered. (iv) No hot-hot or cold-cold matches are allowed. (v) Area cost does not depend on the number of shells. This assumption could be relaxed through the approach of Trivedi et al. (1988).

In the sequel, the proposed retrofit strategy is first outlined. The multiperiod hyperstructure network model is then briefly presented. The model for the controllability considerations of the disturbance propagation linked to the retrofit model follows. An example problem is finally presented to illustrate the trade-offs between cost-flexibility-controllability.

OUTLINE OF THE PROPOSED RETROFIT STRATEGY

The key idea for simultaneously considering flexibility and controllability aspects within a retrofit model for heat exchanger networks is to incorporate the model of the disturbance rejection criterion of Georgiou and Floudas (1989b) within the multiperiod retrofit model of Papalexandri and Pistikopoulos (1992). The proposed retrofit strategy can then be summarized in the following steps (see Figure 1).

STEP 1 : Having specified a flexibility target for the existing heat exchanger network, a flexibility analysis is performed to determine whether the structure is flexible or not. The flexibility analysis model is based on the active set strategy (Grossmann and Floudas, 1987) and is represented as a mixed-integer non-linear programming (MINLP) problem (Pistikopoulos and Papalexandri, 1992). The solution of this model defines the "critical" parameter points (periods of operation), that are considered, together with the nominal point, in the multiperiod model.

STEP 2 : A multiperiod hyperstructure network representation is then developed, where the one-to-one (Ciric and Floudas, 1989) and/or the one-to-many (Yee and Grossmann, 1991) relationship between heat exchangers and stream matches is employed.

STEP 3 : The model for the multiperiod hyperstructure network is developed, which explicitly includes a disturbance rejection criterion. The solution of the model results in a retrofitted network structure, which features minimum total annualized cost and total disturbance rejection and is flexible to operate at the specified periods of operation.

The "new" structure is subjected to a new flexibility analysis subproblem and the procedure continues from STEP 1, until a structure is obtained that also features the desired degree of flexibility (see Figure 1).

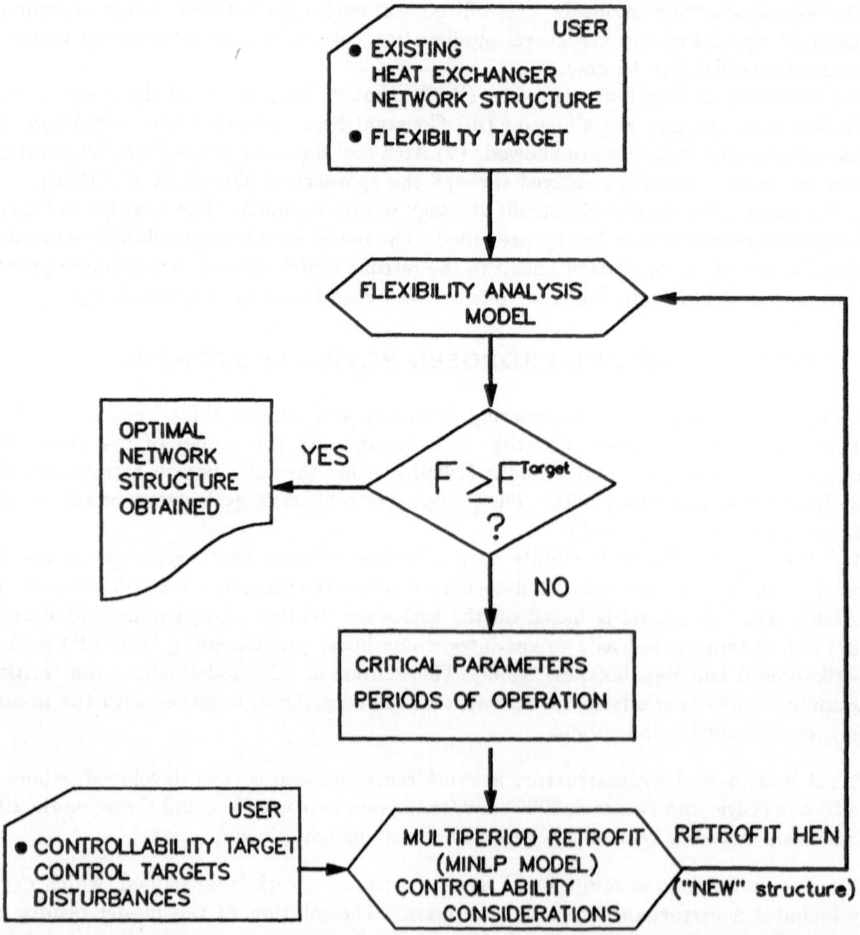

Figure 1. Iterative scheme for improving the operability of HEN

MULTIPERIOD HYPERSTRUCTURE NETWORK REPRESENTATION

The main building element of the proposed retrofit framework is a multiperiod hyperstructure network representation. This can be based either on an one-to-one (Ciric and Floudas, 1989) or an one-to-many (Yee and Grossmann, 1991) relationship between heat exchangers and stream matches. In this work the first representation will be adopted, where the network piping layout will be explicitly modeled through binary variables. The considered multiperiod network hyperstructure is illustrated in Figure 2. As it can be seen, each potential stream match corresponds to a potential exchanger unit, which may be an existing exchanger in the network or a new one to be purchased. Each stream entering the network is split towards each potential heat exchange the stream may contribute. The possibility of multiple heat exchange between two streams is taken into account with multiple potential exchangers being considered for a stream match, whenever ΔT_{min} constraints prevent single heat exchange between two streams. An overall bypass flow is also considered for each process stream. Prior to each heat exchanger a mixer is considered for each stream, where the flow from the initial splitter and the bypass flows from all the other potential exchangers of the stream are merged into a flow towards the exchanger. This flow is further split before entering the exchanger, so that potential bypasses in each exchanger are taken into account and heat transfer area requirements are met. After each heat exchanger unit a splitter is also considered for each process stream, from which a potential flow is driven towards the final mixer of the stream at the network outlet, and bypass flows are considered towards the mixers prior to the potential exchangers of the stream. Binary variables are assigned to all potential stream flows to account for the network piping layout. It must be noted that the structural binary variables, determining the network layout, are defined uniquely for all periods of operation, whereas flows, temperatures and heat amounts may vary from one period to another.

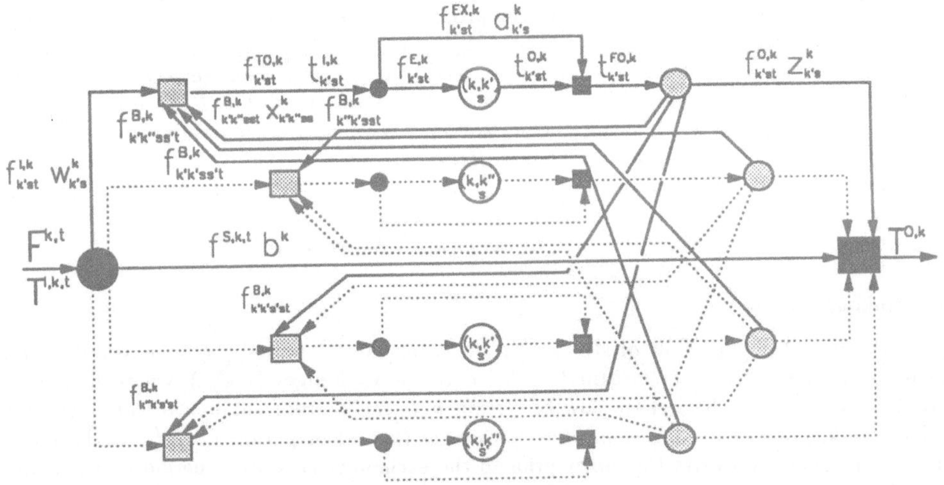

Figure 2. Multiperiod hyperstructure network representation

Let us denote by **HP** and **CP** the set of hot and cold process streams, **HU** and **CU** the set of hot and cold utilitites, respectively, and

$$\mathbf{H} = \mathbf{HP} \cup \mathbf{HU} \qquad \mathbf{C} = \mathbf{CP} \cup \mathbf{CU} \qquad \mathbf{HCT} = (\mathbf{HP}) \cup (\mathbf{CP})$$

To account for potential multiple heat exchange, the temperature range is partitioned into **IT** temperature intervals, based on the inlet temperatures of hot and cold process streams. The temperature difference for this partitioning is considered equal to zero (EMAT=0). If the temperature approach of two process streams is close to the considered ΔT_{min}, potential heat exchange between the two streams is considered in more than one "subnetworks". Denote the set of potential subnetworks in each period of operation by **IS**. Note that the actual utility consumption will be determined by the minimization of the total cost, while multiple heat exchange in the final retrofitted network will be based on the feasibility constraints with respect to ΔT_{min}.

In order to formulate the model for the hyperstructure of Figure 2, the following definitions are necessary:

. Sets:

$R_i^h = \{j \in C / i \in HP \; and \; (i,j,s) \; is \; a \; possible \; match, \; where \; s \in IS\}$
$R_j^c = \{i \in H / j \in CP \; and \; (i,j,s) \; is \; a \; possible \; match, \; where \; s \in IS\}$
$S_{ijs}^h = \{j' \in R_i^h / a \; bypass \; from \; (i,j',s') \; to \; (i,j,s) \; is \; possible, \; where \; s \; and \; s' \in IS\}$
$S_{ijs}^c = \{i' \in R_j^c / a \; bypass \; from \; (i',j,s') \; to \; (i,j,s) \; is \; possible, \; where \; s \; and \; s' \in IS\}$

. Binary variables:

y_{ijs}, decision variable to denote the existence of a match (i,j) in subnetwork $s \in$ IS in the final retrofitted network. $w_{k's}^k$, decision variable to denote the existence of a piping segment between the initial splitter of the process stream $k \in HCT$ and the exchanger (k,k's), where $k' \in R_k$. $z_{k's}^k$, decision variable to denote the existence of a piping segment between the outlet of the exchanger (k,k',s) and the final mixer of the process stream $k \in HCT$, where $k' \in R_k$. $a_{k's}^k$, decision variable to denote the existence of a bypass piping segment at the exchanger (k,k',s) (for the hot or the cold side, $a_{kk's}^h, a_{kk's}^c$ respectively). $x_{k'k''ss'}^k$, decision variable to denote the existence of a piping segment for the bypass flow of process stream $k \in HCT$ from exchanger (k,k",s') to exchanger (k,k',s), where $k' \in R_k$ and $k'' \in S_{kk'}$. b_k, decision variable to denote the existence of an overall bypass for stream $k \in HCT$. d_{ijs}^n, decision variable for the assignment of existing exchanger $n \in X$ to the match (i,j,s). m_{ijs}, decision variable to denote the purchase of a new exchanger unit for the match (i,j,s). v_{ln}, decision variable to denote the assignment of the area of existing exchanger $n \in X$ to the location of exchanger $l \in X$.

. Continuous variables:

f_{kt}^S, is the overall bypass flow of stream $k \in HCT$ at the tth period. $f_{k'st}^{Ik}$, is the flow from the initial splitter of process stream $k \in HCT$ to the exchanger (k,k',s), where $k' \in R_k$ in period t. $f_{k'k''ss't}^{Bk}$, is the bypass flow of stream $k \in HCT$ from the splitter after the heat exchanger (k,k",s') to the mixer prior to exchanger (k,k',s) in period t. $f_{k'st}^{Tk}$, is the flow of stream $k \in HCT$ that exits the mixer prior to the exchanger (k,k',s) in period t. $f_{k'st}^{Ek}$, is the flow of stream $k \in HCT$ that goes through the exchanger (k,k',s) in the tth period. $f_{k'st}^{EXk}$, is the flow of stream $k \in HCT$ that bypasses the exchanger (k,k',s) in period t. $f_{k'st}^{Ok}$, is the flow of stream $k \in HCT$ from the splitter after the exchanger (k,k',s) to the final mixer of the stream in period t. Q_{ijst}, is the total amount of heat exchanged in exchanger (i,j,s) in period t, where $i \in H, j \in R_i^h, s \in IS$. $t_{k'st}^{Ik}$ is the temperature of stream $k \in HCT$ before the exchanger (k,k',s) in period t. $t_{k's}^{Ok}$, is the temperature of stream $k \in HCT$ after the exchanger (k,k',s) in period t. $t_{k'st}^{FOk}$, is the temperature of stream $k \in HCT$ after the mixer, which is considered after the exchanger (k,k',s) in period t. A_{ijst}, is the area required for the heat exchange in exchanger (i,j,s) in period t. AEA_n, is the existing area assigned to exchanger $n \in X$. AA_n, is the additional area required in exchanger $n \in X$. NA_{ijs}, is the total area of an exchanger that needs to be purchased for the match (i,j,s). U_{ijt}, is the overall heat transfer coefficient of match (i,j) in period t. ΔT_{ijmax}^s, is the largest possible temperature drop through the exchanger of the match (i,j,s), equal to $\mathbf{T}^{O,i} - \mathbf{T}^j$.

The mathematical formulation of the hyperstructure model is then as follows (Pistikopoulos and Papalexandri, 1992b):

a. Mass balances for each stream at the initial splitter, the mixer and the splitter prior to each considered exchanger, the splitter after each considered exchanger, for each period of operation. For example, at the initial splitter of stream k and at the mixer prior to the exchanger (k, k', s):

$$\sum_{s \in IS} \sum_{k' \in R_k} f^{Ik}_{k'st} + f^{S}_{kt} = \mathbf{F}^{kt} \qquad k \in HCT,\ t = 1, \ldots, N. \tag{1}$$

$$f^{Ik}_{k'st} + \sum_{k'' \in S_{kk'}} f^{Bk}_{k'k''sst} + \sum_{s' \in IS} \sum_{k'' \in R_k} f^{Bk}_{k'k''ss't} - f^{Tk}_{k's} = 0 \quad k \in HCT, k' \in R_k, s \in IS,\ t = 1, \ldots, N. \tag{2}$$

b. Energy balances at the mixers of the superstructure of each process stream prior to heat exchangers, for each period of operation.

$$\mathbf{T}^k f^{Ik}_{k'st} + \sum_{k'' \in S_{kk''}} f^{Bk}_{k'k''sst} t^{FOk}_{k''st} + \sum_{s' \in IS} \sum_{k'' \in R_k} f^{Bk}_{k'k''ss't} t^{FOk}_{k''s't} - f^{Tk}_{k'st} t^{Ik}_{k'st} = 0$$

$$k \in HCT, k' \in R_k, s \in IS,\ t = 1, \ldots, N. \tag{3}$$

c. Energy balances over the exchangers for the hot and the cold streams, eg for hot streams:.

$$Q_{ijst} - f^{E_i}_{jst}(t^{I_i}_{jst} - t^{O_i}_{jst}) = 0 \qquad i \in HP, j \in R^h_i,\ s \in IS,\ t = 1, \ldots, N. \tag{4}$$

d. Equations that define the log-mean temperature difference between two streams that exchange heat and constraints that define the minimum required area for a heat exchnage between two streams.

e. Energy balances at the mixers of the superstructure of each process stream (i) after the exchanger and (ii) at the final mixer

$$f^{EXk}_{k'st} t^{Ik}_{k'st} + f^{Ek}_{k't} t^{Ok}_{k'st} - f^{Tk}_{k'st} t^{FOk}_{k'st} = 0 \qquad k \in HCT, k' \in R_k, s \in IS,\ t = 1, \ldots, N. \tag{5}$$

$$\sum_{s \in IS} \sum_{k' \in R_k} f^{Ok}_{k'st} t^{FOk}_{k'st} + f^{S}_{kt} \mathbf{T}^{Ikt} - \mathbf{F}^{kt} \mathbf{T}^{O}_{kt} = 0 \quad k \in HCT,\ t = 1, \ldots, N. \tag{6}$$

f. The constraints that account for feasible operation.

$$t^{I_i}_{js} - t^{O_j}_{is} \geq \Delta T_{min} \quad i \in HP,\ j \in R^h_i \qquad\qquad t^{O_i}_{js} - t^{I_j}_{is} \geq \Delta T_{min} \quad i \in HP,\ j \in R^h_i \tag{7}$$

g. Constraint (8) gives an upper bound to the amount of heat that can be exchanged between hot stream i and cold stream j. If a match (ijs) does not exist $(y_{ijs} = 0)$ this heat amount is equal to zero.

$$Q^{ijst} - y_{ijs}\mathrm{UPBND} \leq 0 \quad i \in H, j \in R_i, s \in IS,\ t = 1, \ldots N. \tag{8}$$

h. Constraints in (9) give an upper bound to the flow of a stream to and through an exchanger and ensure that such a flow is not present if the corresponding match does not exist $(y_{ijs} = 0)$. If the match (i, j) is selected then the flow of the hot and the cold stream through the exchanger is bounded from a minimum based on the heat load and the corresponding maximum temperature drop. This is described by the constraints in (14), eg for hot streams:

$$f^{T_i}_{jst} - \mathbf{F}^{it} y_{ijs} \leq 0 \quad f^{E_i}_{jst} - \mathbf{F}^{it} y_{ijs} \leq 0 \quad f^{E_i}_{jst} - \frac{Q_{ijst}}{\Delta T^s_{ijmax}} \geq 0 \quad i \in HP, j \in R_i\ \ s \in IS,\ t = 1, \ldots N. \tag{9}$$

i. Constraints (10)-(11) provide an upper bound to stream flows to and from heat exchangers and bypass flows. When the corresponding piping segments do not exist in the retrofitted network ($w^k_{k's} = 0$, $z^k_{k's} = 0$, $a^k_{k's} = 0$, $x^k_{k'k''s} = 0$, *or* $b_k = 0$) these flows are forced equal to zero.

$$f^{Ik}_{k'st} - w^k_{k's}\text{UPBND} \leq 0 \qquad f^{Ok}_{k'st} - z^k_{k's}\text{UPBND} \leq 0 \qquad f^{EXk}_{k'st} - a^k_{k's}\text{UPBND} \leq 0$$
$$f^{Bk}_{k'k''st} - x^k_{k'k''s}\text{UPBND} \leq 0 \quad k \in HCT,\ k' \in R_k,\ s \in IS,\ t = 1, \ldots, N \tag{10}$$

$$f^S_{kt} - b_k\text{UPBND} \leq 0 \qquad k \in HCT,\ t = 1, \ldots, N \tag{11}$$

j. Logical constraints that ensure well-defined structures.

Constraints (12) and (13) ensure that each process stream exchanges heat at least once and enters at least one exchanger. Constraints in (14)-(15) ensure that if a match does not exist ($y_{kk's} = 0$) the piping segments connected with the corresponding exchanger do not exist either.

$$\sum_{j \in R_i} \sum_{s \in IS} y_{ijs} \geq 1 \quad i \in HP \qquad \sum_{i \in R_j} \sum_{s \in IS} y_{ijs} \geq 1 \quad j \in CP \tag{12}$$

$$\sum_{k' \in R_k} \sum_{s \in IS} w^k_{k's} \geq 1 \quad k \in HCT \qquad \sum_{k' \in R_k} \sum_{s \in IS} z^k_{k's} \geq 1 \quad k \in HCT \tag{13}$$

$$w^k_{k's} - y_{kk's} \leq 0 \quad z^k_{k's} - y_{kk's} \leq 0 \quad a^k_{k's} - y_{kk's} \leq 0 \quad k \in HCT,\ k' \in R_k,\ s \in IS \tag{14}$$

$$x^k_{k'k''ss'} - y_{kk's} \leq 0 \quad x^k_{k''k's's} - y_{kk's} \leq 0 \quad k \in HCT, k' \in R_k, k'' \in S_{kk'}, s, s' \in IS \tag{15}$$

Constraint (16) prevents cycling piping between two heat exchangers of the same process stream, while constraint (17) prevents the presence of multiple bypasses when the process stream exchanges heat only once.

$$x^k_{k'k''ss'} + x^k_{k''k's's} \leq 1 \qquad k \in HCT,\ k', k'' \in R_k \tag{16}$$

$$\sum_{k' \in R_k} \sum_{s \in IS} a^k_{k's} + b_k + (1 - \sum_{k' \in R_k} \sum_{s \in IS} y_{kk's})\text{UPBND} \leq 1 \qquad k \in HCT \tag{17}$$

Constraints in (18)-(19) ensure that if a process stream constibutes to a heat exchange (k,k',s) then the stream enters and exits the exchanger.

$$w^k_{k's} + \sum_{s' \in IS} \sum_{k'' \in S_{kk'}} x^k_{k'k''ss'} - y_{kk's} \geq 0 \qquad k \in HCT,\ k' \in R_k,\ s \in IS \tag{18}$$

$$z^k_{k's} + \sum_{s' \in IS} \sum_{k'' \in R_k,\ k' \in S_{kk''}} x^k_{k''k's's} - y_{kk's} \geq 0 \qquad k \in HCT,\ k' \in R_k,\ s \in IS \tag{19}$$

Constraint (20) ensures that if a hot or cold stream exchanges heat with a cold or hot utility respectively, then it leaves the network only from that exchanger.

$$z^k_{k's} - y^k_{k''s'} \leq 0 \qquad k \in HP \cup CP,\ k' \in R^k \cap (HP \textit{ or } CP), k'' \in R_k \cap (HU \textit{ or } CU) \tag{20}$$

k. Match - Exchanger assignments

Equation (21) ensures that for a match (i,j), in a subnetwork $s \in IS$ either an existing unit will be assigned or a new one will be purchased. When a match (i,j,s) is not present ($y_{ijs} = 0$) then neither an existing unit is assigned to it ($d^n_{ijs} = 0$) nor a new one is purchased for it ($m_{ijs} = 0$).

$$\sum_{n \in X} d^n_{ijs} + m_{ijs} - y_{ijs} = 0 \qquad i \in H,\ j \in R_i,\ s \in IS \tag{21}$$

Equation (22) defines the total existing area assigned to the location of an exchanger n.

$$AEA_n = \sum_{l=1}^{X} (EA_l v_{nl}) \qquad n \in X \tag{22}$$

Constraints in (23) define the area of a new exchanger to be purchased for the match (i,j,s). If such an exchanger will not be purchased ($m_{ijs} = 0$), then this area is zero, otherwise it must suffice for the heat exchange taking place. Constraint in (24) gives a lower bound to the additional area that must be purchased for an existing exchanger.

$$NA_{ijs} + (1 - m_{ijs})UBND \geq A_{ijs} \qquad NA_{ijs} - m_{ijs}UBND \leq 0 \qquad i \in HP, j \in R_i, s \in IS \tag{23}$$

$$AA_n + AEA_n + (1 - d^n_{ijs})UPBND \geq A_{ijs} \qquad n \in X;\ i \in HP, j \in R_i, s \in IS \tag{24}$$

Constraints in (25) define the assignment of the area of an existing exchanger either to a match or to the location of another existing exchanger.

$$\sum_{l=1}^{X} v_{ln} \leq 1 \qquad \sum_{i \in HP} \sum_{s \in IS} \sum_{j \in R_i} d^n_{ijs} + \sum_{l \neq n} v_{ln} \leq 1 \quad n \in X \tag{25}$$

l. Retrofit cost - Objective function

Equations (1)-(25) define the hyperstructure model, where all the possible alternatives for the existing network and the existing equipment are embedded. The retrofit of heat exchanger networks is driven then from the minimization of an objective function, which is the total annualized operating and retrofit investment cost. The terms of the objective function are: (i) Operating cost of hot and cold utility consumption, (ii) Cost of purchasing a new exchanger, which includes the standard cost of a new exchanger and the cost of its area, (iii) Cost of purchasing additional area for an existing exchanger, (iv) Cost of reassigning an existing exchanger to a different match, (v) Cost of assigning the area of an existing exchanger to the location of another existing exchanger, (vi) Cost of making a new match, (vii) Repiping cost, which is assumed to be independent of exchanger reassignments. It must be noted that the retrofit model simultaneously accounts for operating and investment cost (retrofit model *without decomposition*).

Therefore, the total cost of structural modifications including operating cost to be minimized will be:

$$OBJ = min\{\sum_t \{ \sum_{i \in HU} \sum_{j \in R^i} \sum_{s \in IS} CHU_i Q_{ijst} + \sum_{j \in CU} \sum_{i \in R^j} \sum_{s \in IS} CCU_j Q_{ijst}\} +$$

$$min\{\sum_i \sum_j \sum_s \{\{aNA_{ijs}{}^b + NEm_{ijs} + \sum_n aAA_n{}^b + (NEA + C^n_{ij})d_{ijst} + \sum_{l \neq n} CAA_{ln} v_{ln}\} + CNM_{ijs}$$

$$+ \mathbf{CP}[w^i_{js} + \sum_{k' \in S_{ij}} \sum_{s' \in IS} x^i_{jk'ss'} + z^i_{js} + b_i] + \mathbf{CP}[w^j_{is} + \sum_{k' \in S_{ij}} \sum_{s' \in IS} x^j_{ik'ss'} + z^j_{is} + b_j]\} y_{ijs}\} \quad (26)$$

The above equations along with the objective function in (26) define a mixed-integer nonlinear programming (MINLP) model for the retrofit of heat exchanger networks. The solution of this model can be achieved through decomposition techniques, e.g. Generalized Benders Decomposition (Geoffrion, 1972, Floudas, 1990) or Outer Approximation (Duran and Grossmann, 1986). However, due to the presence of non-convexities in the formulation, global optimality of the solution can not be guaranteed, unless global optimization methods (Floudas and Visweswaran, 1991) are employed. In the next section, it will be shown how operability considerations can also be incorporated within the multiperiod retrofit model.

STRUCTURAL CONTROLLABILITY ASPECTS

As a structural controllability consideration, a disturbance rejection criterion based on the structural concept of *disturbability* (Georgiou and Floudas, 1989a) is used in this work. A *variable structural* or *incidence matrix* is introduced, the entries of which are either zero or take arbitrary values depending on whether a variable participates in an equation or not. Since only the disturbance input variables are of interest, the manipulated variables, such as utility consumption, do not participate in the structural matrix. The outlet temperatures of process streams are usually considered as control targets, while the stream flowrates and process stream inlet temperatures are considered as disturbance variables.

The objective is to determine a structure that features total disturbance rejection, given a set of control targets and disturbance variables. In general, a variable x_i is disturbable from an input d_j, when the generic rank of the matrix M that results from the network structural matrix after deleting the column corresponding to x_i is equal to the number of the (active) equations-rows of the network (see Georgiou and Floudas, 1989b).

For the multiperiod heat exchanger network hyperstructure, a variable structural matrix is developed, where the existence of each "row" depends on the existence of the corresponding stream match in the retrofitted network. Furthermore, the existence of each variable depends on the existence of the corresponding stream match, and in particular on the existence of a associated piping segment, in the final network. Binary variables (q in general) are introduced (i) for each equation to denote whether an equation is redundant or not from a structural point of view, (ii) for each variable to denote whether the corresponding column in the network structural matrix is redundant or not, (iii) for each variable and equation to denote whether this variable is an output variable of the corresponding equation. These new binary variables, q, are related to the original binary variables, Y, defining the network structure (by Y, we denote the set of all the binary variables that define the network layout, i.e. stream matches, piping segments, etc) in the following way:

$$q - Y \leq 0 \quad (27)$$

This set of constraints ensure that if a continuous variable or an equation (energy or mass balance) does not exist in the retrofitted network then it is not active in the network structural matrix either. For example, if $q^{Ik}_{k's}$ is assigned to the flow $f^{Ik}_{k's}$ and $w^k_{k's}$ is the decision variable corresponding to the existence of the piping segment between the initial spliiter of stream k and the exchanger (k, k'), the constraint becomes $q^{Ik}_{k's} - w^k_{k's} \leq 0$. If $w^k_{k's} = 0$, the corresponding piping segment does not exist and the flow $f^{Ik}_{k's}$ does not exist as degree of freedom in the network either ($q^{Ik}_{k's} = 0$).

The explicit formulation for the determination of the generic rank of a matrix within the retrofit model can then be achieved as follows: The maximization of the number of output variables is expressed introducing a scalar variable U, which is forced to be less than the generic rank of the structural matrix, using the single input connectability formulation (Georgiou and Floudas, 1989a). This scalar variable is also introduced in the objective function with a negative sign and a large positive coefficient, so that the minimization of the objective variable ensures the proper determination of the generic rank. A constraint is also included forcing the generic rank of the chosen network structure to be strictly less than the number of the network active equations. The number of the active rows of the network representation model can be expressed as a function of the structural decision variables as follows:

$$\text{No of active rows} = 2*\text{card(HP)}+2*\text{card(CP)}+6*\sum_{i\in HP}\sum_{j\in R^i}\sum_{s\in IS} y_{ijs}+6*\sum_{j\in CP}\sum_{i\in R^j}\sum_{s\in IS} y_{ijs} \quad (28)$$

When multiple control objectives are considered, the disturbance rejection criterion is explicitly taken into account for each one of the control objectives.

It should be noted that the introduction of the disturbance rejection criterion does not introduce any extra nonlinearities, but it increases significantly the size of the problem, in terms of number of rows and especially of number of binary variables. When decomposition techniques are applied to solve the retrofit model, however, the binary variables that correspond to the disturbability criterion can be relaxed as continuous in the Primal subproblem. This is due to the total unimodularity of the remaining matrix for these variables once the "structural" binary variables have been fixed to integer values (Georgiou and Floudas, 1989a). Then, the "disturbability" variables will automatically take integer values at the optimal solution.

ILLUSTRATIVE EXAMPLE

The proposed retrofit framework to improve the operability of heat exchanger networks will be illustrated with the heat exchanger network shown in Figure 3a. It involves one hot and two cold process streams and a single hot utility (steam). Stream data are given in Table 1. The flowrate heat capacity of the hot stream is considered to vary :

$$11kW/K \leq \mathbf{F}^H \leq 16kW/K$$

A $\Delta T_{min} = 10K$ is also given. Heat exchange areas are not taken into account in this analysis.

A flexibility analysis of the existing network yields a flexibility index FI=0.67, defining the uncertainty range, where the network can feasibly operate, to be $12kW/K \leq \mathbf{F}^H \leq 15kW/K$.

Based on the cost data shown in Table 2 (a large piping cost is considered to avoid extensive splitting of the streams and to implicitly take into account pressure drop considerations) and considering two periods of operation (the nominal one and the critical one defined by the vertex of the uncertainty range $\mathbf{F}^H_v = 16kW/K$) the multiperiod MINLP retrofit model is developed, where controllability considerations are also included. The oulet temperature of the hot stream and of the cold stream $C2$ are considered to be the controlled variables and the heat capacity flowrates of the process streams the disturbance variables. The multiperiod MINLP retrofit model with the disturbance rejection criteria involves 729 rows and 1022 variables, 826 of which are binary. The model has been solved with the application of GBD through GAMS modelling system. After the relaxation of the "disturbability" variables the Primal subproblem featured 618 rows and 951 variables while the Master problem 113 rows and 72 variables, 68

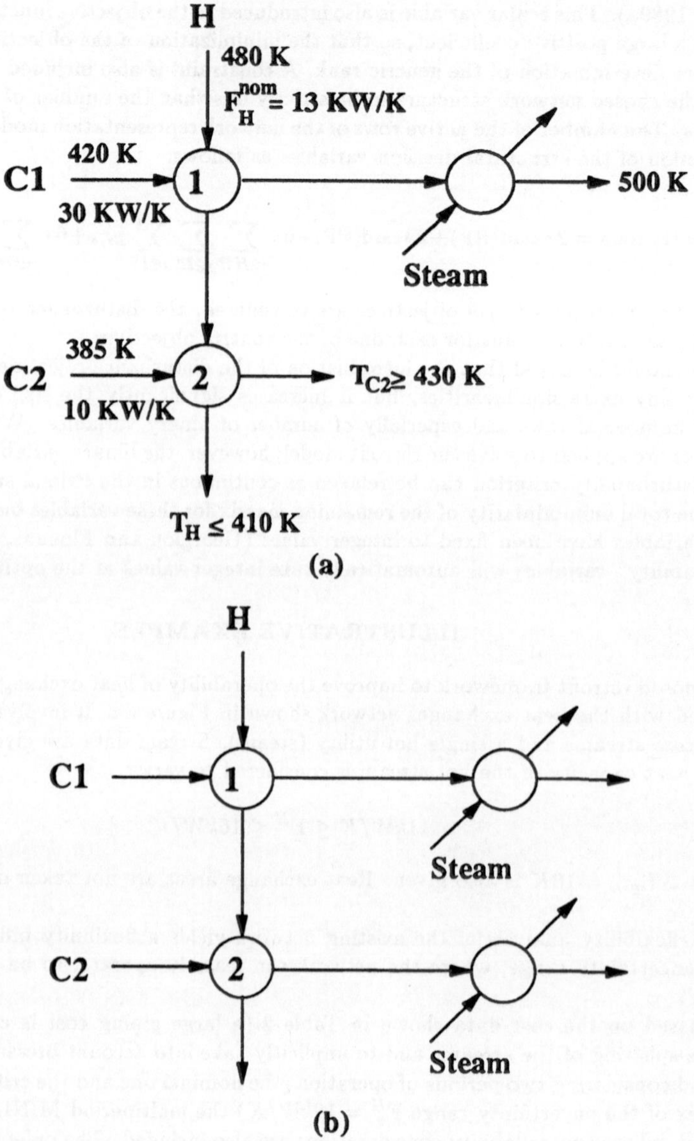

Figure 3. Existing and retrofitted network structures

of which are binary. The model was solved in 3 GBD iterations. The resulting retrofitted network is illustrated in Figure 3b. It features an additional heater and an annualized total cost of $\$41613.75yr^{-1}$. The network is proven to be fully flexible in the whole range of uncertainty of $\mathbf{F}_H$ and it features total disturbance rejection. However, it should be noted that global optimality can not be guaranteed (due to non-convexities).

STREAM No.	T_N^{IN} (K)	T^{OUT} (K)	FCP^N (KW/K)
H	480	≤ 410	nom. 13
C1	420	500	30
C2	385	≥ 430	10

Table 1: Stream data.

Steam cost	$\$11.05KW^{-1}yr^{-1}$
Cost of new exchanger	$\$2000yr^{-1}$
Piping cost	$\$10000\ yr^{-1}$

Table 2: Cost data.

CONCLUSIONS

A retrofit framework has been presented in this paper in order to improve the operability of heat exchanger networks. An iterative procedure was proposed to derive flexible network structures, while a disturbance rejection criterion was explicitly incorporated within the retrofit model, to reject networks structures that feature disturbance propagation. The proposed strategy is based on a multiperiod MINLP retrofit model, where the existing equipment is explicitly accounted for. The resulting network features minimum total annualized cost and minimal disturbance propagation or total disturbance rejection. With the proposed retrofit framework, trade-offs of flexibility -disturbance rejection - operating cost - investment cost are properly assessed.

REFERENCES

1. Brooke,A., D. Kendrick and M. Meeraus, *GAMS: A User's Guide.* Scientific Press, Redwood City, 1988.

2. Cerda, J., M.R. Galli, N. Camussi and M.A. Isla, Synthesis of Flexible Heat Exchanger Networks-I. Convex Networks. Comp. & Chem. Eng., 1990, **14**.

3. Cerda, J. and M.R. Galli, Synthesis of Flexible Heat Exchanger Networks - II. Nonconvex Networks with Large Temperature Variations. Comp. & Chem. Eng., 1990, **14**.

4. Colberg,R.D. and M. Morari, Analysis and Synthesis of Resilient Heat Exchanger Networks, Advances in Chemical Engineering. J. Wei. Ed., 1988, **14**.

5. Duran,M.A. and I.E. Grossmann, An Outer Approximation Algorithm for a Class of Mixed Integer Nonlinear Programs. Math. Program, 1986, **36**.

6. Floudas, C.A., OASIS: Discrete/Continuous Optimization Approaches in Process Systems. Computer Aided Systems Laboratory. Department of Chemical Engineering. Princeton University, 1990.

7. Floudas,C.A. and A.R. Ciric, Strategies for Overcoming Uncertainties in Heat Exchanger Network Synthesis. Comp. & Chem. Eng., 1989, **13**.

8. Floudas,C.A. and I.E. Grossmann, Synthesis of Flexible Heat Exchanger Networks for Multiperiod Operation, Comp. & Chem. Eng., 1985, **10**.

9. Floudas,C.A. and I.E. Grossmann, Automatic Generation of Multiperiod Heat Exchanger Network Configurations. Comp. & Chem. Eng., 1987, **11**.

10. Floudas,C.A. and V. Visweswaran, A Global Optimization Algorithm (GOP) for certain classes of Nonconvex *NLP's-I & II*. Comp. & Chem. Eng., *1990*, **14**.

11. Galli,M.R. and J. Cerda, Synthesis of Flexible Heat Exchanger Networks - III. Temperature and Flowrate Variations. Comput. Chem. Eng., 1991, **15**.

12. Geoffrion, A.M., Generalized Benders Decomposition, Jl. of Opt. Theory and Appl., 1972, **10**.

13. Georgiou,A. and C.A. Floudas, Optimization model for generic rank determination of srtuctural matrices. Int. J. Control, 1989, **49**.

14. Georgiou,A. and C.A. Floudas, Simultaneous Process Synthesis and Control: Minimization of Disturbance Propagation in Heat Exchanger Networks. FOCAPD, Colorado, 1989.

15. Grossman,I.E., A.W. Westerberg and L.T. Biegler, Retrofit Design of Processes. Proc. of 1st Int. Conf. on Found. of Comp. Aided Proc. Oper. (FOCAPO), Park City, Utah, July 1987.

16. Gundersen,T., Retrofit Process Design - Research and Application of Systematic Methods. FOCAPD - 89, Colorado, 1989.

17. Gundersen,T., Process-Syntese for Design og Regulering av Prosessanleg Arsmoteseminar i Norsk Forening for Automatisering, Soria Moria Konferansesenter, Oslo, 1992.

18. Kotjabasakis,E. and B. Linnhoff, Sensitivity Tables in the Design of Flexible Processes: 1. How much Contingency in Heat Exchanger Networks is Cost - Effective?. I. Chem. Eng., Chem. Eng. Res. Des., 1986, **24**.

19. Luyben,M.L. and C.A. Floudas, A Multiobjective Optimization Approach for analyzing the Interaction of Design and Control-I. Submitted for publication, 1992.

20. Mathisen,K.W., S. Skogestad and T. Gundersen, Optimal Bypass Placement in Heat Exchanger Networks. AIChE Spring National Meeting, New Orleans, 1992.

21. Papalexandri,K.P. and E.N. Pistikopoulos, A Multiperiod MINLP model for improving the flexibility of Heat Exchanger Networks. ESCAPE-2, 1992, for publication.

22. Pistikopoulos, E.N. and K.P. Papalexandri, An MINLP Retrofit Approach for Improving the Flexibility of Heat Exchanger Networks. IRC Report, Imperial College, 1992.

23. Saboo,A.K. and M. Morari, Design of Resilient Processing Plants: IV. Some New Results on Heat Exchanger Network Synthesis. Chem. Eng. Sci., 1984, **39**.

24. Viswanathan,J. and I.E. Grossmann, DICOPT++ - Version 2.1. Engineering Design Research Center, Carnegie Mellon University, Pittsburgh, 1991.

25. Yee T.F. and I.E. Grossmann, Optimization Model for Structural Modifications in the Retrofit of Heat Exchanger Networks. Proc. of 1st Int. Conf. on Found. of Comp. Aided Proc. Oper. (FOCAPO), Park City, Utah, July 1987.

26. Yee, T.F. and I. Grossmann, A Screening and Optimization Approach for the Retrofit of Heat-Exchanger Networks, Ind. Eng. Chem. Res., 1991, **30**.

NEW CRITERIA FOR THE DESIGN OF OPTIMAL HEAT EXCHANGER NETWORKS

Dr. IOANNIS S. ANDREOU
Chemical Engineer / Safety and Hygiene Section Head
Hellenic Aspropyrgos Refinery SA, 54 Amalias Ave, 10558 Athens

ABSTRACT

This paper presents a new algorithm based on thermodynamics for design purposes. It uses all the necessary thermodynamic criteria which can lead to a family of near optimal heat exchanger networks as for as the utilities consumption is concerned. The new method is called "the pseudo-pinch point method" [1].

PREVIOUS WORK

Several optimal networking methods are available. Those methods are either based on linear programming techniques or on heuristics based on thermodynamics. From the above techniques the best known one is "the pinch point method" [2], which is already applied to many industrial applications with good results. The basic aspects of this method are the definition of two separate thermodynamic areas around the pinch point and the analysis based on the problem definition table (Composite curves) [3],[4] which defines the minimum requirement of hot and cold auxiliary loads. Also the method under consideration evaluates the minimum number of heat exchangers (Euler Theorem) [4].

DEVELOPMENT OF THE MODEL

The new algorithm is based on the following new arguments :

1.- The full driving forces utilization by means of two new diagrams, [T_{hot}-DT] & [T_{cold}-DT], used as selection criteria.

2.- The new criterion called "The heat capacity surplus criterion". This principle combined with the maximum driving forces utilization, ensures that the remaining pieces of streams after a new match, may also be networked.

3.- The "pseudo-pinch point" thermodynamic heuristic. This handles the design problem away from the pinch point vicinity, under the same criteria, valid near the pinch point.

The above heuristic is the basis of the new algorithm.

4.- A data bank of prototype matches is presented. These matches follow the new thermodynamic criteria-heuristics and are used for the design done near and away from the pinch point.

The general form of the system of composite curves [4], is shown in figure 1.

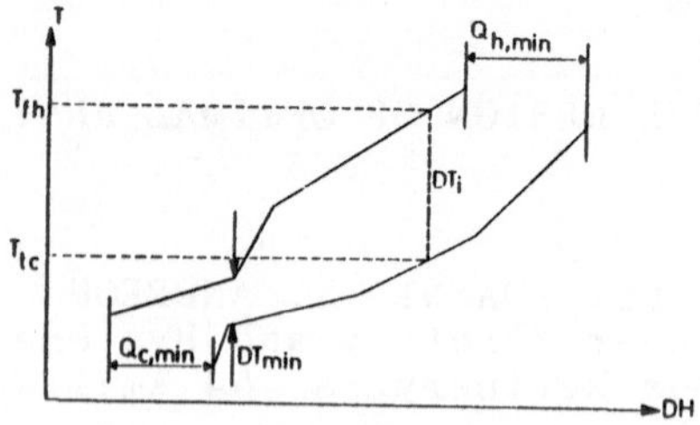

Figure 1. The general form of the Composite Curves System

By using a very simple algorithm, we produce the curves shown in Fig. 2, 3. We do that by combining the $-T_{fh}-$ or $-T_{tc}-$ temperatures with the temperature differences $-DT_i-$, as shown with the interrupted line of Fig. 1 (Diagrams [T_{hot}-DT], [T_{cold}-DT] correspondigly).

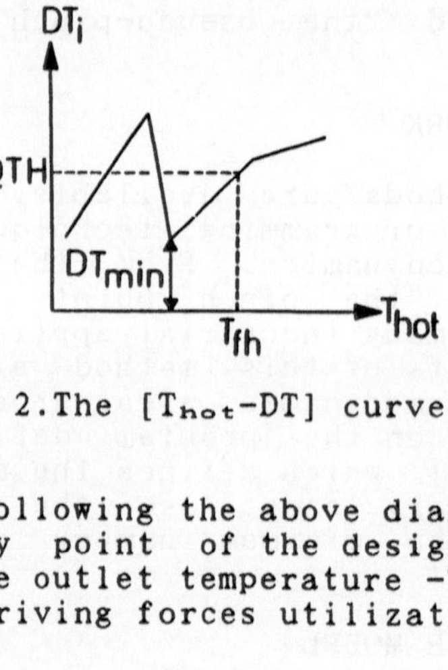

Figure 2.The [T_{hot}-DT] curve

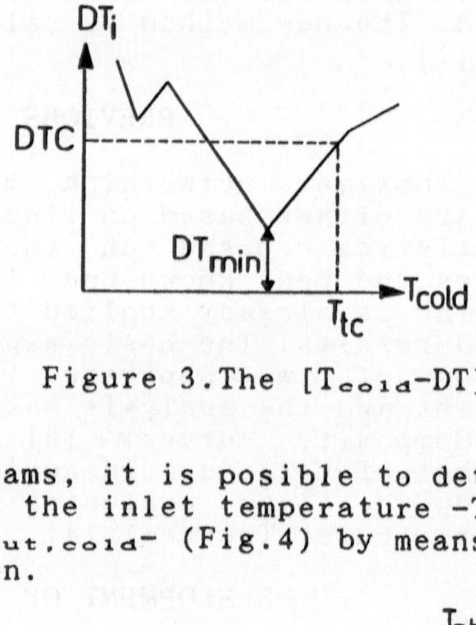

Figure 3.The [T_{cold}-DT] curve

Following the above diagrams, it is posible to determine, at any point of the design, the inlet temperature $-T_{in,hot}-$ and the outlet temperature $-T_{out,cold}-$ (Fig.4) by means of the full driving forces utilization.

$T_{in,hot}=T_{tc}+DT_i$
$T_{out,cold}=T_{fh}-DT_i$

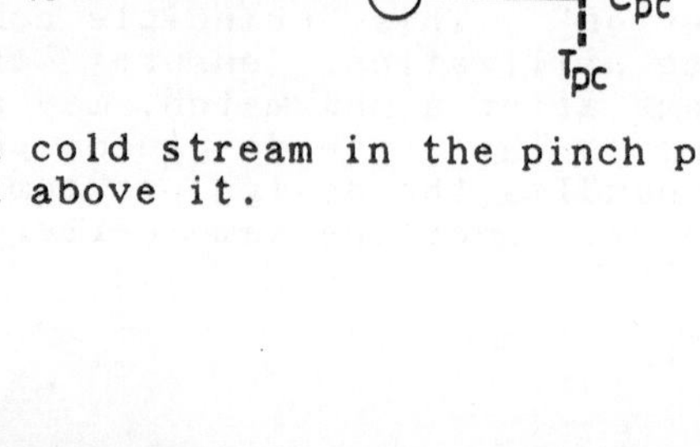

Figure 4.The match of hot and cold stream in the pinch point region and above it.

At any point of the design, we use either the [T_{hot}-DT] diagram or the [T_{cold}-DT] one, which depends on the streams temperatures and their thermal capacities.

The "Heat capacity surplus criterion" simplifies the evaluation of all the possible matches between a hot and a cold stream, but also, it points out the necessity of the division of the streams (if the criterion is not valid).

It is also used for the specification of the fraction of streams when they have to divide into pieces.

The "Heat capacity surplus criterion" is : "The match between a hot -i- and a cold -j- stream which satisfies the inequality :

$$C_{pc,j} \geqslant C_{ph,i}$$

is acceptable if :"

$$\sum_{j=1}^{NC} C_{pc,j} - \sum_{i=1}^{NH} C_{ph,i} \geqslant C_{pc,j} - C_{ph,i} \qquad (1)$$

For example if there are 5 hot and 5 cold streams at the pinch point vicinity, then 120 different combinations must be done, each of which includes 5 matches. Then we must select the most proper match.

With the application of the new criterion, only 14 matches have to be evaluated (in the worst case), independently of the divisions of the main streams. The proof of the above criterion (1) is shown in Appendix A.

The "pseudo-pinch point" thermodynamic heuristic handles the design problem away from the pinch point vicinity under the same criteria, valid near the pinch point.

A typical representation of the remaining (after the pinch point matches) problem is shown in Fig. 5.

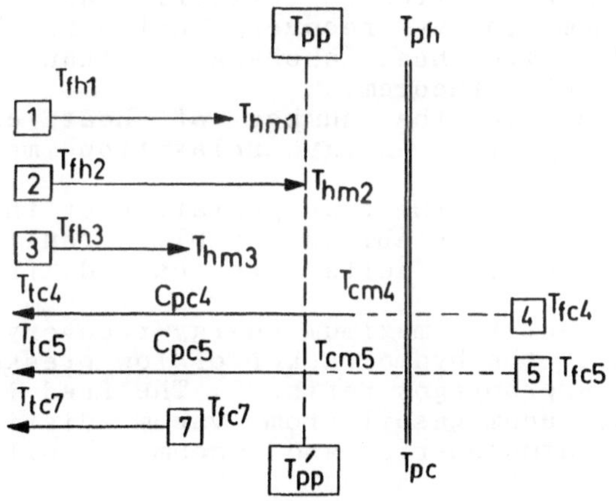

Figure 5. The remaining problem after the pinch point matches

If T_{hmi}, T_{cmj} are the minimum temperatures of hot and cold streams, which remain after the pinch point matches, then:

$$T_{hmi} - T_{cmj} = DT \geqslant DT_{min} \quad (2)$$

By definition : $T_{hm2} = \min (T_{hmi}) \;:\; i=1,2,3$ (3)

and $T'_{pp} = T_{hm2} - DT_{min}$ (4)

where T_{pp} is the pseudo pinch point temperature

It is clear that $T'_{pp} \geqslant T_{cmj} \;:\; j = 4,5$ (5)

The cold streams with j=4,5 can give their thermal charges to any hot stream without restrictions. We define the **"free thermal charge, QF"**, as the charge of the cold stream being distributed in the temperature interval **[T_{cmj} - T'_{pp}]** and we calculate it with the equation (6).

$$QF_j = \sum_{j=1}^{N} C_{pcj} * (T'_{pp} - T_{cmj}) \quad (6)$$

Next, the QF charge is transferred to the higher temperature interval defined with the temperatures T'_{pp} and T_{tcj}. This transfer becomes with a fictitious increasing of the thermal capacity -C_{pcj}-. The new -C'_{pcj}- is given from the equation (7).

$$C'_{pcj} = C_{pcj} + QF_j \;/\; (T_{tcj} - T'_{pp}) \quad (7)$$

After that, the lower temperature of the cold streams will be the T'_{pp}. Then at the cold end of the remaining problem, the temperature difference will become again equal to DT_{min} and the same criteria (as for the design at the pinch point) are valid.

All the above heuristics - criteria are used in the design of maximum energy recovery networks. Usually these networks include more heat exchangers than the networks defined from the Euler theorem.

The reduction in the number of heat exchangers is achieved thanks to the energy relaxation methods (loop breaking).

At the same time, the transportation of thermal charges along the flow paths, reestablish the DT_{min} faults.

Usually the DT_{min} faults are caused from the loop breaking.

Next, we present the maximum energy recovery design for a real industry unit, the hydrocracker of low pressure (U-4000) of the Hellenic Aspropyrgos refinery. The feed is gasoil from visbreaking unit, vacum gasoil from vacum distillation unit and mixture of atmospheric and vacum gasoil from storage tanks.

The design basis of the low pressure hydrocracker unit is given in Figure 6 and the thermophysical data of the streams are presented in Table 1.

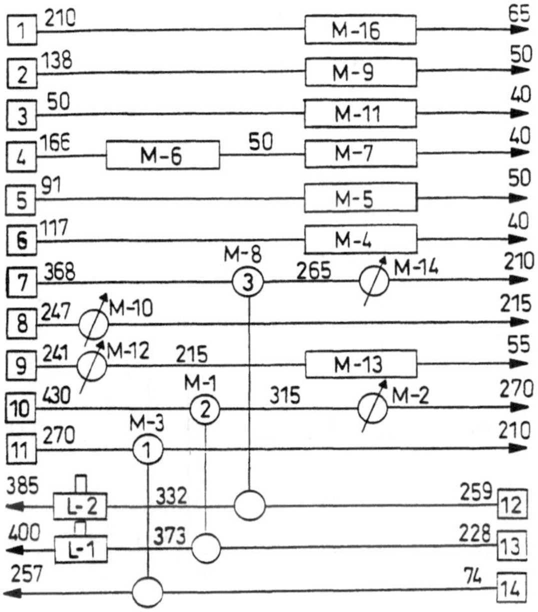

Fig. 6 : Design basis of Hydrocracker of low pressure

TABLE 1
Hydrocracker of low pressure : Thermophysical Data

Serial No	Temperature Deg C	C_p Kcal/hr C
1	210 - 65	5.669
2	138 - 50	75.807
3	50 - 40	1.100
4	166 - 50	27.233
	50 - 40	13.700
5	91 - 50	168.195
6	117 - 40	1.818
7	368 - 265	75.660
	265 - 210	67.582
8	247 - 215	93.750
9	241 - 215	25.769
	215 - 55	21.663
10	430 - 315	177.574
	315 - 270	154.422
11	270 - 210	44.950
12	259 - 332	106.753
	332 - 385	132.385
13	228 - 373	140.834
	373 - 400	156.593
14	74 - 257	14.738

The problem definition table, gives the following results : T_{ph} = 247 deg C, $Q_{c,min}$ = 2546599 Kcal/hr and $Q_{hot,min}$ = 1055.83 Kcal/hr

U_{min} = 8 (Minimum number of heat exchangers - Euler theorem)

In fig.7 the design of the maximum energy recovery network, is given.

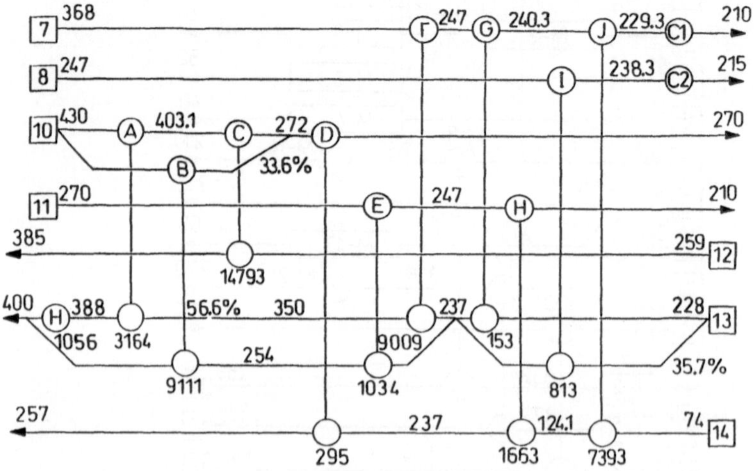

Figure 7.The maximum energy recovery network for the low pressure Hydrocracker.

A new technique for the construction of a matrix ("**DESIGN MATRIX - DM-**") is introduced. This technique is the thermodynamic representation of the maximum energy recovery networks.

We present the DM (-DESIGN1-) matrix as follows :

$$
DESIGN1=\begin{bmatrix}
7 & 368 & 0 & 0 & 0 & 0 & 0 & F & G & 0 & 0 & J & 210\\
8 & 247 & 0 & 0 & 0 & 0 & 0 & 0 & 0 & 0 & I & 0 & 215\\
10 & 430 & A & B & C & D & 0 & 0 & 0 & 0 & 0 & 0 & 270\\
11 & 270 & 0 & 0 & 0 & 0 & E & 0 & 0 & H & 0 & 0 & 210\\
12 & 385 & 0 & 0 & C & 0 & 0 & 0 & 0 & 0 & 0 & 0 & 259\\
13 & 400 & A & B & 0 & 0 & E & F & G & 0 & 0 & 0 & 228\\
14 & 257 & 0 & 0 & 0 & D & 0 & 0 & 0 & H & I & J & 74
\end{bmatrix} \qquad (8)
$$

The lines of the above matrix include :

The serial number of the streams, their initial and final temperatures and the places (non zero elements) of the heat exchangers.

A loop is called a "**n-th class loop**", when the maximum number between hot and cold streams which exchange heat loads, is -n-.

So the "**SIMPLE LOOP -SL-**" (Fig. 8a) is characterized as a 1-st class loop and the "**SIMPLE TRANSITIONAL LOOP -STL-**" (Fig. 8b and 8c) as a 2-nd class loop.

In the above matrix -DESIGN1- we recognize only the two following simple loops :

[SL1] = [A,B] & [SL2] = [F,G]

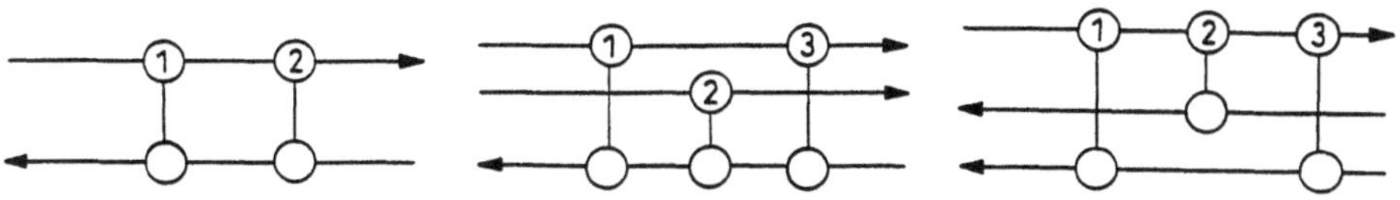

(a) -SL-　　　(b) -STL-　　　(c) -STL-

Figure 8. Schematic form of -SL- and -STL-

After recognizing all the -SL- & -STL- loops, the above matrix takes the following form (matrix DESIGN2), a form which helps us to find all the n-th class loops (n>2)

$$DESIGN2 = \begin{bmatrix} 0 & 12 & 13 & 14 & 0 \\ 7 & 0 & SL2 & J & CU1 \\ 8 & 0 & I & 0 & CU2 \\ 10 & C & SL1 & D & 0 \\ 11 & 0 & E & H & 0 \\ 0 & 0 & HU1 & 0 & 0 \end{bmatrix} \quad (9)$$

The above matrix includes both the places of hot (HU) and cold (CU) auxiliary loads. In order to find all the 2-nd class loops, we investigate the non zero different elements (i,j), of the DESIGN2 matrix.

We define two numbers -k-, -l- (k < NH and l < NC) in such a way that the parallelogram [(i,j),(i,l),(k,j),(k,l)] has non zero apexes.

All these different parallelograms define 2-nd class loops. In the above DESIGN2 matrix we recognize the following loops.

[SL2-CU1-CU2-I]　　[SL2-SL1-D-J]　　[SL2-J-D-SL1]

[SL1-D-H-F]　　[SL2-F-H-J]

To recognizing all the different loops, it is a very difficult step. Two loops will be different when their breaking leads to different design networks.

We define as **"INCORPORATED ELEMENT -IE"** the heat exchanger with the lower heat load.

To make a breaking of a loop, we add and substract the heat load of (IE) to the loads of the other elements, so that the enthalpy balance won't be changed.

With this procedure and starting every time from a different non zero element, the algorithm produces families of near optimum heat exchangers networks

With the same procedure (parallelogram method) the algorithm searches for the n-th class loops (n > 2).

In figure 9 these loops are presented.

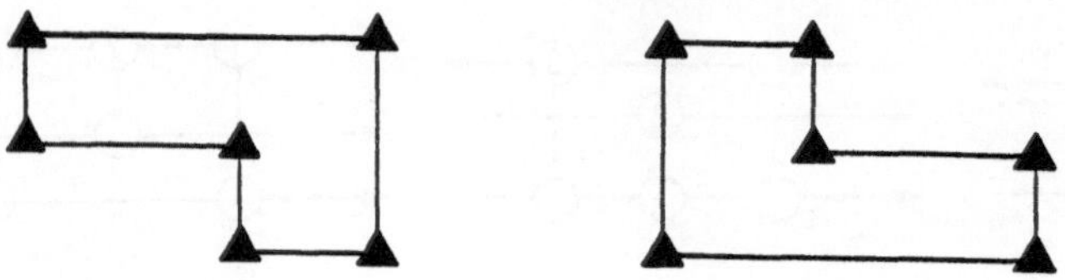

Figure 9. n-th class loop with n > 2

The final algorithm step is the correction of the DT_{min} faults. This correction can be completed with the stream splitting and the transportation of thermal charges along the flow paths. In fig. 10 a simple network is presented.

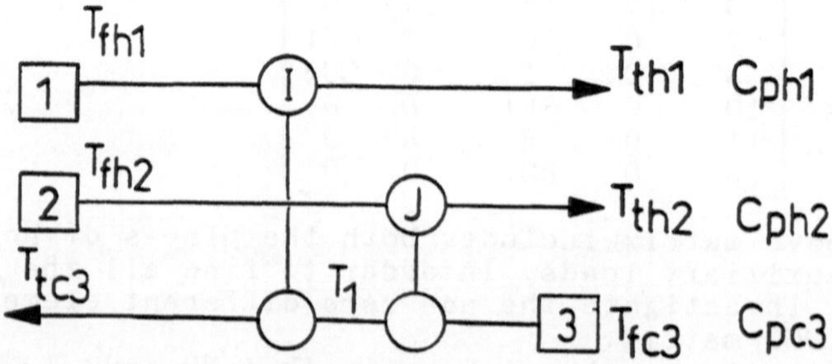

Fig. 10 : Representation of a simple network

The outlet temperature $-T_1-$ is given by the equation (10)

$$T_1=(T_{fh2}-T_{th2})*(C_{ph2}/C_{pc3})+T_{fc3} \quad (10)$$

If $T_{th1} < T_1$ then the rule : "The hot stream must always be kept hotter than the combined cold stream", is not valid (DT_{min} fault at heat exchanger -I-) and it is necessary (for thermodynamic reasons) to divide the cold stream (3) into two pieces (Fig. 11)

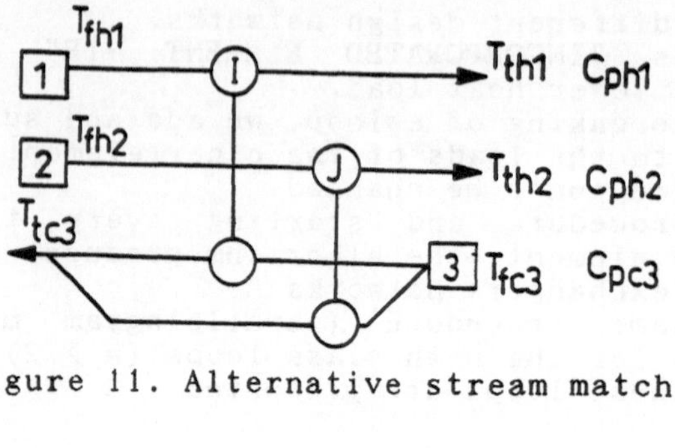

Figure 11. Alternative stream matching

Figure 12 shows the "FLOW PATH" between streams, heat exchangers and the auxiliary loads. These loads can be of the same (Fig. 13a) or of different kind (Fig. 13b).

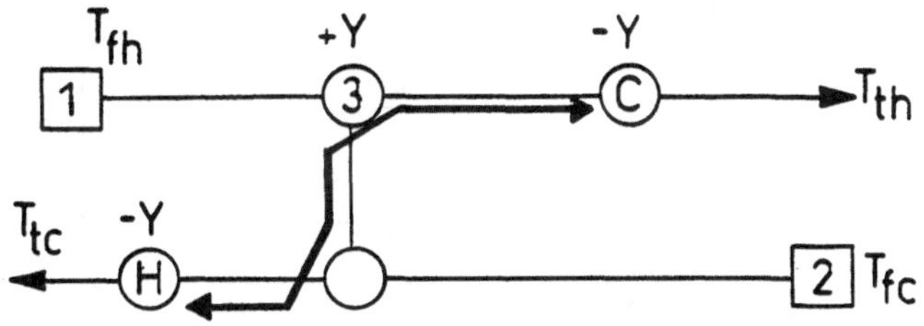

Figure 12. Flow Path definition

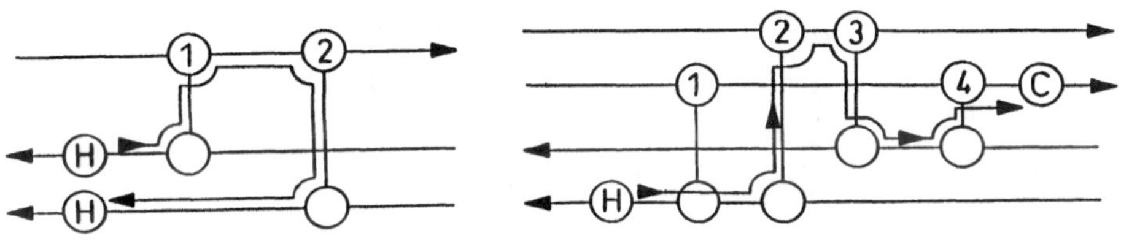

(a) (b)

Figure 13. Flow path between the same (a) or of different (b) kind of auxiliary heat loads

It is true that if we have a reduction of -Y- thermal units of the auxiliary hot load (H), then we must increase the charge of the No 3 heat exchanger (Y thermal units). At the same time, the auxiliary cold load must be reduced with the same amount (Y thermal units). This transportation of thermal loads along the flow paths, is very important as it corrects all the DT_{min} faults.

All the above techniques, finally lead to the following family (Fig. 14) of near optimum heat exchanger networks for the industrial problem of hydrocracker of low pressure. Table 2 presents the duties of the above networks (family) of heat exchangers.

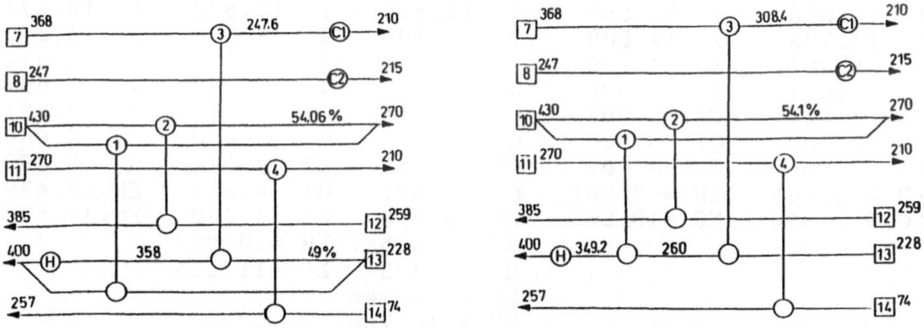

Proposed Network 1 Proposed Network 2

Fig. 14 : The proposed networks for the low pressure Hydrocracker industrial problem

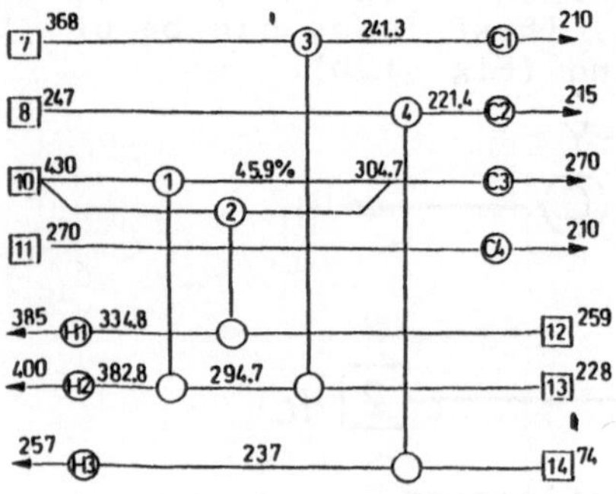

Proposed Network 3

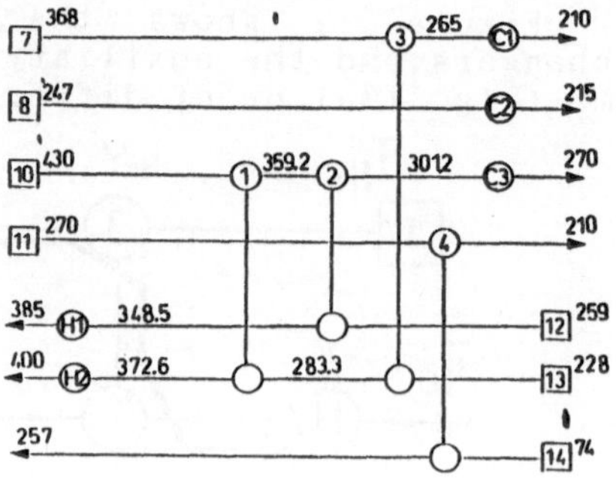

Proposed Network 4

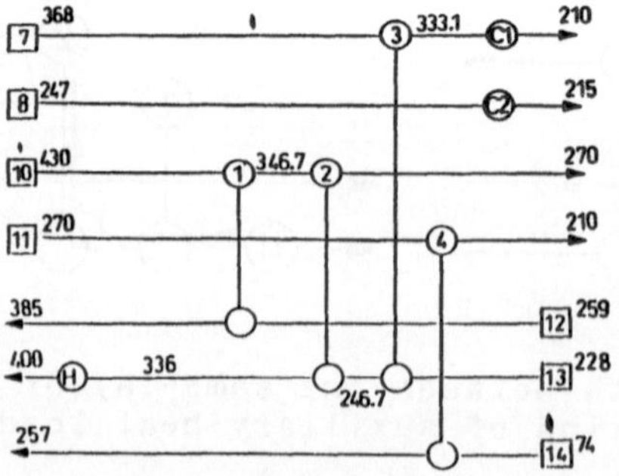

Proposed Network 5

Figure 14. (Cont.) The proposed networks for the low pressure Hydrocracker industrial problem

TABLE 2
Thermal Charges (MMKcal/hr) of the low pressure Hydrocracker proposed networks 1,2,3,4 & 5

NETWORK 1	NETWORK 2	NETWORK 3	NETWORK 4	NETWORK 5
1 12.570	1 12.560	1 12.560	1 12.570	1 14.793
2 14.793	2 14.809	2 2.402	2 9.975	2 12.570
3 8.967	3 4.506	3 9.450	3 7.793	3 2.634
4 2.697	4 2.696	4 9.392	4 2.697	4 2.697
C1 2.544	C1 7.003	C1 2.115	C1 3.717	C1 8.877
C2 3.000	C2 3.000	C2 0.597	C2 3.000	C2 3.000
H 3.106	H 3.106	C3 5.358	C3 4.818	H 9.439
ΣH = 3.106	ΣH = 7.581	C4 2.697	H1 4.818	ΣH =9.439
ΣC = 5.554	ΣC =10.003	H1 5.358	H2 4.280	ΣC=11.877
		H2 2.695	ΣH = 0.098	
		H3 0.295	ΣC =11.535	
		ΣH = 8.348		
		ΣC = 10.768		

ΣH = Hot auxiliary load
ΣC = Cold auxiliary load

CONLUSIONS

In the above proposed design method, we have attempted a combination of the Linnhoff method advantages with these of the new heuristics - criteria. The analysis of the problem, the determination of the auxiliary loads and the pinch point, are done following the procedure of the "problem definition table" as this has been proposed by Linnhoff [4].
Based on the above characteristics a design algorithm has been developed able to face not only the bibliographic problems but the real industrial problems as well.
The choice of matches is done under the three following conditions :

1.- Orthological use of available driving forces (DT)
2.- Production of simple networks (the lower number of stream divisions)
3.- Networks with the maximum degree of flexibility (ergonomic networks)

The application of the proposed algorithm, both in bibliographic and real industrial problems, has shown that it can achieve these goals quickly and safely.
Finally, it should be stressed, that an important element for the success of this method as well as of every other method, is the good knowledge of the specific characteristics and special demands of the problem under consideration.

NOMENCALTURE

C_{ph} = Thermal capacity of hot stream
C_{pc} = Thermal capacity of cold stream
C'_{pc} = Fictitious thermal capacity of cold stream
DT_{min} = Minimum allowable temperature approach
DTH, DTC = Driving forces. Diagrams [T_{hot}-DT], [T_{cold}-DT]
DT, DT_i = Temperature difference between hot and cold streams
NH, NC = Number of hot and cold streams respectively
N = Number of heat exchangers of the design basis
$Q_{hot,min}$ = Minimum hot auxiliary load
$Q_{cold,\ min}$ = Minimum cold auxiliary load
T_{fh}, T_{th} = Initial and final temperatures of hot streams
T_{fc}, T_{tc} = Initial and final temperatures of cold streams
T_{ph}, T_{pc} = Pinch Temperature for hot and cold streams ($DT_{min}=T_{ph}-T_{pc}$)
T_{pp} = Pseudo pinch point temperature
U_{min} = Minimum number of heat transfer units

REFERENCES

1.- Andreou I. : "Design of optimal heat Exchanger Networks", PhD Thesis, National Technical Univercity of Athens (NTUA) 1989
2.- Linnhoff B. : "Thermodynamic analysis in the design of process Networks", PhD thesis, University of Leeds, 1979
3.- Linnhoff B., Hindmarsh E. : "The Pinch method for heat exchanger networks", Chemical Engineering Science, Vol. 38, No 5, 1982, pp 745-763
4.- Linnhoff B. : "A user guide on process integration for

efficient Use of energy". Institution of Chemical Engineers, 1982

APPENDIX A

The heat capacity surplus Criterion Proof for the above pinch region

When one hot -i- and one cold -j- stream are matched in the vicinity of the pinch point, then :

$$C_{pc,j} \geqslant C_{ph,i} \qquad (A.1)$$

Adding all the above inequalities, we take the following one :

$$\sum_{j=1}^{NC} C_{pc,j} - \sum_{i=1}^{NH} C_{ph,i} \geqslant 0 \qquad (A.2)$$

After the first match the above expression (A.2) must be also valid, so :

$$\left[\sum_{j=1}^{NC} C_{pc,j} - C_{pc,j}\right] - \left[\sum_{i=1}^{NH} C_{ph,i} - C_{ph,i}\right] \geqslant 0 \qquad (A.3)$$

or finally :

$$\sum_{j=1}^{NC} C_{pc,j} - \sum_{i=1}^{NH} C_{ph,i} \geqslant C_{pc,j} - C_{ph,i} \qquad (A.4)$$

PARADIGMS IN PROCESS INTEGRATION
A TECHNICAL AND EPISTEMOLOGICAL APPRAISAL

GEORGE J. PROKOPAKIS[1] and VASSILIKI KINDI[2]
[1]SPEC Ltd, 75 Patission, GR-10434, Athens
[2]Philosophy Department, Deree College, GR-15342, Aghia Paraskevi

ABSTRACT

Two main schools of thought are identified within the field of Process Integration: (a) Pinch Technology and (b) Mathematical Programming Approach. Their development is briefly presented, along with their basic methodological characteristics. It is argued that, in addition to their technical-methodological differences, the two trends are characterised by different notions, criteria, and committed followers to the extent that each one may be considered what philosophers of science call a Paradigm. The consequent scientific, technical, and sociological issues are examined.

INTRODUCTION

Within the field of Chemical Engineering and in response to the oil crises of the 70s and 80s and the consequent need for more efficient and economical designs of process plants, the research activities in Process Integration flourished, to the extent that a sub-field has emerged. Currently, two main currents may be identified within the field of Process Integration: (a) pinch technology and (b) mathematical programming methods. Although it appears that the former enjoys greater acceptance from the industrial end-users, the relating research activities are to a large extent localised (primarily in the UK). The latter is the subject matter of the research activities in numerous research establishments in the USA and to a lesser extent in Europe. Furthermore, since the early 80s, when the mathematical programming techniques made their entrance, the communication between the two schools of thought has been very limited.

Within each current, one may identify a framework of specific notions, theories, applications, lexicon, along with explicitly stated methods, rules, and criteria. This framework is complemented with tacit presuppositions, prejudices, and committed followers. Each school of thought may be considered what the philosophers of science call a **Paradigm**. In this work, a technical appraisal of the two approaches is given, along with an epistemological account.

HISTORY

Pinch technology (PT) made its formal appearance in the literature in 1978, [1]. Ever since, almost all relating activities bear the seal of Professor Bodo Linnhoff and his co-workers. The fundamental contribution of the Linnhoff approach is the recognition of the existence of the pinch point, the "point" in the temperature versus heat load (or similar coordinates) diagram, through which no heat is transferred at maximum energy recovery. The pinch point decomposes the system (plant) into sub-systems as far as the transfer of energy is concerned. Following this observation, techniques have been developed to manipulate the energy picture to identify feasible matches for the exchange of energy and obtain near optimal solutions to a large number of process integration problems.

Although Linnhoff and his co-workers were the ones who developed the pinch concepts to the extent that we now talk about *Pinch Technology*, their origin may be traced to German Mechanical Engineering of the post war era. The first use of the composite curves (name given later by the Linnhoff group) was made by Hohman in 1969. Subsequently, Umeda et al., [2], made use of the composite curves to locate side reboilers and condensers in distillation columns. The real maturity of the technology came after the formation of the Centre for Process Integration at UMIST (1982) and of Linnhoff March Ltd. The former plays an important role in supporting R&D activities, disseminating the technology, and providing channels for the industrial feedback, and the latter in commercialising the technology.

Predecessors of mathematical programming (MP) approaches were early attempts in the mid-70s to solve the complex, combinatorial in nature, problem of designing heat exchanger networks. These efforts aimed at solutions by using sheer computer power, and were exercises in futility. The break came after it was realised that the problem has its equivalent in the field of Operations Research, namely it is a problem of resource allocation. Interestingly enough, the first paper in this direction in the Chemical Engineering literature is co-authored by Linnhoff, [3]. Four milestones may be recognised in the area: (a) the introduction of the transshipment model, [4]; (b) the introduction of the notion of the *superstructure*, [5]; (c) the use of the MINLP formulation as a viable tool for the solution of process integration problems, [6]; (iv) the interface of integration methodologies with process simulators for detailed calculations, [7].

THE METHODOLOGIES

The Nature of the Problem

Process integration deals with complex optimisation problems characterised, in general, by two key aspects:

1. selection of structurally different alternatives, introducing combinatorial characteristics,
2. optimisation of each one of the alternatives with respect to the design parameters.

In many instances, the problem may have multiple solutions and/or need to be dealt with subject to different, often competing, criteria.

The Pinch Approach

The approach is "physical" in the sense that, the problem is decomposed into sub-problems, by first determining the optimal energy requirements allocation (it can be dealt within a thermodynamic framework) and, then, by using tools and insights developed on an ad hoc basis establishes targets for the capital expenditures. The approach may be characterised as evolutionary, since it is based on the solution of a core problem around which a complete solution is built. The approach is largely dependent upon the intervention of the engineering practitioner in the solution procedure and cannot guarantee optimality.

The Mathematical Programming Approach

The approach is "mathematical-numerical", since a complete model is developed and its rigorous solution is attempted by numerical means. It calls for the construction of a *superstructure* (i.e., a super flowsheet in which all structural alternatives are present) involving discrete variables, to account for the combinatorial decisions (i.e., a process unit exists or not in the solution), and continuous variables (design variables of process units). Then, the problem is split into two, a master problem involving the discrete variables and a linearised model for the flowsheet, and a rigorous non-linear optimisation model involving the process parameters only. The iterative solution of the two problems is theoretically guaranteed to lead to the optimum in a finite number of iterations. The quality of the solution depends on the completeness and soundness of the model.

CROSSFIRE

It is very interesting to realise that the cross-reference of the published work between the two schools of thought is minimal. References are regularly made only to compare solutions obtained by either approach. MP references in PT publications are virtually non-existent. PT references in MP publications are made almost exclusively on basic concepts (e.g., pinch point, temperature intervals). Although either one is fully aware of the other's developments, each current appears self-sufficient. Each trend has committed followers who tend to minimise the other's contribution. In the following the basic arguments will be reconstructed.

PT followers tend to consider MP as brute force approaches utilising computing power and tools to solve problems that may be solved rather easily using the pinch concepts. Although this argument holds true for the early attempts, it is completely unfounded for the later developments (after ca 1983). In the authors' view, it is exactly the incorporation of sound thermodynamic relationships in the MP approaches that allows for the efficient and effective use of modern computer technology and makes viable the solution of problems that cannot be solved in any other way.

MP followers tend to deny any originality to the PT. It is characteristic that "leaders" of the current often refer to the composite curves (a PT term generally accepted) as Hohman or Umeda curves! Furthermore, PT is accused of incompleteness in its methodological approach, since the emerging design depends on the subjective judgement and the experience of the practitioner. Finally, MP tend to attribute the success of PT to the commercialisation efforts of CPI and Linnhoff March. However, the litmus test for all development efforts is their successful integration in engineering practice; if it were only one contribution of the PT to be recognised, it should be the catalytic industrial role it has played and still plays.

AN EPISTEMOLOGICAL ACCOUNT

Having established the framework of coexistence of the two separate communities, let us resort to the field of philosophy of science in our attempt to shed some light to the issue. T.S. Kuhn, in his pivotal work *The Structure of Scientific Revolutions*, [8], introduced the term *normal science* to mean "research firmly based upon one or more past scientific achievements, achievements that some particular scientific community acknowledges for a time as supplying the foundation for its further practice." Furthermore, he introduced the, closely related to normal science, term *Paradigm*, to mean "achievements that are (a) sufficiently unprecedented to attract an enduring group of adherents away from other modes of scientific activity, and (b) sufficiently open-ended to leave all sorts of problems for the redefined group of practitioners to resolve."

With these two terms and the ensuing analysis, Kuhn means to suggest that some accepted examples of actual scientific practice - examples that include law, theory, application, and instrumentation together - provide models from which spring particular coherent traditions of scientific research. The characteristics of *normal science* are:

- the problems the practitioners face **do not** aim at major novelties, conceptual or phenomenal; the results gained in normal research are significant because they add to the scope and precision with which the paradigm may be applied;
- the solution to these problems is anticipated, the way to achieve this outcome, however, is very much in doubt;
- the research activities are actually *puzzle-solving* activities;
- the scientific community, having accepted a paradigm, acquires a criterion for choosing problems, which, for as long as the paradigm is considered valid, are assumed to have a solution;
- the practitioners are concentrated on problems, whose solution depends on their ingenuity;
- the scientific community and the corresponding paradigm are characterised by special notions, theories, applications, and lexicon.

For those familiar with the developments within the two currents over the last 15 years, the applicability of each one of the above characteristics to each one of the currents is common place. Apart from the first papers of the PT, no major novelty has been produced, although the body of work has established the approach as a major industrial tool. Furthermore, from paper to paper the next development was anticipated, either as an extension of the previous, or as an apparent industrial need. Similarly, after the introduction of the notion of superstructure and the establishment of the general framework within which MP approaches attempt the solution of problems, the research programmes revolved around extending the applicability of the concepts, validation of solutions, etc. Even the lexicon is different: notions such as composite and grand composite curves, central to the PT, are irrelevant within MP. Similarly, the notion of the superstructure or the master problem, central to the MP, are irrelevant within PT.

Science or Art?

To attempt an answer to the title question we resort to Karl Popper, another pivotal figure of the philosophy of science. Popper, [9], claims that, a theory deserves to be called scientific if it is *falsifiable* (i.e., it supplies in advance the conditions that, if satisfied, will render it false). Applying this criterion of scientificity to the issue at hand, we might claim that the pinch approach cannot by its nature determine in advance the conditions that would lead the adherents of the theory to reject it. If the achievement of a prespecified goal is the analogon of scientficity in the case of technology, we observe that the pinch approach does not have any means to evaluate the outcome of the design activity vis a vis the goal (the optimum solution).

The solution may be compared to other (previous) designs only. Hence, the approach fails the falsifiability criterion. On the other hand, the mathematical programming approach, by theoretically guaranteeing the optimum, satisfies the Popperian criterion.[1]

The scientific status of a theory is not the only factor that affects its acceptance. It also depends on factors external (at least in the strict sense) to the scientific community. In practical terms, the distinction established in the previous paragraph, bears very little significance, since all engineering research is called upon to serve industrial needs. In that respect, pinch technology has succeeded in transferring applied academic research to industrial applications in a very effective way. However, this distinction helps to understand a number of issues, both technical and sociological, when one compares the two approaches.

Technical Issues

The resolution of a number of technical issues within PT required the execution of specific research programmes, while they never became an issue within MP. We attribute this difference to the more coherent scientific nature of the latter. Examples follow:

- *sensitivity:* linear sensitivities are a keystroke away within MP after the solution is available, for any commercial linear programming package;
- *forbidden matches:* may be accounted for within MP by suppressing a variable;
- *stream splitting:* accounted for within MP by adding variables and constraints;
- *multiple utilities and grand composite curves:* right at the outset, [4], the utilities were treated as process streams;
- *area targets:* within the theoretical MP framework the notion bears no significance; recently, [10], targets were used as a means to facilitate the solution.

Sociological Issues

In this section we refer to the external issues that affect the direction of scientific research. Examples follow:

- to the authors' knowledge, not a single academic research project has been funded by the USA National Science Foundation in PT research since 1982; numerous projects have been funded in MP research; as a result, almost no academic research is being conducted in the USA in PT;
- at least three MP researchers have been named Presidential Young Investigators[2] in the USA since 1984, when the award was established; however, all of them needed at least

---

1 In this analysis it is assumed that the problem definition (including the model) in the two cases is identical

2 An award that provides for $25,000 of governmental funds annually for five years, plus an additional $37,500 per annum, provided that the researcher can secure matching funds from the industry, for a grand total of $100,000 per annum to be used at the researcher's discretion

eight months to secure the industrial matching funds; the Centre of Process Integration has received the prestigious Queen's award from the UK government and enjoys the continuous support of some 20 leading process industries;

- up to the recent JOULE II call for proposals, only one PT project had been supported from the CEC, in a topic marginal to the main PT line of research; CEC has supported one project on the development of a common environment for the application of MP and PT;
- PT can produce hundreds or even thousands of references of industrial applications; MP's actual industrial applications are limited to very large corporations with in-house computer support groups.

The above suggest that, although MP enjoys the academic recognition, the PT is winning the golden apple of industrial acceptance.

An Explanation

The above paradox may be attributed to the following reasons:

- MP emerged as a viable alternative too late; the process industry responded quickly to the oil crises, by making the required investments in process integration, using to a large extent engineering judgement and the then emerging PT know-how, leaving, thus, little room for significant improvements in the areas where MP could be more effective;
- the core model to MP, the transshipment model, has been solved in the Operations Research literature since 1963; it took twenty years for the transfusion to take place;[3]
- MP, being the result of academic research and without very much effort put in the technology transfer issues, was available to the practicing engineer as a monstrous amalgam of huge optimisation problems and computer jargon; on the other hand, PT offered simple tools, reduceable in many instances to back-of-the-envelope calculations;
- MP relies very much on theories and tools imported from a field (Operations Research) other that of the targeted end-user (Chemical Engineering); PT relies on basic thermodynamic concepts, drawn from the end-user's field of expertise;
- MP is virtually impossible to become part of any Chemical Engineering undergraduate curriculum, as the latter stand now; PT is not only an excellent subject to teach upperclassmen, but it is a tool to integrate the process related concepts taught in basic courses (material and energy balances, thermodynamics, process engineering);

---

3 Two comments are appropriate here (a) the basic notions of the pinch point and the zero energy flux across the pinch have their exact equivalents in the transshipment model, in the max/min flow and min/max cut of the network, these are known since 1963 as well, (b) in a similar case, the SQP algorithm for nonlinear optimisation was introduced by Powell in an obscure conference in late 1977, and by 1979 it had already become the Chemical Engineering standard

- until the tools and environments of MP become friendly and familiar to the practicing engineer, no widespread acceptance of the technology may be expected.

CONVERGENCE?

It should be noted that very recently there is evidence that the picture painted above may be changing. Publications from the MP side appear to start making explicit use of PT developments (e.g., [10], [11]). More particularly, the notion of energy and area targets starts becoming part of the culture of MP. The reason is twofold: (a) a practical one, i.e., the insights gained from the PT targeting analysis may be used to facilitate the search for the optimum; (b) a marketing one, i.e., the industrial penetration is facilitated by the use of a lexicon in terms familiar to the client. It is not the first time that such a transfusion is made. In fact, it was the very notion of the temperature interval of Linnhoff that helped the MP researchers to realise the applicability of the transshipment model in process engineering. However, the extent of the use and the significance of the targeting concepts in the latest MP developments is such that, one may be optimistic about an eventual convergence of the two schools.[4]

The efforts by research groups to integrate tools from both approaches in a common environment should be mentioned as another form of convergence. Having realised the relative advantages of the two approaches, tools and methodologies applying either MP or PT are available to the user. The decision on the tools to be used is left with the designer. A notable example in this direction is *SYNEP*, [12].

CONCLUSIONS

In this paper we presented an account of the two competing schools of thought, both from a technical and an epistemological viewpoint, with emphasis on the latter. It is the authors' firm belief that both the field and industry have only to gain from a convergence of the two schools of thought. The "methodological superiority" of mathematical programming is not by itself sufficient to discard the pinch technology, moreover, since the latter has deep physical insights to contribute. The industrial acceptance of the pinch technology is not, in turn, by itself sufficient to discard mathematical programming.

If the two lines of research do not converge, the authors predict that they will coexist ad infinitum. The MP will become more manageable by the end-user and will gradually gain more industrial acceptance. However, for a long time to go it will remain the *rich man's tool*, i.e., will be used by industrial concerns that (a) can afford the overhead resulting from the

4 We cannot resist the temptation to mention that one of the co-authors of the latest MP papers has a solid background in PT.

required in-house support; and (b) deal on a regular basis with problems that MP is clearly superior (e.g., large scale heat integration and optimisation with detailed process simulation). On the other hand, PT will enjoy its leading advantage, but as the computer and optimisation culture spreads wider, will start becoming the *poor man's tool.*

In the authors' view, there is a great challenge and opportunity for all parties concerned (i.e., researchers, funding organisations, industry) for developments in the direction of the unification of the two approaches, making best use of the advantages of both. The CEC DG XII, with its recourses, its established networks, and its leadership in the direction of community-wide research, may play a pivotal role in meeting this challenge.

LITERATURE CITED

[1] Linnhoff, B. and Flower, J.R., Synthesis of heat exchanger networks: Part I. Systematic generation of energy optimal networks. AIChE J, 1978, **24**, 633

[2] Umeda, T., Hirai, A., and Ichikawa, A., Synthesis of optimal processing systems by an integrated approach. Chem Eng Sci, 1972, **27**, 795

[3] Cerda, J., Westerberg, A.W., Mason, D., and Linnhoff, B., Minimum utility usage in heat exchanger network synthesis - A transportation problem. Chem Eng Sci, 1983, **38**, 373

[4] Papoulias, S. and Grossmann, I.E., A structural optimization approach in process synthesis. Parts I, II, III. Comp & Chem Eng, 1983, 7, 695

[5] Andrecovich, M.J. and Westerberg. A.W., An MILP formulation for heat-integrated distillation sequences synthesis. AIChE J, 1985, **31**, 1461

[6] Duran, M.A. and Grossmann, I.E., A mixed-integer non-linear programming approach for process systems synthesis. AIChE J, 1986, **32**, 592

[7] Lang, Y.-D., Biegler, L.T., and Grossmann, I.E., Simultaneous optimization and heat integration with process simulators. Comp & Chem Eng, 1988, **12**, 311

[8] Kuhn, T.S., ***The Structure of Scientific Revolutions***, Chicago Univ. Press, 1962

[9] Popper, K., ***The Logic of Scientific Discovery***, Basic Books, New York, 1959

[10] Yee, T.F., Grossmann, I.E., and Kravanja, Z., Simultaneous optimization models for heat Integration - Parts I, II, III. Comp & Chem Eng, 1990, **14**, 1151

[11] Gundersen, T. and Grossmann, I.E., Improved optimization strategies for automated heat exchanger network synthesis through physical insights. Comp & Chem Eng, 1990, **14**, 925

[12] Marechal, F. and Kalitventzeff, B., *SYNEP1*: A methodology for energy integration and optimal heat exchanger network synthesis. Comp & Chem Eng, 1989, **13**, 695

PRACTICAL ASPECTS OF PROCESS INTEGRATION AND THEIR IMPLICATIONS FOR DESIGN

E.KOTJABASAKIS and I.D.GREMOUTI

Centre for Process Integration
Department of Chemical Engineering
University of Manchester Institute of Science and Technology (UMIST)
Manchester M60 1QD
United Kingdom

ABSTRACT

Pinch Technology has proved to be effective in developing the best integrated process designs for both new plants and retrofits. The cost benefits, in terms of both energy and capital, have been outstanding. However, there is a wide-spread belief that such significant savings, achieved through integration, must by necessity impair process flexibility.

Application experience has shown the opposite. Use of the recently developed techniques ensures that flexibility is incorporated into the integrated process design in the most cost-effective way. In this paper, two basic approaches for the design of integrated processes are discussed; Multiple Base Cases (MBC) and Parameter Ranges. Both approaches systematically explore the opportunities present during integration and therefore lead to designs, which are easier to operate and less expensive in terms of energy and capital than the less integrated designs.

A major spin-off from the work is a new approach to the design of heat exchanger networks subjected of fouling. Large potential savings have been identified.

INTRODUCTION

Flexibility is one of the main concerns during process design. For example, plants have to deal with a range of throughput, a range of product specifications, seasonal variations, and with such problems as catalyst deactivation and fouling. At the design stage some degree of flexibility must be introduced to ensure that the plant will be able to cope with uncertain parameters during operation.

A large part of the academic research on flexible process design has focused mainly on heat exchanger networks. This is probably so because the overall processes are considered too complex. In addition, temperature control is a major factor determining overall process operability. Thus, the design of flexible heat exchanger networks can yield valuable results at less than the full complexity.

The trend towards more highly integrated processes has fostered concern that operability and flexibility will suffer. A wide-spread assumption is that better integration means worse operability.

Experience has shown the opposite. Not only are highly integrated processes being operated, but also many turn out to be cheaper to build, cheaper to run, and easier to operate than their less integrated counterparts. Experience has also shown that process optimization should involve three aspects: capital costs, operating costs (mainly energy) and operability costs (throughput, product purity, etc.). All of these aspects need to be quantified and compared.

In this article, the work on flexibility conducted at UMIST for the past nine years is summarized. It will be shown that integrated processes do not need to give flexibility problems. Rather, they can offer opportunities. The procedures introduced are aimed at establishing the trade-off between energy, capital and flexibility. Two basic approaches have been adopted:

(i) Multiple Base Cases

(ii) Parameter Ranges

Unfortunately, it is rare, in practice, that the overall flexibility requirements of a process can be expressed by either the first or the second approach. These flexibility requirements are usually a mixture of both. Therefore, a design procedure has been proposed to consider this more general problem and it is also discussed in this paper.

MULTIPLE BASE CASES

With this approach, different cases of operation, known as Multiple Base Cases (MBC), are considered. Each base case operates for a specified period of time, has its own heat and material balance and its own values for all process variables. Even though this definition might not correctly reflect actual operational requirements, it has, however, the merit that the problem is well defined in easy terms and a solution can be attempted with reasonable effort, providing that the number of base cases is not excessive. It is also commonly used in industrial practice.

The design objective in multiple base case HENs is, as in a single base case, to minimise total cost (energy and capital). In addition, we should have feasible operation, which means that all stream target temperatures are at their desired values for all base cases.

Until recently, such targeting and design techniques were restricted to one base case. Jones[6] developed a targeting procedure for the design of heat exchanger networks under multiple base case operation. Total cost targets are calculated using targets for the utility consumption, the heat transfer area and the number of heat transfer units.

We then explore the energy and capital trade-off to find the minimum total cost target. This minimum total cost can be calculated for any combination of base cases. In this way, we can predict the cost implications of operating additional base cases. Flexibility targets, in terms of total cost, can now be easily and rapidly established.

The method not only optimises flexibility, but also helps us initialise the network design. As a result, the HEN so derived will require little or no evolution to find the minimum total cost.

MBC design types

MBC designs are classified into three types: conventional, resequence and repipe. The different structures exist because there are choices regarding the utilisation of heat exchangers across base cases. These choices are:

(i) In conventional designs exchangers remain between the same two streams and each stream encounters exchangers in the same order during the operation of all base cases.

(ii) In resequence designs exchangers are kept between the same two streams during the operation of all base cases, but streams no longer have to encounter exchangers in the same order. Resequencing exchangers give us the freedom to make better utilisation of capital.

(iii) In repipe designs no constraints are placed upon heat exchanger matching. Heat exchangers can be between different streams during the operation of different cases. This type of design gives us more freedom to utilise capital. Therefore, the total cost requirements are less than in resequence or conventional design types.

For each of the three design types discussed above, targets are calculated for the energy, the area and the number of units.

Energy targets

The energy targets for MBC operation, at a given set of permissible driving force (ΔT_{min}) values, can be calculated for each individual base case using the composite curves. The energy target is independent of the design type.

Area targets

The area target for a single base case is based on the vertical heat transfer model[5]. An **area matrix**[1] is used to show the distribution of area between streams. The sum of all the area contributions will always equal the area target. The area target in MBC networks is developed by the cross-comparison (using LP) of the area matrices across base cases. Each design type has its own area target. Details can be found elsewhere[1].

Units target

If a stream is located either totally above or totally below the pinch, then it must have at least one match associated with it. However, if a stream crosses the pinch, it must have at least two matches associated with it, for a maximum energy recovery design. To extend the argument to MBC, all that it is required are the stream contributions of all base cases. Each design type has its own units target. The algorithm is fully explained in reference 1.

Total cost target

The individual targets for energy, area and units are used to calculate the total cost targets for each design type. To determine the minimum total annualized cost, we use a multi-dimensional search. This involves a trade-off of energy against capital cost and identifies a set of optimum ΔT_{min}s, one for each base case. Each design type has a different capital/energy trade-off and therefore, different optimum ΔT_{min}s (see Fig.1).

Design initialisation

The optimum targets provide the required information for the correct initialisation of the design (see Figure 1). Starting from this initialisation, we can use new design tools that consider the design of all base cases simultaneously. Consequently, we can now systematically and relatively rapidly develop MBC designs. The MBC Design Method ensures that the targets will be achieved during design. Applications have shown that the total cost targets predict the minimum total cost of the optimum design with an error of less than 10%.

Flexibility targets

The total cost targeting procedure can be used to establish flexibility targets. For instance, consider the problem of optimising the flexibility when we have three possible operating scenarios:

1) Case 1 and Case 2.
2) Case 1, Case 2 and Case 3.
3) Case 2 and Case 3.

We can now rapidly target for the minimum HEN total cost of each operating scenario, assess the additional costs and choose the best operating scenario. These flexibility targets, shown in Figure 2, allow us to optimise the flexibility correctly in a realistic time scale ahead of detailed design.

PARAMETER RANGES: SENSITIVITY TABLES

The MBC approach only considers a limited number of the many possible combinations of circumstances. In practice, it may be necessary to provide flexibility for parameters that vary over defined ranges. We need a tool to examine and evaluate changes in sensitivity. Sensitivity Tables[2] is such a tool.

A realistic approach to the problem of designing for flexibility requires operating ranges of some parameters to be specified, rather than several "stiff" base cases. The main problem with this approach was that it is dependent on good initialisations to achieve optimal designs.

Sensitivity tables

Sensitivity Tables give the correlation between stream supply temperatures, flowrates, exchanger parameters, such as area, overall heat transfer coefficient, etc., and their effects throughout the network. They can be used for the rapid screening of sensitivities. The information contained shows the engineer, which exchangers in his process govern sensitivities, what limits are, and where the most cost effective locations for the additional investment are.

Three kinds of tables are required for the full analysis. They are:

T(TS) Sensitivity Table
T(CP) Sensitivity Tables
and T(UA) Sensitivity Tables

Tables 1, 2 and 3 show the Sensitivity Tables for a simple example. The T(TS) and T(CP) tables are used to establish the network response, whilst the T(UA) tables are used "backwards" to rectify design changes.

Discussion

The approach proposed is different from the multiperiod approach both in procedure and objective. Sensitivity Tables do not attempt to lead to rigorously best solutions. They give an approximate (but sufficiently accurate) overview of the non-linear sensitivities involved during the design of the flexible heat exchanger networks. They show which exchangers in the process govern flexibility and what the limits and sensitivities are. The engineer gains insight. This helps to quickly and efficiently screen out the cheapest option(s).

Change in the network structure cannot be readily identified using this approach. Evidently changes in structure, where appropriate, can only be identified using the MBC approach, and this is discussed next.

DESIGN PROCEDURE

A non-rigorous, yet practical, design procedure for flexibility is emerging. It recognises the advantages of both the parameter ranges and MBC models, and it is illustrated in Figure 3. The different process operating conditions must be first identified. The different base cases are then set up. These multiple base cases are used to optimise the three-way cost trade-off. The design so developed must be checked for feasible operation under all possible parameter ranges conditions. Small infeasibilities are eliminated by adding contingency: additional area for heat exchangers, additional plates for distillation columns, etc. If we have gross infeasibilities, then we must reassess the set up of the original base cases and produce a new design. Once all possible parameter ranges are feasible, we have a flexible, cost effective design.

REDUCING THE COST OF FOULING IN HEAT EXCHANGER NETWORKS

Fouling influences the way in which process plants are designed and operated. Practically every heat exchanger manufactured includes an allowance for fouling through the "fouling factor". All methods for reducing and predicting the effects of fouling are based on the implicit assumption that fouling is counteracted by overdesigning the fouling equipment.

An alternative approach has recently been developed(3). It starts with the observation that alternatives often exist for the selection of individual equipment sizes and trade-offs. The designer may specify one piece of equipment to be smaller at the expense of another one being larger. There are many good reasons why, if a piece of equipment is subjected to fouling, it should be made smaller rather than larger. This can be done by increasing the size of other pieces of equipment, which are not subject to fouling.

In other words, we advocate the following simple concept: we need to oversize to compensate for fouling but let us oversize the non-fouling equipment!

In heat-exchanger networks, installation of additional area at one point forces temperature changes and hence, changes the performance of the other units in the network elsewhere. The same amount of additional area installed in different exchangers has different effects, depending on network sensitivities. There is always a most cost effective location for additional investment. Sensitivity Tables are ideally suited to assess the effect of fouling on network performance and to decide on appropriate design changes. Changes in (UA) can be used to express fouling variations as much as area changes.

We can illustrate the approach with a simple example(3). A heat exchanger network is shown in Figure 4. Stream 3 contains a foulant, which is active once the temperature exceeds 125°C and as a result, only exchanger 1 is affected. With or without fouling, we have to achieve the final temperature of 175°C on stream 3. Employing the standard practice of oversizing the fouling exchanger, we find that exchanger 1 requires an additional 148 m^2 of surface area. An alternative form of direct compensation is the installation of a new heater at the outlet of the fouling exchanger.

The Sensitivity Tables show that oversizing exchanger 3 instead of the fouling exchanger 1 could also have the desired effect of maintaining this temperature. Calculation shows that no more than an additional 103 m^2 will fully compensate for the fouling occurring in exchanger 1. This is 30% less additional area than would be required for direct compensation in exchanger 1. Furthermore, exchanger 3 has a beneficial downstream effect on the heater on stream 4. As a byproduct of oversizing exchanger 3 rather than exchanger 1, the overall utility requirement is reduced by 15%.

We have demonstrated with this simple example how to compensate for fouling by exploiting heat exchanger network sensitivities. The gains achieved can be significant. Initial experience shows that use of this approach can save as much as one half of the total cost of fouling[3].

CONCLUSIONS

As mentioned in the Introduction, there is a wide-spread belief that integration must cause operability problems. This is no doubt true-if integration is specified non-expertly.

Integration and operability are certainly linked. There is a clear call for integrating not only our processes but also our approach in design to energy cost, capital cost and flexibility. The incentives are significant. Practical cases are reported elsewhere of reduced energy costs, reduced capital costs and improved flexibility[2,4].

REFERENCES

1. Jones, P.S., (1991). Targeting and Design for Heat Exchanger Networks under Multiple Base Case Operation. *PhD Thesis*, UMIST, England.

2. Kotjabasakis, E., and B. Linnhoff (1986). Sensitivity Tables for the Design of Flexible Processes (1). *Chem.Eng.Res.Des.*, 64, 197-211.

3. Kotjabasakis, E., and B. Linnhoff (1987). Better System Design Reduces Heat-Exchanger Fouling Costs. *Oil & Gas Journal*, Sept 28, 49-56.

4. Linnhoff, B., D.W. Townsend, D. Boland, G.F. Hewitt, B.E.A. Thomas, A.R. Guy, and R.H. Marsland (1982). *User Guide on Process Integration for the Efficient Use of Energy.* IChemE, Rugby, England.

5. Townsend, D.W., and B. Linnhoff (1984). Surface Area Targets for HeatExchanger Networks. *IChemE Annual Research Meeting*, Bath, U.K.

T(TS) - Sensitivity Table

	$\hat{TS}_1$	$\hat{TS}_2$	$\hat{TS}_3$	$\hat{TS}_4$
$\hat{T}_1$	0.333	0.334	0.0	0.333
$\hat{T}_2$	0.091	0.293	0.121	0.495
$\hat{T}_3$	0.0	0.333	0.0	0.667
⋮				

Example of use: If $\hat{TS}_1$ = +10°C, then, $\hat{T}_1$ = (0.333) x (+10°C) = +3.33°C.

Table 1: The "T(TS) Sensitivity Table" gives the correlation between changes in stream supply temperatures $(\hat{TS}_i)$ and the resulting changes of network temperatures $(\hat{T}_i)$.

T(CP) - Sensitivity Tables

- One for each stream

$\hat{CP}_i$	....	-20%	-10%	0%	+10%	+20%	....
$\hat{T}_1$		-11.2	-5.4	0.0	+4.9	+9.3	
$\hat{T}_2$		-9.9	-4.8	0.0	+4.3	+8.4	
$\hat{T}_3$		0.0	0.0	0.0	0.0	0.0	
⋮							

Example of use: If $\hat{CP}_i$ = -20% then, $\hat{T}_1$ = -11.2°C, $\hat{T}_2$ = -9.9°C,

Table 2: The "T(CP) Sensitivity Table" for stream i gives the correlation between changes in its heat capacity flowrate $(\hat{CP}_i)$ and the resulting changes of network temperatures $(\hat{T}_i)$.

T(UA) - Sensitivity Tables

- **One for each exchanger**

$\hat{UA}_i$	....	-20%	-10%	0%	+10%	+20%	....
$\hat{T}_1$		-12.6	-6.4	0.0	+3.9	+7.3	
$\hat{T}_2$		+2.9	+1.8	0.0	-4.3	-8.4	
$\hat{T}_3$		0.0	0.0	0.0	0.0	0.0	
⋮							

Example of use: If $\hat{UA}_i$ = +10% then, $\hat{T}_1$ = +3.9°C, $\hat{T}_2$ = -4.3°C, etc.

Table 3: The "T(UA) Sensitivity Table" gives the correlation between changes in (UA_i) of a given heat exchanger i and the resulting changes of network temperatures ($\hat{T}_i$) .

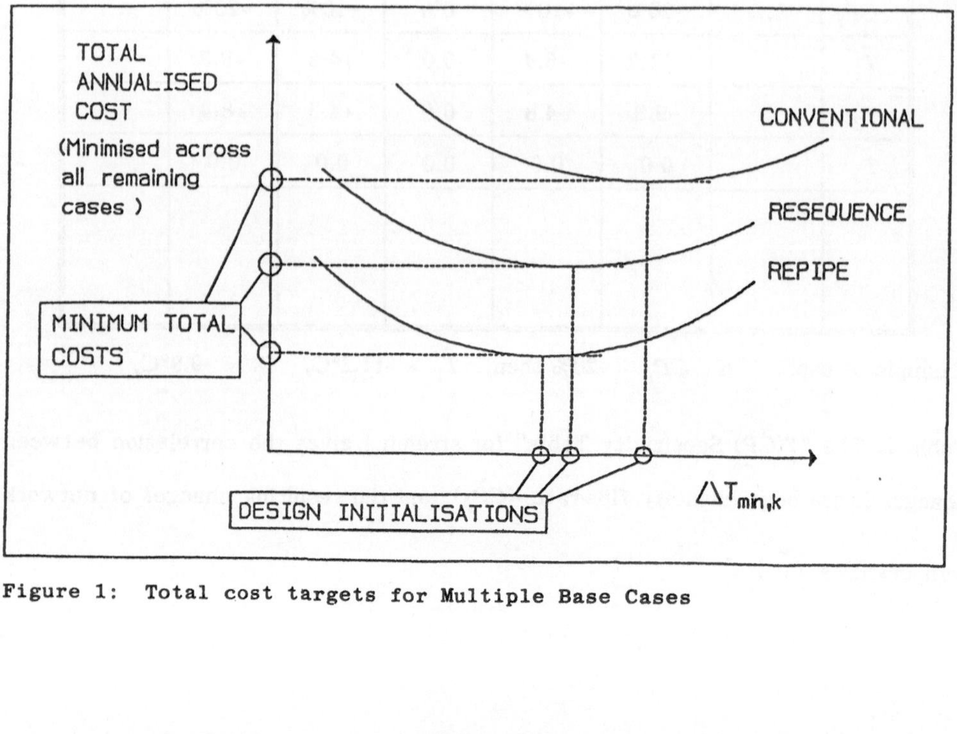

Figure 1: Total cost targets for Multiple Base Cases

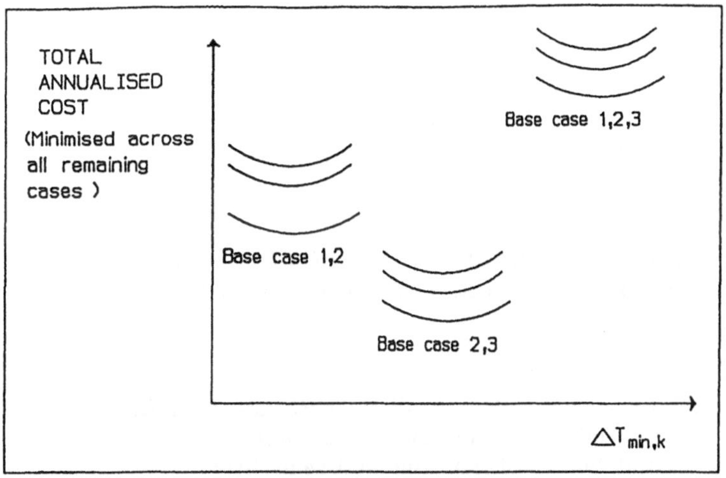

Figure 2 : Flexibility targets allow us to optimize the flexibility requirements ahead of design.

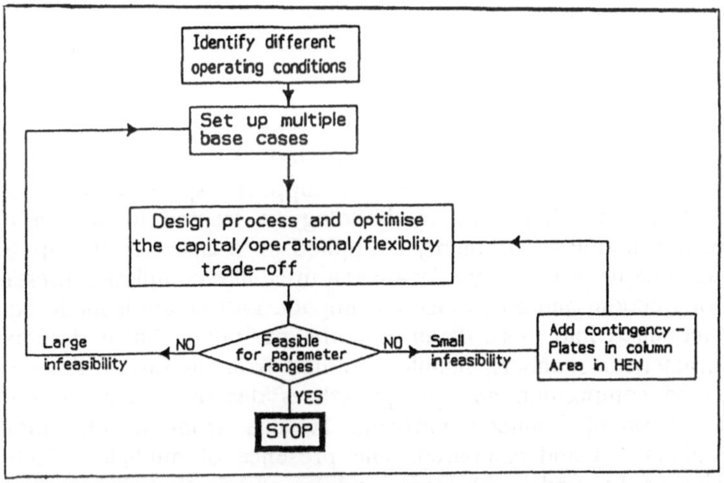

Figure 3 : A flexible process design procedure.

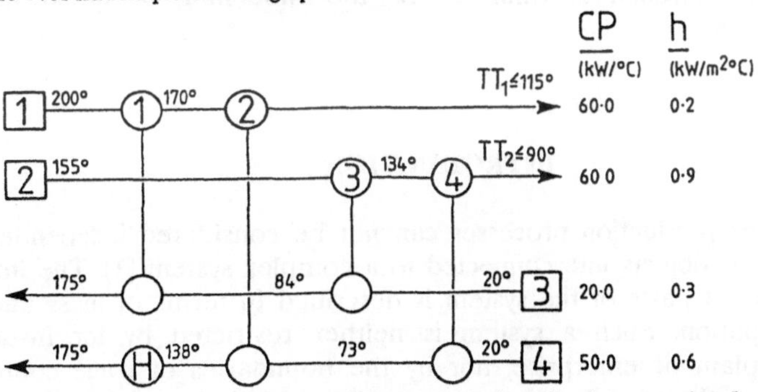

Figure 4: The heat exchanger network for a simple example

QUANTITATIVE MODELLING OF PRODUCTION CHAINS WITH RESPECT TO ENVIRONMENTAL EFFECTS

A.J.D. LAMBERT, W.T.M. WOLTERS and J. CLAUS
Eindhoven University of Technology,
Grad. School of Ind. Eng. and Man. Science,
P.O.B. 513, 5600 MB Eindhoven, NL

ABSTRACT

This study is devoted to mathematical modelling of production chains. Production chains are decomposed into modules that are treated in a similar way, based on mass- and energy flows. This is useful in decision-support, especially in the field of environmental performance. A general methodology is treated to have a proper interface between the modules. The theory is applied on a case, viz. the production chain of pork. This case involves many relevant features, like complexity, presence of internal degrees of freedom and environmental impact. Distinction is made between the core that is modelled into more detail, and the periphery that is described by more "static" characteristic numbers, mainly obtained from literature. As a core a combination of feed composition and pig growth is described, focused on cost, energy use, and flows of bounded nitrogen. Results from weakly non-linear optimisation are generated and compared. The presence of multiple objectives is discussed. Extension of the model with more modules and features is indicated. It is claimed that this approach is valuable for the modelling of broad classes of production processes.

INTRODUCTION

For many purposes production processes can not be considered independently, but should be treated as objects interconnected to a complex system [1]. The interaction between the different parts of the system is described in terms of mass and energy flows, and information. Such a system is neither restricted by for instance the boundaries of a plant or enterprise, nor by the boundaries of some country. The resulting concept is the so-called chain-approach, which is increasingly encountered in topics on industrial engineering as logistics, quality control etc.

The growing concern on environmental and ecological questions highlights another field of application for the chain approach. With respect to the effect of production on the environment, it is meaningless to make too sharp distinctions between different enterprises or production steps. It is necessary to analyze the effects on the environment of the production chain. This implies the study of the material flows over the whole trajectory (from source to sink, thus incorporating a number of production systems), starting at the extraction of the required resources and ending at the emission of all applied materials from, respectively towards the environment.

METHODOLOGY

To gain insight in the mass flows that are involved in the production system of the commodity to be considered, the status-quo is described. Because this is done on a large time scale, e.g. one year, a continuous description is used. It is build up by an alternating series of activities and commodities (fig. 1a).

Activities are (aggregated) production processes that transform flows of principal input materials with respect to place, time and quality, using flows of auxiliary materials including energy carriers. Thus flows of principal products and by-products are generated (fig. 1b).

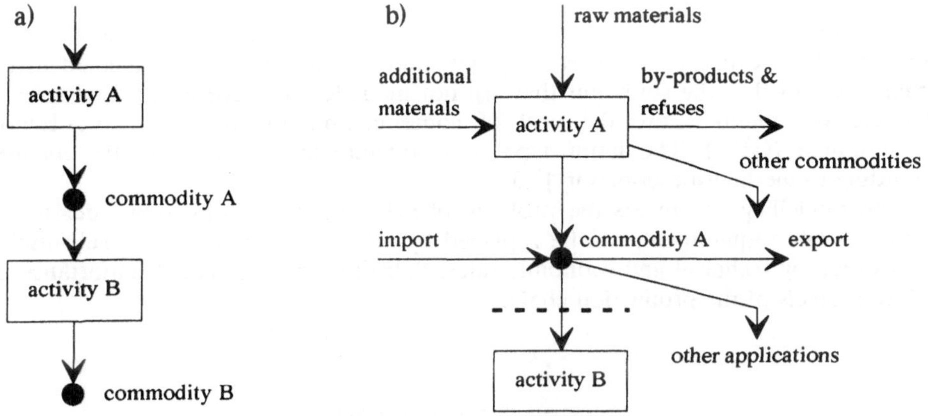

Figure 1. The alternating sequence of activities and commodities.

Commodities are treated as nodes. Incoming flows are production and import (from outside the system boundary). Outgoing flows are applications inside and outside the system boundary, including export.

Because of the strict separation between activities a modular structure of the model is obtained. The modules are coupled via nodes represented by the commodities. This makes it simple to extend the model or to change its structure.

Since mass conservation is valid, in both cases incoming and outgoing flows should be matched, apart from possible accumulation (neglected in this study).

The mass flows that occur in the production system are aggregations of various substances, with their own characteristics. In general these mass flows should be decomposed according to the purpose of the description. Because mass flows are interfaces between different activities, problems may arise. In the next section it is demonstrated how to cope with such problems in a specific situation. An additional complication is the fact that for the individual constituents of a mass flow generally no conservation exists. Chemical reactions, phase transitions and so on may take place that should be described conscientiously.

The activities impose a transfer function, i.e. a relation between input and output flows. To describe a status quo, this relation can be taken from statistical data ("top-down" approach). To describe changes in the production process, more knowledge about the characteristics of the activity is required in order to express it in a mathematical model ("bottom-up" approach). This should proceed in no more detail than meaningful to the problem to be solved.

Transfer functions are governed by parameters that express the physical state of the production system. Part of these parameters represent degrees of freedom, that can be manipulated according to some definite rules. These rules are mainly based on economic laws. Production takes place in enterprises that take decisions such that profits are optimal. This is translated in the model into maximum added value, which implies that the cost of the involved commodities plays an important controlling role. In the model the activities are described as mathematical enterprises, that optimise the added value, subjected to constraints that are imposed by quality requirements, legislation, environmental requirements and so on.

Information should be attached to the mass flows in the form of characteristic numbers. They express properties of production systems that are connected to the system described in the study but that are not modelled in more detail. Examples of characteristic numbers are the cost of some commodity and its Gross Energy Requirement (GER). The latter type of characteristic numbers can be obtained according to methods described in [2,3].

In modelling one meets the problem of substitution. A commodity, designated to meet some requirement, can be replaced by other commodities. This substitution is governed by technical and economic rules. Substitution is of crucial importance on different levels of the production chain.

CASE DESCRIPTION

The choice of the pork production system to demonstrate the methodology described in the previous section, is based on the following considerations:

1. It is a complex system (see fig. 2 for an indication) involving many production processes, often based on process industry.
2. It is a system enclosing large mass flows (fig. 3). In the Netherlands these mass flows are even a substantial part of the mass flows that occur in the total production system.
3. Data on various aspects of the production system are available.

4. It is a system with a large environmental impact. Especially the manure problem and the ammonia emission related to it are considered as restrictive for a sustainable development of the production system.
5. There are many internal degrees of freedom to change the system. Large efforts are devoted to the solution of the environmental problems related to pork production, leading to a decrease of emissions, an enhancement of the quality and applicability of emissions, and treatment of the emissions.

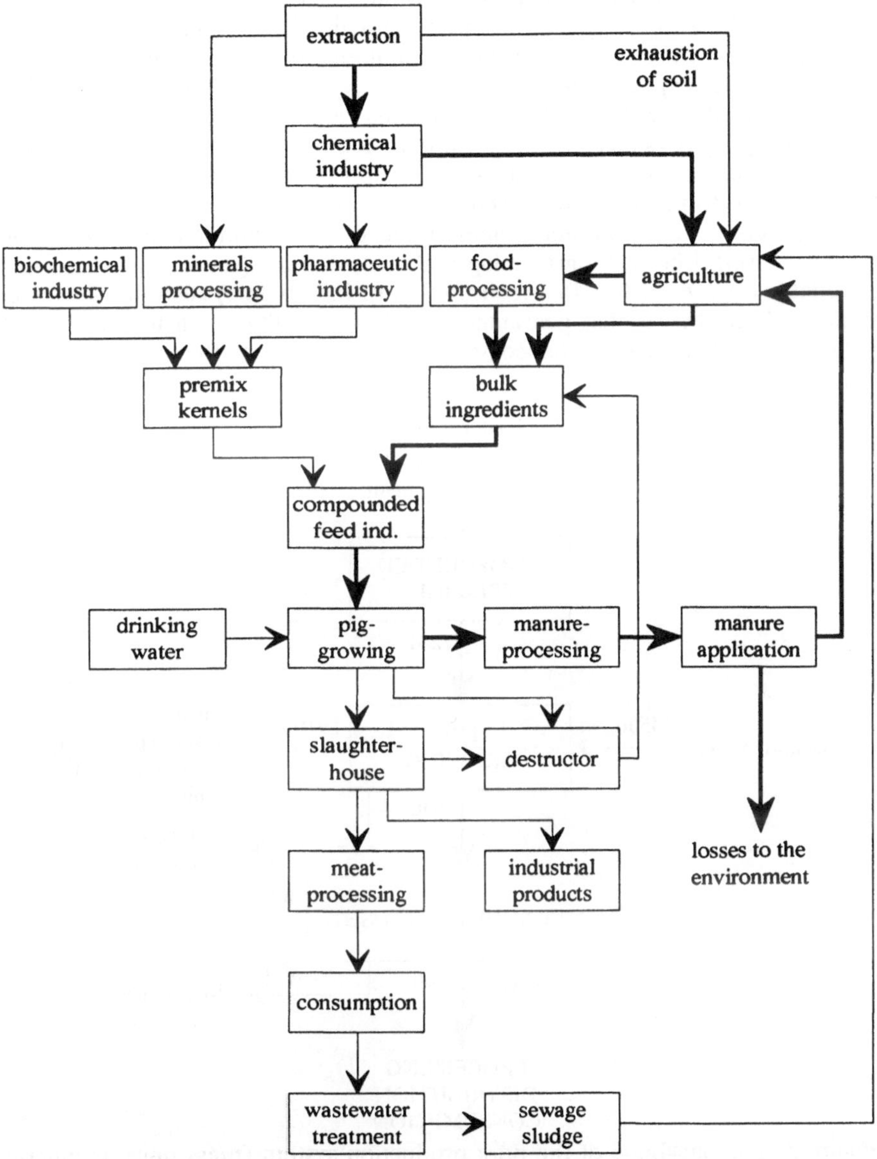

Figure 2. The activities in the pork production system

The modelling of complete networks of related production systems, of which fig. 2 only represents a simplified impression, is a complicated task. It is, however, possible to concentrate on the backbone of such a network. For pork production the backbone is represented in fig. 3. It involves compounded feed production, pig husbandry, and slaughterhouse. Mass flows in ktonnes/yr are indicated. They originate from statistical data.

Pigs are mainly fed by compounded feed. The remaining part consists of products like dried sugar-beet pulp and wet industrial by-products. Macro-ingredients of compounded feed are primary products (agricultural products submitted to simple transformations), secondary products (submitted to complicated and/or energy-intensive operations) and dried industrial by-products. These ingredients may originate from all over the world, and are thus responsible for a large consumption of fossil energy carriers due to transportation. Additives, especially vitamins, synthetic amino acids and enzymes, are of increasing importance. By using small quantities of them, some constraints may be relaxed.

In practice, the optimum composition of compounded feed is commonly determined using Linear Programming (LP) calculations (see e.g. [4]), manipulating the available degrees of freedom (substitution of ingredients) such that — subject to a number of constraints that guarantee nutritional quality — a feed is produced at minimum cost. This makes the composition change with fluctuations in cost and availability of the ingredient.

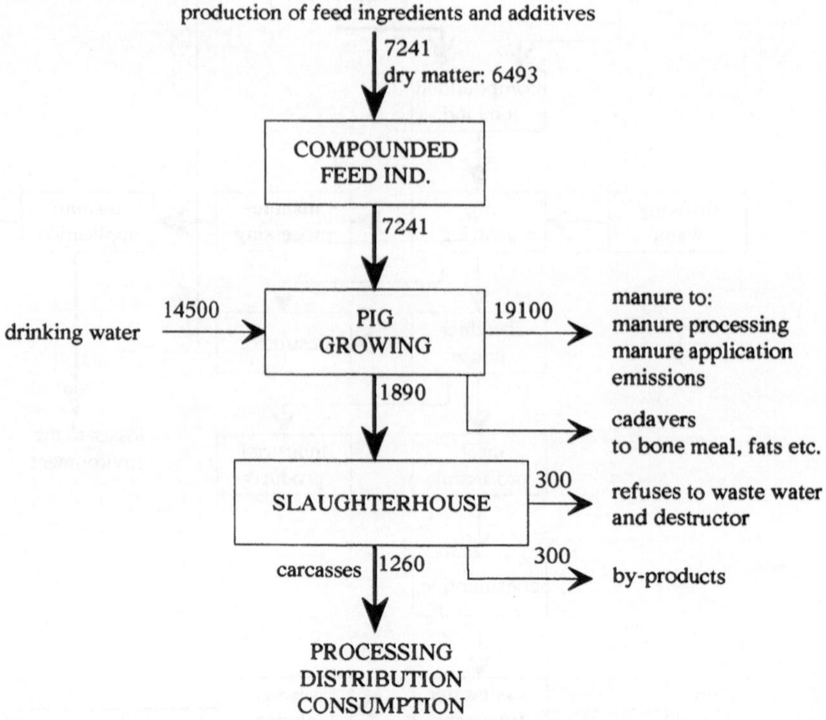

Figure 3. The backbone of the pork production system (mass units: ktonnes/yr).

Pig husbandry is separated in piglet production and pig fattening. Both activities have different characteristics. Detailed modelling of piglet production is difficult, more progress has been made in modelling of pig fattening (weight range: 23-105 kg). This work resulted in computer models, the so-called "pig simulators" like [5] and [6]. They are mainly designed for the farmer's use and return data on retention and slurry composition starting with given pig characteristics, feed quantity and composition. Piglet production is not good modelled till now. In such cases a top-down approach is the only possible approach.

Pig and sow slurries are essentially useful by-products. However, because of their quantity and water content their applicability is restricted. Storage and transportation provide considerable problems. Measures are taken to change manure composition by changing feed composition, minimising water consumption and applying feeds that are adapted to the weight of the fattening pigs (phase feeding). Moreover, there is a tendency towards manure-processing on an industrial scale, producing pellets that are easy to transport and manipulate. This end-of-pipe type of solution, however, is not self-supporting in the field of energy consumption and thus requires additional external energy.

Slaughterhouses are substantial users of water and energy, and they may release considerable amounts of contaminated water. Here a given raw material (living pig) is divided in a lot of mass flows, that are destined to fresh meat, meat products, recycling to feed, petfeed, pharmaceutical products, and other industrial products (brushes, gelatine etc.). Transport of meat and other products often proceeds over impressive distances and continuous refrigeration is required in the most cases.

THE MODEL

The description presented in the foregoing section indicates that it is elaborate to model the total production system. The intention is to build the model gradually and with modular structure. Modules can be combined provided the composition of the mass flows is such that two modules can communicate with each other. The extent of detail depends on the requirements of the user, on the relative importance of the respective module for the final result, and on the availability of reliable data.

With respect to the compounded feed ingredients, characteristic numbers have been taken from studies in energy analysis like [3] and [7]. Dutch statistical data on import and export of feed ingredients have been used [8].

To demonstrate the approach of our study two modules have been modelled and coupled. These modules involve the compounded feed industry and the fattening pig. Algorithms were the LP formulation for the compounded feed industry (aimed at determining a feed composition at the lowest cost), and the linear model of the pig simulator TMV [5] (returning the rate and composition of retention at given feed and pig characteristics). The combination of both models results in a new and flexible optimisation tool. It makes use of the available degrees of freedom to optimise with respect to various objectives including environmental criteria like emissions, occupation of area, and energy use. It is also possible to translate environmental requirements into constraints and, subsequently, to minimise costs.

The combination of TMV and the LP model of the feed composition results in a weakly non-linear model because the objective (e.g. cost related to retention) is a quotient of two linear expressions. The objective function is highly flexible. The model has been written in GAMS simulation language [9].

RESULTS

The model has been demonstrated by study of measures to decrease the amount of bounded nitrogen in pig slurry. These measures involve constraints on nitrogen content of animal feed. Constraints on essential amino acids then may become active. This means that synthetic amino acids should be applied to the feed in order to keep the solution feasible. Moreover, even in the case of a feasible solution, application of synthetic amino acids may lead to lower cost of the feed. Due to (bio)technological developments there is a tendency towards decreasing prices of synthetic amino acids, making it economically feasible to diminish the protein content of the feed. However, not the protein content of the feed is an objective, but rather the effectiveness of protein conversion. This leads to non-linear problems like minimising the ratio of the nitrogen content in manure to the one in retention (inverse protein effectiveness). Table 1 presents some results on optimising according to various objectives. The composition of feed A and B changes with the fluctuations in the prices of the ingredients. Feeds C through F were determined by optimising with respect to technical objective functions. Consequently they may be considered as theoretical limits.

TABLE 1
Optimum feed with respect to different criteria.

	Cost of feed (*f*/kg)	Cost of retention (*f*/kg)	Growthrate (g/day)	Inverse protein effectiveness	GER (MJ/kg feed)	GER (MJ/kg retention)
A	**0,345**	1,060	947.944	3.083	3.729	11.560
B	0,353	**1,052**	983.558	3.083	3.985	11.888
C	1,027	2,947	**1030.084**	3.083	13.828	40.004
D	0,543	1,639	976.531	**1.083**	4.205	12.738
E	0,505	1,581	937.491	2.526	**2.675**	8.385
F	0,517	1,537	987.167	2.405	2.799	**8.320**

GER = Gross Energy Requirement
Bold = optimisation criterion
*f*1.0 ≈ US$ 0.5

Table 2 shows the compositions of the optimum feeds presented in table 1. Note that these compositions are for fattening pigs at a weight of 60 kg. In the lower weight range (20-35 kg) and in the higher weight range (from about 90 kg upwards) there are significant deviations from these solutions.

TABLE 2
Composition of optimum feeds from table 1

Ingredient (g/kg)	A	B	C	D	E	F
Tapioca	400	400	400	400	-	-
Soybean meal	57	86	-	11	-	-
Barley	-	-	354	547	650	610
Peas	132	74	48	-	150	150
Maize glutenmeal	100	100	100	-	-	-
Sunflowermeal	100	100	-	-	-	-
Hominy feed	150	150	-	-	150	150
Molasses	50	50	-	-	50	50
Animal fats	11	40	40	40	-	40
Synth. lysine	-	-	2	2	-	-
Synth. methionine	-	-	-	0.0686	-	-
Synth. threonine	-	-	56	0.0137	-	-
Synth. tryptophan	-	-	0.283	0.360	-	-
Synth. isoleucine	-	-	0.0649	-	-	-

INTENDED EXTENSIONS

The model described here involves optimisation of feed composition, growth rate and composition of the retention of fattening pigs. The completion of the model is in progress and a first step is combination with calculation of the manure composition for fresh manure and for manure composition after storage. This proceeds using, among others, algorithms taken from [6]. The manure composition is starting point for manure application, including industrial manure processing. Modules for several methods of industrial manure processing are developed, with available models [10] as a starting point. These models, however, have fixed degrees of freedom and in the optimisation calculations described here, the possibility of variation of process parameters will be included. Moreover, we intend to go into more detail and focus on energy use, emissions, and applicability of the final product, i.e. pellets.

The slaughterhouse, and the use of the parts the pig is divided in, will also be included. Finally, to enable the evaluation of the sacrifices with respect to natural resources that have to be made to produce a specific product, in this case pork or pork products, a multiple criterion evaluation of a discrete number of feed compositions will be performed.

GENERALISATIONS AND CONCLUSIONS

To define the environmental performance of a product, one usually defines it by a set of characteristic numbers, like the GER, the cost or an indicator for the amount of emissions.

It is demonstrated that computer models of production chains are a powerful extension of this concept. They offer the possibility to study the influence of changing process parameters on these "static" characteristic numbers. The process parameters are in turn controlled by scientific-technological progress, by economic and market features, by prescriptions, and by legislation. Moreover, they can be considered as in instrument to study the consequences of new, e.g. environmental, measures to the production chain as a whole.

Because the construction of a complete model is complicated and reliable data are not always available, as a starting point a partial model has been described. It is indicated that the -always present- lack of knowledge and/or completeness can be by-passed using static characteristic numbers for the modules that are not modelled into detail. In an example it is demonstrated that optimisation according to different objectives leads to diverging values for the independent variables. Sensitivity analysis and multiple-criterium evaluation are the appropriate instruments to find optimum solutions that may meet all relevant criteria and objectives in a satisfying way.

The method described here is demonstrated on pork production. This production process, however, includes the features that are relevant to a broad class of other production processes.

REFERENCES

1. Mesarovic M.D. et.al., Theory of Hierarchical Multilevel Systems, Academic Press, 1970.

2. IFIAS, Energy Analysis, Report of: Energy Analysis Workshop on Methodology and Conventions, IFIAS, Stockholm (S), 1974

3. Boustead I. and Hancock G.F., Handbook of Industrial Energy Analysis, Ellis Horwood, Chichester (UK), 1979

4. Katzman I., Journal of Farm Economics, 1956, **38**, 420-429.

5. IVVO (Institute for Animal Feed Research) and others, Technisch Model Varkensvoeding (TMV-Technical Model Pig Nutrition), Rosmalen (NL), 1991 (Computer Program and Manual, in Dutch)

6. Aarnink A.J.A., and Ouwerkerk, E.N.J. van, Model for Estimating Volume and Composition of Slurry from Fattening Pigs, IMAG (Institute for Mechanisation, Labour, and Buildings), Wageningen (NL), 1990 (Computer Program and Manual, in Dutch)

7. Leach G., Energy and Food Production, IPC Science and Technology Press, Guildford (UK), 1976

8. CBS (Netherlands Central Bureau of Statistics), Agricultural Data 1990, Voorburg (NL), 1990

9. Brooke A. et.al., GAMS a User's Guide, Scientific Press, Redwood City (USA), 1988

10. Have P.J.W. ten, Een Drietal Mestverwerkingsroutes doorgerekend ten aanzien van Materiaal- en Energiestromen (Three Methods of Manure Processing, computed with respect to Material- and Energy-flows), IMAG-note 339, Wageningen (NL), 1988 (in Dutch)

DESIGN AND ENERGY ANALYSIS OF ABSORPTION-DRIVEN MULTIPLE EFFECT EVAPORATORS

STYLIANOS C. BIKOS and JOHN R. FLOWER
Department of Chemical Engineering, The University of Leeds,
Leeds LS2 9JT, U.K.

STAVROS YANNIOTIS
Department of Agricultural Industry, Agricultural University of Athens, Iera Odos 75,
GR-118 55 Athens, Greece.

ABSTRACT

Absorption-driven multiple effect evaporators have been reported to achieve significant energy savings compared to conventional systems. This paper explores limiting economic circumstances and heat integration options which complicate the decision between absorption-driven and steam-driven systems. A simple way of performing preliminary feasibility studies is illustrated through a low boiling point elevation example. A situation is presented where absorption-driven evaporators can never cost less than steam-driven ones.

NOTATION

A	heat transfer surface area, m^2
C	cost coefficient, (monetary units/yr)/(m^2 or kW)
CU	cold utility consumption, kW
DT	approach temperature
HU	hot utility consumption, kW
K	total cost of a system, monetary units/yr
K_{cu}	composite cooling cost coefficient, (monetary units/yr)/kW
K_{ab}	cost of absorptive solution, monetary units/yr
m_v	flow rate of vapour absorbed in absorber, kg/h
M_f	flow rate of absorptive solution fed to the absorber, kg/h
Q	heat load transferred from the absorber to the main evaporator, kW

U overall heat transfer coefficient, W/m^2/degC
V_{cold} vapour flow rate from coldest main evaporator effect, kg/h
x weight mass fraction of solute in aqueous solution, dimensionless

Subscripts

a refers to surface area of evaporator effects
ab absorptive solution
ad absorption-driven system, HEN excluded
at absorption-driven system, HEN included
c condenser
cu cold utility
hu hot utility
hx refers to surface area of a heat exchanger
sd steam-driven system, HEN excluded
st steam-driven system, HEN included
v vapour

Superscripts

h heat exchange network

INTRODUCTION

Absorption-driven (AD) multiple effect evaporators exploit the principle of a heat pump to reduce steam consumption. The hottest effect is heated by a concentrated hygroscopic aqueous solution which absorbs water vapour generated in the coldest effect. The dilute absorptive solution is itself regenerated in a multiple effect steam-driven (SD) evaporator. Figure 1 shows an example of an AD system with its regenerator (only the hottest and coldest effects of the main evaporator are shown).

Recent experimental work [1] reportedly identified savings of 30% to 45% in steam consumption and 50% in cooling water when compared to a similar SD unit [2] *"depending on the efficiency of sensible heat recovery in the regenerator, the heat losses in the regenerator, the flow rate of the absorptive solution and the overall heat transfer coefficient of the absorber"*. The problem was also identified of balancing the vapour requirement of the absorber and the capacity of the regenerator.

In this paper, the question of sensible heat recovery is seen in the context of total costs rather than energy savings only. A simple criterion for choosing between AD and SD evaporators is then suggested, based on results for various serial configurations of the main evaporator.

METHOD OF CALCULATION

Based on previous work [3], a computational model has been built to simulate absorption driven multiple effect evaporator systems. The principles and modelling of isobaric, diabatic absorption in a falling film absorber have been discussed by Yanniotis and Le Goff [2]. The range of operation of the absorber is limited by equilibrium, heat and mass transfer Coupling the absorber with an evaporator introduces tighter limits. A robust computational model of the process needs to obey these limits.

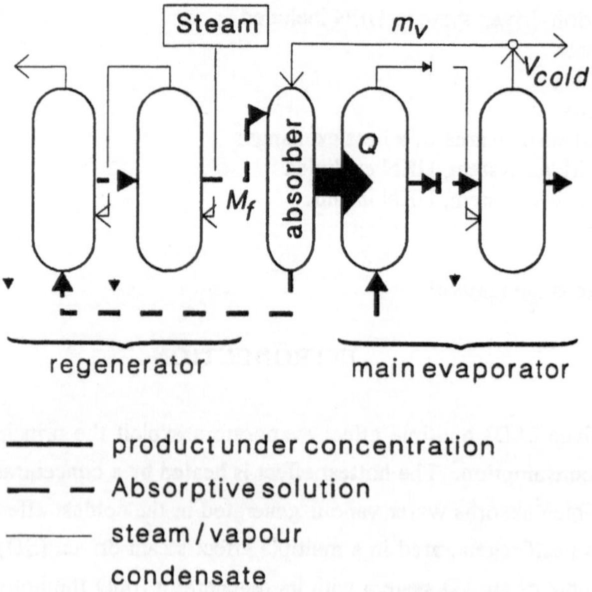

Figure 1. A simplified absorption-driven multiple-effect evaporator

Design calculations

The design algorithm of a multiple effect evaporator is based on the iterative solution of heat and mass balances [3]. Absorption calculations are necessary for the estimation of the area of the hottest effect. The surface area used for heat transfer across the walls of the falling film absorber is directly proportional to the height of the absorber, which is determined using the "Mass Transfer Unit" concept [2]. The area is estimated subject to a process-dependent overall heat transfer coefficient and respecting the limits discussed above. After the main evaporator calculation, the evaporation task of the regenerator is fully determined.

Sensible Heat Recovery

Sensible heat can be recovered from hot liquid streams flowing between evaporator effects and hot condensate streams flowing out of calandrias. This heat can be used to heat up cold liquid streams flowing between evaporator effects, including possibly a cold feed stream. The resulting heat exchanger network (HEN) introduces new hot and cold utility requirements and additional capital investment.

In this work the Problem Table Algorithm of Linnhoff and Flower [4] is used to identify the minimum energy requirements of the HEN. Heat transfer area targets are calculated by the methods of Ahmad and Smith [5]. A certain percentage of utilities is added to the minimum requirements to achieve practical heat transfer. This approach stems from the ideas for the use of a minimum energy flux [6] instead of a minimum approach temperature and provides explicit control of the energy consumption.

Cost Equations

To avoid the dependency of cost-related results on time and location, the rest of the discussion is directed towards identifying limiting economic scenarios which make the absorption driven evaporators cheaper, therefore preferable. Results are based solely on process related quantities and the reader is expected to feed his/her own economic data.

For simplicity, the total cost of a multiple evaporator system has been expressed as a linear expression of surface areas and utility requirements. The coefficients are the corresponding annualized marginal costs of these quantities.

If an SD evaporator is considered alone, its total annual cost is:

$$K_{sd} = C_{hu}HU_{sd} + K_{cu}CU_{sd} + C_{a}A_{sd} \tag{1}$$

In the absorption-driven systems the corresponding costs of the regenerator are included:

$$K_{ad} = C_{hu}HU_{ad} + K_{cu}CU_{ad} + C_{a}A_{ad} + K_{ab} \tag{2}$$

Other cost contributions are neglected as constant. The highest marginal costs are likely to be C_{hu} and C_a. Their ratio provides a convenient criterion for comparison between steam driven and absorption driven systems. The two types of system cost the same if:

$$\frac{C_{hu}}{C_a}^{*} = d_1 + \frac{K_{cu}}{C_a}d_2 + \frac{1}{C_a}d_{ab}$$

$$d_1 = \frac{A_{ad} - A_{sd}}{HU_{sd} - HU_{ad}} \qquad d_2 = \frac{CU_{ad} - CU_{sd}}{HU_{sd} - HU_{ad}} \qquad d_{ab} = \frac{K_{ab}}{HU_{sd} - HU_{ad}} \tag{3}$$

The annual cost of absorptive solution is neglected in the rest of the analysis as virtually constant. If the AD system consumes less hot utility than the SD system, $(C_{hu}/C_a)^*$ is the minimum ratio between steam and surface area costs that makes the AD system cheaper. Otherwise, $(C_{hu}/C_a)^*$ is the maximum such ratio.

If the heat exchanger network that accompanies an evaporator is also taken into account, the cost equations become:

$$K_{st} = C_{hu}HU_{st} + K_{cu}CU_{st} + C_a A_{sd} + C_{hx}A^h_{st} \quad (4)$$

and

$$K_{at} = C_{hu}HU_{at} + K_{cu}CU_{at} + C_a A_{ad} + C_{hx}A^h_{at} + K_{ab} \quad (5)$$

By neglecting K_{ab}, the ratio $(C_{hu}/C_a)^*$ is now given by:

$$\frac{C_{hu}}{C_a}^* = d_3 + \frac{K_{cu}}{C_a}d_4 + \frac{C_{hx}}{C_a}d_5 + \frac{1}{C_a}d'_{ab} \quad (6)$$

$$d_3 = \frac{A_{ad} - A_{sd}}{HU_{st} - HU_{at}} \qquad d_4 = \frac{CU_{at} - CU_{st}}{HU_{st} - HU_{at}} \qquad d_5 = \frac{A^h_{at} - A^h_{st}}{HU_{st} - HU_{at}} \qquad d'_{ab} = \frac{K_{ab}}{HU_{st} - HU_{at}}$$

RESULTS

The Example

The economic trade-offs when comparing absorption-driven and steam-driven evaporators are illustrated by a simple example (Table 1). Design calculations were carried out using the computational schemes outlined above. The inlet temperature of the absorptive solution was always taken at the middle of its feasible range. All systems were restricted to equal-area effects. The regenerator was taken in backward feed.

Case Studies

Each steam-driven system can have a number of corresponding absorption-driven systems where the regenerator part varies. In the following case studies the SD system is one where *no* sensible heat is recovered from liquid streams between effects.

Evaporator/regenerator systems: Assume an AD system where the liquid stream flowing between the two regenerator effects is heated up outside the regenerator (by other process heat or steam). The hot utility consumption of the SD system is larger than that of the equivalent AD system (Figure 2) for all the flow patterns examined. The AD system consumes roughly 40-50% less steam! For the AD system to be preferable in the widest possible range of economic conditions $(C_{hu}/C_a)^*$ must attain its lowest possible value. If that value happens to be negative, the AD system is *always* cheaper.

TABLE 1
Example data.

Main Evaporator	
Feed Flowrate, kg/h	200
Feed composition, w/w	0.20
Feed temperature, degC	70
Product composition, w/w	0.40
Product temperature, degC	40
Boiling point elevation, degC	5
Heat transfer coefficient, W/m²/degC	8.52(T+273)-1440
Hottest effect U, W/m²/degC	600
Regenerator	
Hygroscopic solution	NaOH
Inlet composition, w/w	0.50
Outlet composition, w/w	0.49
Heat exchanger network	
Hot utility temperature, degC	177
Hot utility film coefficient, W/m²/degC	2500
Cold utility temperature, degC	30-35
Cold utility film coefficient, W/m²/degC	1000
Process stream film coefficient, W/m²/degC	1300
Energy penalty, % of minimum required	10

Figure 3 gives the critical value of (K_{cu}/C_a) for each flow pattern examined, for which $(C_{hu}/C_a)^*=0$. All values are positive meaning that for each flow pattern there exists a critical ratio (K_{cu}/C_a) such that the AD system is *always* preferable to the SD system.

The coefficient K_{cu} is a lumped expression of both capital and running costs of the condenser: $K_{cu}=C_{hx}/(U_c DT_c)+C_{cu}$. Even if the cost of cooling water and heat exchanger surface area are given, the approach temperature in the condenser can be independently decided to achieve at least the critical ratio K_{cu}/C_a.

Total systems: Consider now the previous systems together with the accompanying HENs. The AD system consumes more steam by 3-78% (Figure 4). For the AD system to become cheaper, it is now necessary for the steam to be as cheap as possible compared to capital costs. Relation (6) provides an upper bound on the ratio C_{hu}/C_a and this tends to favour the AD systems as it moves to higher values. If this upper bound turns out non-positive, the AD system is *never* cheaper than the SD system, however cheap steam may become. This does not happen in this case unless too high approach temperatures are employed in the condensers.

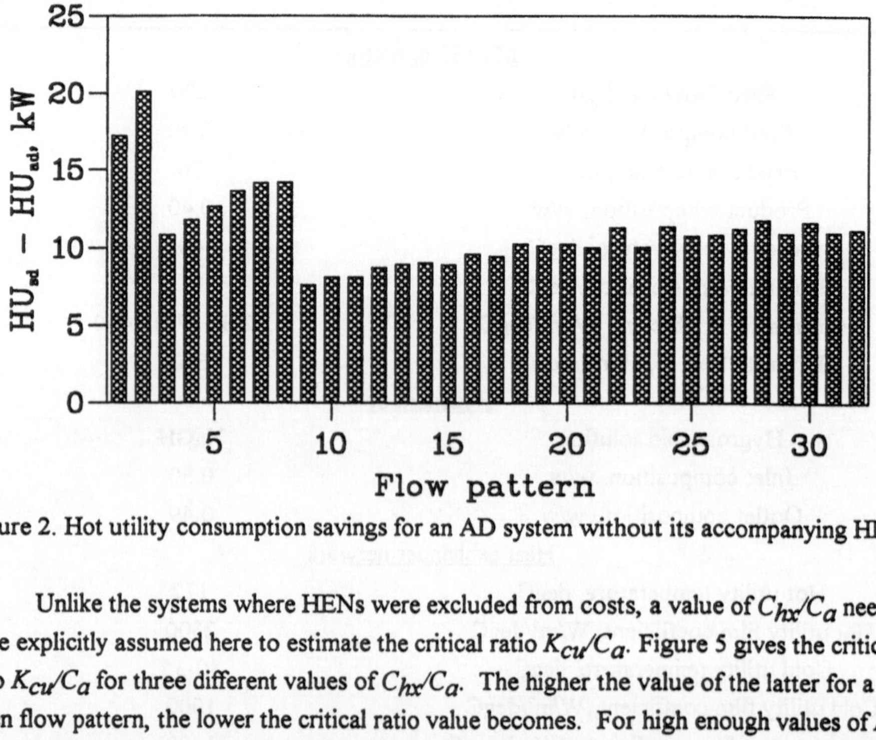

Figure 2. Hot utility consumption savings for an AD system without its accompanying HEN.

Unlike the systems where HENs were excluded from costs, a value of C_{hx}/C_a needs to be explicitly assumed here to estimate the critical ratio K_{cu}/C_a. Figure 5 gives the critical ratio K_{cu}/C_a for three different values of C_{hx}/C_a. The higher the value of the latter for a given flow pattern, the lower the critical ratio value becomes. For high enough values of DT_c and/or cheap cooling water the AD system is *never* cheaper than the SD system. Note that the ratio C_{hx}/C_a relates to the complexity of an evaporator effect (moving parts, special materials) as compared to an ordinary heat exchanger.

Less heat recovery: Assume now an AD system, where the liquid stream between the two regenerator effects is heated up in the regenerator itself. Figure 6 presents the difference in steam consumption for the total systems, HENs included, and Figure 7 presents the critical ratio (K_{cu}/C_a) for three different values of the C_{hx}/C_a ratio. The critical ratio is here *always* negative. Since this is an upper bound for which the AD system costs less, an AD system is *never* economically preferable to an SD one within the examined range of C_{hx}/C_a!

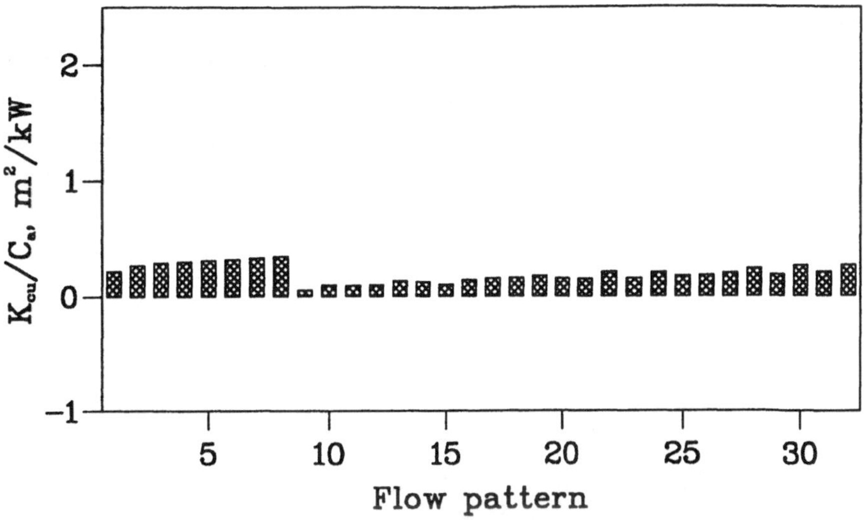

Figure3. Critical ratio (cooling cost/effect area cost) for an AD system without its accompanying HEN.

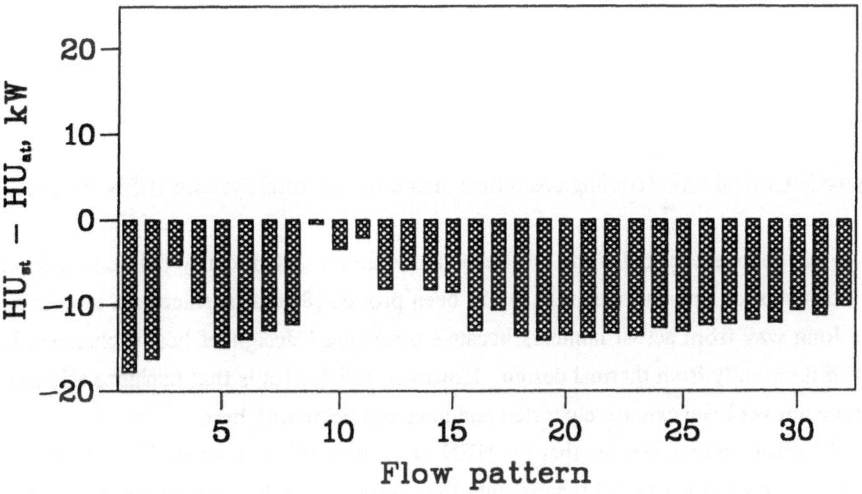

Figure 4. Hot utility consumption difference between the SD and AD systems (HENs included).

DISCUSSION

The analysis above provides a measure for comparison between absorption driven and steam driven systems independent of time or place. If the process of interest is simulated once, the feasibility of one or the other configuration (flow pattern, AD or SD, heat recovery) can be easily assessed by using the relevant economic data and the equations above. This flexibility seems useful at the design stage where various economic scenarios may need to be explored early on. The idea of expressing the criterion of feasibility in terms of the ratio of the most significant cost coefficients C_{hu}/C_a agrees with recent, similar studies for reactor-separator systems [7].

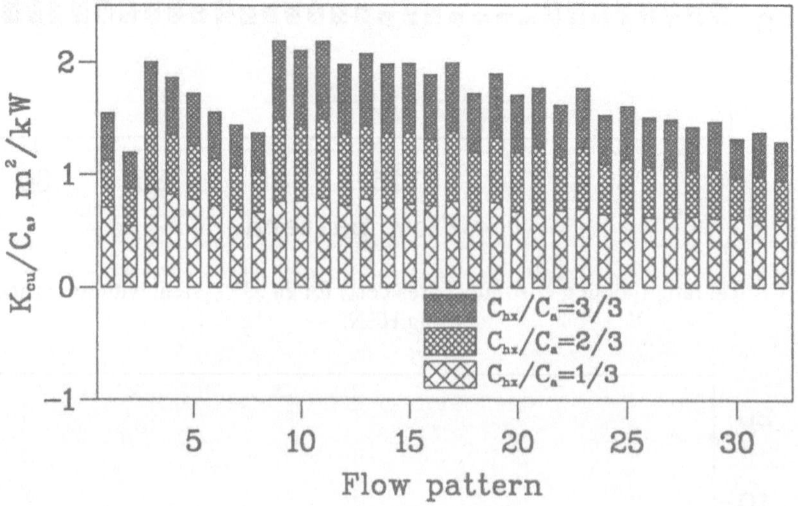

Figure 5. Critical ratio (cooling cost/effect area cost) for total systems (HENs included).

Heat recovery calculations are deliberately simple and identify only target levels of utility consumption and surface area. It has been proved [8] that surface area targets can lie quite a long way from actual numbers because mechanical design of heat exchangers is approached differently from thermal design. However, reliable tools that minimize this discrepancy have not yet been extensively tested and have not been used here.

All results above assume that the HEN of each system consumes 10% more hot/cold utility (whichever is the largest) than required for zero approach temperature. Different such percentages can be tested to provide a favourable result in the feasibility study.

The sensible heat required in the regenerator of AD systems seems to be quite significant in a system where hot streams dominate. It is not surprising that its inclusion in the heating requirements of the regenerator completely changes the cost picture and even makes the AD systems more expensive at all instances. This agrees with earlier warnings about the importance of sensible heat recovery [2].

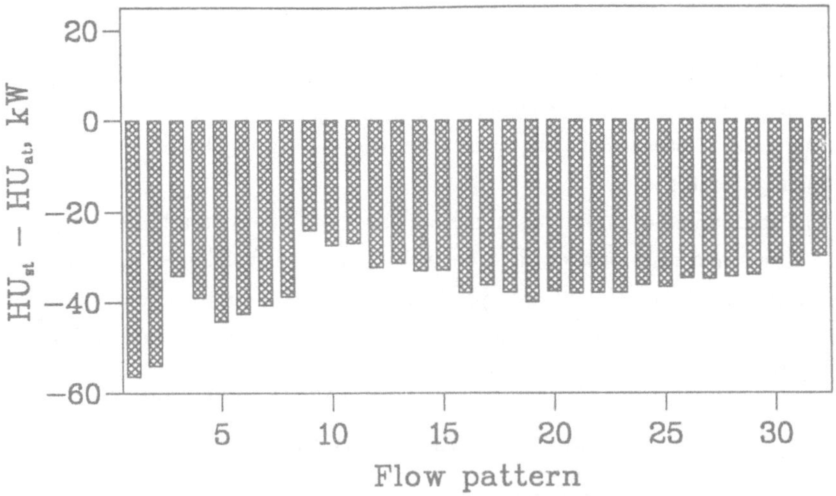

Figure 6. Hot utility consumption difference between total systems (HEN included): the case of less heat recovery in the regenerator.

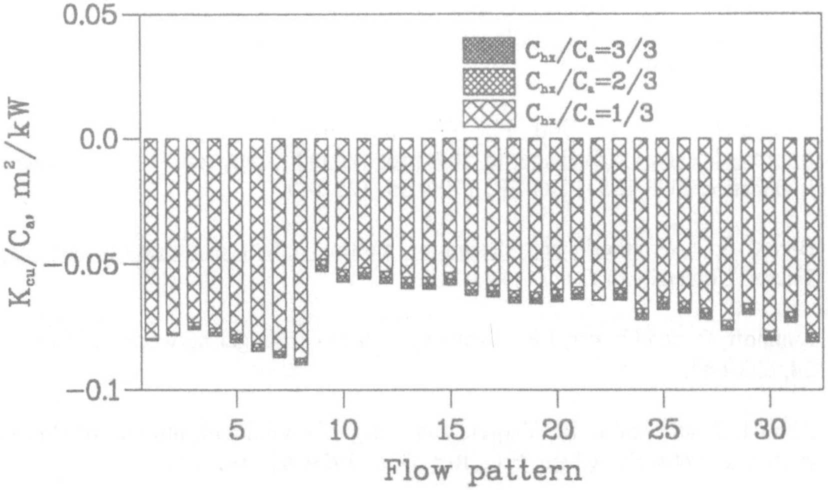

Figure 7. Critical ratio (cooling cost/effect area cost) for total systems: the case of less heat recovery in the regenerator.

CONCLUSIONS

In this paper, the issue of choosing between absorption-driven and steam-driven evaporator systems has been addressed in terms of process-related quantities only.

It has been found that economic circumstances exist where absorption-driven evaporators are always preferable and others where this is never the case. Capital costs can outweigh steam savings, especially if all the evaporator streams are considered in a heat exchanger network. Steam savings may even never occur if sensible heat is not recovered sensibly in the regenerator.

The calculations of both types of system have been based on a Newton-type iterative solution of heat and mass balances and on well established techniques for heat exchanger network targeting. In the case of absorption-driven evaporators, the choice of design variable values particular to the absorber has been subjected to rigorously established bounds that eliminate infeasibilities.

ACKNOWLEDGEMENTS

The authors wish to thank Dr. P.A. Pilavachi for his encouragement to produce this work.

REFERENCES

1. Yanniotis, S. and Moschovitis, P., Absorption-driven multiple effect evaporators. In Industrial processes, Proceedings of a contractors' meeting, ed. P.A. Pilavachi, CEC Publication EUR 12246 EN, 1988, pp. 121-130.

2. Yanniotis, S. and Le Goff, P., Absorption-driven multiple effect evaporators. In Improved energy efficiency in the process industries, ed. P.A. Pilavachi, CEC Publication EUR 13541 EN-FR, 1990, pp. 129-140.

3. Bikos, S.C., Synthesis and design of multiple effect evaporation systems, Ph.D. Thesis, University of Leeds, 1991.

4. Linnhoff, B. and Flower, J.R., Synthesis of heat exchanger networks, AIChE J., 1978, **24**, 633-642.

5. Ahmad, S. and Smith, R., Targets and design for minimum number of shells in heat exchanger networks, Chem. Eng. Res. Des., 1989, **67**, 481-494.

6. Fraser, D.M., The use of minimum flux instead of minimum approach temperature as a design specification for heat exchanger networks, Chem. Eng. Sci., 1989, **44**, 1121-1127.

7. Drage, M.J., Economic trade-offs in process design, PhD Thesis, University of Nottingham, 1992.

8. Polley, G.T. and Panjeh Shahi, M.H., Interfacing Heat Exchanger Network Synthesis and Detailed Heat Exchanger Design, Chem. Eng. Res. Des., 1991, **69**, 445-457.

APPENDIX

The N effects of an evaporator can be linked by liquid and vapour streams in N! permutations if no stream splits are allowed (serial configuration). In a double effect evaporator number the effects 1 and 2 in the order the liquid under concentration visits them. Then, the forward feed arrangement is one where vapour visits the two effects in the same order, 1 and 2. We denote this flow pattern 12. Similarly, the backward feed of a four effect evaporator is flow pattern 4321. For simplicity, graphs of the main text quote flow patterns 1, 2, 3, ..., 32. These correspond to permutations 12, 21,123,132,...,4321 respectively.

SESSION 10:

Process Integration
Continued

Chairman: Mr J. Kosmadakis

SIMULATION AND OPTIMIZATION OF AN INTEGRATED INDUSTRIAL ENERGY SYSTEM

Dr. A. I. LYGEROS
Manager of Research Development and Planning
Hellenic Aspropyrgos Refinery S.A.
Amalias Ave 54, GR-105 58 Athens, Greece

C. MERENTITIS
Chemical Engineer
Kolokotroni Str. 40
GR-136 71 Acharnai, Athens, Greece

ABSTRACT

This work demonstrates how the integrated energy system of a complex industrial plant can be simulated, linearized and optimised by Linear Programming (LP). This approach greatly simplifies the analysis of the parameters involved and reveals energy conservation potentials. The energy system of Hellenic Aspropyrgos Refinery S.A. (HAR) a complex 130.000 BPD petroleum refinery, is used as a case study. After a very brief description of HAR's production scheme, the refinery's energy system (consisting of the energy supply, energy conversion and energy utilisation subsystems) is presented and analysed. The mathematical equations that describe the system are presented and the objective function which gives the total energy cost is formulated. Since HAR's energy system is a rather typical situation, the approach, analysis and methods demonstrated here can be used in many other industrial applications.

INTRODUCTION

There are many ways to improve the efficiency of energy utilisation in the process industries. Most of the efforts in this area have been devoted to energy saving techniques aiming at the reduction of heat losses (improvement of thermal efficiency of fired heaters and steam boilers, better insulation etc.), waste heat recovery, heat exchanger network optimization, more efficient use of energy in unit operations, 2nd law thermodynamic analysis, better control of industrial processes etc. The work presented here is concerned with a different approach for improving energy efficiency in a large industrial plant. It is based on the optimization of the energy supply and distribution system and the allocation of available energy sources to energy uses in a way that minimises overall energy consumption.

More specifically, the purpose of this presentation is to demonstrate how the integrated energy system of a large scale industrial plant can be simulated, linearized and optimised by Linear Programming (LP). This approach greatly simplifies the optimization procedure and reveals energy conservation potentials. The energy system of Hellenic Aspropyrgos Refinery S.A. (HAR), a complex petroleum refinery, is used to demonstrate the approach.

HAR's energy system is by no means a unique situation. Similar systems and problems are encountered in all large refineries and in fact in all large industrial installations. Thus, the analysis, methods and solutions presented here can very well be used in many other industrial applications. In this respect, emphasis has been given to the methodology and techniques used rather than to the specific results obtained from the particular simulation model of HAR. The presentation focuses on the philosophy of this approach, the analysis required, the general guidelines for building and operating such a model and the evaluation of the results.

THE REFINERY

Hellenic Aspropyrgos Refinery is a state owned refinery located about 16 km South West of Athens. HAR was established in 1958 and since then has undergone several expansions. The refinery today produces about 5.8 million metric tons of petroleum products per year and covers approximately 55% of Greece's needs. The last - about 600 million dollars - expansion and modernisation project, was completed in 1987. The main purpose of this project was to construct cracking units for converting residuals into distillate products.

The refinery has two crude distillation units of a total capacity of 130.000 BPD. The distillation bottoms (reduced crude) of these units go to a vacuum distillation unit the overhead of which (VGO) after desulfurization is fed to an FCC unit where it is cracked into lighter products, mainly gasoline. The refinery has a Continuous Catalyst Regeneration reformer (CCR), gasoil desulfurization unit, C_3/C_4 polymerization units, MTBE production, amine H_2S removal units, Clauss sulfur recovery units, LPG, gasoline and kerosene MEROX units, hydrogen production etc. It is a refinery of high complexity with a total of 25 integrated process units. In the last expansion project mentioned above, a distributed control system (DCS) for the entire refinery was also installed.

Energy cost in a petroleum refinery is one of the largest components of its total operating cost. For a complex refinery of the size of HAR the energy cost amounts to about 30 million dollars per year at today's prices. Hence, energy conservation by optimization of energy utilisation is a key factor for minimising the total operation cost. This is particularly important in today's hard business environment of the petroleum refinery industry.

THE ENERGY SYSTEM

The energy needs of H.A.R. amount to about 400,000 tons of FOE/year or approximately 7 wt % on the crude oil processed.

These needs are satisfied from the following energy sources:

	%
• Electricity from the National Grid	.3
• Fuel Gas (F.G.)	49.5
• Propane	5.7
• Fuel Oil (F.O.)	31.2
• Coke of the FCC unit	13.3

Figure 1 shows a simplified diagram of the refinery's energy system. As shown, the energy system is fully integrated. Primary energy sources (i.e. Fuel Gas, Fuel Oil and Coke) are converted to electricity, steam or heat for the refinery's process units. Thus the refinery's energy system can be considered as consisting of three subsystems, namely:

- Energy Supply
- Energy Conversion, and
- Energy Utilisation

A brief description of each of the subsystems follows.

Energy Supply

The Energy Supply System of the refinery consists of the Electricity Supply from the National Electricity Grid, the Fuel Gas System, the Fuel Oil System, Propane and Coke of the FCC Unit, as primary energy sources for the refinery.

Electricity Supply: The refinery has its own electricity production system . The electricity production capacity is today greater than the refinery's electricity needs and gives the possibility of exporting electricity to the National Grid. In special cases the refinery may

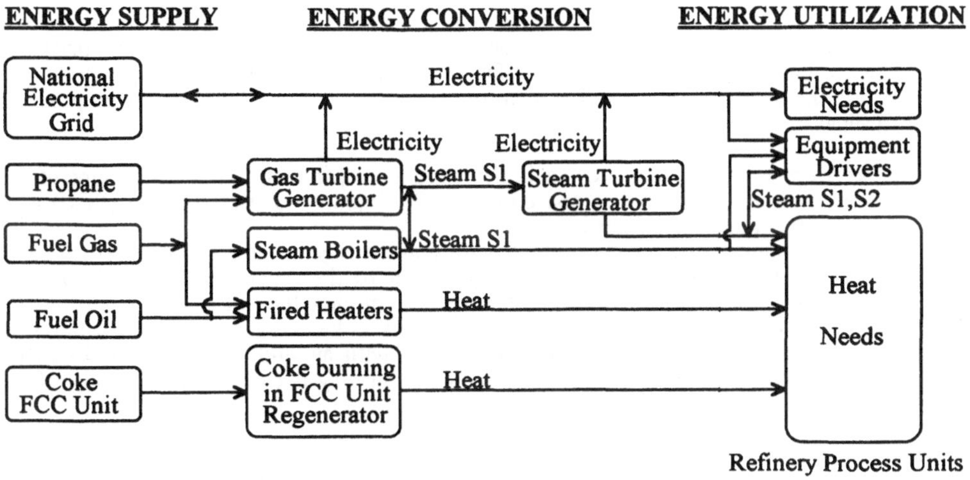

Figure 1. Energy system of Hellenic Aspropyrgos Refinery.

need to import electricity from the National Grid. Thus the refinery is connected to the National Grid for two reasons: safety and possibility of exporting electricity.

Fuel Gas Supply: Fuel Gas (F.G.) is the largest primary energy source in HAR. Fuel Gas is by-product of the refinery's process units. It consists of light hydrocarbons, methane (C_1) to butane (C_4) and a small percentage of hydrogen (about 5 vol. %). The F.G. has, in the average, a molecular weight of about 17 kg/kmole.

The F.G. comes from the light hydrocarbons contained in crude oil (separated in the crude distillation units) and from cracking of heavier hydrocarbons in the refining processes, mainly Gasoline Reforming, Fluid Catalytic Cracking and Mild Hydrocracking.

The F.G. is used in the fired heaters and in 2 gas turbines for co-generating electricity and steam. There are 18 fired heaters in H.A R. which can use Fuel Oil, Fuel Gas or both simultaneously and 5 fired heaters which use only F G

Fuel Gas cannot be stored. A small inventory in the accumulation vessels and piping distribution system does exist (depending on pressure) but there is no storage for F.G. Thus, the rate of consumption of F.G must always equal the rate of production. Any unbalance is automatically directed, by the refinery's safety systems, to the flares where it is burned. The amount of F.G. burned in the flares is a total loss for the refinery. This means that the balance between F.G. production and consumption is a very important factor for energy conservation and optimization.

Fuel Oil Supply: The second largest primary energy source in H.A.R. is Fuel Oil (F.O.). The F.O. used in HAR is commercial industrial grade Fuel Oil (370°C+, 900 kg/m^3, 370 cSt at 50°C max) of special low sulfur content (0.7 wt % max) because of the strict environmental regulations in the greater Athens area.

The F.O. is used in the fired heaters and in the steam boilers. In addition to the 18 dual firing (F.G./F.O.) heaters mentioned above, there are also 4 steam boilers which use only F.O.

Propane Supply· Propane is only used as a supplement of F.G. burned in the gas turbine generators if such need arises or economics favour it. Propane is sellable final product. Its use as a fuel in the refinery depends on propane storage availability and its selling price. As mentioned above, the refinery's energy system is fully integrated. Thus use of propane in the gas turbine generators results in diverting F.G. from the gas turbine generators to fired heaters substituting F.O. This means that there is actually a trade-off between F.O. (which is also a sellable product) and propane and the use of one or the other depends on their selling prices.

The propane used as a fuel in the refinery is pumped from liquid storage tanks to a vaporizer and the gas from the vaporizer is mixed with the refinery's fuel gas stream going to the gas turbines.

Coke from the FCC unit: Coke is neither a final product nor a by-product of the refinery. It is continuously produced and burned in the Fluid Catalytic Cracking (FCC) unit, covering part of the unit's energy needs.

The FCC unit consists mainly of the reactor and the regenerator. Hot catalyst (in a fine powder form) flows continuously from, the regenerator to the reactor and from the reactor back to the regenerator in a closed loop. In the reactor entrance the hot catalyst is mixed with the liquid oil feed. Due to the cracking chemical reactions, coke is deposited on the catalyst. In the regenerator the coke on the catalyst is burned off with air and the regenerated catalyst flows back to the reactor. The heat released from coke burning is carried by the catalyst from the regenerator to the reactor.

Energy Conversion

The primary energy sources (F.O., F.G. and Propane) are converted to more usable forms of energy such as heat, steam and electricity in order to be utilised. The conversion systems are described below.

Steam Production: In order to satisfy the variety of steam needs of the refinery's process units in the most economical way, the refinery uses 4 different grades of steam. The main properties of these steam grades are summarised in table 1.

TABLE 1
Steam grades used in the refinery

Grade Designation	Pressure Bar	Temperature oC	Enthalpy kJ/kg
S1	42.4	410	3233
S2	13.5	320	3084
S3	3.7	150	2755
S4	4.7	160	2769

Steam is produced in the boilers, in the gas turbine electricity generators (producing electricity in a co-generation scheme) and in the waste heat boilers of the process units.

The refinery has 4 boilers of total production capacity of 130 MT/h of grade S_1 steam. The two gas turbine generators at full production can produce 60 MT/h of S1 steam. The waste heat boilers of the process units produce various kinds of steam, mainly S1 and S2.

Steam is used for:

- Thermal needs of the process units in heaters, reboilers etc.
- Electricity production in the steam turbine
- Stripping in distillation columns
- In steam turbines, used as drivers of pumps and compressors

Steam S1, used in the steam turbine generator and in the various equipment drivers of the process units, is degraded to lower grades, desuperheated (by condensate injection) and used in the process units for heating purposes. A balance between production and consumption of the various steam grades must always be maintained in order to avoid degrading higher grades into lower grades in expansion stations (without mechanical energy production). A simplified diagram of the refinery's steam system is shown in figure 2.

Electricity Production: Electricity is produced in two gas turbine generators, with a capacity of 18 MW each and in a steam turbine generator with a capacity of 16 MW. As mentioned before the refinery's total electricity production capacity is today greater than its needs. Hence it is not necessary to operate the gas turbine and the steam turbine generators at their maximum production level. This allows optimization of electricity production by selecting the operating level of each machine. Taking into account that the gas turbines produce steam while the steam turbine consumes steam (affecting the overall steam balance and the required production of the steam boilers) one can appreciate the need for overall optimization of the refinery's integrated energy system. In addition, as mentioned above, electricity can be exported to National Electricity Grid, depending on prices, thus adding another parameter to the optimization problem.

Heat Production: As described above fuel Gas and/or Fuel Oil is used in the refinery's fired heaters, in which the process fluids are heated to the desired temperature for reaction, distillation etc.

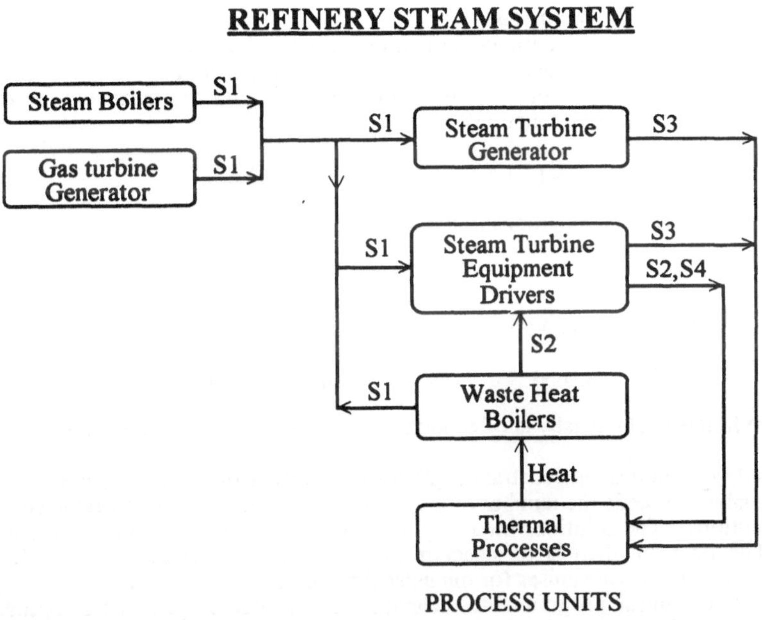

Figure 2. Refinery steam production and utilisation system.

Energy Utilisation

The refinery's energy needs can be classified as:

- General electricity needs (instruments, lighting etc.)
- Steam or electricity for equipment drivers (pumps and compressors)
- Heat (steam or direct burning of F.G. or F.O.)

These needs are satisfied through the energy supply and conversion systems described above. From the point of view of energy optimization, it is important to note that the large pumps and compressors have two drivers, a steam turbine and an electric motor, of 100% capacity each. Either one of the drivers can be used, depending on steam and electricity availability and cost.

Also, as mentioned above, most of the large fired heaters (18 out of a total of 23 heaters) can use either F.G. or F.O. or both fuels simultaneously. Selecting one or the other fuel depends on the overall optimization of the energy system.

THE OPTIMIZATION PROBLEM

HAR's Energy System (as of any large modern refinery) is very intricate and involves a great number of interrelated parameters. Table 2, summarises the main factors that have to be taken into account for the optimization of the Energy System. It is obvious that optimization of a system of that size and complexity cannot be achieved by the common-sense methods, inspired guesswork or systematic trial-and-error procedures. In this case a more scientific method must be adopted. One approach is to form a mathematical model which takes into account all the interrelated factors and to compute the optimum solution from the model.

The idea is by no means new. It is an approach that since the 2nd World War has been known as Operations Research, a branch of science concerned with the application of scientific methods and techniques to decision-making problems. The technique most commonly associated with operations research is linear programming.

The Linear Programming Technique

In mathematical terms, linear programming is a procedure whose central theme is finding the point where a linear function defined on a convex polyhedron assumes its maximum or minimum value. A typical linear programming problem is to find the maximum (or minimum) of a linear expression (objective function):

$$C_1X_1 + C_2X_2 + \dots\dots\dots + C_nX_n \quad (1)$$

subject to linear side conditions (constraints):

$$X_j > 0$$

$$\begin{aligned} a_{11}X_1 + \dots\dots\dots\dots + a_{1n}X_n &\geq b_1 \\ \dots\dots\dots\dots\dots\dots\dots\dots\dots\dots\dots & \\ a_{m1}X_1 + \dots\dots\dots\dots + a_{mn}X_n &\geq b_m \end{aligned} \quad (2)$$

The objective function (1) usually is of economic nature (profit to be maximised or cost to be minimised).

The side conditions or constraints (2) form a system of linear equations. When the number of variables exceeds the number of equations the system will in general have an infinite number of solutions. The solutions involving negative values of one or more variables X_j are discarded. The problem then is to select from the remaining set of feasible solutions the optimum, i.e. the one that maximises (or minimises) the objective function.

To apply the linear programming technique to the specific problem of optimising the energy system of HAR (or any other industrial plant) we must proceed by the following steps:

(a) Define the goal and objective function

(b) Describe the limitations or constraints under which the energy system must operate (energy and material balances). Linear programming requires that the objective function and constraints be stated in the form of a matrix.

(c) Solve the problem, that is, find a way to arrive at the best solution out of the many feasible solutions of the problem. This may be done using a commercial LP computer program.

(d) Make a sensitivity analysis, that is, explore the effect of changes of key parameters on the best solution.

FORMULATING THE MODEL OF HAR'S ENERGY SYSTEM

The Objective Function

The goal is to minimise energy cost. For given operation of the process units, the energy cost of HAR consists of the sum of costs of F.O., F.G., propane, coke of the FCC unit, electricity imported from the National Grid, minus the value of electricity exported to the National Grid.

F.G. and coke are produced in certain amounts which depend on the operation of the process units and they are totally consumed. The LP model is instructed (in the constraints) to give fixed values to the variables of F.G. and coke, equal to their production rates in every case. Their unit cost is also constant in a given study period. Therefore the costs of F.G. and coke are not decision variables, i.e. their value does not affect the optimum solution. They are included however in the objective function in order to have the true total cost of the energy consumed.

TABLE 2

Key parameters for energy system optimization

A. Operation of gas turbine generators-steam turbine generator and boilers

1. Electricity (within certain limits) can be produced either by the gas turbines or by the steam turbine generator.

2. Gas turbine generators produce steam at the same time. Thus, increased gas turbine level of production results in increase of steam availability reducing the required production of steam by the steam boilers.

3. Steam turbine generator consumes steam. Increasing steam turbine generator's level of production results in reduced steam availability thus increasing the required production of the steam boilers.

B. Use of electricity or steam

4. Operation of the large compressor and/or pump drivers with electricity or steam affects electricity and steam needs and operation of the gas turbine generators, steam turbine generator and boilers as per 1, 2 and 3 above.

5. Electricity can be exported to the National Electricity Grid (depending on availability and prices) thus affecting the operation of the gas turbines, steam turbine and boilers as per 1, 2 and 3 above.

6. Production and consumption of the various steam grades must be kept in balance to avoid degrading higher steam levels to lower levels at a loss (i.e. without production of mechanical work).

C. Use of F.G. or F.O. or Propane

7. Most fired heaters can be operated either with F.G. or F.O. or both. Choice of level affects F.G. availability and may make it necessary to consume propane as a supplement. It may also affect the operation level of the gas turbine generators, steam turbine generator and boilers as per 1, 2 and 3 above.

8. Surplus quantities of propane, in connection with its price as compared to F.O. price must be taken into account in considering its share in covering energy needs.

9. Fired heaters have different efficiencies with F.G. and F.O. This must be taken into account in the choice of fuel as per 7 & 8 above.

10. F.G. cannot be stored. F.G. consumption must always equal F.G. production, otherwise excess F.G. will be burned in the flares at a total loss.

The objective function for HAR's total energy cost is then formulated as follows:

MIN [(FO) * (FO unit cost) + (FG) * (FG unit cost) + (Prop.) * (prop. unit cost) +
+ (Coke) * (coke unit cost) + (electr. import) * (electr. unit cost) -
- (electr. export) * (electr. price)].

where the quantities of F.O., F.G., propane and coke are in MT/h, the electricity in kWh/h, the unit costs in \$/MT and \$/kWh, and the total energy cost in \$/h.

The quantities of F.O., F.G. etc. are calculated by the model in order to satisfy the refinery's energy needs and the energy and mass balances specified in the constraints. The unit costs are parameters which are given by the user for the period under study.

The Constraints

Having defined the objective function, the next step is to specify the conditions (constraints) that must be satisfied in order to have a feasible solution. In this case the constraints are energy and material balances. From the energy and material balances the model will calculate the required quantities of F.O., F.G., propane, coke and electricity which shall be used in the objective function to calculate the total energy cost. The formulation of constraints is as follows.

Fired Heaters Operation: The model takes into account 23 fired heaters. From these 18 heaters can use F.O. or F.G. or both simultaneously and 5 heaters use only F.G. Each heater has different thermal efficiencies with F.O. and F.G. The heat balance for heater i is:

$$(FO)_i * (eff)_{FOi} * (HV)_{FO} + (FG)_i * (eff)_{FGi} * (HV)_{FG} = H_i$$

where:

$(FO)_i$, $(FG)_i$	: are the FO & FG consumed in this heater in MT/h
$(eff)_{FOi}$, $(eff)_{FGi}$	: are the corresponding efficiencies of the heater (with FO & FG)
$(HV)_{FO}$, $(HV)_{FG}$	: are the heating values of FO & FG in kcal/MT
and H_i	: is the required process heat duty of the heater.

There are 23 such equations, for i=1 to 23. For the heaters that use only FG the FO terms are zero.

The total FO and total FG that are used in the fired heaters are:

$$THFO = \sum_i (FO)_i \quad \text{and} \quad THFG = \sum_i (FG)_i$$

Steam Boiler Operation: The steam boilers consume FO and produce steam S1. Within certain limits of production (in which an average overall boiler efficiency can be used) the steam boiler FO consumption (SBFOC, MT/h) is related to the steam boiler steam S1 production (SBS1P, MT/h) by a relation of the form:

$$SBFOC = \alpha 1 * SBS1P$$

where $\alpha 1$ is an average coefficient derived from actual operating data.

The maximum total steam production of the boilers is 130 MT/h which introduces one more constraint:

$$SBS1P \leq 130$$

Gas Turbine Generator Operation: The gas turbine generators consume F.G. and produce electricity and steam of grade S1. The relationship between F.G. consumed and electricity and steam produced (i.e. the overall efficiency of the gas turbine generator) depends on the level of production, the pressure of the gas and the ambient air temperature. These relationships are given by the gas turbine generator manufacturer in operating curves. With some assumptions and simplifications linear equations of the following form were derived:

$$GT1FGC = \alpha 2 + \beta 2 * GT1EPP$$

and

$$GT2FGC = \alpha 3 + \beta 3 * GT2EPP$$

where GT1FGC, GT2FGC	: are the consumptions of FG in the gas turbines 1 and 2 in MT/h
and GT1EPP, GT2EPP	: are the productions of electricity from the gas turbine generators 1 and 2 in MW.

Similar equations can be derived for the gas turbine steam S1 production (GT1S1P, MT/h) as a function of electricity production (GT1EPP, MW).

$$GT1S1P = \alpha 4 + \beta 4 * GT1EPP$$

and

$$GT2S1P = \alpha 5 + \beta 5 * GT2EPP$$

The maximum electricity production of each gas turbine generator is 18 MW, while for practical reasons minimum production is 8 MW. These limits introduce the following additional constraints:

$$GT1EPP \geq 8 \ , \ GT1EPP \leq 18$$

$$GT2EPP \geq 8 \ , \ GT2EPP \leq 18$$

Steam Turbine Generator Operation: The steam turbine electricity generator consumes steam of grade S1 and produces electricity and steam of grade S3. The manufacturer of the steam turbine generator provides operating curves from which for a given consumption of S1 steam, we can calculate the production of electricity for various levels of production of S3 steam. In order to use them in the LP model these production curves must be linearized by regression analysis. In general, for a certain production of S3 steam of the steam turbine (STS3P, in MT/h) the consumption of S1 steam by the turbine (STS1C, MT/h) is related to the electric power production of the turbine (STEPP, MW) by an equation of the form:

$$STS1C = \alpha 6 + \beta 6 * STEPP$$

The coefficients α6 and β6 were calculated from the manufacturer's production curves. These coefficients depend on the production of steam S3.

The maximum production of electricity from the steam turbine is 16 MW and for practical reasons the production cannot fall below 6 MW. These introduce two new constraints:

$$STEPP \geq 6 \quad \text{and} \quad STEPP \leq 16$$

Process Equipment Drivers Operation: For each large pump there is a second (spare) installed of 100% capacity. One of the two pumps is driven by an electric motor and the other by a steam turbine. Only one pump is in operation every time.

In order to let the model know which of the two pumps is in operation, we used integer linear programming and more specifically "0-1 programming". To do this we defined two integers that can be either 0 or 1, integer JSijk for the steam turbine and integer JEijk for the electric motor.

JSijk is 1 when the steam turbine is in operation.
i is the grade of the steam entering the turbine
j is the grade of the steam leaving the turbine
k is the designation number of the pump
JEijk is 1 when the electric motor is in operation (for uniformity i,j,k have the same values as for the steam turbine, although i and j have no significance for the electric motor).

At all times : $JSijk + JEijk = 1$

The power consumption of pump k at any time is given by:

$$JSijk * A_k + JEijk * B_k$$

where A_k is the steam of grade i consumed and steam of grade j produced, in MT/h and B_k is the electricity consumed in kW.

Energy and Material Balances: After the equations describing the operation of all equipment have been written as above, the overall energy and material balances for the entire plant are formulated. By summing-up individual consumptions, the total consumptions of F.G., F.O., propane and electricity (to and from the National Grid) are calculated. These consumptions are used in the objective function in search of the minimum total energy cost of the refinery. The entire set of equations is written in the form of a matrix as required by LP. For the specific problem of the energy system of HAR, the matrix consists of 78 rows (equations) by 101 columns (variables).

A Note on Accuracy: In order to be able to include all the variables into one working model and to express by linear functions all the relationships involved, some assumptions and simplifications have necessarily to be made. However these simplifications do not limit the use of the model. Linear programming is usually used to indicate directions rather than to calculate exact values at least for some of the variables. Not to mention that even if the best solution was known in every detail, in reality in a large industrial installation very seldom this could be fully implemented.

RESULTS AND ANALYSIS

The model described above defines a linear programming problem which can be solved by a commercial LP program. The program calculates the optimum value of the objective function (which in this case is the minimum total energy cost) and determines the values of all the variables involved in order to achieve this optimum. From the values of the variables we get information on how to operate the various components of the energy system. Such information includes:

- what is the optimum operation level of the gas turbine and steam turbine generators
- what is the optimum steam S3 extraction from the steam turbine generator
- what is the optimum operation level of the steam boilers
- what is the optimum amount of electricity to be exported or imported to and from the National Grid
- what is the optimum amount of propane to be used
- how is every one of the fired heaters to be operated (with F.G. or F.O. and in what percentage each)
- how are the large pumps to be operated (steam or electricity driven)
- what are the consumption of F.G., F.O., propane each grade of steam and electricity to each section of the energy system

etc.

By adjusting the operation according to this information an overall optimization of the energy system is achieved, which results in the minimum total energy cost as calculated by the objective function.

From the model we can also calculate unit costs of the utilities produced i.e. each grade of steam and electricity. These prices are not directly involved in the optimization procedure but are useful in allocating the total energy cost to the various process units in order to estimate their operating cost.

Sensitivity Analysis

The solution obtained as described above depends of course on the values of the parameters introduced as constants in the model, i.e. the unit price of F.O., propane and electricity, the available quantity of F.G. etc. By varying the values of these parameters we can establish the range of their value in which the optimum solution is still valid. In other words we can establish how sensitive to the various parameters is the optimum solution. This information is very useful in order to know in advance the effect of changes of the parameters on the optimum solution and determine if it is required to make operating adjustments in order to stay in the optimum.

The Practical Value of the Energy Optimization Model

The model built for the energy system of HAR proved in actual practice to be a very useful tool. Several cases studied for various production schemes (depending on availability and operating level of the refinery's process units, refinery product needs, availability of energy sources etc.) revealed that substantial savings can be achieved through the optimization of the operation of the energy system as described above. In every case the model selects the optimum operating scheme and calculates the operating parameters for each of the components of the energy system. Unit production costs as well as marginal costs for steam, electricity etc. are in every case calculated by the program, providing useful information for other studies and management decisions.

The savings that can be achieved in every operating case by optimising the energy system by the model, certainly depend on the operation before using the model, established by habit or experience. When operation remains constant for long periods of time, experience and trial and error procedures can perhaps achieve an acceptable level of optimization (although this certainly requires much greater time, effort and cost). In today's real world in the refining industry however (especially for single independent refineries as HAR) operating scheme must change almost every day in order to adjust operation to the ever changing conditions of crude oil supply, product demand and product prices. In addition, in large industrial complexes, changes have very often to be made due to internal reasons for unscheduled maintenance needs. In these cases the energy optimization model is indispensable for getting operating guidelines for the energy system and achieving the best possible results quickly and inexpensively.

REFERENCES

1. Merentitis, Ch., Optimization of the distribution and utilisation of fuel gas and fuel oil in a petroleum refinery by linear programming. Diploma Thesis, Department of Chemical Engineering, National Technical University of Athens, November 1991. (in Greek language)

2. Dano, Sven, Linear Programming in Industry, Springer-Verlag, Wien-New York, 1974.

3. Lygeros A.I., Magoulas K., Marinos-kouris D., Actual Blending Linear Programming model is developed ,Oil and Gas Journal ,1988, **86**, July 18, 44-48.

4. Bouiloud Ph. Compute Steam Balance by LP, Hydrocarbon Processing, 1969, **48**, No 8, 127-128.

5. Tsai T.C., Microcomputer program optimizes steam systems, Oil and Gas Journal, 1986, **84**, Feb. 24, 67-72.

6. Seinfeld J.H. , McBride W.L., Optimization with multiple performance criteria, Industrial and Engineering Chemistry, 1970, **9**,1, Jan., 53.

HEAT INTEGRATION OF A NEW FORMIC ACID PULPING AND RECOVERY PROCESS

ESA MUURINEN and JORMA SOHLO
University of Oulu
Department of Process Engineering
Linnanmaa
SF-90570 Oulu
Finland

ABSTRACT

A formic acid recovery process was preliminary designed for the MILOX pulping method. The effect of various process parameters on the energy consumption of the process was studied using computer simulation. When suitable process conditions were found the overall process was heat integrated using pinch technology. Significant savings can be achieved by adjusting the amount of water fed to the process, the liquor-to-wood ratio in pulping and the concentration of formic acid in pulping liquor. The energy consumption of the process can be further decreased by about 40 % using pinch technology in heat integration.

INTRODUCTION

The search for less polluting processes has initiated the development of several new pulping methods. Many of these new methods are so called organosolv processes using organic solvents as delignifying agents. One of the most promising organosolv methods is the MILOX process [1] using formic acid and peroxyformic acid as solvents in a totally sulphur and chlorine free process.

Our previous work [2,3,4] has clearly shown that the energy costs of the chemical recovery are critical when total operating costs of the formic acid pulping and recovery process are considered. The purpose of the present work is for that reason to cut down the energy consumption of the process.

THE MILOX PROCESS

Pulping and bleaching

MILOX pulping is a three-stage delignifying method (fig. 1) where hydrogen peroxide is added in the first and the third stage to form peroxyformic acid. All pulping stages can operate at atmospheric pressure, but a slight pressurizing of the second stage improves the results. The pulp obtained from birch has a kappa number of 5...10 when about 2 % (on wood) hydrogen peroxide is used. The pulp yield is 43 % and the viscosity is about 1300 dm^3/kg. The birch pulp produced by the MILOX process can be bleached to 90% brightness with alkaline hydrogen peroxide.[5]

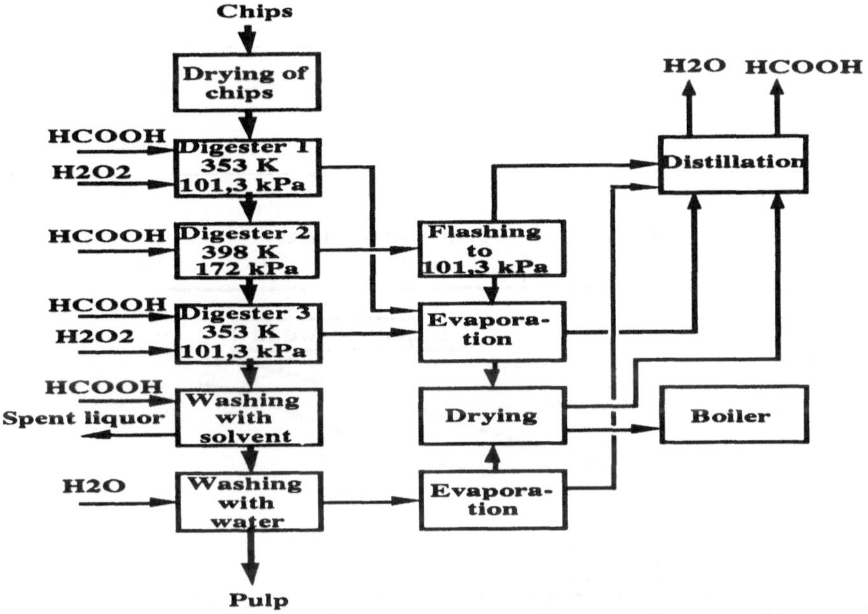

Figure 1. The flowsheet of the pulping and recovery process.

Recovery of chemicals

A recovery cycle based on distillation (fig. 1) was designed for the MILOX process. Pulp is washed with fresh solvent before the traditional washing with water to prevent lignin from condensing on the surfaces of the fibres. Spent liquor from solvent washing can be used as pulping liquor. All chemicals are volatile and they are recovered from spent liquors by multistage evaporation and drying. The solid fraction from drying is burned to produce energy. The acid concentration in pulping has to be about 85% by weight and the formic acid containing vapours collected from evaporation and drying must be concentrated by e.g. pressure shift distillation before they can be used as pulping liquor.

THE EFFECT OF SOME PROCESS PARAMETERS ON THE ENERGY CONSUMPTION OF THE MILOX PROCESS

The material balances for the pulping stages were calculated using the results of black liquor analysis. The energy and material balances for the recovery cycle (fig. 1) were calculated using computer simulations. The energy consumptions of the MILOX process was calculated with several values of chip moisture content, liquor-to-wood ratio and formic acid concentration in pulping (table 1).

When large amounts of water are fed to the process with chips the distillation unit has to produce more concentrated formic acid causing an increase of the energy costs (fig. 2).

TABLE 1
Conditions used when testing the effects of process parameters

Variable	Chip moisture (% by weight)	Liquor-to-wood (kg/kg)	Formic acid concentration (% by weight)
Chip moisture	5...50	4:1	83.5
Liquor-to-wood	20	3:1...11:1	83.5
Formic acid con.	20	4:1	75...90

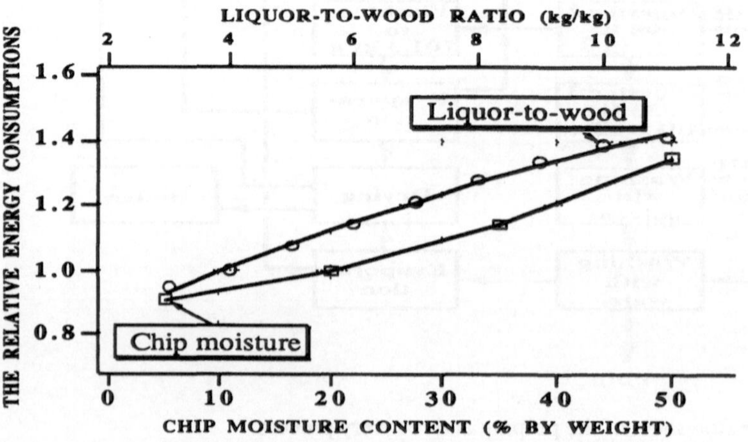

Figure 2. The effect of the chip moisture content and the liquor-to-wood ratio on the energy consumption.

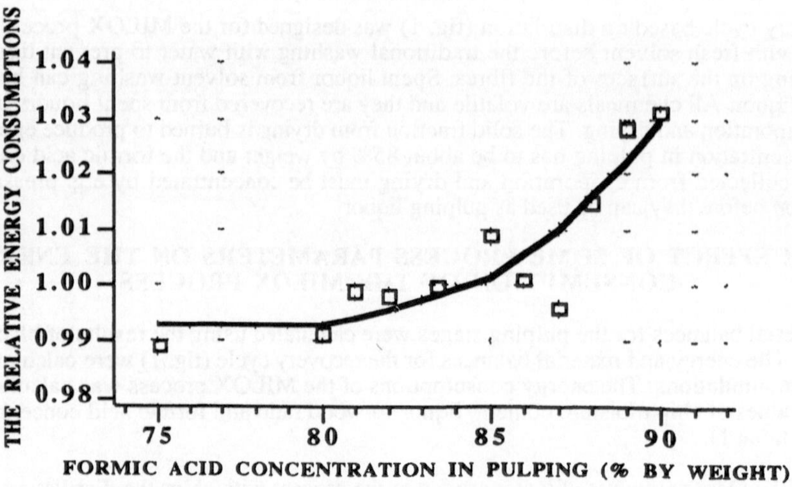

Figure 3. The effect of the formic acid concentration on the energy consumption.

The increasing liquor volume is causing an increase in the energy costs when the liquor-to-wood ratio increases (fig. 2). When the liquor-to-wood ratio is increased the formic acid concentration 83.5 % can be reached with less concentrated acid. This makes the increase of energy costs less steep with high liquor-to-wood values.

When the formic acid concentration in the first delignifying stage increases, the energy consumption of the distillation unit also increases but the decreasing water volume decreases the energy costs in other parts of the process (fig. 3). These effects compensate each others and the energy costs are nearly constant with concentrations 75...90 %.

HEAT INTEGRATION

A heat exchanger network for the MILOX pulping and recovery process was designed using pinch technology. The unit operations connected in the network are chip drying, pulping, evaporation, drying of concentrate and distillation.

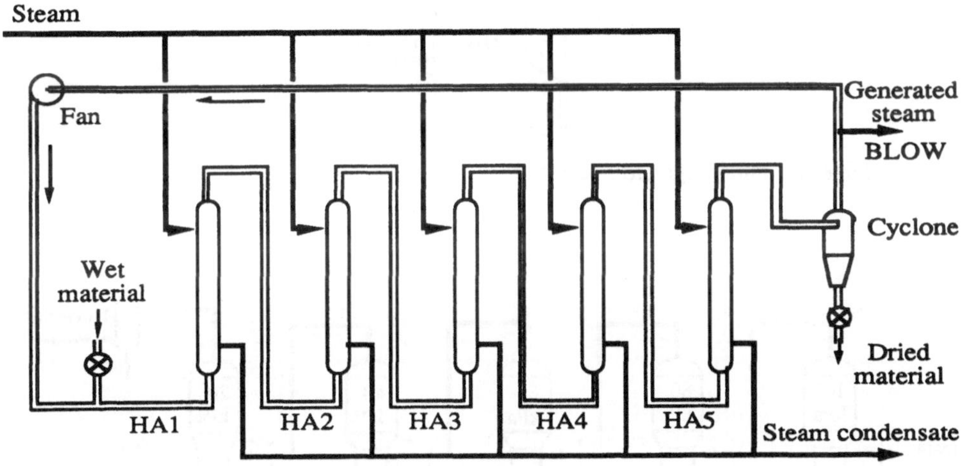

Figure 4. Steam dryer and streams to be integrated.

Chips are dried prior to pulping in a five-stage steam dryer [6]. Pulping is carried out in three digesters. The second stage operates at a pressure of 170 kPa. Spent liquors from pulping and pulp washing are concentrated in a five-stage evaporation plant. When concentrated in the evaporation plant, the liquor is dried in a double-drum dryer to improve the recovery rate of formic acid. Formic acid containing vapours collected from evaporation and drying are concentrated in a pressure shift distillation unit to break the azeotrope formed by formic acid and water (77.5 % HCOOH at 101 kPa). The integrated streams are shown in figures 4...8 and table 2.

The composite curves and the grand composite curve are presented in figures 9 and 10, when the minimum temperature difference is 10 K. The minimum hot utility of the process is 367 energy units and the minimum cold utility is 296 energy units. The consumptions of thermal energy in the unit operations before and after integration are presented in table 3 and the amounts of energy to be transferred from the process in table 4. The amount of heat to be transferred to the process can be decreased with 42 % and the amount of heat to be transferred from the process with 47 % by using pinch technology to design the heat exchanger network.

The costs of separating byproducts (acetic acid etc.) have not been taken into account in these calculations.

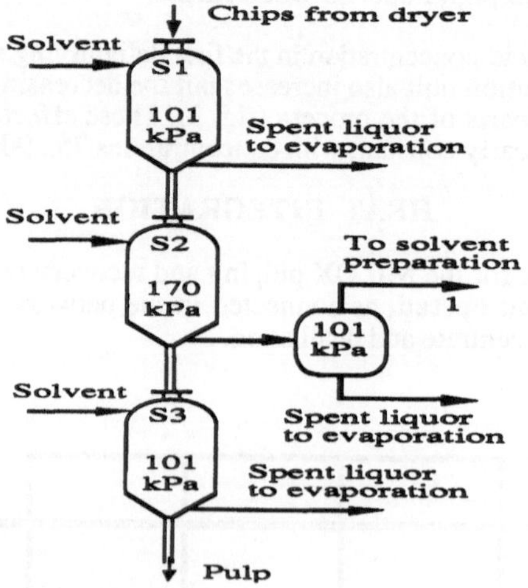

Figure 5. Digesters and integrated streams.

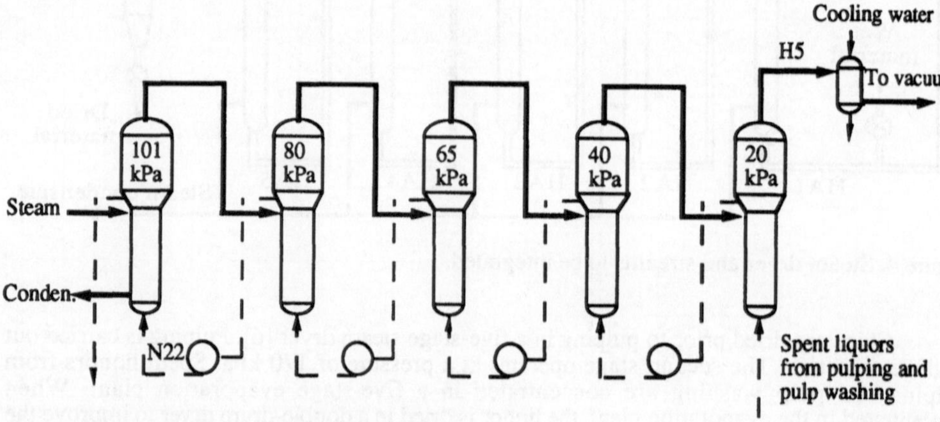

Figure 6. Evaporation and integrated streams.

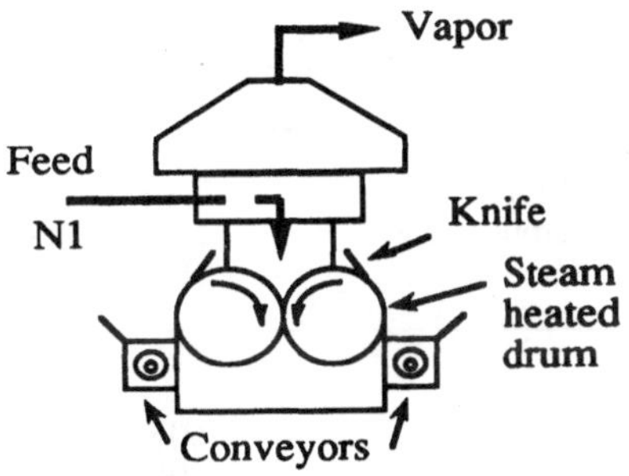

Figure 7. Double-drum dryer and integrated streams.

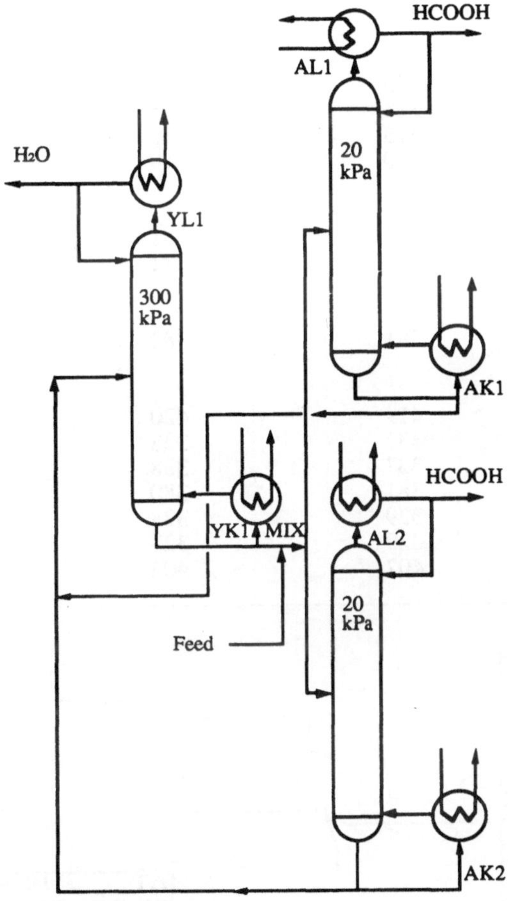

Figure 8. Distillation and integrated streams.

TABLE 2
Stream data

Stream number and type		Supply temperature (K)	Target temperature (K)	Relative stream heat duty
DRYING OF CHIPS				
HA1	cold	407	408	4.0
HA2	cold	407	408	4.0
HA3	cold	407	408	4.0
HA4	cold	407	408	3.9
HA5	cold	407	408	4.0
BLOW	hot	407	406	-14.2
PULPING				
S1	cold	293	353	6.5
S2	cold	313	398	59.1
S3	cold	313	353	4.6
1	hot	380	379	-3.2
EVAPORATION				
N22	cold	373	381	24.6
H5	hot	332	331	-20.2
DRYING OF LIQUOR				
N1	cold	381	382	8.0
DISTILLATION				
YK1	cold	419	420	268.8
AK1	cold	337	338	119.6
AK2	cold	337	338	119.6
MIX	hot	381	380	-40.9
AL1	hot	329	328	-143.2
AL2	hot	329	328	-143.2
YL1	hot	407	406	-194.6

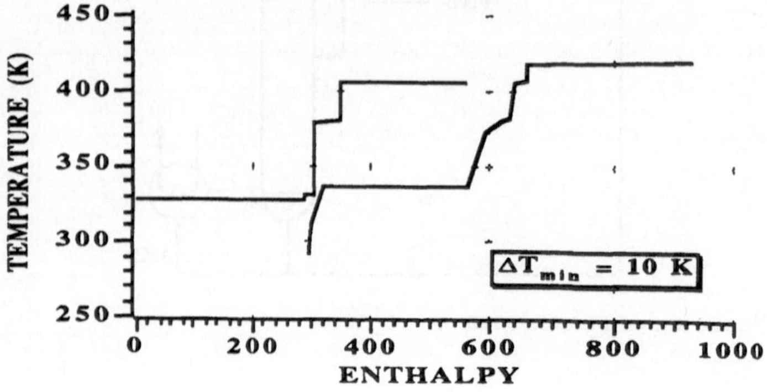

Figure 9. The composite curves.

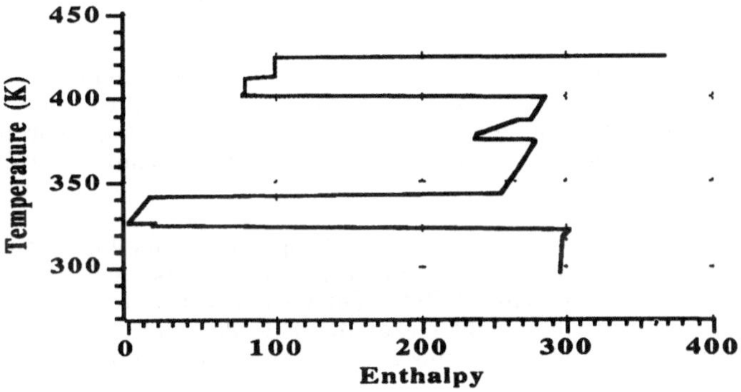

Figure 10. The grand composite curve.

TABLE 3
The relative consumptions of thermal energy in the unit operations before and after heat integration

Unit operation	Relative consumptions of thermal energy		Change %
	Before integration	After integration	
Drying of chips	20	20	0
Pulping	74	58	-22
Evaporation	25	25	0
Drying of liquor	8	0	-100
Distillation	509	269	-47
	636	372	-42

TABLE 4
The relative needs of cooling in the unit operations before and after heat integration

Unit operation	Relative needs of cooling		Change %
	Before integration	After integration	
Drying of chips	14	0	-100
Pulping	3	0	-100
Evaporation	20	18	-13
Drying of liquor	0	0	0
Distillation	522	277	-43
	559	295	-47

CONCLUSIONS

The main factors affecting the energy costs in the MILOX pulping and recovery process are the amounts of liquor and particularly water fed to the process. Minimising of both decreases mainly the energy consumption of the recovery section. The formic acid concentration of the solvent being between 75 and 90 % does not have any significant affect on the energy consumption of the process.

The heat integration of the MILOX process using pinch technology decreases the need of thermal energy to be transferred from the environment to the process by about 40 %. The need of cold utility decreases by almost 50 %. The energy consumption of particularly the distillation plant is relatively high. This makes the heat integration of the unit operations important if the MILOX process has to compete with traditional pulping methods.

ACKNOWLEDGEMENTS

The authors are indebted to the Technology Development Centre of Finland (TEKES) for financial support.

REFERENCES

1. Poppius, K., Sundquist, J. and Wartiovaara, I., Chemical pulping of birch and pine chips by the three stage peroxyformic acid method. In Wood processing and utilization, ed. J.F. Kennedy, G.O. Phillips and P.A. Williams, Ellis Horwood Ltd, Chichester, 1989, pp. 87-92.

2. Muurinen, E. and Sohlo, J., Recovery of chemicals from organosolv wood-pulping processes. In Wood processing and utilization, ed. J.F. Kennedy, G.O. Phillips and P.A. Williams, Ellis Horwood Ltd, Chichester, 1989, pp. 81-6.

3. Muurinen, E., Sohlo, J., Vanhanen, T. and Kivelä, E., Recovery of chemicals from organosolv woodpulping processes. In 1989 ISWPC Poster Sessions, Tappi Press, Atlanta, 1989, pp. 223-5.

4. Muurinen, E. and Sohlo, J., Recovery of chemicals from peroxyformic acid pulping. In 6th ISWPC Proceedings Volume 2, Appita, Melbourne, 1991, pp. 169-73.

5. Sundquist, J., Bleached pulp without sulphur and chlorine chemicals by a peroxyacid/alkaline peroxide method - an overview. Paperi ja Puu, 1986, **68**, 616-20.

6. Svensson, C., Industrial applications for new steam drying processes in forest and agricultural industry. In Drying'85, Washington DC, 1985, pp. 415-9.

INTEGRATION OF ABSORPTION HEAT TRANSFORMERS IN THE PROCESS INDUSTRY - APPLICATIONS IN THE OLEOCHEMICAL, PULP AND PAPER INDUSTRIES

ALY, G., ABRAHAMSSON, K. AND JERNQVIST, Å.
Department of Chemical Engineering 1, University of Lund,
P.O. Box 124, S-221 00 Lund, Sweden

ABSTRACT

Incorporating a heat transformer unit in an existing process to increase its energy efficiency was studied in this work. Two different applications; an evaporation plant in the pulp and paper industry and a fat hydrolysis unit in the oleochemical industry, were investigated.

In the first application, optimal energy conservation strategies were investigated using a heat transformer system incorporated with the evaporation plant of the pulping process. A process configuration, designed for the largest energy consumer unit in the evaporation plant, with a heat transformer boosting the temperature of the last vapor stream by 31°C, would reduce the amount of live steam used in this unit by 18.5 %. The pay-off period for this case was calculated as 4.4 years.

In the second application saturated water vapor at 100°C, available from two flash vessels in a fat hydrolysis unit, has previously been condensed in a dump condenser and discharged. Incorporating a heat transformer unit would enable the recovery of almost half this heat as saturated steam at 135°C and would result in a pay-off period of 1.4 years.

One of the main practical features of these applications is that installing the heat transformer unit requires minimum changes in the existing plant. Operating data from two industrial plants were used in developing the different process configurations.

ENERGY CONSERVATION IN THE PROCESS INDUSTRY

The total energy used by the Swedish industry in 1990 was 139 TWh, which is equivalent to 12.0 Mtoe. The pulp and paper industry, which is conventionally the largest member of the industrial sector in Sweden, used about 44% of that amount, i.e. 63 TWh. It should be noticed however that spent liquors, internally produced by the different mills, are utilized as prime energy source and thus 29 TWh were produced that particular year. Examples within the Swedish pulp and paper industry include a current average specific process energy consumption of 6.0 MWh/ton pulp (basis: 90 % dry solid content), for bleached sulfate pulp, corresponding to 21.8 GJ/ton. The corresponding values for bleached sulfite pulp are 4.49 MWh/ton and 16.4 GJ/ton respectively [1].

Evaporation and drying, together with distillation, are the most energy intensive unit operations in the process industry. For instance, the average annual energy used in 1987 by all evaporation units in the Swedish process industry was about 11.6 TWh, 89.4 % of which was used within the pulp and paper sector alone. The corresponding figures for drying were 21.2 TWh and 76.9 %, respectively. Distillation plants used approx. 5.8 TWh, and a very small fraction of this amount is assigned to the pulp and paper industry, since condensate stripping is the main distillation process. On the other hand, heating of process water, pulp digesting and pulp bleaching are the third largest energy intensive unit operations in that sector.

A portion of the waste heat is theoretically recoverable to produce steam or to supply other process heating requirements. Incorporating various types of heat pumps within different units in a process burdened by waste heat streams having appropriate temperature levels is an efficient method for energy recovery. A literature survey revealed, however, that relatively few applications have so far been implemented in the process industry. Heat pumps can generally be divided into mechanical compressor heat pumps, thermal compressor heat pumps, and absorption heat cycles. The latter group can be subdivided into absorption heat pumps AHP and absorption heat transformers AHT. The coefficient of performance COP is often used for the evaluation of an absorption heat cycle. However, this is not always the best criterion to express the effectiveness of this type of heat pump systems [2-3]. A more detailed description of absorption heat transformers can be found elsewhere [1-4].

The main objective of this study was to investigate the feasibility of incorporating a heat transformer system in an existing process to increase its energy efficiency. Two different applications; an evaporation plant in the pulp and paper industry and a fat hydrolysis unit in the oleochemical industry, were investigated.

LIQUOR EVAPORATION IN THE PULP AND PAPER INDUSTRY

In the magnefite pulping process, the spent sulfite pulping liquor is treated conventionally by evaporation and subsequent burning to produce steam and electric power and to recover the chemicals for re-use in the cooking process. Figure 1 shows a simplified block diagram for the existing evaporation plant consisting of three units having a total evaporation capacity of 208.4 ton evaporated water/h. According to the current mode of plant operation, the evaporation capacities of the three units are 25.9 % (53.9 ton/h) for the pre-evaporation unit,

51.4 % (107.2 t/h) for Line 2, and the balance for Line 1. The thin liquor is fed to a modern pre-evaporation unit, having five falling-film evaporators, where its concentration of the solid contents is increased to about 20 weight %. About 30% of this liquor is assigned to Line 1 and the balance to Line 2. The evaporators in these two units are of the rising-film forced-circulation type. Depending on the production scheme, each unit can be operated either in a 5- or 6-effect mode.

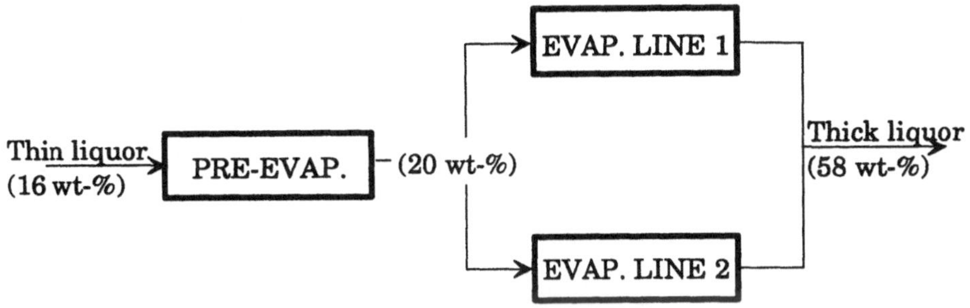

Figure 1. A simplified block diagram of the evaporation plant.

The evaporation plant is also utilized to preheat three external streams of cooking liquor, process water and boiler water. A total amount of heat equivalent to 10.39 MW is extracted from the evaporation plant for preheating purposes. The pre-evaporation unit delivers 32 % of the required amount of heat while the contribution of evaporation lines 1 and 2 is 48 % and 20 %, respectively. The following boundary conditions must be fulfilled:

- to maintain the preheating duties as discussed above,
- to avoid major changes in the existing plant when the heat transformer system is incorporated,
- to avoid any losses in the condensate and boiler feed water streams,
- to maintain the current temperature levels in some components of the existing plant.

FAT HYDROLYSIS IN THE OLEOCHEMICAL INDUSTRY

The initial process for obtaining fatty acids from fats and oils is hydrolysis, whereas the triglycerides are split, by means of water, into diluted glycerine solution and mixed fatty acids. This mixture is purified by hydrogenation and distillation, or via separation into individual fatty acids of different chain lengths by fractional distillation. Figure 2 shows a simplified flowsheet for a fat hydrolysis process, both with the existing dump condenser and the proposed absorption heat transformer.

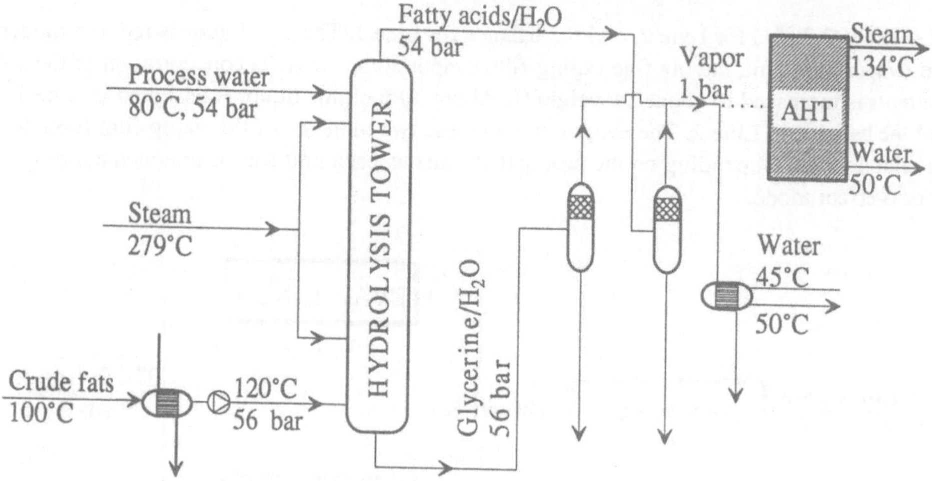

Figure 2. Flowsheet of a continuous fat hydrolysis process.

The pre-treated fats are first preheated and then fed at the bottom of the hydrolysis tower. The hydrolysis reaction is usually carried out continuously using deionized and deaerated water at a temperature of 250 - 255 °C and a pressure of 50 - 55 bar. The high temperature and pressure employed assures a higher water solubility in the fat phase, but the use of high pressure is primarily to keep water in the liquid phase at temperatures above its boiling point. Injection of direct high-pressure steam in the hydrolysis reactor is frequently practiced to retain the reaction temperature and increase the agitation intensity of the reacting phases. The hydrolysis reaction yields a diluted aqueous glycerine solution and a mixture of different fatty acids. These pressurized streams are fed to separate flash vessels, kept at atmospheric pressure, to reduce their pressure and temperature. Saturated water vapor leaving the flash vessels at 100°C is condensed in a dump condenser where the temperature of process cooling water is increased by 5°C.

COMPUTER PROGRAMS

Two computer program packages, INDUNS and SHPUMP, were developed at our department for the simulation of general multiple-effect evaporation processes and different absorption heat pump systems, respectively. A brief presentation of both packages is given below.

The Computer Program INDUNS

This program is based on the principle of modular composition of evaporation plant flowsheets using a so-called unit cell [5,6]. The unit cell is designed to provide maximum flexibility, as well as maximum simplicity of the flowsheet data input procedure. Each unit cell consists of an evaporation vessel, a thermo-compressor, liquor and condensate flash vessels, a condensate level controller, heat exchangers, liquid mixing/splitting vessels and vapor mixing/splitting vessels. This configuration, together with a connection matrix, would enable the user to easily assemble evaporation flowsheets of arbitrary complexity. The main function of the connection matrix is to inform INDUNS about the flowsheet of the plant.

A comprehensive data base comprising the physical properties of black liquor, Ca-based sulphite liquor, milk, sugar solutions and seawater is attached to INDUNS. Three special subfunctions for the estimation of the overall heat transfer coefficients are also incorporated. The first subfunction is used for black liquor in LTV evaporators. The second one for forced circulation evaporators and heat exchangers. The third subfunction is used for falling-film evaporators.

The Computer Program SHPUMP

This is a flow-sheeting program constructed for both design and evaluation calculations of seven basic and well established absorption cycles [7]. These are the heat pump (HP), heat transformer (HT), double-stage HP, double lift HP, double-stage HT, double lift HT, and heat pump transformer. Nine different component modules are incorporated to provide the user with maximum flexibility in constructing any absorption cycle. The nine modules represent an evaporator, absorber, generator, condenser, heat exchanger, mixer, splitter, pump and valve. The program delivers pressure, temperature, concentration, enthalpy, and flowrate profiles for the absorption cycle under consideration. For the heat exchanger, either the overall heat transfer coefficient, U, and the heat transfer area, A, or their product, U·A, will be calculated together with the minimum temperature difference which can be attained. Being menu-driven, together with an interactive user manual, the program is designed to provide maximum flexibility, as well as maximum simplicity of the flowsheet data input procedure.

RESULTS AND DISCUSSION

Liquor Evaporation

The three evaporation units were first simulated using actual operating data to calculate complete temperature, concentration and overall heat transfer coefficient (U-values) profiles. The U-values and apparent temperature differences for the forced circulation effects of Line 2 were 0.8 - 1.7 kW/m^2 K and 5.1 - 18.4°C, respectively. Different system configurations were simulated to determine the optimum location and size of the heat transformer for evaporation Line 2, which is the major energy consumer of the evaporation plant. The best results were obtained by the configuration shown in Figure 3. The heat transformer boosts the temperature of the vapor leaving the last effect from 70 to 101 °C. This vapor is to be mixed with the vapor leaving the third effect, and used as the heating medium in the fourth effect. It should be mentioned that installing a heat transformer according to this configuration requires minimal changes in the existing plant.

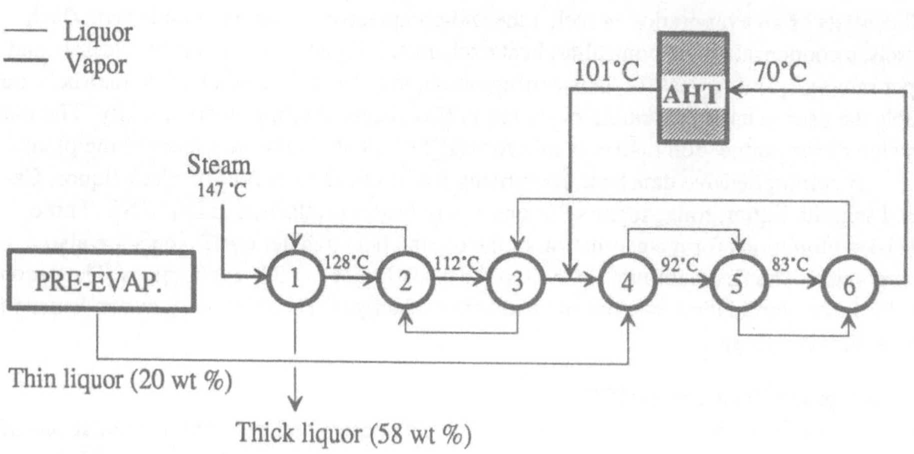

Figure 3. A heat transformer incorporated in the evaporation plant.

The heat transformer would reduce the amount of live steam consumed in Line 2 by 18.5 %, corresponding to 3.09 MW. Compared to the total thermal energy required for the evaporation plant, this saving would be 12.0 %. Figure 4 shows the different components of the heat transformer which would deliver an output of 6.3 MW. The basic design parameters for the AHT system are displayed in Table 1. Based on H_2O-NaOH as the working pair, the operating pressure range would be 3 - 25 kPa, while the concentration would be 51.2 and 48.6 weight % for the strong and weak solutions, respectively.

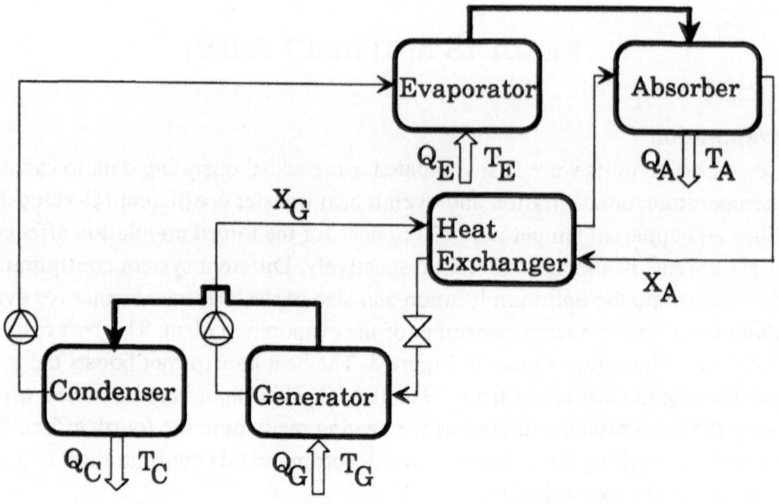

Figure 4. Schematic flowsheet for the absorption heat transformer system.

An economic evaluation was performed to determine the economic feasibility of incorporating heat transformer systems in this type of evaporation plants. Based on the information obtained for the Hoogovens plant [8], the installed equipment cost was assumed to be 410 $/kW. The pay-off period of this type of investment depends mainly on the local steam cost, heat transformer efficiency and also on how much of the existing equipment in the evaporation plant that can be utilized as components in the heat transformer. In this work,

Table 1: Basic design data for the liquor evaporation heat transformer system

Variable		Evaporator	Generator	Absorber	Condenser	H. Ex.
Effect, Q	(kW)	7,500	6,326	6,314	7,512	-
Flowrate	(kg/h)	11,520 [a]	9,720 [a]	10,080 [a]	709,200 [b]	-
Temperature, T	(°C)	70	70	101	19	-
LMTD	(°C)	5.0	6.1	6.3	8.8	10.1
Pressure	(kPa)	25	3	25	3	-
Conc., X	(wt% H_2O)	100	48.6	51.2	100	-
U·A	(kW/°C)	1,500	1,033	1,003	856	563

[a] Steam [b] Cooling water

the third factor was not considered in the economic evaluation and a COP-value of approx. 45% was assumed. The annual plant operation time and steam cost were taken as 8,000 hrs and 6.6 $/GJ, respectively. The pay-off period for this case was calculated as 4.4 years. It should be noticed however that this steam cost, corresponding to 14 $/ton, is considered to be relatively low compared to a comparable Scandinavian pulp and paper mill having the same capacity. For a steam cost of 9.7 $/GJ, corresponding to 20.6 $/ton, the pay-off period would be reduced to 3.0 years.

Following a successful 2 year operation of a 10 kW pilot plant AHT with self-circulation, a relatively larger unit has recently been built as an industrial demonstration plant. The self-circulation, which eliminates the need for circulation pumps, is attained according to the thermosiphon principle, whereas the pressure difference of the working fluid in the different components of the transformer can be achieved and maintained through a difference in hydrostatic pressures [9,10]. The host site is the pulp and paper industry involved in this work and the AHT unit, utilizing H_2O-NaOH as working pair, will be incorporated in one of the evaporation plants in the mill. Sodium hydroxide solutions are more propitious to the pulp and paper industry compared to other working pairs. The unit will be operated for 12-18 months which would allow for a comprehensive field testing under realistic operational conditions. It will intentionally be subjected to substantial perturbations using thermodynamic properties and flowrates of the waste heat sources, and the corresponding responses will be monitored.

Fat Hydrolysis

A comprehensive energy mapping of an industrial fatty acid production plant was performed in co-operation with a major Swedish oleochemical manufacturer. The results concerning the fat splitter, with a capacity of 3400 kg/h, are summarized in Table 2. As can be observed, the total energy input and output for this hydrolysis process is 3.37 and 2.93 MW, respectively. Thus, the energy losses were estimated to be 13% corresponding to about 440 kW. The energy mapping revealed that a total of 500 kg/h of saturated water vapor is produced at 100°C from four flash vessels used to depressurize condensate streams emerging from different processing units in the plant. It should be mentioned that the flowrate of steam withdrawn from both flash vessels in the fat splitting unit is 434 kg/h as indicated in Table 2. The balance, 66 kg/h, represents steam withdrawn from two other flash vessels used to depressurize different condensate streams coming from the fatty acid distillation, glycerol evaporation, and glycerol distillation units. This vapor has a heat content of 314 kW and is currently condensed in a dump condenser and discharged. Incorporating an AHT system, using H_2O-NaOH as the working pair, would enable the recovery of almost half this energy and a temperature lift of 34°C can be achieved. Refering to Figure 4, the AHT in this case is supplied by a total of 314 kW of waste heat corresponding to 500 kg/h of saturated water vapor at a temperature of 100°C. The heat energy delivered by the transformer system is 153 kW corresponding to a flowrate of 255 kg/h of saturated steam at a temperature of 134°C. This 3 bar steam can be utilized in different locations in the fatty acid plant such as evaporation of crude glycerol and distillation of crude fatty acid mixtures.

Table 2: Energy mapping of the fat hydrolysis unit

Stream	Flowrate (kg/h)	Energy (kW)
Raw fat	3,373	151
Hydrolysis water	2,361	276
High pressure steam	809	625
Low pressure steam	130	97
Cooling water	34,000	2,170
Electricity	-	53
Raw fatty acids (1 bar)	3,237	186
Diluted glycerol (1 bar)	2,872	322
Condensate (1 bar)	434	40
Low pressure condensate	130	17
Cooling water	34,000	2,367
Thermal losses	-	438

Using the pay-off analysis, an economic evaluation was performed to determine the economic feasibility of incorporating AHT systems in this type of chemical plants. Based on a derived cost function [11], the installed equipment cost was assumed to be 535 $/kW of delivered heat. The annual operation time and steam cost of this particular plant were taken as 7,200 hrs and 14.25 $/GJ, respectively. Consequently, the total installed cost of the AHT system and the corresponding annual saving of steam (3 bar) would be approximately 81,900 and 56,500 $, respectively. The pay-off period for this heat transformer system would then be 1.45 years.

Table 3: Basic design data for the fat hydrolysis heat transformer system

Variable		Evaporator	Generator	Absorber	Condenser	H. Ex.
Effect, Q	(kW)	161	153	153	161	-
Flowrate	(kg/h)	256[a]	244[a]	253[a]	27,780 [b]	-
Temperature, T	(°C)	100	100	134	50	-
LMTD	(°C)	3.0	6.7	4.9	5.7	6.5
Pressure	(kPa)	90.9	14.7	90.9	14.7	-
Conc., X	(wt% H_2O)	100	50.1	53.4	100	-
U·A	(kW/°C)	53.6	22.7	31.2	28.2	16.5

[a] Steam [b] Cooling water

The amount of heat transferred as well as the flowrate, temperature and concentration of the different streams are given in Table 3, together with design data for the different components of the AHT system. The investment cost of each component of the AHT system is an increasing function of its heat transfer area A (m^2), which is proportional to the amount of heat transferred Q (kW). However, the heat transfer area depends on the design characteristics of the component, but the product U·A (kW/°C) is only a function of both the driving force, normally expressed as the logarithmic mean temperature difference LMTD (°C), and the amount of heat transferred Q (kW). As shown in Table 3, the LMTD and U·A-values of the AHT vary in the range of 3.0-6.7°C and 16.5-53.6 kW/°C, respectively.

Due to phase change in both the evaporator and condenser, the overall heat transfer coefficient U of these components are rather high. Design values of 2.5-3.0 kW/m^2 °C for the evaporator and 1.5-2.5 kW/m^2 °C for the condenser can easily be realized. On the other hand, dealing with rather viscous solutions in both the absorber and generator, the U-values of this type of heat transfer equipment are relatively low and are normally in the range of 0.6-1.0 kW/m^2 °C. Similarly, the solution heat exchanger can be designed with a U-value in the range of 1.0-1.6 kW/m^2 °C. Based on these values, the AHT system investigated in this study would require a total heat transfer area of 93-147 m^2, depending on the geometric configuration of each component.

ACKNOWLEDGMENTS

The financial support of the Swedish National Board for Industrial and Technical Development and the Swedish Council for Building Research is gratefully acknowledged.

REFERENCES

1. Abrahamsson, K., Aly, G. and Jernqvist, Å., Heat transformer systems for evaporation applications in the pulp and paper industry. Nordic Pulp and Paper Research J., 1992, **6**, 9-16.

2. Jernqvist, Å., Abrahamsson, K. and Aly, G., On the efficiencies of absorption heat transformers. J. Heat Recovery Systems & CHP, 1992, In press.

3. Jernqvist, Å., Abrahamsson, K. and Aly, G., On the efficiencies of absorption heat pumps. ibid, 1992, In press.

4. Gidner, A. and Jernqvist, Å. Energy Conservation in the Sugar Industry. Accepted for publication, International Conference on Energy Efficiency in Process Technology, Athens (1992).

5. Bolmstedt, U. and Jernqvist, Å., Simulation of the steady-state and dynamic behaviour of multiple-effect evaporation plants. Part I: Steady-state simulation. Computer Aided Design, 1976, **8:3**, 142-148.

6. Bolmstedt, U. and Jernqvist, Å., Simulation of the steady-state and dynamic behaviour of multiple-effect evaporation plants. Part II: Dynamic simulation. ibid, 1977, **8:1**, 29-40.

7. Abrahamsson, K. and Jernqvist, Å., Modeling and simulation of absorption heat pump cycles. Submitted for publication (1992).

8. Rinheat OY, Finland, private communication (1991).

9. Eriksson, K. and Jernqvist, Å., Heat transformers with self-circulation: Design and preliminary operational data. Int. J. Refrig., 1989, **12**, 15-20.

10. Abrahamsson, K. and Jernqvist, Å., Design features of different components for heat transformers with self-circulation. Submitted for publication (1992).

11. Aly, G., Abrahamsson, K. and Jernqvist, Å., Application of absorption heat transformers for energy conservation in the oleochemical industry. Submitted for publication (1992).

ENERGY CONSERVATION IN THE SUGAR INDUSTRY

GIDNER, A. and JERNQVIST, Å.
Department of Chemical Engineering I, University of Lund,
P.O. Box 124, S-221 00 Lund, Sweden

ABSTRACT

The technical and economic feasibility of incorporating an absorption heat transformer (AHT) unit was investigated to increase the energy efficiency of an existing evaporation-continuous crystallization plant in a sugar mill. The evaporation plant consists of a single-effect pre-evaporator connected in series with a quadruple-effect unit. The plant is operated in a co-current mode using live steam. To increase the energy efficiency of the plant, an extensive number of heat exchangers is used and secondary vapour streams are withdrawn to provide the leaching and crystallization plants with the thermal energy needed.

Using operating data from a major Swedish sugar mill, three different process configurations have been simulated using two flow sheeting programs. Two waste heat streams were utilized in the AHT where the generator is supplied by vapour leaving the crystallization unit, while waste heat from the evaporation plant is supplied to the evaporator of the AHT. The results of these simulations revealed that the total amount of live steam currently used in the plant can be reduced by about 20%. Based on an operation time of 2,000 hrs/year, an economic analysis revealed that a pay-off period of four to five years would be obtained. Installing the heat transformer requires minimum changes in the existing plant.

INTRODUCTION

Unit operations such as evaporation, distillation and drying are the most energy intensive in the process industry due to the phase conversion from liquid to vapour. For instance, in 1990 the amount of energy consumed in all Swedish evaporation plants was about 11.6 TWh, which is equivalent to 1 Mtoe. This is to be compared with the total energy used by the Swedish industry which was 139 TWh [1].

A portion of the waste heat is theoretically recoverable to produce steam or to supply other process heating requirements. The heat recovered would reduce the primary energy input into the process. Incorporating various types of heat pumps within different units in a process burdened by waste heat streams having appropriate temperature levels is an efficient method for energy recovery. However, relatively few applications have so far been implemented in the process industry.

The main objective of this work was to investigate the technical and economic feasibility of incorporating an absorption heat transformer unit to increase the energy efficiency of an existing evaporation plant in a sugar mill having an annual production capacity of 85,000 ton.

THE HEAT TRANSFORMER

Absorption heat pumps (AHP) are based on the principle of utilization of the enthalpies of evaporation-condensation of solutes in appropriate working fluid mixtures. These systems consume very little electric energy and use mainly a high-temperature (primary heat) and a low-temperature (waste heat) heat source to recover useful heat at an intermediate temperature as illustrated in Fig. 1. Some electric energy is used only for the purpose of recirculating the working fluid.

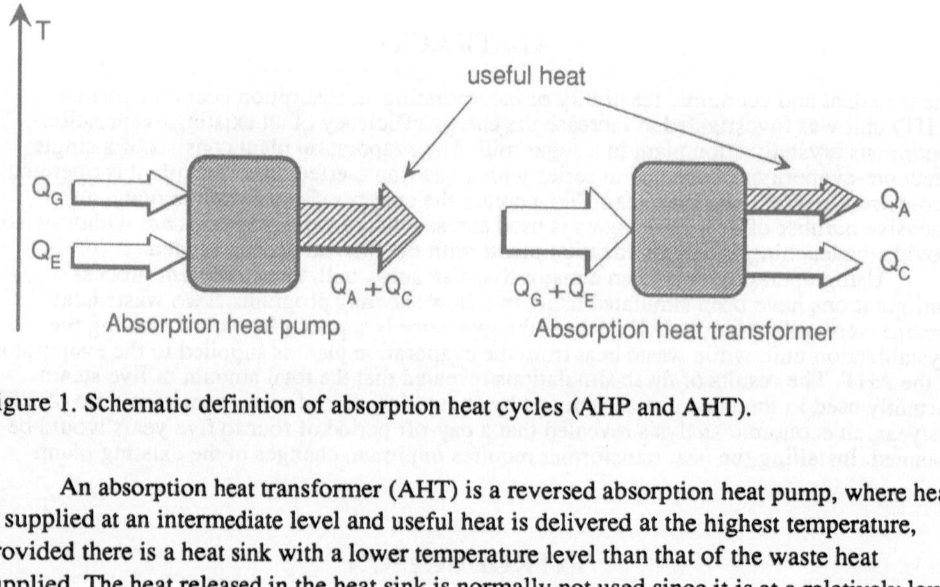

Figure 1. Schematic definition of absorption heat cycles (AHP and AHT).

An absorption heat transformer (AHT) is a reversed absorption heat pump, where heat is supplied at an intermediate level and useful heat is delivered at the highest temperature, provided there is a heat sink with a lower temperature level than that of the waste heat supplied. The heat released in the heat sink is normally not used since it is at a relatively low temperature.

The heat transformer, which consists of four main components, is normally operated at three temperature and two pressure levels as can be seen in Fig. 2. Besides those four components, the system usually includes one heat exchanger and two pumps. The absorber and the evaporator operate at a higher pressure level than the generator and condenser. The waste heat input is supplied to the generator and evaporator, at a middle temperature level, while about half of the heat input is produced at a higher temperature in the absorber and the other half is dissipated at a lower temperature in the condenser. The available useful energy output of the AHT is accordingly about 50% of the waste heat input. However, an AHT does not consume high-grade thermal energy and only a small amount of electric energy is used to recirculate the liquid streams. The circulation pumps can be totally eliminated if the heat transformer is operated in the self-circulation mode [1].

A working fluid pair is circulated between the absorber and the generator to provide the heat pump loop. Absorption occurs in the absorber and desorption occurs in the generator. Working fluids consist of a refrigerant and an absorbent. When heat is added to the working fluid mixture, the refrigerant evaporates, leaving behind a weak working fluid. As the refrigerant is condensed, the latent heat is given off to the cooling water. When the refrigerant is vaporized in the evaporator and rejoined with the weak working fluid, heat is liberated to the process.

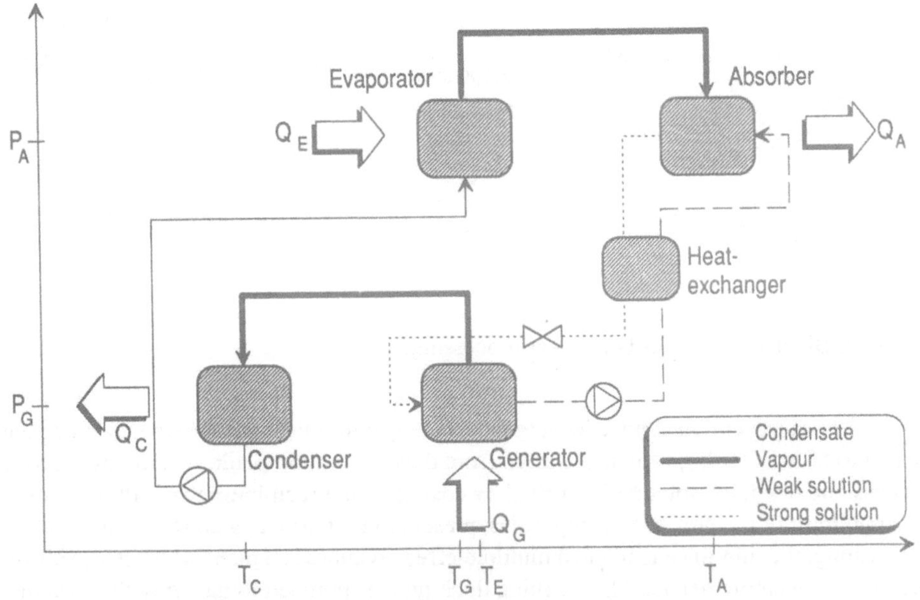

Figure 2. Schematic flowsheet for the absorption heat transformer system.

THE BEET-SUGAR PROCESS

Figure 3 shows a simplified flowsheet of the beet-sugar process. The washed beets are sliced into cossettes which are leached at 70-75°C using hot water. The water percolates counter-currently through the cossette mass, where about 98% of the sugar is usually leached. The sugar-depleted cossettes, known as pulp, are run to pulp presses where the moisture content is reduced from about 95% to between 76 and 84%. The water pressed from the wet pulp is usually sterilized by brief heating and recycled as a reflux stream to the leaching equipment. After the addition of molasses, the pressed pulp is dried to a moisture content of about 10%, cooled and pelletized before final shipping as ruminant animal feed.

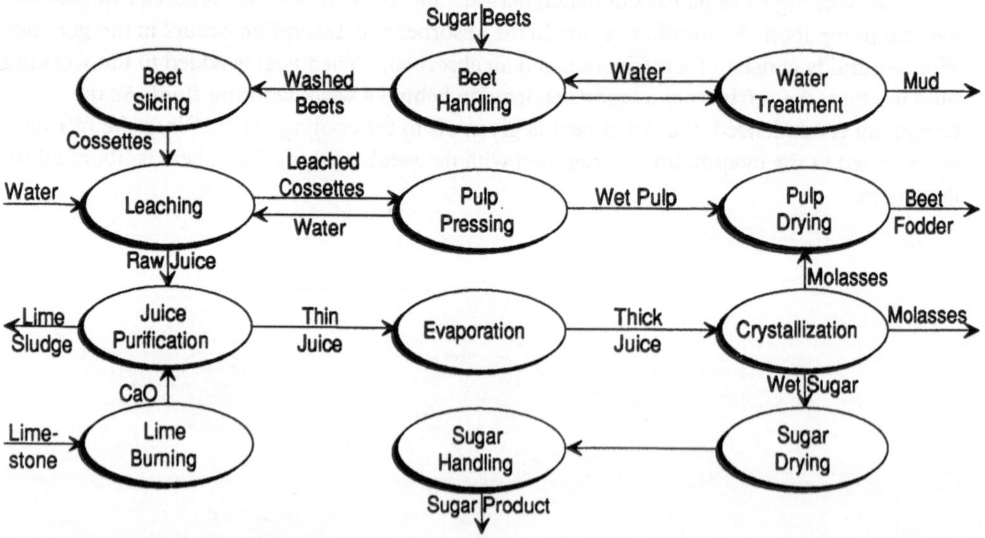

Figure 3. Block diagram for beet-sugar processing.

The sugar-enriched juice leaving the leaching equipment is first screened and then heated to 80-85°C using vapour extracted from the evaporation plant. In a liming-carbonation process, the raw juice solution is purified by coagulation, precipitation and filtration to separate most of the nonsucrose impurities, present in both true and colloidal solution. After pre-heating, the thin juice is fed to a multiple-effect evaporation plant which is described below in some more details. The resulting thick juice is pumped to the crystallization plant where a saturation environment is maintained to yield an optimum crystal growth rate initiated by an embryo crystal population. The fillmass, a mixture of sugar crystals and mother liquor, is fed to centrifugals and following one or two brief washes with pure hot water, the liquid phase, called syrup, is recycled to the crystallization plant. The wet white sugar crystals are sent to a dryer-granulator-cooler unit where the moisture content is reduced from 1- 2% to 0.02 - 0.03%. Finally, the granulated sugar is screened and sacked.

THE EXISTING EVAPORATION PLANT

Figure 4 shows the evaporation plant which consists of a single-effect pre-evaporator (effect # 5) connected in series with a quadruple-effect unit. The plant has a total heat transfer area of 17,600 m^2 and is operated in a co-current mode using steam from a back pressure turbine.

Three effects are of the falling-film type (effects # 1, 3 and 4), while the second effect consists of four Robert-type evaporators. The pre-evaporator is also of Robert-type. The falling-film and Robert evaporators have a heat transfer area varying from 2,500 to 5,000 m^2 and 900 - 2,000 m^2, respectively. To increase the energy efficiency of the plant, an extensive number of heat exchangers is used and secondary vapour streams from both third and fourth effects are withdrawn to provide the leaching and crystallization plants with the thermal energy needed. The crystallization plant consists of three units currently operated in a batchwise mode. The first unit will shortly be replaced by a modern crystallizer for continuous mode of operation. The thermal energy needed in the crystallization plant is supplied using the main part of the vapour streams extracted from the evaporation plant.

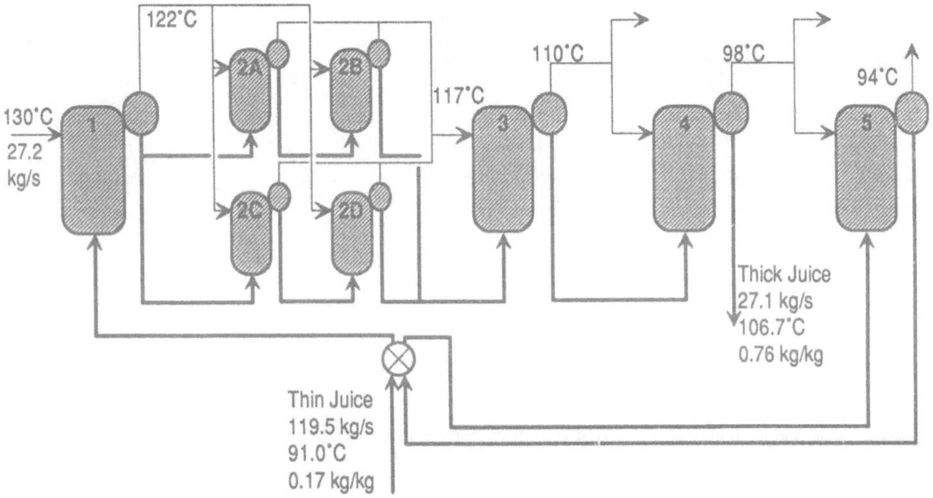

Figure 4. The existing evaporation plant.

SIMULATION RESULTS

SHPUMP and INDUNS are two computer program systems which have been developed for the simulation of different absorption heat cycles and general multiple-effect evaporation processes, respectively [1,2]. A large number of computer runs were performed to establish both local and global optimal energy conservation strategies for the evaporation plant investigated in this work.

The evaporation plant was first simulated using actual operating data, provided by a major Swedish sugar mill, to calculate complete temperature, concentration, and overall heat transfer coefficient (U-values) profiles. The total live steam consumption in the current mode of operation is 27.2 kg/s which is equivalent to 59.1 MW. The U-values and apparent temperature differences were in the range 1.5 - 3.0 kW/m^2K and 3.5 - 7.3 °C, respectively.

Different system configurations were simulated to determine the optimum location and size of the heat tranformer. To find a solution with considerable energy conservation, the continuous crystallization unit, which is supplied with vapour extracted from the fourth effect, had to be incorporated in the calculations.

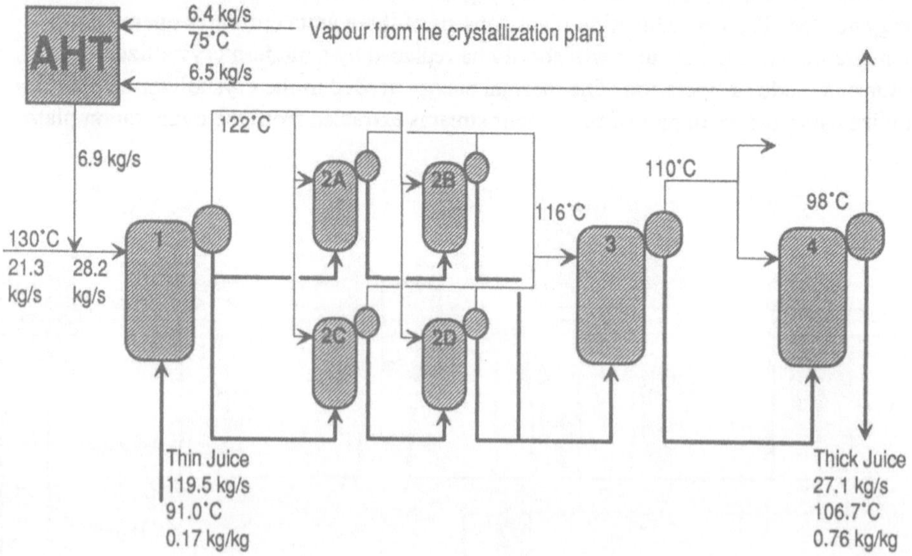

Figure 5. The AHT system incorporated in the evaporation plant according to configuration I

Three main process configurations were simulated and analysed. For all process configurations investigated in this work, an additional heat exchanger was incorporated in the condensate loop between the condenser and the evaporator components of the AHT system. This heat exchanger would utilize a considerable portion of the heat content of the condensate stream leaving the evaporator to preheat the unsaturated condensate stream entering the evaporator. This modification was found necessary in order to maximize the utilization of the vapour stream extracted from the evaporation plant to increase the useful thermal energy delivered by the AHT.

Figure 5 shows a simplified flowsheet of configuration I where the pre-evaporator, effect # 5, would be disconnected. The vapour stream leaving the evaporation plant from the fourth effect will now be splitted into three streams where 55.7% of the total flow rate is extracted to the continuous crystallization unit, 6.9% to preheat different juice streams, and the balance would be fed into the evaporator component of the AHT. The heat transformer system in this case was designed to deliver vapour compatible with the heating steam used in the first effect of the evaporation plant. This can be achieved by utilizing the vapour leaving the crystallization plant, at 75°C, as heat input in the generator component of the AHT while 37.4% of the vapour stream leaving the fourth effect, at 98°C, will be used in its evaporator component. The simulation results revealed that the heat transformer system in this configuration would reduce the amount of live steam needed in the evaporation and crystallization plants from 27.2 to 21.3 kg/s, i.e. a reduction of 22%, corresponding to 12.8 MW thermal energy. Table 1 displays the basic design parameters for the different components of the heat transformer which would deliver an output of 15.1 MW. Based on H_2O - NaOH as the working pair, the operating pressures would be 3.8 and 78.5 kPa, while the concentrations would be 51.7 wt% for the strong soulution and 48.2 for the weak one.

Table 1.
Basic design data for the absorption heat transformer of configuration I.

Variable		Evaporator	Generator	Absorber	Condenser	H. Ex.
Effect, Q	(MW)	16.4[a]	14.8	15.1	16.1	17.0
Flow rate	(kg/s)	6.5[b]	6.4[b]	6.9[b]	766[c]	
Temperature, T	(°C)	98	75	130	23	
LMTD	(°C)	5	6.6	7.1	7.2	7.5
Pressure	(kPa)	78.5	3.8	78.5	3.8	
Conc., X	(wt% H_2O)	100	48.2	51.7	100	
U·A	(kW/°C)	2940	2240	2130	3220	2150

[a] Including 1.7 MW from the additional heat exchanger [b] Steam [c] Cooling water

Although this configuration results in a substantial energy conservation, it assumes a rather high saturation temperature (75°C) of the vapour leaving the continuous crystallization unit, where the fillmass normally has a boiling point elevation of approximately 10°C. There is a certain critical temperature above which the sugar content in the fillmass tends to caramelise, causing a loss of sucrose and a colouration which will persist to the final crystals of sugar [3]. If the pressure in the continuous crystallization unit decreases, there will be a lower pressure of the vapour delivered to the generator. This pressure is to low to attain the desired pressure of the vapour leaving the absorber, unless it is possible to increase the pressure in the evaporator of the AHT.

This problem was solved in configuration II by connecting, on the vapour side, the pre-evaporator in parallel with the fourth effect, as illustrated in Fig. 6. Since the concentration of the juice in the pre-evaporator is much lower, it is possible to obtain a higher saturation pressure of the vapour leaving this effect than the corresponding pressure from the fourth effect. Supplied by vapour from the pre-evaporator effect, the evaporator of the AHT can therfore be operated at a relatively higher pressure than in the first configuration. Consequently, this arrangement would make it possible to decrease the pressure in the crystallization plant and still allow the AHT to deliver vapour, to the first evaporation effect, at the desired temperature of 130°C. This simple modification would also result in some saving in live steam consumption since pre-heating the raw juice solution can be achieved using vapour from the pre-evaporator, now maintained at a higher temperature level, instead of vapour from the third effect. Installing the AHT system according to this configuration requires very few changes in the existing plant, and would reduce the amount of live steam from 27.2 to 22.5 kg/s, i.e. a reduction of 17%, corresponding to 10.2 MW.

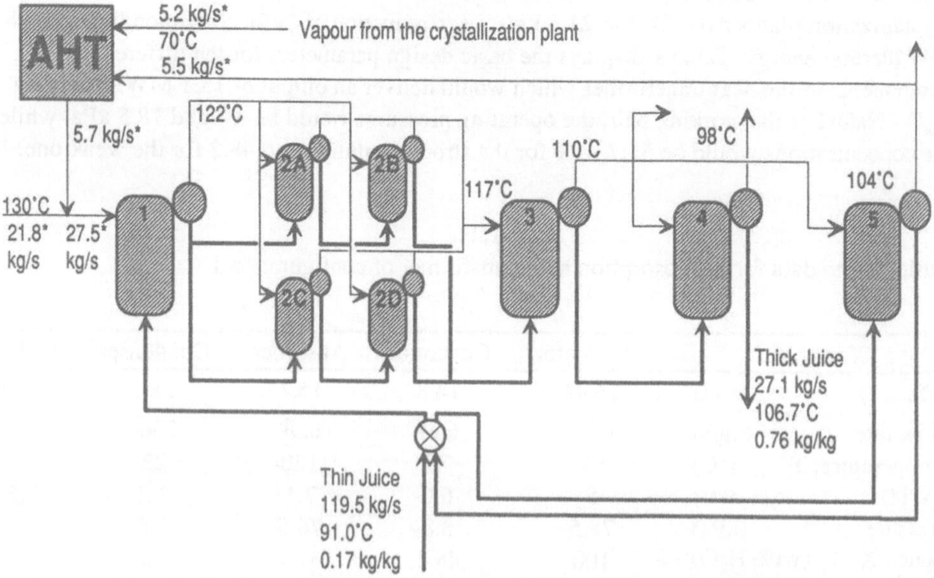

Figure 6. Schematic flowsheet for configuration II and III. *Flows valid only for configuration III

Table 2 displays the basic design parameters for the different components of the heat transformer which would deliver an output of 10.8 MW. Based on H_2O - NaOH as the working pair, the operating pressure range would be 4 - 98 kPa, while the concentration range would be 55.8 and 52.6 weight % for the strong and weak solutions, respectively.

Table 2.
Basic design data for the absorption heat transformer of configuration II.

Variable		Evaporator	Generator	Absorber	Condenser	H. Ex.
Effect, Q	(MW)	12.0[a]	10.6	10.8	11.8	14.1
Flow rate	(kg/s)	4.8[b]	4.5[b]	5.0[b]	469[c]	
Temperature, T	(°C)	104	70	130	24	
LMTD	(°C)	5	6.5	6.8	7.6	7.5
Pressure	(kPa)	98	4.0	98	4	
Conc., X	(wt% H_2O)	100	52.6	55.8	100	
U·A	(kW/°C)	2160	1620	1580	2360	1880

[a] Including 1.2 MW from the additional heat exchanger [b] Steam [c] Cooling water

The main objective of configuration III was to increase the availability of the vapour leaving the pre-evaporator effect under different operating conditions since this vapour is the main source of thermal energy to the evaporator of the AHT. This would result in an increase of the high pressure vapour delivered by the AHT, taking into account the surplus in vapour leaving the continuous crystallization unit. Optimization of the thermal energy needed for pre-heating the raw juice stream, according to Fig. 7, would save some vapour from the pre-evaporator. Keeping the same design features of the AHT as in configuration II, the live steam consumption in this case can be reduced from 27.2 to 21.8 kg/s, i.e. a reduction of 20%, corresponding to 11.7 MW. As indicated in Table 3, the thermal output delivered by the heat transformer would be 12.4 MW.

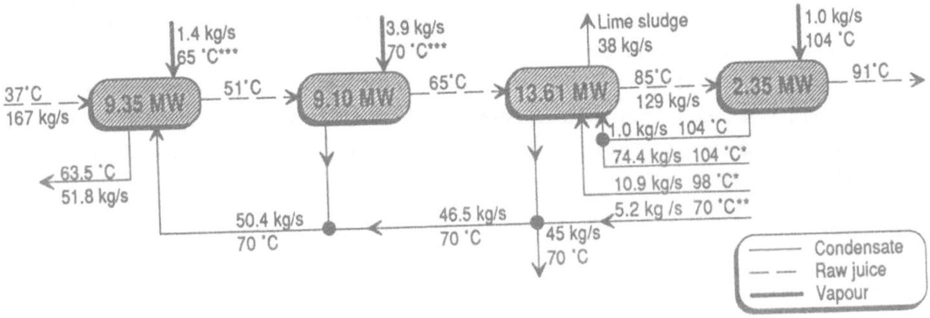

Figure 7. Optimization of the juice pre-heating scheme. *Condensate from the evaporation plant. **Condensate from the generator. ***Vapours from the crystallizers.

It should also be pointed out that incorporating the AHT system, according to configuration III, would require minor physical changes in the existing evaporation plant as clearly illustrated in Figures 5 and 6.

Table 3.
Basic design data for the absorption heat transformer of configuration III.

Variable		Evaporator	Generator	Absorber	Condenser	H. Ex.
Effect, Q	(MW)	13.8[a]	12.1	12.4	13.5	16.2
Flow rate	(kg/s)	5.5[b]	5.2[b]	5.7[b]	537[c]	
U·A	(kW/°C)	2470	1860	1810	2710	2150

[a] Including 1.4 MW from the additional heat exchanger [b] Steam [c] Cooling water

ECONOMIC EVALUATION

An economic evaluation was performed to determine the economic feasibility of incorporating a heat transformer in this type of evaporation plants. Based on cost information recently published [4], the installed equipment cost for the heat transformer system was assumed to be 260 $/kW of delivered heat. The pay-off period of this type of investment depends mainly on the local steam cost, heat transformer efficiency, operation time, and also on how much of the existing equipment in the evaporation plant that can be utilized as components in the heat transformer. The steam cost was taken as 8.7 $/GJ, corresponding to 110 SEK/ton, and a thermodynamic efficiency of approximately 47% was assumed. In this case the third factor is most important due to the seasonable characteristics of this type of industry. In this work, the annual operation time was taken as 2,000 hrs/year while the fourth factor was not considered in the economic evaluation.

The pay-off period for the heat transformer in the last and most interesting case (configuration III) would be 4.4 years. It should be noticed, however, that the pay-off period would be reduced to one year for a conventional process industry with an annual operating time of 8,000 hrs/year.

ACKNOWLEDGMENTS

The authors are grateful to Dr Claes Gudmundsson and Mrs Karin Forsberg of the Swedish Sugar Company for their cooperation, supplying the necessary process data, and for their kind permission to publish the results.

The financial support of the Swedish National Board for Industrial and Technical Development and the Swedish Council for Building Research are gratefully acknowledged.

REFERENCES

1. Aly, G., Abrahamsson, K. and Jernqvist, Å., Integration of absorption heat transformers in the process industry - applications in the oleochemical, pulp and paper industry. Accepted for publication, International Conference on Energy Efficiency in Process Technology, Athens (1992).

2. Bolmstedt, U. and Jernqvist, Å., Simulation of the steady-state and dynamic behaviour of multiple-effect evaporation plants. Part I: Steady-state simulation. Computer Aided Design, 1976, **8:3**, 142-148.

3. McGinnis, R.A., Sugar Beet Technology, 3rd ed., Beet Sugar Development Foundation, Colorado (1982).

4. Scheihing, P.E. and Cuervo. L.A., Market opportunities of industrial chemical heat pumps in the United States. IEA Heat Pump Centre Newsletter, 1990, **8**, 16-19.

SESSION 11:

Dynamic Simulation and Batch Processes

Chairman: Prof. D. Tsahalis

SIMULATION AND CONTROL OF FAST TRANSIENTS IN PROCESS AND UTILITY PLANT

H. Yeung[1], A. Guilbert[1], G. Heyen[2], B. Kalitventzeff[2], K. Murphy[3], D. Marchio[3], M. Gill[4]

1. School of Mechanical Engineering, Cranfield Institute of Technology, Bedford MK43 OAL (UK)
2. L.A.S.S.C. Universite de Liege, Sart Tilman B6, B4000 LIEGE (Belgium)
3. Centre d'Energetique, Ecole des Mines de Paris (France)
4. W.S. Atkins, Woodcote Grove, Asley Road, Epsom, Surrey KT18 5BW (UK)

ABSTRACT

A dynamic simulation program, FASTran, is underdevelopment capable of modelling fast transients and control systems in rotary machines and utility networks. Improving energy efficiency and safety is the incentive to study the behaviour of such systems in operation ranges close to instabilities (e.g. compressor surge). The program is based on the concept of Object Oriented Programming methodology. The most important features of OOP such as data independence, reusability and inheritance are implemented. Lumped parameter models have been developed to simulate dynamic behaviours of plant equipments. The models are based on one dimensional approximation of continuity momentum and energy equations. Controller design tools are embedded in FASTran with easy transfer of information between simulation and controller design. This permits the results of simulation to be used for controller design and optimisation.

INTRODUCTION

This paper describes part of a research programme, supported by the JOULE Programme, where Cranfield Institute of Technology (U.K.), University of Liege (B) and Ecole des Mines de Paris (F) join efforts to develop modelling and control design tools. W.S. Atkins(UK) and Thomassen International(N) are the two industrial associated partners of the project. A first goal of the project is modelling of fast transients (time scale from 10^{-3} to 10s) in rotodynamic machines and utility distribution networks. Improving energy efficiency is the incentive to study the behaviour of such systems in operation ranges that are closer to instability than usual operation mode. For instance, the maximum polytropic efficiency of compressors is usually located close to the surge limit.

Precise modelling of the process behaviour in the domain of interest should allow the reproduction of the occurrence of instabilities or even their prediction at the design stage. Next control algorithms can be developed in order to allow safe operation with improved energy efficiency.

This paper gives an overview of the FASTran dynamic simulator, the theoretical basis of dynamic simulation and the controller design method. The simulation results are compared with test data obtained during the commissioning of a cogeneration plant.

THE FASTran PROGRAM

A new dynamic simulation package, FASTran, capable of modelling fast transients and control systems in rotary machines and utility networks is under development, see Heyen et al (1,2). Some important features of the program are described below.

The program has been coded in FORTRAN with portability in mind. This allows the reuse large portions of existing codes (e.g. the physical property package) with only minor adaptations.

Unlike most programs, its architecture does not rely on PLEX data structures, which allow all process information to be shared between the various subsystems of the simulator. Whilst the PLEX architecture is memory efficient, it may lead to difficulty in debugging and maintenance of the code, since any subroutine has access to the whole data base. FASTran is based on the Object Oriented Programming methodology. The most important features of OOP such as data independence, reusability and inheritance are implemented.

The main advantage of object based architecture is better structuring of data. Each module has the capability to access and structure its own data in the most convenient way. For instance, reading data from the input file can be done differently for each object if this looks easier for the user,(or the programmer). This allows easier interfacing with other data structures, such as data base management systems or fancier user interfaces.

A typical illustration is found in the way heat exchanger unit has been implemented. Objects describing fluid flow in pipes and storage in vessels are available. The simulation equations take into account an enthalpy transfer term, whose magnitude is not specified in the model. This heat transfer term is evaluated in a "wall" object, that has access to the various heat transfer correlations. Thus a heat exchanger can be modelled by merging two "fluid flow" elements (e.g. a pipe and a mixing vessel) and one "wall" element. The wall element will obtain geometric information and physical properties from the fluid flow elements. Different wall elements can be developed, to handle various types of heat exchange (e.g. radiation, convection). The wall objects and the heat flow objects can be developed independently, and any improvements in one of them can be made available at no cost to any combination of objects.

Five main object classes have been defined.

1. Equipments correspond to basic processing operations e.g. compressor, piping element, valve, expander, heat exchanger, combustor, etc). Controllers are also available in the simulator.

2. Streams correspond to the state of any material, energy or information transfer between equipments. Shaft power, electricity or heat are treated as stream objects as well as material flows.

3. Components correspond to the chemical species, or to identifiable materials whose properties can be characterised (e.g. a petroleum fraction).

4. Mixture objects handle the physical property models that allow the estimation of thermodynamic properties and phase equilibria for any stream containing a given list of components.

5. Reactions objects are able to handle mass and energy balance in reacting systems, to evaluate chemical equilibria or kinetic expressions.

A computer model of a plant under consideration is formed by a collection of equipments connected by streams. This information is read in via data files. Simulation begins with the initial conditions as defined after successive initialisation, verification and checking. Perturbations and forcing function may be defined as step, ramp or sine functions or smooth transitions between two states(with continuous first derivative with respect to time). A sparse matrix adaptation of the DASSL package (3) is used to obtain the solution of the resulting large set of algebraic and differential equations.

MATHEMATICAL MODEL FOR FAST TRANSIENTS

To achieve a concise model, the flow in the system is assumed to be one dimensional. Using the assumption, the conservation equations can written for a general system element (see Fig. 1), as follows:

Conservation of mass

$$\frac{d}{dt}\int_x \rho A dx = M_1 - M_2 \tag{1}$$

Conservation of linear momentum

$$\frac{d}{dt}\int_x M dx = (PA + Mu)_1 - (PA + Mu)_2 + Fnet \tag{2}$$

Conservation of energy

$$\frac{d}{dt}\int_x \rho \left(e + \frac{u_2}{2}\right) A dx = (Mh_o)_1 - (Mh_o)_2 + \dot{E}net \tag{3}$$

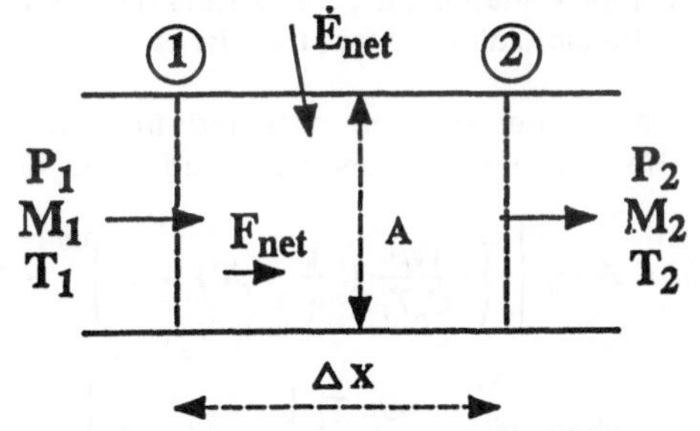

Figure 1. General System Element

The equations can be further simplified by use of some assumptions which are generally valid in the flows of interest in process and utility plants. In many cases, the Mach number of the flow remains low enough for terms involving Mach number to be neglected. Also, by replacing the terms under the integral in the left hand side of the equations by average values (the lumped parameter approximation), we have

$$V \frac{d\bar{\rho}}{dt} = M_1 - M_2 \tag{4}$$

$$\Delta x \frac{d\bar{M}}{dt} = (PA)_1 - (PA)_2 + Fnet \tag{5}$$

$$V \frac{d\,(\overline{\rho e})}{dt} = (Mh_o)_1 - (Mh_o)_2 + \dot{E}net \tag{6}$$

The terms Fnet and Ėnet represent the force and energy input into the element. These are estimated by assuming that the steady state data remain valid under transient conditions. The equations form the basis of all the fluid handling models. Various approximations can be incorporated provided that they are justifiable and are appropriate to the problems under investigation. Some of the equipment models developed are given below.

For pipes, Fnet is related to frictional pressure loss. For isothermal flows, Eqns (4),(5) and (6) become

$$V \frac{dP_2}{dt} = \gamma\, ZRT_1\, (M_1 - M_2) \tag{7}$$

$$\Delta x \frac{dM_1}{dt} = (PA)_1 - (PA)_2 + Fnet \tag{8}$$

$$\text{where } Fnet = A_1\, Cf\, \frac{\Delta x}{D} (\tfrac{1}{2}\, \rho_1\, u_1^2)$$

$$T_2 = T_1 \tag{9}$$

It should be noted that the pipe equations allow the transmission of pressure waves at the speed of sound enabling the simulation of fast transients.

For compressors, Fnet and Ėnet can be obtained from the pressure and power characteristics (Fig 2). The derivation has been detailed in Heyen et al(2).

$$Fnet = A\, P_1 \left\{ \left(\frac{Wp}{C_p T_1} \left(\frac{1}{\eta} + X \right) + 1 \right)^{1/n} - 1 \right\} \tag{10}$$

$$\text{where } W_p = \frac{C_p\, T_1}{\frac{1}{\eta} + X} \left\{ \left(\frac{P_2}{P_1} \right)^n - 1 \right\}$$

$$n = \frac{ZR}{Cp}\left(\frac{1}{\eta} + X\right)$$

$$\dot{E}net = \frac{W_p\ M_1}{\eta}$$

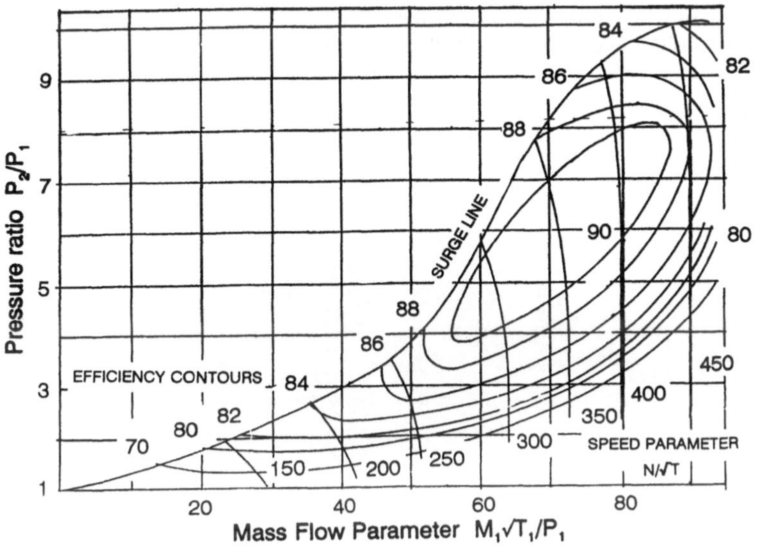

Figure 2. A Typical Compressor Characteristic

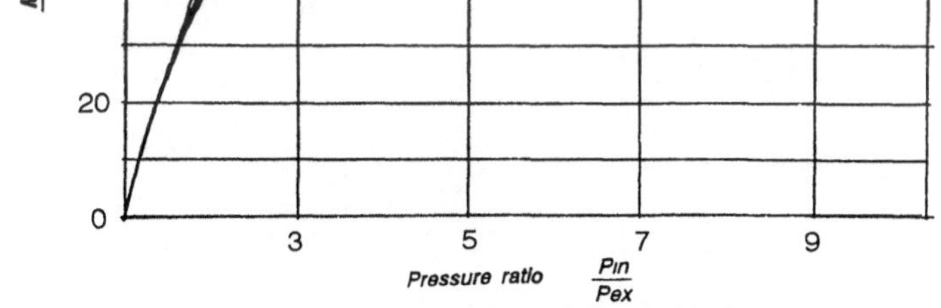

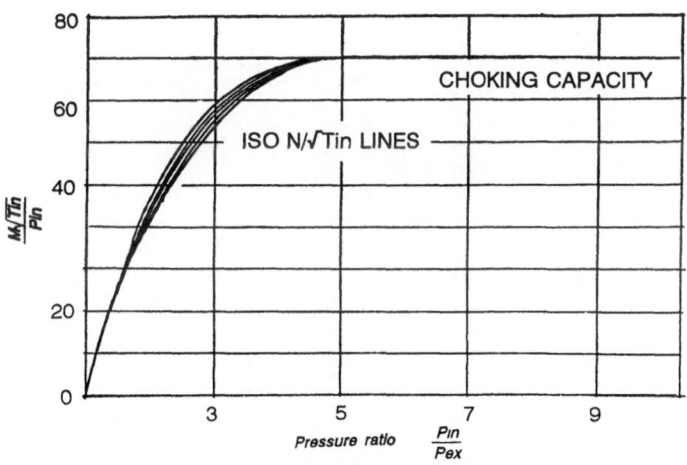

Figure 3. A Typical Turbine Characteristic

The lumped parameter model for turbines is similar. The performance characteristic of a typical axial flow turbine is shown in Fig 3. For control valves, $\dot{E}net$ can be set to zero while Fnet can be calculated from the valve characteristics (pressure drop versus valve opening relationship).

Fig 4 shows the a typical simulated output from the FASTran program for a step change in the outlet valve of a compressor system. It reveals the movement of pressure waves in the pipe between the compressor and valve. Close comparisons between simulated and experimental results have been obtained, see Heyen et al (2).

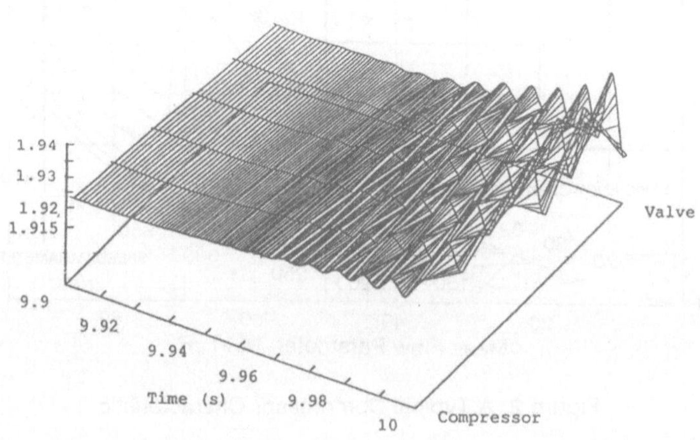

Figure 4. FASTran Simulation Result (Pressure Profile)

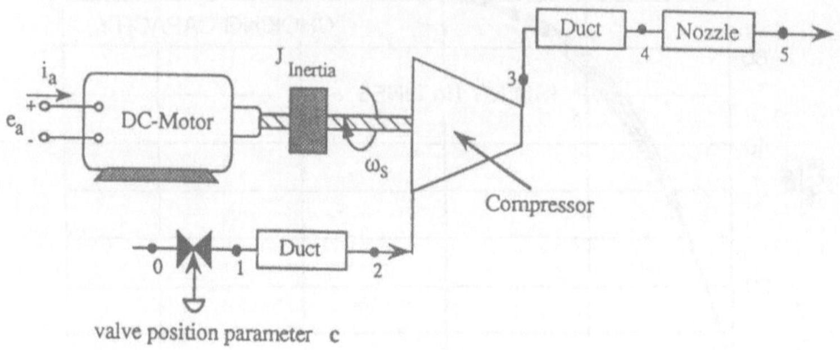

Fig. 5. Motor Driven Compressor System

CONTROLLER DESIGN WITH REFERENCE TO A CENTRIFUGAL COMPRESSOR SYSTEM

Controller design procedures are embedded in the FASTran with easy transfer of information between simulation and controller design. The user may easily switch from dynamic simulation to controller design tools. This permits the results of simulation to be used for controller design and optimisation.
The motor compressor system configuration illustrated in Fig.5 is used to outline the controller design methodology. The compressor is driven by a DC electric motor. An inlet throttling valve is connected upstream of the inlet duct, and a nozzle is connected to the outlet duct. The compressor characteristic is known for different speeds. The surge line separates a stable operating region, where small perturbation create damped oscillations, from an unstable region characterized by violent oscillations in mass flow. It is desired to regulate the outlet pressure P_3 and inlet mass flow rate M_2 to the compressor. The control variables are motor armature voltage e_a and a valve position parameter c_2.

The Small Signal Linearisation (ASME Book (4)) process is used to develop linear stat-space models which are used for the controller design. The equations that describe the plant are reorganised in the form

$$x_p = f(x,z,u,w) \tag{11a}$$

and

$$0 = z - g(x,u,w) \tag{11b}$$

where the vectors x,z,u and w are defined as:

$$x = [M_1, P_2, M_2, P_3, M_3, P_4, \omega_s, i_a]^T \quad \textit{state variables}$$

$$z = [M_o, P_1, T_1, T_2, T_3, M_4, T_4, M_5, T_5)^T \quad \textit{algebraic variables}$$

$$u = (C_2 e_a)^T \quad \textit{control variables}$$

$$w = [T_o, P_o, P_5]^T \quad \textit{disturbance variables}$$

Let xp, up and wp be defined as small perturbations of the state x, control u and disturbance w from some operating point x^{opt}.

$$\dot{x}_p = A_p x_p + B_p u_p + \Gamma_p w_p + h(x^{opt}, u^{opt}, w^{opt}, x_p, u_p, w_p)$$

The Jacobian matrices Ap, Bp and Γ_p are evaluated at the operating points x^{opt}, u^{opt}, w^{opt} and h is the remainder term.

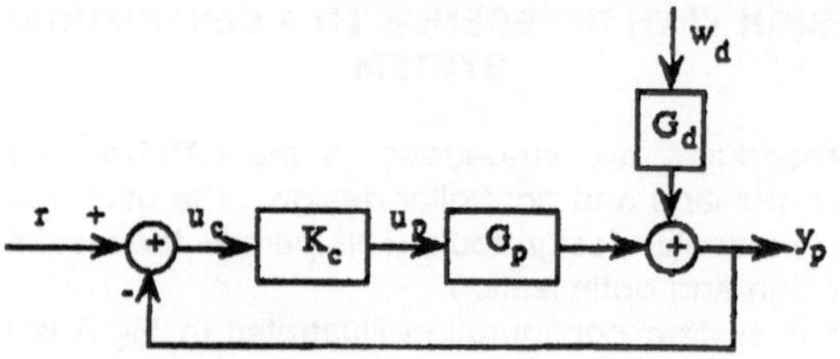

Fig. 6 Controller Block Diagram

The step responses of the resulting small signal models are compared by simulation experiments against the original nonlinear model in transient and steady state at other operating points. The small signal model is then used for the controller design when it accurately describes the nonlinear behaviour. To complete the description of the P-LTI (perturbed linear time invariant) model of the motor compressor model, a measurement vector y_p whose elements are presumed to be linear combinations of the state-variables, is added to the P-LTI model

$$y_p = C_p x_p$$

The Objective of the closed loop controller design is to regulate the inlet mass flow and the outlet pressure of the compressor. The closed loop controller structure is illustrated in Fig.6. The reference signals r are the respectively the desired values for input mass flow and the outlet pressure of the compressor. The transfer functions $G_p(s)$ and $G_d(s)$ are

$$\text{Plant:} \quad G_p(s) = C_p(sI - A_p)^{-1} B_p$$

$$\text{Disturbance:} \quad G_d(s) = C_p(sI - A_p)^{-1} \Gamma_p$$

The controller Kc has been built to be block diagonal, and for each input-output channel is a PI controller. The two input and output channels are respectively the regulation errors on M2, P3 and the variations of c and e_a. The controller Kc is similarly described by a LTI system of the form:

$$\dot{x}_P = A_C x_C + B_C u_C$$

$$y_c = C_c x_c + D_c u_c$$

$$u_c = r - y_p$$

The design procedure is composed of four steps:

1. compute the Jacobian matrices Ap,Bp,Γ_p and form the P-LTI model
2. check to see if the matrix pair (Ap,Bp) of the small signal model is stablizable,
3. check to see if the matrix pair (Cp,Ap)
4. tune the controller parameters such that the closed loop system is stable and the transient behaviour is satisfactory.

Fig 7 illustrates the action of the tuned controller to step changes of the mass flow and pressure setpoints. At time t=0, the two setpoints are changed and the controller acts simultaneously on voltage to adapt the speed of the compressor and on the valve position. The controller gives good regulation for the chosen setpoints of mass flow and pressure. The simultaneous responses on voltage and valve position, are shown to be absolutely coupled. The simulation illustrated a dominance of the controller's valve position over that of the controller's motor armature voltage.

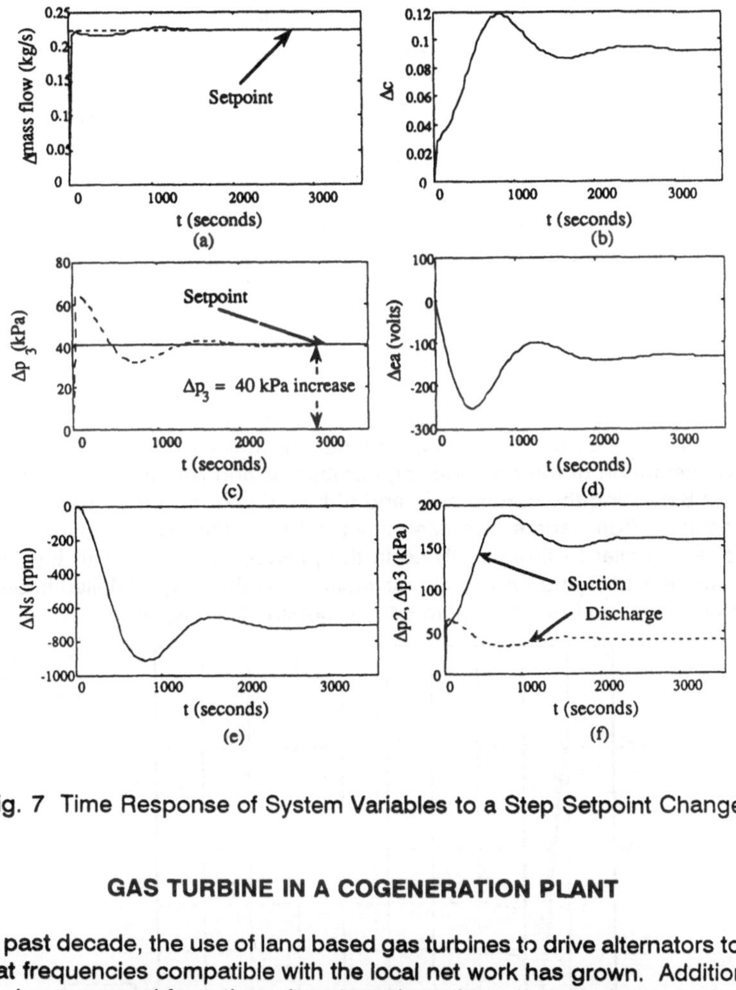

Fig. 7 Time Response of System Variables to a Step Setpoint Change

GAS TURBINE IN A COGENERATION PLANT

During the past decade, the use of land based gas turbines to drive alternators to produce electricity at frequencies compatible with the local net work has grown. Additional useful energy can be recovered from the exhaust gas by using a heat exchanger. Often steam is generated which can be for a process, as in the case of cogeneration, or can be used to generate additional electrical power in a steam turbine.

In FASTran two single shaft gas turbine models are provided. The first model (Stage 1)

is based on the simplified mathematical representations by Rowen(5). Fig. 8 shows a schematic of a gas turbine showing its essential components. In this model the gas turbine and its control are represented by transfer functions. It has been developed for GE (General Electric) heavy duty gas turbines ranges from 18MW to 106MW. The model is intended for power system stability investigations, dispatching strategy and contingency planning for system upsets. The model is limited to 95 -107 percent of the rated speed.

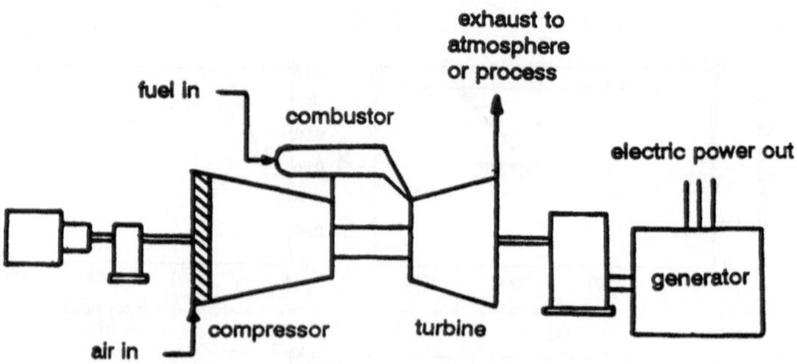

Fig. 8 Schematic Diagram of a Gas Turbine

The Stage 1 model does not include the simulation of gas dynamics even though the exhaust temperature can be calculated. The second model (Stage 2) includes a more detailed representation of the compressor, combustor and gas turbine. The fuel flow is calculated by balancing the compressor and turbine works and load requirement rather than interpolated from some predetermined tables. The compressor and turbine 'submodels' are similar to those outlined in the previous sections with the inclusion of appropriate compressor and turbine characteristics. In this way, detailed information on gas dynamics in any part of the system can be obtained at any time.

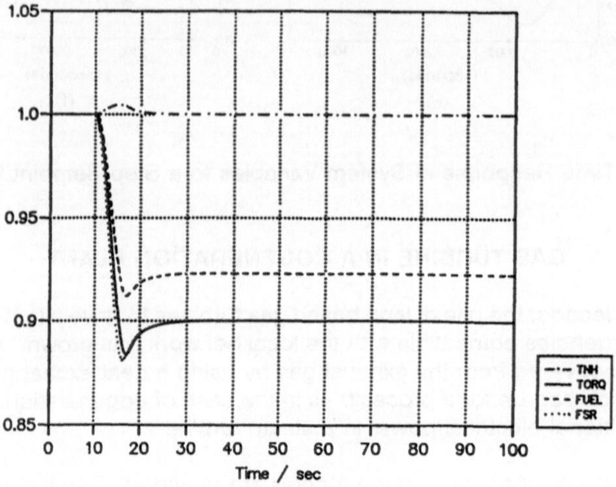

Figure 9a System response for 10% load change - Stage 1 Model

Fig.9 compares the response of a gas turbine to load change. One can see that both Stage 1 and Stage 2 give similar behaviour for rotational speed, fuel requirement and output torque. The Stage 1 model predicts an increase in compressor air mass flow with load while the Stage 2 model predicts a decrease in mass flow but an increase in discharge pressure. Rowan (6) mentioned that higher load is associated with higher compressor discharge pressure and air flow. However, an inspection of compressor characteristics (see Fig. 2) reveals that a higher pressure ratio is associated with a lower mass flow.

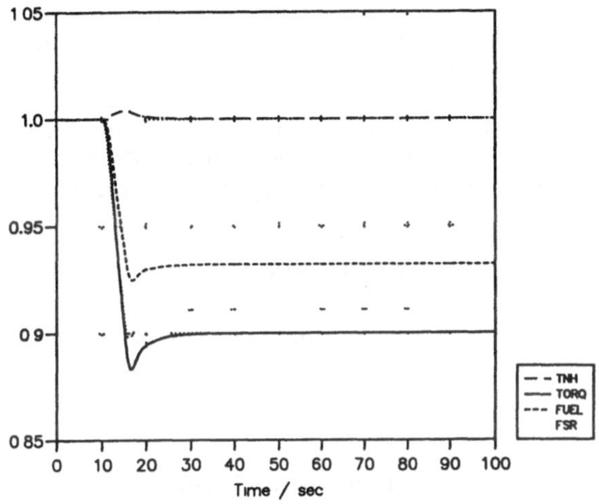

Figure 9b. System response for 10% load change - Stage 2 Model

Load	Compressor Outlet temp (°C)		Compressor Outlet Pressure (bar)		Turbine Exhaust temp./(°C)	
	site data	simulated	site data	simulated	site data	simulated
15.14	602	622	8.58	9.46	650	655
20.48	609	626	8.94	9.66	697	705
30.26	623	629	9.99	9.99	836	797

Table 1: Comparison of part load simulated and site data

Table 1 compares the simulated (Stage 2 model) compressor outlet temperature, pressure and exhaust temperature for various load with some site test data. The gas turbine is part of a cogeneration plant that is installed by Thomassen International and is being commissioned in India. The calculated results agrees fairly well with the measured values for a range of load conditions. The discrepancy is attributed to that the compressor and turbine characteristics are extrapolated from one single data point. With detailed characteristics, accurate performance simulation is expected. Work is presently undertaken to simulate the shut down behaviour of the gas turbine.

CONCLUSION

A general simulation program based on the concept of Object Orientated Programming has been produced. The program includes equipment models which can simulate fast transient behaviour, e.g. compressor surge. The models have been validated with test data. The simulation of a plant operation shows how the model may be used to study the system response to control actions and hence provide a design aid for control system specification. Robust control design procedures for a class of compression and gas turbine systems is given using linear approximation models embedded in the code. These design tools are coupled to the simulator and the user may easily switch from dynamic simulation of a plant to the controller design tools. Validation against real plant data proved that the approach can be used to study complex plant behaviour.

ACKNOWLEDGEMENT

The authors thank the Commission of European Communities (Joule Programme) for supporting the work described. The contributions of J F Mackie of Thomassen International bv are also gratefully acknowledged.

NOMENCLATURE

A	Cross sectional area
C	Valve position parameter
C_f	Friction coefficient
C_p	Specific heat
D	Diameter
e	Specific internal energy
e_a	Armature voltage
E	Energy
F	Force
h	Specific enthalpy
i_a	Armature current
M	Mass flow
P	Pressure
R	Gas constant
T	Temperature
u	Velocity
u	Control variable
V	Volume
w	Disturbance variable
x	Length
x	State variable
X	Compressiblity function
z	Controlled variable

Z	Compressibility factor
ρ	Density
w_s	Shaft speed

REFERENCES

1. Heyen G, Kalitventzeff B, Hutchinson P, Yeung H, Gill M.,"Simulation of Fast Transients in Fluid Transport Equipments and Utility Networks" ESCAPE-1, May 24-29, 1992(a), Elsinore, Denmark

2. Heyen G, Murphy K, Marchio D, Kalata P, Kalitventzeff B., "Dynamic Simulation and Control of Gas Turbines and Compressor Systems" To be presented, ESCAPE-2, October 5-7, 1992(b), Toulouse, France

3. Petzold L.R. "A description of DASSL: A Differential Algebraic System Solver" SAND 82-8637, Livermore USA, 1982.

4. ASME Book "Nonlinear System Analysis and Synthesis, Dynamic Systems and Control", vol.1, Chapter 1, New York, 1978

5. Rowen W.I., "Simplified Mathematical Representations of Heavy-Duty Gas Turbines" ASME Journal of Engineering for Power, Oct 1983 Vol 105 pp865-869

6. Rowen W.I. "Operating Characteristics of Heavy Duty Gas Turbines in Utility Service" ASME paper 88-97-150, Gas Turbine and Aeroengine Congress, Amsterdam, The Netherlands, June 6-9, 1988

RELIABILITY ASPECTS OF ENERGY EFFICIENT OPERATION OF TECHNICAL SYSTEMS

G. BECKER, A. BEHR, G. BARTSCH
Department for Process and Energy Technology
Technical University of Berlin
Marchstraße 18
1000 Berlin 10, FRG

ABSTRACT

A method is presented to determine abstract reliability related costs of technical systems. The approach makes use of dynamic simulation of the system to find the consequences of equipment failure as a function of calendar time and the duration of the failure. An inhomogenous Markovian model is used then to find expected cost functions related to failed equipment. The evaluation of the Markov process can be performed numerically, applying the same simulating software, which has been used to determine the deterministic consequences before. The approach is introduced with the aid of a simple example involving a batch process requiring external heating.

INTRODUCTION

Operating costs of process plants frequently have a complex structure. This is especially true, if costs related to reliability are to be considered. These latter are strongly influenced by stochastic factors; in fact, reliability related costs are (in a mathematical sense) random variables, as they depend on (random) life times and repair times of the equipment [1]. A suitable way to quantify this type of cost is an expected value function, which is defined as the average cost for some given observation period. These costs are not necessarily to be understood in terms of currency, but in any unit desired. Possible types of cost are: heat which is not delivered to the end user, emissions due to defective off gas processing systems etc.

To determine these costs, the response of the system due to component failure has to be evaluated. In most cases, this response is a dynamic one. Considering the stationary balance equations is in many cases sufficient to describe the system in its normal operating state; however, as soon as equipment fails, dynamic behavior will be significant. This is why reliability analysis almost necessarily is related to dynamic systems simulation, as soon as consequences (abstract costs) are within the scope of the analysis.

Within the context of a big research and development project on the simulation of large and complex energetic systems at the Technical University of Berlin, a method is being implied to estimate the average effects (costs) of non-standard operation, as it may be caused by unreliable equipment. Basically, the approach is to determine primary cost functions using the deterministic simulating tool. This will result in a cost structure, which is valid for a given accident scenario. This cost structure may depend on major influence factors, which are the duration of the failure, and the time, when failure occurs. The actual random behavior is modelled in a second step by applying a suitable stochastic model. Experience shows, that inhomogenous Markov processes, though they have not been used frequently in practise [3], are a good compromise between generality, and ease of application. Their use implies solution of a system of differential equation [2], which in most cases is well within the capabilities of the simulation tool, which has also been used for the first stage of the analysis. The result is an expected cost function, which may serve as a tool for planning the allocation of resources, and for decision making between concurring designs. The tool is developed with the intention to handle complex systems up to complete plants and more. In the following, however, for the purpose of clarification of the methodology, a simple example shall be considered throughout the rest of this paper.

A simple problem

As an example, a reaction is considered, which takes place in a reaction vessel operating in a batch-wise manner. The reaction requires a certain temperature, which is provided by a heat exchanger, which delivers heat at virtually no cost, as there are other exothermic processes in the plant. Batches are started at regular intervals, if process heat is available. For some reasons, the temperature must be kept above a given limit until a batch is finished. Two alternative heat sources are proposed to provide a redundancy to process heat: Either, an additional supply of steam may be installed. This would imply high initial costs of installation, as some additional piping is required. Alternatively, an electric heater may be provided, which is inexpensive to install, though it causes comparatively high costs, if it has to be used as process heat become unavailable during a batch. This heater is switched on, as soon as the temperature in the vessel falls below a critical value. Now, as the reaction vessel has some heat capacity, temperature will decay slowly. Only, if process heat remains unavailable for longer than a critical time called the tolerable down time, external heat will be consumed until the end of the batch.

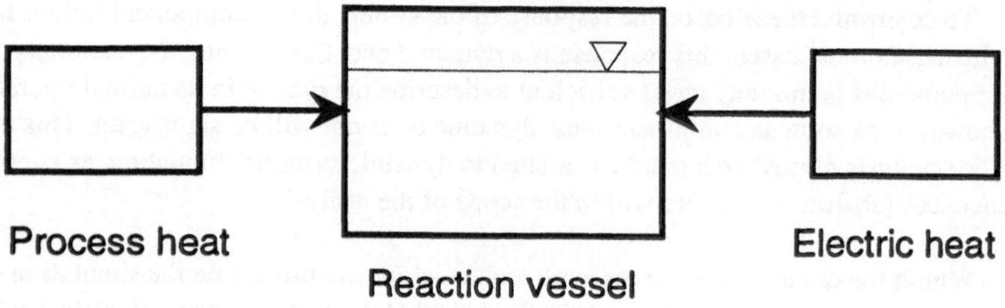

Figure 1. Schematic of the example under consideration

Given this scenario, a question of interest for decision making may be: how much (electrical) energy will be required on an average due to unavailability of the process heat source ? To answer this question, clearly a combination of deterministic (physical-chemical) and stochastic modelling is required. By a deterministic simulation, the tolerable down time for the process heat supply is determined, whereas a stochastic model is used to find the probability, that the heater is required at some time t. The integral over this probability is proportional to the expected cost of electrical energy required.

DETERMINISTIC SIMULATION

The question to be answered by deterministic simulation is: After what time T_{Tol} will the temperature in the vessel fall below some critical value θ_{crit}, which necessitates electrical heating, if the primary heat source fails. An accurate solution for this question will involve thermodynamics and reaction kinetics, as the enthalpy required at some time will depend on the temperature of the vessel, and on the degree, up to which the reaction is finished at the time of failure of the primary heat source. Also, losses of heat to the surrounding have to be considered. For the purpose of this example, the deterministic model has been kept simple. The reaction enthalpy has been neglected, and the reaction vessel has been modelled by a single heat capacity and a single heat conductivity. Under these assumptions, the time from the removal of the heat supply until θ_{crit} has been reached will not depend on the time of failure, as the resulting balance equation is homogenous w.r.t. time. This simplification, however, is not a restriction of the approach; if the reaction kinetics are modelled more accurately, the resulting tolerable down time will be a function of time, which will result in a time dependent transition rate in the associated Markov model. As this latter is inhomogenous for other reasons, a time dependent transition rate would cause no principle difference. The resulting temperature after loss of heating is given in fig. 2. It is indicated there, that the critical temperature (65 ^{o}C) will be reached after approx. 3 hours, which will be used as a tolerable down time for the stochastic part of the model presented below.

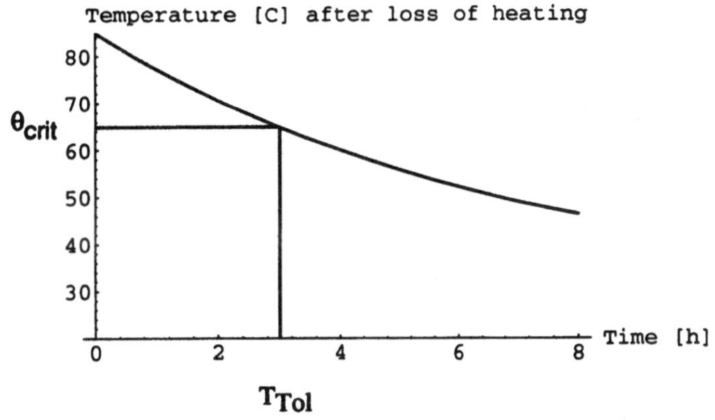

Figure 2. Temperature in the reaction vessel after loss of primary heat supply

AN INHOMOGENOUS MARKOVIAN MODEL

Essentially, a Markovian model is represented by a number of disjoint states and the transitions between these states [2]. The states have to be chosen in a suitable way, such that for some time t, the future behavior of the system will depend only on the state, the system occupies at time t, but not on the past, i. e., the way, this state has been reached. For the given example, a model of six states is given in fig. 3, which are defined as follows:

State	Description
1	primary heat supply intact and batch processing active
2	primary heat supply defect and θ_{crit} has not been reached yet
3	primary heat supply defect and θ_{crit} has been reached
4	primary heat supply intact and no batch process active
5	primary heat supply defect (unnoticed) and no batch process active
6	primary heat supply under repair and no batch process active

States 5 and 6 have been distinguished, as a failure of the primary heat supply is not noticed immediately, if no heat is consumed, i. e., if no batch is active. However, the policy is to check before starting a batch, whether primary heat is available. If it is, the batch is started, otherwise, state 6 is assumed; the batch is omitted, and repair action starts.

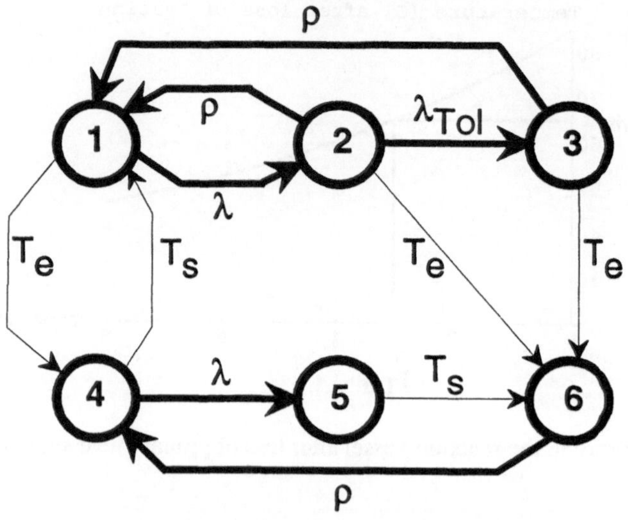

Figure 3. A Markov state graph for the loss of primary heat supply

Between the states, the following (random) transitions are possible:

From state 1 to 2: transitions occur with the failure rate of the primary heat supply (λ).
From state 2 to 1: transitions occur with the repair rate of the primary heat supply (ρ).
From state 2 to 3: transitions occur where the rate is the reciprocal of the tolerable down time T_{Tol} of the primary heat supply (λ_{Tol}).
From state 3 to 1: transitions occur with the repair rate of the primary heat supply (ρ).
From state 4 to 5: transitions occur with the failure rate of the primary heat supply (λ).
From state 6 to 4: transitions occur with the repair rate of the primary heat supply (ρ).

The following transitions will not occur randomly, but rather at fixed points of time (the starting time and the end time of the batches. This results in an inhomogenous model.

From state 1 to 4: transitions occur, when a batch terminates.
From state 2 to 6: transitions occur, when a batch terminates.
From state 3 to 6: transitions occur, when a batch terminates.
From state 4 to 1: transitions occur, when a batch starts.
From state 5 to 6: transitions occur, when a batch (is supposed to) start.

Electric heat will be consumed only, if the process is in state 3. Thus, p_3 (t), the probability, that the process is in state 3 has to be determined. Integrating p_3 over some time interval will yield the expected time T_{el}, the system is in state 3 during this interval. As it has been assumed, that the power of the electric heater is constant, the amount of electric heat required will be proportional to T_{el}.

This modelling needs some commenting. First, it can be seen, that the operating policy of the plant under consideration can be taken into account in detail. This makes the approach an invaluable tool to assess different operating policies for purposes of decision making. For example, if the primary heat supply is not checked before a batch is initiated, there would be no state 6, but rather a transition from state 5 to 3, when a batch starts. Or, if on the other hand, the state of the primary heat supply would be made observable by some additional instrumentation, there would be no state 6, but rather a random transition from state 5 to 4 with the repair rate (ρ). In practise, this type of analysis is only possible, if the reliability analyst closely collaborates with either plant designers or (preferably) plant owners. This, however, is true for nearly all types of reliability analysis.

As a second issue, the underlying assumptions of modelling have to be considered. In an ordinary Markovian model, all random transitions must be governed by rates, which may possibly depend on calendar time, but not on the length of the duration, the system was in some given state. If the random variables cannot be described by this type of rates, this imposes the necessity to include more states. Whereas a constant failure rate for some device is a realistic assumption, a constant repair rate (i. e. exponentially distributed repair times) may be questioned. It is, however, possible, to approximate any distribution by a multi-Erlang distribution, which may be represented by additional states in the process. This is not a restriction of the modelling approach. Another simplification is the modelling of the tolerable down time by a rate. Actually, the transition from state 2 to 3 should take place, if the process has been in state 2 for longer than a fixed value T_{Tol}. A more precise modelling would lead to an inhomogenous SemiMarkovian model, which presently is not implemented. Modelling by a constant rate leads to slight conservativities.

EVALUATION OF MARKOVIAN MODELS WITH DISCRETE TRANSITIONS

Evaluation of a Markov process [2] involves (within the context of this paper) the determination of the state probabilities p_j (t) for a given set of initial conditions p_j (0), where

$$p_j(t) = \Pr\{\text{state } j \text{ is held at time } t\} \tag{1}$$

The process given before includes two different types of transitions, which are represented by two types of edges in the Markovian state graph. Thus, the state graph can be described formally as $G = (v, e_1, e_2)$. The set of vertices v shall represent the states of the process. There are two sets of directed edges e_1 and e_2 defined on $v \cdot v$. The elements of e_1 represent ordinary transitions and are labeled with the transition rates λ_{ij}. The elements of e_2 represent transitions occurring due to discontinuities and will subsequently be referred to as discrete transitions. They are labeled with pairs of the type (R_{ij}, q_{ij}). q_{ij} represents the probability, that a transition occurs into state j, given that the process is in state i at the time t_k, which is a fixed time point of a discontinuity. R_{ij} is some rule to determine the value(s) of

t_k for the edge under consideration. In practise, these rules are quite simple, as batches occur and terminate at regular time intervals, with possibly an exception for the time of the first batch. In this case, R can be given as

$$R_{ij}: t_k = T_{IO} + n \cdot T_I \qquad (2)$$

where n is a natural number (including 0),
T_{IO} is the beginning of the first batch,
T_I is the interval between two successive batches.

Without loss of generality, assume, that R renders at a given time t the largest value of t_k less than t.

This process strictly exhibits the Markovian property. This can be seen easily by noting, that it can be modelled equivalently by an inhomogenous Markov process with time dependant transition rates for e_2. Such transition rate could be given by

$$\lambda_{ij}(t) = \sum_{(k)} \delta(t - t_k)\, q_{ij} \qquad (3)$$

where $\delta(\cdot)$ represents a Dirac impulse.

This type of discontinuous time dependent transition rate requires special treatment [4]. When the process is started at t = 0, it will behave exactly like the reduced process $G^* = (v, e_1)$, until the first transition time is reached. It is not necessary to consider the set e_2. At the time point of a transition, i.e. during the interval (t_k, t_k+dt), the probability of transition via e_1 is of order dt and thus negligible with respect to the finite probability of transition via e_2. Thus, the behavior during a discontinuity is completely represented by the reduced process $G^{**} = (v, e_2)$. Afterwards, the process is continued with G^* until the next discontinuity.

The solution of G^* is straight forward [2] by solving the related system of differential equations for the state probability p_j (t). This system of differential equations can be given a

$$\dot{P}_j(t) = \sum_{i=1, i \neq j}^{m} \lambda_{ij} \cdot P_i(t) - \sum_{i=1, i \neq j}^{m} \lambda_{ji} \cdot P_j(t) \qquad (4)$$

where: λ_{ij} = transition rate from state i to j

G^{**}, on the other hand, is solved by the following reasoning: During a small time interval dt, at most one transition may occur. Thus, the state probability immediately following a transition, which shall be represented by $p_j(t_k{+}dt)$ can be given easily, if the state probabilities immediately before the transition, which will be written as $p_j(t_k)$, are known.

$$p_j(t_k+dt) = p_j(t_k) \cdot \left(1 - \sum_{(i \in e_2^*)} q_{ji}\right) + \sum_{(i \in e_2^*)} q_{ij} \cdot p_i(t_k) \quad (5)$$

Here, $e_2{}^*$ is that subset of edges of e_2, which is valid at the time of discontinuity t_k, which is determined by the associated rule R.

This equation may be interpreted in the following way [4]: The process will be in state j after a discontinuity at t_k, if either, it is in state j before the discontinuity and has not left it during the discontinuity (first term), or, if it has been in some state i before, and a transition occurred into state j (second term).

For small processes (up to three or four states), evaluation may occur analytically. For larger processes, like the one given in the example, numerical methods for solving systems of stiff differential equations have to be applied. Suitable methods are implemented in the simulation package; thus, it can be used to determine also the stochastic results.

PROBABILISTIC SIMULATION

The process presented before has been solved numerically with the following parameters.: The failure rate of the primary heat device has a value of 5e-04 / h, which corresponds to apr. 4 failures per year. Its average repair time is 3 h. A batch lasts 8 h, and there is one batch per day. The resulting function $p_3(t)$ is given in the following fig 4

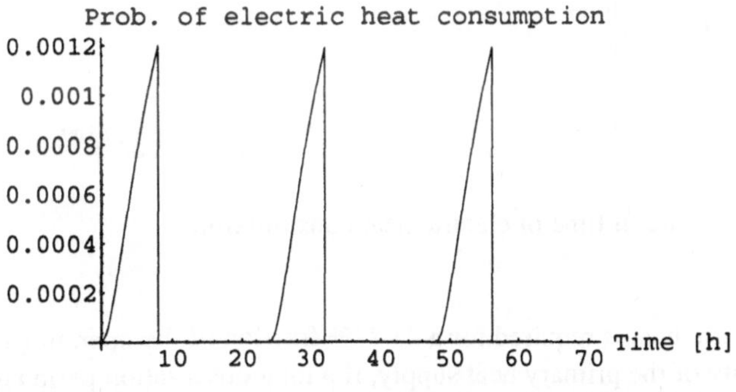

Figure 4. Resulting probability of electric heat consumption over time

Clearly, there may be consumption of electric heat during a batch, though not between batches. As the policy is to start a batch only, if primary heat is available, the probability of electric heat consumption is zero, when a batch starts, and increasing during the batch. The integral, which is the average time of electric heat consumption, has the following form:

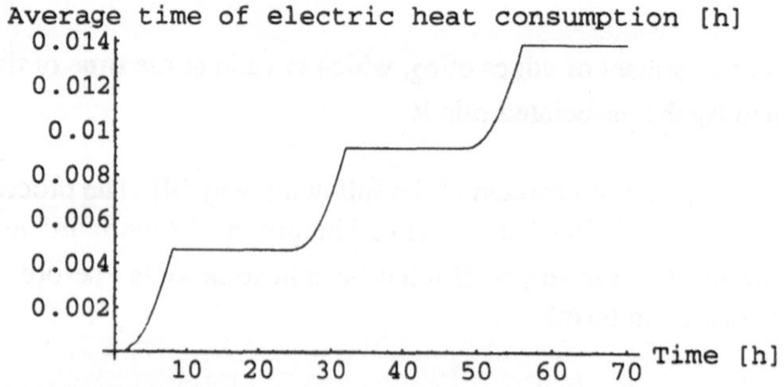

Figure 5. Resulting average time of electric heat consumption

From the periodic character of fig 4, it can be concluded, that the relative share of T_{el} to calendar time will converge to a constant limit. From the resulting plot, this limit can be found as 0.0002 p.u. apr.

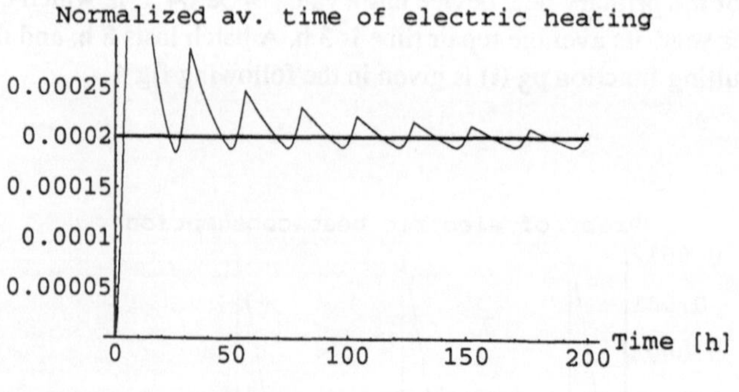

Figure 6. Average share of time of electric heat consumption

Thus, electric heat is required for a 0.02 % fraction of the operating time of the plant due to unreliability of the primary heat supply, if a long observation period is considered. With this result, a decision can be obtained easily, if the investment costs and the costs per unit heat energy are known for the electric heater, and for the additional steam supply. It is obvious, that for very long periods of time, a steam heater will always always be the cheaper alternative. However, as a basis for a decision, it is necessary to know, how long this time is going to be.

CONCLUSIONS

Reliability costs contribute to overall costs. Due to their stochastic nature, they are not easy to control. After a long time of virtually no reliability related cost, there may be suddenly a large contribution due to failure of some sensitive equipment. This method to estimate reliability related energy costs in terms of a mean value function is considered to be of great benefit for purposes of decision making, planning, and comparison of concurring design alternatives. Compared with other approaches, like crude Monte-Carlo techniques, considerably less computational effort is required. Especially, if an analysis of this type is performed for the overall costs of system reliability with contributions of many components, the expected value function will be a useful indication for future reliability costs.

The collaboration between groups interested in simulation and a group concerned about systems reliability has synergetic effects. A simulation code under development has been extended to suit probabilistic problems and to be able to determine expected cost functions on the basis of primary costs determined by the simulating software. In times, where systems show a tendency to become more and more interconnected and dependent on each other, where complexity increases more and more, reliability is no more an obvious matter, which the engineer may decide via his thumb, but rather a field of systematic research and design effort.

REFERENCES

[1] Barlow, R. E., Proschan, F., Mathematical Theory of Reliability, New York (etc): John Wiley & Sons 1965.

[2] Howard, R. A., Dynamic Probabilistic Systems, 2 vols, New York (etc): John Wiley & Sons 1971.

[3] Smotherman, M., Zemoudeh, K., A Non-Homogenous Markov Model for Phased Mission Reliability Analysis, IEEE Transactions on Reliability, vol R38, 1989, pp 312 - 313.

[4] Becker, G., Camarinopoulos, L., Mixed discrete and continuous Markovian Models for Components and Systems with Complex Maintenance and Test Strategies, Proc. of the European Safety and Reliability Conference 1992 in Kopenhagen, Denmark, June 10 - 12, 1992, Elsevier: 1992.

DESIGN OF ENERGY EFFICIENT BATCH PROCESSES

DR GREG ASHTON
Principal Consultant,
Linnhoff March Limited,
Tabley Court, Moss Lane, Over Tabley, Knutsford,
Cheshire, WA16 OPL

ABSTRACT

This project is aimed at developing industrially useful procedures for the reduction of batch plant energy costs. These procedures address all categories of batch plants, namely; single product, multiproduct and multipurpose.

The research covers all stages in the synthesis and analysis of batch processes, including:

- design and scheduling optimisation
- process integration
- dynamic simulation
- utility system optimisation

Industrial studies on factories manufacturing resins and pigments have illustrated how the new procedure can be applied. These studies have shown how energy savings between 8% - 20% of process heating costs can be achieved.

Future work will develop these procedures further and address the link between energy efficiency and environmental issues in batch plants. This work will also be supported by the CEC JOULE programme. Further industrial case studies will be carried out.

INTRODUCTION

Linnhoff March is currently co-ordinating research work into the 'Design and Operation of Energy Efficient Batch Processes'. This 'Eurobatch' project is supported by the CEC JOULE Programme. The report on the first phase of this project will be submitted to the CEC this year [1].

The project has considered all of the three main categories into which batch process plants can be divided, namely:

- single purpose plants, in which equipment is dedicated to production of a single product.

- multi purpose plants, in which batches are processed as and when equipment becomes available without a common pattern.

- multi product plants, in which equipment can be operated flexibly to produce different products in parallel trains of equipment.

The main objective of the project is to develop industrially useful and accepted procedures for the design and operation of energy efficient batch processes. To achieve this objective, researchers have combined to develop an overall procedure aimed at optimising batch plant design and operation. The main responsibilities of these co-workers can be summarised, as follows:

Universitat Politecnica de Catalunya (UPC); development of computer aided procedures for design and scheduling of batch operations in multiproduct plants.

Universite de Liege; dynamic process simulation of batch equipment to enable modelling of start-up and operation and as the basis for detailed equipment specifications.

SPEC Ltd, Athens; synthesis of multipurpose batch plants and testing of the developments with industrial case studies.

University of Manchester Institute of Science and Technology (UMIST); appraisal and improvement of existing procedures for the energy analysis of batch process and the development of a computer programme which utilises these methods.

Linnhoff March Ltd, Manchester; project coordination and testing of procedures by application to industrial case studies. AEA, Harwell have also assisted Linnhoff March and UMIST to develop and test process integration procedures.

University College, Dublin (UCD); optimisation and control of batch plant utility systems for heating and cooling batch equipment.

BATCH PROCESS OPTIMISATION PROCEDURE

The results of the project can be summarised in terms of an overall procedure which is applicable to new plant design or retrofit. In general, the two most important stages in the design and optimisation of a batch plant are:

- process synthesis
- process analysis

The 'batch onion' diagram in Figure 1 shows the hierarchy in which these stages need to be addressed in a new plant design or in an existing plant retrofit project.

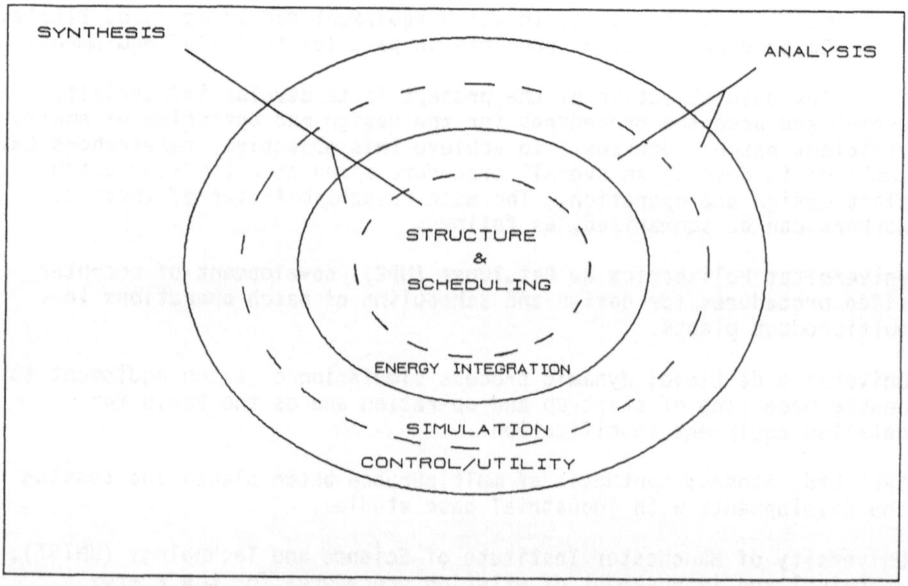

Figure 1: The Batch Onion

Starting at the centre of the onion, the first task in plant synthesis is to determine the structure and scheduling of the batch plant. For a given production capacity and process route, these procedures can be used to optimise the arrangement and scheduling of batch operations. This is analogous to optimising the sequence of reactors and separators in a continuous plant.

The second task in process synthesis is to apply process integration procedures. While one objective of a process integration study is usually the reduction of energy costs directly these techniques are also useful to identify process modifications which will give other important benefits that indirectly reduce specific energy consumption. These other benefits include debottlenecking and utility peak shaving.

For this reason, process integration should be applied at the process synthesis stage, so that the results can be fed back to further improve the structure and scheduling optimisation.

When the process synthesis stage is complete, the resulting flowsheet can be analysed to optimise and specify equipment sizes, process control and utility system operation. The first step in process analysis is usually process simulation. Because of the time-dependency of batch operations, a dynamic simulation facility is important to enable optimisation of these parameters. The design of the process-utility system interface is also an important part of batch process synthesis.

In this project the results of each research group have been incorporated into a suite of computer programmes. The individual programmes have been linked by a common menu system so that the appropriate software can easily be selected by the plant designer. The resulting software will enable state of the art procedures to be applied at each stage in the synthesis and analysis of batch processes. Figure 2 shows the order in which the software can be applied according to the 'Batch Onion' hierarchy.

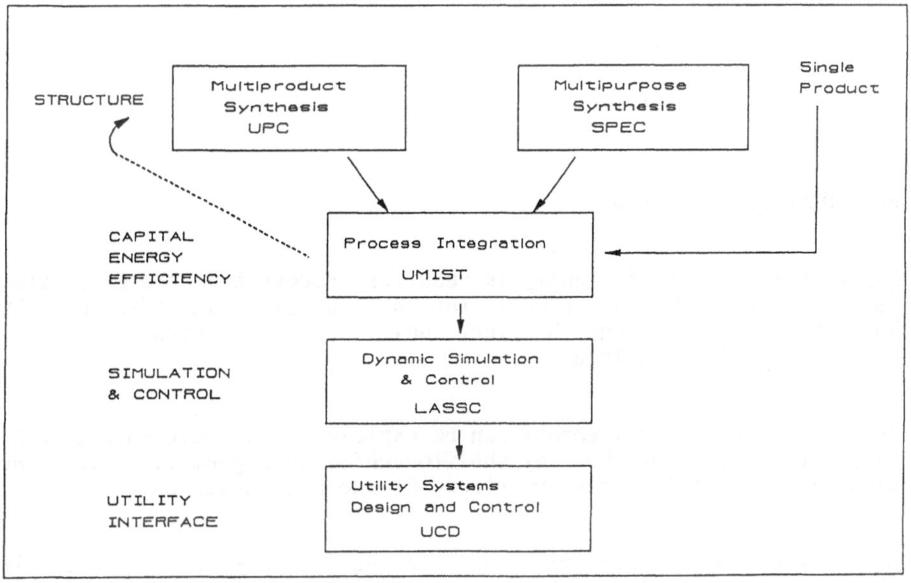

Figure 2: Procedure for Retrofit or New Design

BATCH PROCESS INTEGRATION

A key stage in improving the energy efficiency of batch processes is to carry out a process integration analysis. The technology developed by UMIST and Linnhoff March in cooperation with AEA, Harwell, is focused on this aspect of the procedure.

Early procedures for batch process integration were based on applications experience in Linnhoff March [2] and were aimed primarily at energy saving by heat integration. These procedures were subsequently developed in cooperation with UMIST [3]. A parallel research programme at Harwell [4] independently developed related procedures aimed at optimising energy storage in batch operations.

The current work has brought all of these developments together and further developed the batch process integration models and software. The resulting state-of-the-art process integration techniques are based on the simultaneous application of three process models as follows:

- 'Time Average'
- 'Time Event'
- 'Time Slice'

What Data are Required?

Any process integration analysis requires process heat and mass balance data to be provided as the basis for the analysis. In order to analyse batch process operations, however, additional information as the batch schedule is also required.

The process data requirements can be explained with reference to the simple example plant shown by the flowscheme in Figure 3. This shows a plant which uses two batch reactors, R1 and R2, in sequence.

A convenient way to summarise the data requirements for this plant is shown in Table 1. This table defines the process temperatures, heating and cooling duties and the batch schedule in terms of the time (hrs), temperature (°C) and utility consumption (kWh) for each batch operation.

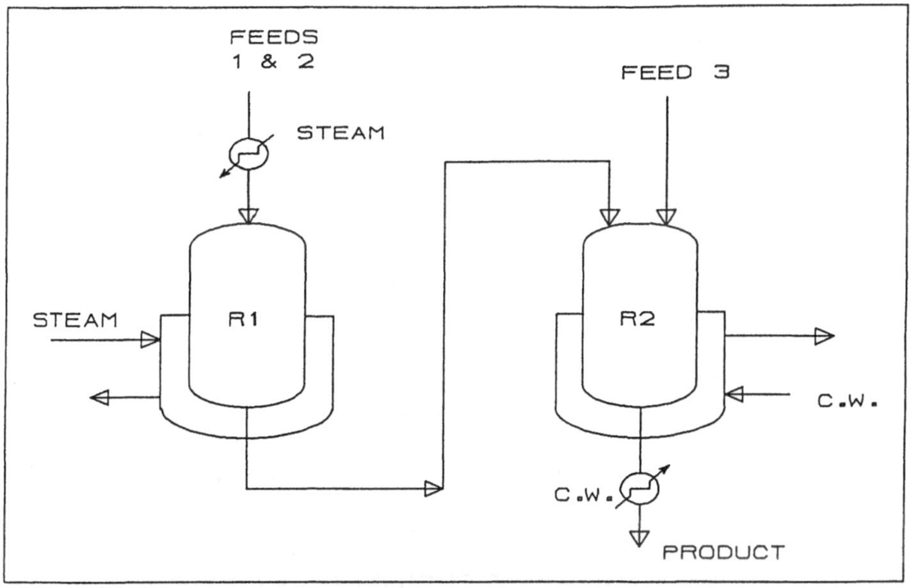

Figure 3: Flowscheme for Example Process

Reactor R1:

Operation	Time (hrs)	Temperature (°C)	Utility (kWh)
R1.1 Preheat and Charge Feeds 1 & 2	1	25-80	50 (Steam)
R1.2 Reaction (Endothermic)	1	80-120	150(Steam)
R1.3 Heating	1	120-150	90 (Steam)
R1.4 Hold Temperature to Maximise Yield	2	150	0
R1.5 Transfer to R2	1	150	0
Reactor R2:			
R2.1 Transfer from R1 and test	1	150	0
R2.2 Cool to Quench Reaction	1	150-100	150 (Cooling Water)
R2.3 Add Feed 3 Slowly (Exothermic)	1	100	75 (Cooling Water)
R2.4 Discharge and Cool Product	1	100-40	175 (Cooling Water)

Table 1: Example Process Data

These data can be expressed in terms of input data for the 'BATCH' process integration programme developed in the JOULE project. This is shown in Table 2. In this table, all of the batch operations associated with energy flows are specified as hot (H) or cold (C) streams. For each stream the starting temperature (TS), the ending temperature (TT), the enthalpy change and the start and finish times in the batch schedule are specified.

Stream		TS	TT	CP	Enthalpy Change	Start time	Finish time	Stream Name
1	C	25.0	80.0	0.909	50.0	0.00	1.00	R1 Step 1
2	C	80.0	120.0	3.75	150.	1.00	2.00	R1 Step 2
3	C	120.0	150.0	3.00	90.0	2.00	3.00	R1 Step 3
4	H	150.0	100.0	3.00	-150.	1.00	2.00	R2 Step 2
5	H	100.0	99.0	75.0	-75.0	2.00	3.00	R2 Step 3
6	H	99.0	40.0	2.97	-175.	3.00	4.00	R2 Step 4

Table 2: 'Batch' Programme Input Data

Time-Average Model

The time average model is used to calculate the minimum energy consumption target and the process pinch temperature. It is analogous to the composite curves used to analyse continuous processes. The main difference is that for batch plants, heating and cooling duties are 'time averaged' over a convenient period.

Figure 4 shows the time average curves for the example plant. With a minimum temperature difference of 20°C, the process heating target is 90 kWh/Batch. This is 200 kWh/Batch less than the heating duty for the existing design, equivalent to potential energy savings of 69%.

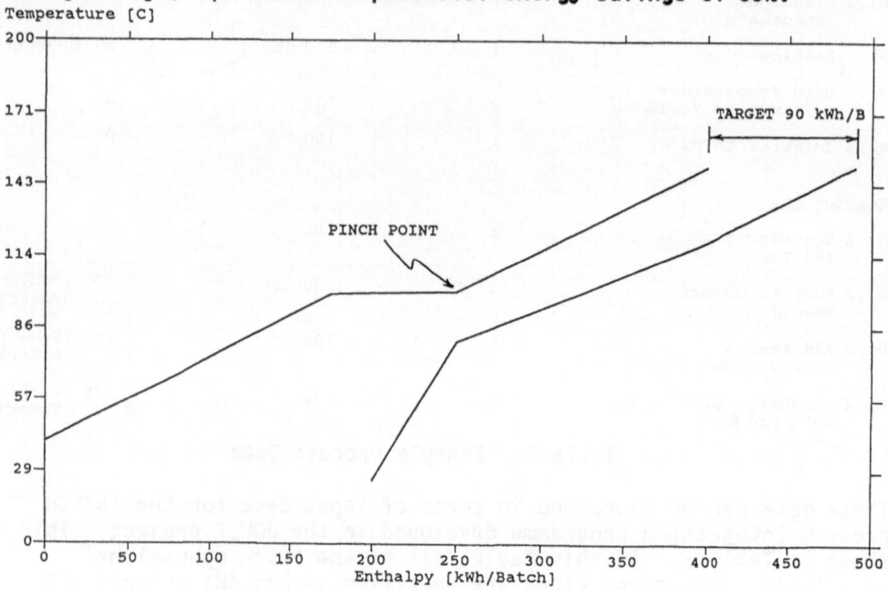

Figure 4: **Time Average Composite Curves**

The fundamental significance of 'Time Average' process heating target is that it is based on a rigorous thermodynamic model of the process. It will always be possible to make an integrated design which will achieve this target by applying the well known principles of pinch technology [6].

The main limitation of the Time Average model is that it does not convey any information about the batch schedule. Additional information on scheduling is important to enable understanding of the practical an economic constraints on energy integration in batch processes. For this reason, it is not appropriate to use the Time Average model alone in batch process integration studies.

Time-Event Model

The time event model defines the batch plant schedule using a Gannt chart. It is used to identify the batch operations on the critical path which limit production and to measure the 'free time' in operations which are not on the critical path. When the time event model is applied in conjunction with the time average and time slice models, it can be used to identify the best opportunities for debottlenecking, energy recovery and peak shaving of utility loads.

The objectives of debottlenecking, or time saving, and utility peak shaving, are important in terms of improving, the energy efficiency of batch plants from three points of view: Firstly, the specific energy consumption of the plant is reduced by enabling more product to be manufactured in the same equipment so that energy losses are minimised; secondly, by creating opportunities for more efficient design and operation of the plant utility systems; and thirdly, by increasing the total value of the benefits attributable to energy saving projects.

An example of a project in the third category could be a new heat exchanger aimed at saving energy by preheating the charge to a batch reactor by heat recovery. If the value of energy savings alone is taken into account, this project may not give an acceptable return on investment. However, if it can be shown that this project also saves time in the reactor, which is on the process critical path, then there is an additional benefit due to increased production. The value of this secondary benefit can increase the return on investment to the point where project implementation can go ahead.

The objective of developing batch process integration procedures which can be used to maximise the secondary benefits attributable to energy saving projects is key to the design of energy efficient batch processes because it maximises the rate of implementation of energy efficient features in the plant design.

In the context of this research project, the objectives of debottlenecking and utility peak shaving are also addressed by the complementary procedures for schedule optimisation, dynamic simulation and utility system design and control.

Figure 5 shows the time event model for the example process with the existing schedule. The batch cycle time of 6 hours is limited by reactor, R1, which is on the critical path. Clearly, the schedule of this reactor must be modified to achieve any capacity increase. The time-event model also shows that reactor, R2, has two hours free time in every batch cycle. This could also be exploited to reduce energy consumption by making heat recovery more economical to implement and/or by peak shaving utility demand.

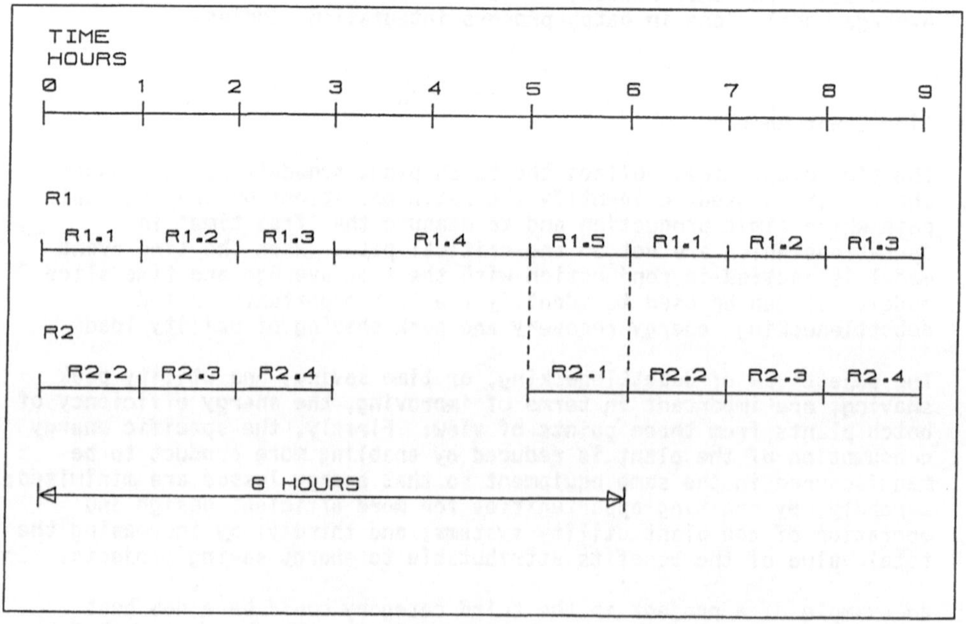

Figure 5: Time Event Model Existing Schedule

Time-Slice Model

To calculate the time slice model, the batch cycle is divided into time slices where each time slice is bounded by the time at which a batch operation starts or finishes. This is analogous to the 'problem table' algorithm for calculating composite curves in pinch technology, where the temperature range is also divided into intervals according to where streams start and finish [6].

It is now possible to construct composite curves separately for each individual time slice in the process, using only hot and cold streams which co-incide in time during that time interval. These time slice composite curves show how much energy saving could be achieved by heat recovery between coincidental streams, so-called direct heat recovery.

Residual heating and cooling duties in each time slice, after direct heat recovery has been maximised, are defined by the corresponding time slice grand composite curves. If there is excess heat which cannot be used for direct heat recovery, this is termed a 'heat source'. It may still be possible to save energy by integrating a heat source in the plant, but this would have to be by indirect heat recovery where the heat is stored over time.

Heat deficits within each time slice are similarly termed 'heat sinks'. It is now possible to define all heat sources and sinks in the plant in a similar way to process streams, that is, in terms of a starting and finishing temperature and a corresponding enthalpy change. These streams can then be combined to calculate composite heat source and heat sink curves for the whole plant. The resuling curves are called the Batch Cascade Composite Curves.

The Batch Cascade Curves for the example plant with the existing schedule are shown in Figure 6. It is important to realise that, unlike the 'time average' curves, these curves are now a function of the plant schedule. They are used specifically to analyse what constraints the schedule imposes on heat recovery opportunities. This overcomes the main shortcoming of an approach which uses the time average model alone.

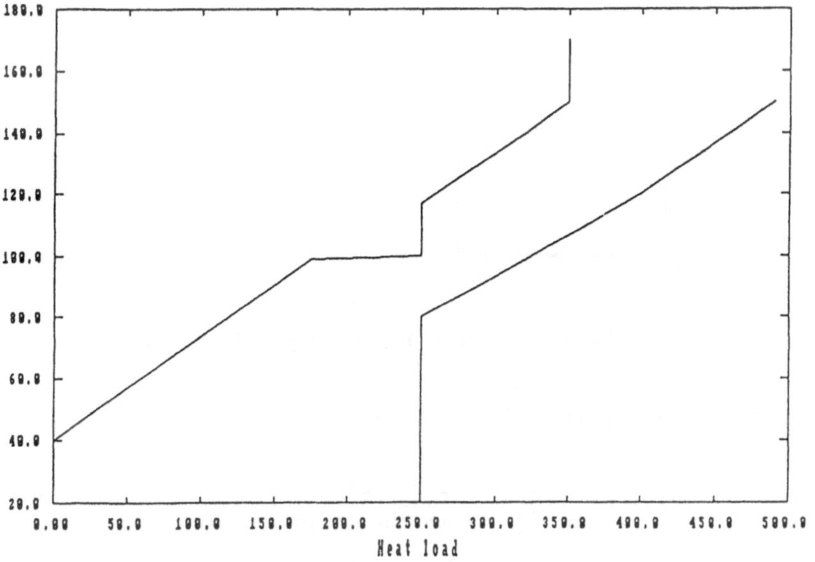

Figure 6: Batch Cascade Curves

The heating target for the example plant measured from the batch cascade curves is 140 kWh/Batch which is higher than the time average target of 90 kWh/Batch. The difference of 50 kWh/Batch can be regarded as the energy penalty due to constraints on heat recovery imposed by the plant schedule. Furthermore, the overlap between the batch cascade curves measures the indirect heat recovery that would be necessary to reach this target.

The time slice model can also be used to calculate the utility demand profiles for the plant. An example showing the steam demand profile is given in Figure 7. It shows that the steam demand peak of 150 kW occurs in the interval from 1 to 2 hours in the batch cycle. This feature of the time slice model is used specifically to investigate opportunities for peak shaving of utility demands, by rescheduling batch operations and/or by heat recovery.

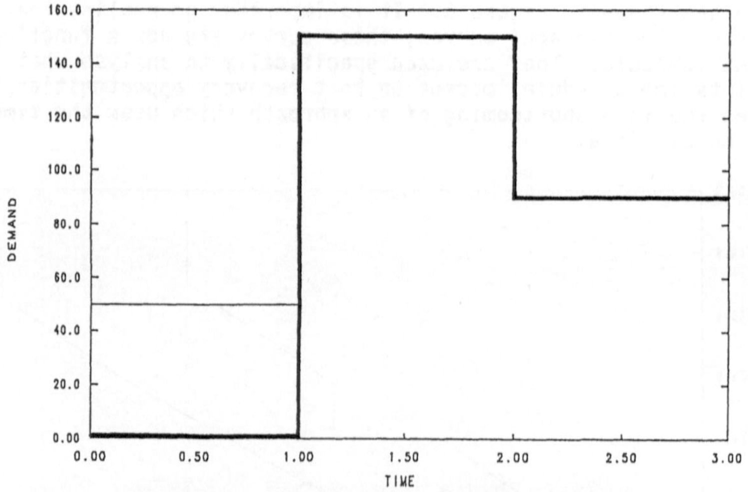

Figure 7: Hot Utility Demand Profile

Process Design Improvements

In this example, the process integration approach has been used to develop an improved design, with the following features:

- the batch schedule was revised to debottleneck the plant. The time event model indicated that free time in R2 could be exploited. The revised time-event diagram showed that the batch cycle time could be reduced to 5 hours. The equivalent increase in production capacity is 20% and there is a consequent reduction in specific energy consumption.

- the batch cascade curves for the revised schedule showed that the time average target of 90 kWh/Batch can be achieved with the revised schedule. It is now possible to specify more cost effective heat recovery projects to achieve this target.

- the steam demand profile for the improved plant design showed that the peak steam demand can be reduced from 150 kW to 90 kW.

In this example, therefore, the process integration approach was used to illustrate how the batch plant design could be improved to achieve three important benefits, namely:

- debottlenecking with reduced specific energy consumption
- maximum energy recovery
- utility load smoothing, leading to more efficient utility system operation

JOULE PROJECT CASE STUDY RESULTS

Two industrial case studies have been completed as part of the Joule project to demonstrate the application of the new batch process integration techniques.

The first project was carried out for Cray Valley Total's factory at Machen, South Wales. This factory manufactures a range of resin products by batch processes.

The study applied the time average, time event and time slice models to the batch process operations. As a result, a strategy for investment into energy savings on this site has been proposed.

A package of projects which will save up to 8% of the resin plant energy costs has been recommended. Short term initiatives included in this package could achieve about 3% savings at under 1 year payback. Importantly, other benefits, including debottlenecking and utility load smoothing have also been identified. It is expected that these other benefits will increase the implementation rate for energy savings and reduce the future investment that is necessary in utility plant.

The second demonstration project was carried out on a site manufacturing pigments, and a total of four single product plants were studied on this site. An important feature of these processes was the high energy consumption associated with heating water for use in product filters.

When the pigment plants were analysed individually, there was found to be some scope for energy saving by recovering waste process heat into water. However, the energy savings were not great enough to justify investment in the new equipment that was necessary.

The main energy saving opportunities on this site were identified when the process integration analysis was extended to consider energy integration between plants. This can easily be done by combining the batch utility curves for all plants into a single model. In this case, this analysis showed increased scope to save up to 20% of process heating costs by heat integration between plants. The analysis also showed how this could be achieved by indirect heat recovery using a hot water system to store heat over time until it is required.

CONCLUSIONS AND FUTURE WORK

Procedures for the design of energy efficient batch processes have been developed as part of a research project supported by the CEC's JOULE programme. These procedures have been incorporated into software which is already developed to the stage where it is industrially useful.

The existing procedures will be further developed and improved in the current project. This future work will focus on research into the interaction between energy and environmental issues in the design and operation of batch plants.

A part of the future work will be to carry out further industrial case studies to demonstrate how this technology can be applied.

REFERENCES

1. Draft Report on the 'Design and Operation of Energy Efficient Batch Processes'. CEC Contract JOUE-0043 (SMA).

2. Cost Reductions on an Edible Oil Refinery identified by a Process Integration Study at Van den Berghs and Jurgens Ltd. UK Energy Efficiency Office Report RD/14. Oct 1986.

3. Linnhoff, B., Ashton, G.J. and Obeng, E.D.A., Process Integration of Batch Processes. AIChE Annual Meeting, New York, 1987.

4. Kemp, I.C. and Macdonald, E.K., Application of Pinch Technology to Separation, Reaction and Batch Processes. IChemE Symposium Series 109, pp 239-257, 1988.

5. Gremouti, I.D., Integration of Batch Processes for Energy Saving and Debottlenecking. MSc Thesis, UMIST, 1991.

6. Linnhoff, B. et al. User Guide on Process Integration for the Efficient Use of Energy. Published by IChemE.

A SIMPLE TECHNIQUE FOR ANALYZING WASTE-HEAT RECOVERY WITH HEAT-STORAGE IN BATCH PROCESSES

S. Stoltze, B. Lorentzen, P. M. Petersen, B. Qvale
Laboratory for Energetics,
Technical University of Denmark,
Building 403, DK-2800 Lyngby, Denmark.

ABSTRACT

The aim of the work has been to arrive at theoretical and practical approaches to the design of heat-storage systems for waste-heat recovery in batch processes.

A simple procedure for determining the number of heat-storage tanks and heat-storage temperatures necessary to achieve the maximum energy saving targets (as calculated by the Pinch-Point Method) is described.

Qualitative arguments justifying the procedure are presented. Through case studies, two of which are presented, it is found that a theoretically large number of heat-storage tanks (twice the number of process streams) can be reduced to just a few tanks. It is found, that waste-heat recovery systems with three heat-storage temperatures (a hot store, a cold store and a heat store at the Pinch Temperature) most often achieve the maximum energy saving targets.

The simplifications are effectuated by introducing several steps that each will lead to larger heat exchangers than if a process - by - process match were used. To some extent the simplifications introduced compromise the economics. However, cases that have been studied show that a two- and a three-tank heat-storage systems are economically attractive while, recovering up to 100 % of the maximum energy saving target.

INTRODUCTION

Use of Heat Storage

A significant percentage (about 50%) of industrial processes are so-called batch processes. Some of the processes are carried out in batch operation because of limited supply or demand (dairy), because of tradition (brewing), or because of special requirements to the process (brewing). Some processes may even improve thermodynamically by having them carried out in a batch mode (for example heating water by use of heat pumps).

Because of todays high demand for peak electricity, economic considerations favour batch operation of the heat generation from public cogeneration plants.

The use of heat storage in connection with batch processes will in its turn often improve the operation of the processes both with respect to economics, controllability and energy utilization. The use of heat storage in connection with cogeneration and district heating systems is already quite common in Denmark. The use of storage of heat in industrial processes is not always done explicitly, but a container used for storage of the material to be processed will often in effect also serve as a store of energy.

Process integration

Integration of industrial processes is often problematic because of the sheer magnitude of the computational work. An industrial process may consist of a significant number of processes (10 to 100) during which energy is added or removed. The chief aim of process integration will be to make use of the heating or cooling surplus from a given part process in other part processes. If all possibilities are to be evaluated, the computational work will be quite large, and often beyond practical computational capacity. If, furthermore, the processes do not occur simultaneously, the heat or cold will have to be stored in energy stores for later use.

For each of the heat-recovery processes, one extra heat exchanger and two heat stores will be used, and the computational work would be expected to increase compared to simultaneous processes.

In order to achieve a reasonable degree of success with process integration, it is therefore very important that some systematic and efficient procedure for reducing the magnitude of the computational work is found. This is the chief aim of the work that is described in the present paper.

Earlier studies

There are only few studies that are concerned specifically with the problems connected with batch operation.

The usual approach consists of averaging the energy interaction over the whole of the batch period. This procedure is simple to apply, but the expectations to the potential for energy recovery is usually exaggerated.

Recently, papers addressing more complex situations such as time varying temperatures (1) and rescheduling (2, 3) have been published.

Also recently three papers have been published by Kemp and Deakin (4, 5, 6) that address batch processes expressly. It is characteristic that the resulting procedures are quite comprehensive and therefore both lengthy and complex. Other authors that have contributed to this development are B. Linnhoff, G. J. Ashton and E. D. A. Obeng and E. K. Macdonald.

These studies will not be discussed further in the present paper. It is our opinion that simplifying procedures will have to be developed in order to make process integration viable for batch processes, and that the procedures that have been published so far have been too complete, lengthy, demanding and complex to be of practical interest for most of the cases that one will encounter.

The present paper
In the present paper, the background for the present simple method is presented. The work is limited to the utilization of variable-mass constant-temperature stores. The application of the method is illustrated by two case studies. The results show that it is justified to use the simple method.

EVALUATION OF THE PROBLEM AND THE METHOD

The chief problem associated with the synthesis of networks is the enormity of the computational work if all possibilities are going to be explored.

It has in general been expected that the introduction of the possibility of energy storage into the problem would lead to increased complexity and a significant increase of the magnitude of the computational work. The chain of arguments that will be presented below will contradict this.

Let us concider a brewhouse example with N_H = 10 hot streams and N_C = 6 cold streams (The case B discussed below). In order to simplify the argument, we assume that one stream completely exchanges heat with one other stream and that any leftover energy in a given stream will remain unused. The total number of heat exchangers (including the physically impossible) will then be

$$N = N_H \cdot N_C = 10 \cdot 6 = 60$$

Then we have to investigate the degree of energy recovery and economics of introducing 1, 2, 3, etc. up to 6 heat exchangers in the system.

The number of permutations for $N_H > N_C$ will be:

$$N_{HX} = (10 \cdot 6)(9 \cdot 5)(8 \cdot 4)(7 \cdot 3)(6 \cdot 2)(5 \cdot 1) = \frac{N_H!}{(N_H - N_C)!} \cdot N_C! = 1.09 \cdot 10^8$$

If we now consider the corresponding energy-storage problem, the maximum numbers of heat stores to be considered, N_S, will be twice the number of process streams. In the case under consideration this will be N_S = 32. A number of these can be eliminated before the onset of the computations. But in the synthesis of the heat-exchanger network discussed above we did not make use of this type of screening. Therefore for the sake of fairness we abstain from this. Now we have:

$$N_{STORE} = \frac{N_S!}{(N_S-2)! \cdot 2!} + \frac{N_S!}{(N_S-3)! \cdot 3!} + \ldots\ldots$$

$$= \sum_{n=0}^{N_S} \frac{N_S!}{(N_S-n)! \cdot n!} - \left(\frac{N_S!}{(N_S-0)! \cdot 0!} + \frac{N_S!}{(N_S-1)! \cdot 1!} \right) = 2^{N_S} - (1 + N_S)$$

For the example under discussion this becomes: $N_{STORE} = 2^{32} - (1+32) = \underline{4.29 \cdot 10^9}$.

However, in practical cases it turns out that only a small number of stores is actually required to achieve the maximum amount of energy recovery (the pinch target). Furthermore, it will be shown below that economical considerations will reduce further the number of stores required to reach an economical optimum.

In both of the two cases shown in the present paper the number of stores required is 3.

If we assume, based on our experience, that the largest number of stores that have to be evaluated is 6, the number of alternatives to be considered will be

$$N_{STORE} = \frac{N_s!}{(N_s-2)! \cdot 2!} + \frac{N_s!}{(N_s-3)! \cdot 3!} + \frac{N_s!}{(N_s-4)! \cdot 4!}$$

$$+ \frac{N_s!}{(N_s-5)! \cdot 5!} + \frac{N_s!}{(N_s-6)! \cdot 6!} = 1.15 \cdot 10^6$$

It can, therefore, be concluded that the problem posed by the introduction of energy storage into the problems is significantly simpler than establishing a heat-exchanger network and will not complicate or expand the problem.

A number of different types of stores may be considered, such as latent heat store and variable-temperature constant-mass store. Because of space constraints the present work is limited to considering Constant-Temperature Variable-Mass Stores only. The conventional two-temperature stratified water tank used in cogeneration systems and the solid-bed high heat-capacity regenerators belong to this group. The use of Constant-Temperature Variable-Mass Stores will require at least two stores and one heat exchanger per process or batch, and thus in reality requires a storage system.

In order to extract heat from a process stream, one low- and one high-temperature store are required, as shown in Figure 1. As mass is moved from the low-temperature store to the high-temperature store, heat is added in a heat exchanger. The only requirement to the temperature levels is that the low-temperature store is below the lowest process temperature and likewise for the high-temperature store. Thus the temperature levels may be lowered from the levels shown without impairing the energy transfer. When heat is added to a stream the stores will have to lie above the corresponding temperatures of the process stream to be heated.

The simple method consists of:

1. Finding all possible store temperature levels given by the end temperatures of the processes, corrected for DT_{MIN} (the minimum temperature difference between the two working media in a heat exchanger).
2. Select the number of stores to be used (start with 2).
3. Try all combinations of the possible temperature levels and calculate the resulting energy flow into and out of the stores.
4. In such systems the operation is cyclical. At the completion of the above procedure some of the stores will have an surplus mass, some will have a deficit. In order to satisfy the condition of cyclical operation, the masses should be evened out. This is firstly done by adding suitable quantities from stores at different temperatures to a given store without changing the temperature of the latter.
5. When the possibilities of this mixing are exhausted, the remaining surplusses or deficits are evened out by moving mass from a store after cooling or heating it to the

desired temperature. This cooling or heating then represents, together with those process streams or parts of process streams, that could not be cooled or heated by the selected stores, the net amount of cold or hot utility required by the processes.

6. Finally, the costs of the investments in tanks, pipes, pumps, heat-exchangers etc. are calculated together with the value of the energy savings caused by these investments. Below, the application of this procedure is illustrated through two case studies.

If composites of several warm and several cold process streams are developed (7) we get the so-called composite curves. A simple illustration is shown in Figure 2 with the, for this case, minimum number of stores for maximum energy recovery. Surprisingly, in many practical cases the maximum possible energy recovery is achieved with only three stores, one below the lowest temperature of the warm composite , one above the highest temperature of the cold composite, and one in the pinch point. (However, in principle, this will depend on the shape of the composite curves, on the chronology of the individual process streams, and on the number of separate batch-process lines.)

CASE STUDIES

Case A: The Simple Pinch Problem
The specifications of this case which is being used in our courses as a teaching example are given in Appendix A.

Table 1. Summary of the results of Case A.

	Stores (Number of, Temperatures, Total Volume)	Heat Exchangers (Number of, Total Area)	Present Value (10^6 Dkr Profit)	Payback Period (Years)	Utilities Hot (Q_H) Cold (Q_C) (GJ/batch)
Economic Optimum	3 25, 55, 140°C 7.90 m^3	3 47.30 m^2	0.509	4.3	Q_H = 3.888 Q_C = 0.504
Maximum Energy Recovery	3 25, 85, 145°C 27.2 m^3	7 86.20 m^2	0.315	7.1	Q_H = 3.384 Q_C = 0
Batch Target					Q_H = 3.384 Q_C = 0
MER, Continuous Operation		3 7.10 m^2	1.124	1.2	Q_H = 3.384 Q_C = 0

With one store at the end of each process stream, using a DT_{MIN} of 5°C, stores could be located at the following temperature levels (°C): 25.0, 55.0, 85.0, 140.0, 145.0, 165.0

By trial, the economically most favorable configuration is found to consist of 3 stores and 3 heat exchangers. The maximum heat recovery, and hence minimum amounts of hot and cold utilities is achieved with 3 stores and 7 heat exchangers (To achieve maximum energy recovery all process streams are divided at the Pinch Point Temperature).

These results are summarized in Table 1. For comparison the same case in continuous operation with simultaneity of the processes over the same total time will require 3 heat exchangers as shown in the last line of Table 1. Since only one of the processes extend over the total time inteval, the size of the heat exchangers will be much reduced in the continuous case.

Case B. Integration of the Processes in the Brewhouse.

The second case to be considered in this presentation, concerns the energy integration of all the processes in the brewhouse, altogether 16 processes. The technical specifications are given in Appendix B. The processes have not been identified by their brewing jargon since this is irrelevant in the present context.

With one store at the end of each process stream, using a DT_{MIN} of 2°C, stores could be situated at the following temperature levels (°C): 9, 10, 28, 39, 53, 63, 69, 72, 76, 78, 82, 96, 98, 99, 102, 103.

By trial, the economically most favorable configuration is found to include 3 stores and 13 heat exchangers. The solution for the maximum heat recovery and hence the minimum amounts of hot and cold utilities coincides with the solution for the economic optimum.

The results are summarized in Table 2.

Even in this case, that is considerably more extensive than the previous case, the economic optimum is found to include only 3 stores.

Some interesting points are illustrated in Figure 3, where the present value and the energy recovery are presented. It is seen that, in this case, it is strictly necessary only to investigate up to and including the use of 4 stores at a time.

Table 2. Summary of the results of Case B.

	Stores (Number of, Temperatures, and Total Volume)	Heat Exchangers (Number of, and Total Area)	Present Value (Million Crowns Profit)	Payback Period (Years)	Utilities Hot and Cold (GJ/batch)
Economic Optimum **and** Maximum Energy Recovery	3 39, 82, 99°C 890.2 m³	13 1622.5 m²	11.29	4.0	Q_h = 30.58 Q_c = 24.74
Batch Target					Q_h = 30.58 Q_c = 24.74

SUMMARY AND CONCLUSIONS

The computational work associated with integration of continuous processes is usually quite extensive, simply because of the large number of processes and components involved. If the processes occur non-simultaneously, or in batches, the size of this work is expected to grow and the complexity of the problems to increase.

However it turns out that this is not neccesarily so. A simple procedure for integration of batch processes is presented. Qualitative arguments and practical experiences from a number of cases show that the introduction of heat storage by this procedure actually lead to a reduction or elimination of difficulties that have been anticipated regarding heat recovery and heat storage.

It is argued that instead of investigating the introduction of heat storage at all the temperature levels that could be considered as candidates, the energy recovery and economics of the system should be investigated introducing stores at a small number of temperature levels. It is found that, in many cases, three temperature levels will be enough, to achieve the energy recovery targets.

Some of the simplifications are realized by sacrificing some of the economical advantages. By sending all the thermal energy to be recovered into storage, the heat exchangers will be significantly larger than if a process - by - process match were achieved. Even so, the economics in the cases studied is quite good, with pay back times of 4.3 and 4.0 years. The favorable economics is explained by the relatively small number of heat exchangers resulting from the application of this procedure. Since the starting cost (at zero capacity) of the components is quite significant, the number of such components is an important parameter in determining the cost of small and medium-size systems. The simplified procedure presented should be very well suited for application to systems in this size range.

REFERENCES

1. Vaselanak, J.A., Grossmann, I.E. and Westerberg, A.W.,
Heat integration in batch processing.
Industrial & Engineering Chemistry. Process design & Development, 1986, 25, pp.357-66.

2. Vaselanak, J.A., Grossmann, I.E. and Westerberg, A.W.,
An embedding formulation for the optimal scheduling and design of multipurpose batch plants.
Industrial & Engineering Chemistry Research., 1987, 26 (1), pp.139-48.

3. Petersen, P.M. and Qvale, B.,
Control and optimization of the refrigeration plant in a brewery.
The proceedings of the 21st Intersociety Energy Conversion Engineering Conference, 1989, pp.1807-12.

4. Kemp, I.C and Deakin, A.W.,
The cascade analysis for energy and process integration of batch processes. Part 1:

Calculation of energy targets.
Chemical Engineering Research and Design, 1989, 67, pp.495-509.

5. Kemp, I.C and Deakin, A.W.,
The cascade analysis for energy and process integration of batch processes. Part 2: Network design and process scheduling.
Chemical Engineering Research and Design, 1989, 67, pp.510-16.

6. Kemp, I.C and Deakin, A.W.,
The cascade analysis for energy and process integration of batch processes. Part 3: A case study.
Chemical Engineering Research and Design, 1989, 67, pp.517-25.

7. Linnhoff, B., Townsend, D.W., Boland, D., Hewitt, G.F., Thomas, B.E.A., Guy, A.R. and Marsland, R.H.,
A user guide on process integration for the efficient use og energy. IChemE, Rugby, 1982.

FIGURES

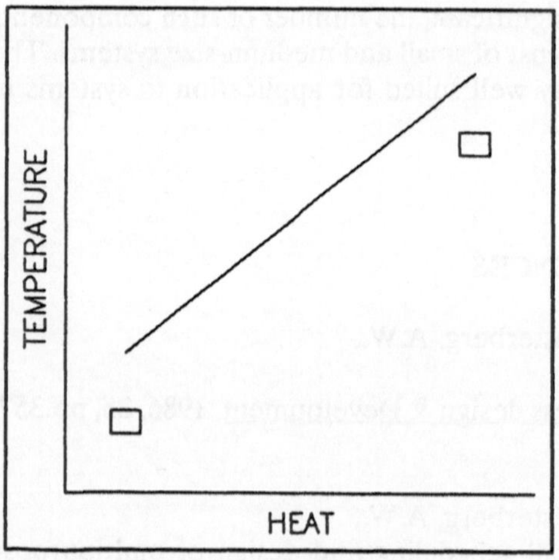

Figure 1. Illustration of two Stores Extracting Heat from a Hot Process Stream.

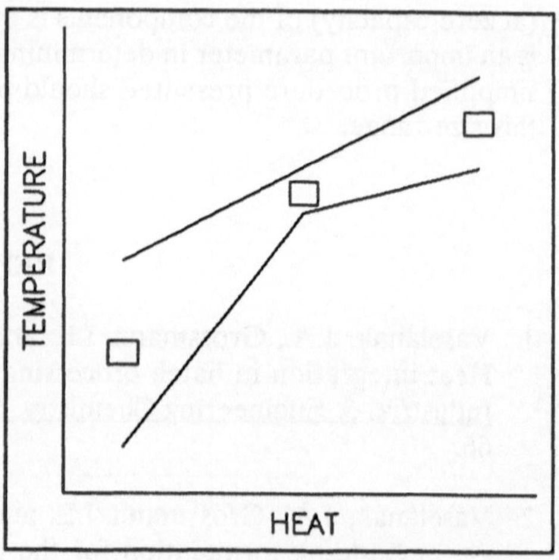

Figure 2. Illustration of the Use of Heat Stores in Connection with Composite Curves

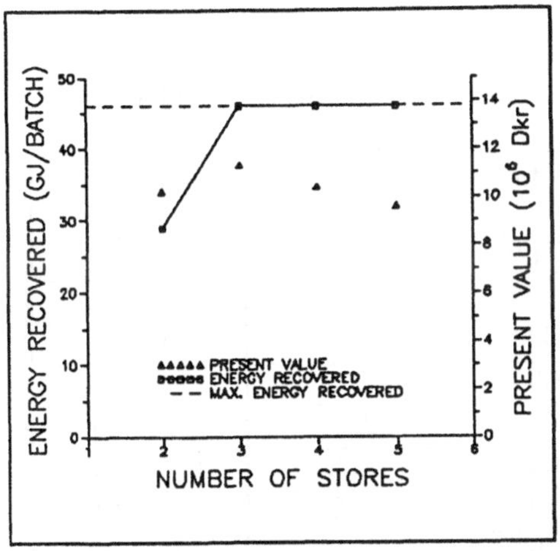

Figure 3. Present Value and Recovered Energy as Function of the Number of Stores (Taken 2, 3, 4 or 5 at a Time).

APPENDICES

Appendix A. Specifications for Case A: The Simple Pinch Problem.

Table A1. Data for the Simple Pinch Problem.

Process stream number	Start temp. [°C]	End temp. [°C]	Mass flow [kg/batch]	Specific heat [KJ/kg°C]	Start and stop time [hours]
1	20	135	14400	1	0.5 - 2.5
2	80	140	64800	1	0 - 3
3	170	60	10800	1	1 - 2
4	150	30	8100	1	1.25 - 2.75

Number of batches per year	2000
Batch period	3 h
Interest rate (real)	6 percent
Write-off period	10 years
Number of stores	2 to 5
Smallest temperature difference DT_{MIN}	5 °C

Appendix B. Specifications for Case B: Integration of Processes in the Brewhouse.

Table B1. Data for a Brewhouse with Batch Size of 1000 hl Cold Wort.

Process stream number	Start temp. [°C]	End temp. [°C]	Mass flow [kg/batch]	Specific heat [kJ/kg°C]	Start and stop time [h]
1	8	51	30000	3.5	0 - 1
2	51	70	30000	3.5	1 - 1.25
3	70	100	30000	3.5	1.5 - 2
4*	100	101	800	2257	2 - 2.33
5	8	37	38000	3.8	1.25 - 1.75
6	61	67	67200	3.7	2 - 2.17
7	67	76	67200	3.7	3.17 - 3.33
8	8	76	64000	4.186	5.75 - 6.75
9	74	100	117200	3.9	5.75 - 6.75
10*	100	101	12000	2257	6.75 - 8.25
11	98	11	105200	4.0	9 - 11
12**	101	100	800	2257	2 - 2.33
13**	100	30***	800	4.186	2 - 2.33
14**	101	100	12000	2257	6.75 - 8.25
15**	100	30***	12000	4.186	6.75 - 8.25
16**	74	30***	14000	3.9	6.75 - 9.75

Number of batches per year	1500
Batch period	9 h 45 min.
Interest rate (real)	6 percent
Write off period	10 years
Number of stores	2 to 5
Smallest temperature difference DT_{MIN}	2 °C

* cooking
** not required by the brewing processes
*** Recommended temperature level below 35°C (environmental concerns).

SESSION 12:

Efficient Production and Use of Electricity

Chairman: Prof. E. Macchi

THE INCREASED ELECTRIC POWER PRODUCTION FROM THE THERMAL POWER STATIONS OF GREECE OPERATING WITH IMPROVED EFFICIENCY

Emm. Kakaras, N. Papageorgiou
National Technical University of Athens
Mechanical Engineering Dept., Thermal Eng. Section
Patission 42, Athens 10682, Greece

ABSTRACT

The paper discusses the possibility of retrofitting the existing lignite fired thermal power plants in northern Greece using a gas turbine as a topping cycle. The necessity of increasing the electric power of these units arises from the continuous decrease of the fuel quality from its nominal value which results to Unit's operational load of 250 MW_{el} instead of 300 MW_{el} of nominal power.

Several possibilities of gas turbine topping of a lignite fired power plant are evaluated in the paper. The increase of the electric power production and the increased efficiency of the total process together with the required modifications of the existing equipment and the availability of the plant are the main criteria for the evaluation of the proposed alternatives.

INTODUCTION

About 50% of the total pollutants emission due to the human activities worldwide are caused by the energy sector which also determines the standard of living in the various countries.

The energy production structure of Greece is based on the exploitation of the local brown coals (lignites) which feature very low calorific value (3500-5500 kJ/kg), high moisture content (55-60%) and medium ash content (10-15%). The lignite fired power plants represent the 70% of the total installed electrical power in Greece and most of them are located in the northern Greece. In comparison to the other fossile fired power plants, the brown coal fired power plants of Greece operate with lower efficiency, higher CO_2 emission and higher pollutants emission (mainly fly ash) due to the poor fuel quality. In order to reduce the environmental impact from these plants a more efficient and environmentally acceptable operational mode is required. This opportunity is offered by the introduction of the natural gas to the energy structure of Greece which will take place in the near future. If a gas turbine is superposed to the system Steam boiler-Turbine-Generator the total electrical power of the system is increased and it is rendered with increased efficiency. For the case of unfired combined cycles with natural gas fired gas turbines the total efficiency can reach 55%.

The combustion gases from the gas turbine feature an O_2 content of 15% and they can be used both for heating and oxidation purposes. The high temperature and the high O_2 content of the gases allow the efficient combustion of almost any fuel with minimum use of air in an additionaly fired combined cycle application. This operational mode leads to lower efficiency than the one of unfired combined cycle but it enhances the utilization of local fuels.

The high energy potential of the gas turbine combustion gases is illustrated in Fig.1 where the adiabatic combustion temperature (which approximates the turbine inlet temperature) is presented in relation to the excess air ratio and the O_2 content of the flue

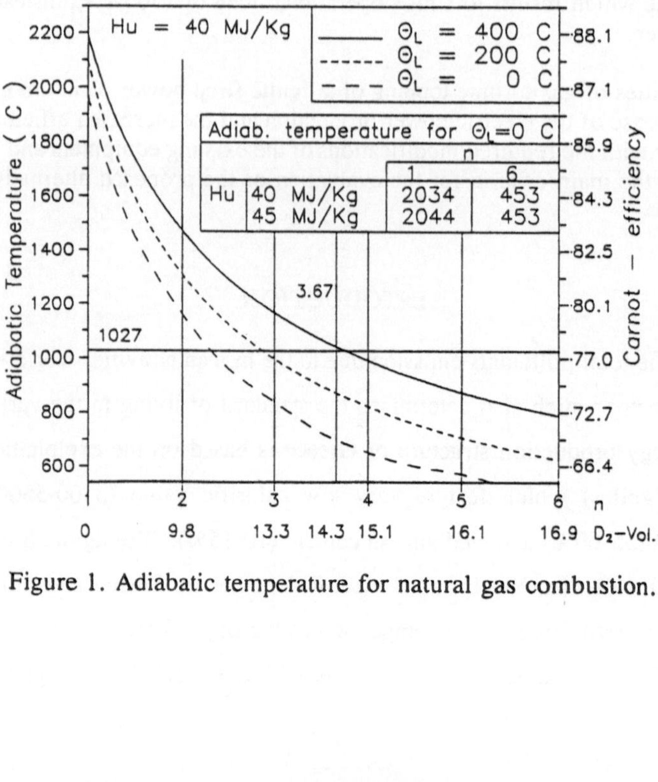

Figure 1. Adiabatic temperature for natural gas combustion.

gas as well as to the Carnot efficiency.

Typical gas turbine characteristics are shown in Table 1 (GT series of ABB). For most cases of retrofiting power plants of 300 MW_{el} (as the greek ones) the gas turbines used are similar the GT 11N.

TABLE 1

Gas Turbine characteristics

Model	GT 8	GT 11	GT 11N	GT 13
Base load power (ISO) [MW_e]	49,1	71,9	81,6	98,2
Peak load power [MW_e]	52,9	78,2	88,9	107,3
Thermal efficiency [%]	32,1	31,1	32,2	32,1
Ambient Temperature [^{0}C]	15	15	15	15
Fuel mass flow [kg/s]	3,12	4,94	5,38	6,92
Thermal power input [MW]	147,05	232,82	253,56	326,14
Flue gas mass flow [kg/s]	183	290	316	406
Turbine Inlet Temperature [^{0}C]	1085	1027	1027	990
Flue gas Temperature [^{0}C]	521	520	515	489

OPERATIONAL PROBLEMS OF THE LIGNITE FIRED GREEK POWER PLANTS

Most of the brown coal fired power plants of Greece consists of 300 MW_{el} Units which are mainly located in Northern Greece. Due to the considerable deviation of the fuel characteristics from its predefined values and the decrease of fuel quality, the usual electrical output of these Units does not exceed 250 MW_{el} with 500 tn/h fuel consumption which represents a considerable financial loss for the Public Power Corporation (PPC) which is the only electricity company in Greece.

The first attempts to increase the load of the above mentioned power plants increasing the fuel's calorific value by mixing lignite with bituminous coal was proved to be not only unefficient but it has caused severe fouling problems on the boiler surfaces. A considerable decrease of the boiler's efficiency due to the increase of unburnt carbon content of the liquid and the fly ash was also detected.

Another alternative which is the increase of the fuel pulverization system capacity does not provide a long term solution to the problem because the use of higher capacity mills will increase the fuel particles velocity inside the boilers furnace leading to extensive erosion problems due to the pyritic nature of the lignite.

In the near future the extensive introduction of the natural gas utilization in the energy structure of Greece is scheduled. Therefore the PPC should examine in depth the possibility of using gas fired gas turbines to increase the electricity production from the thermal power plants. It is obvious that a wide range of alternatives for the combination of gas turbines with existing power plants is available. In this paper, three indicative test cases of brown coal fired power plant repowering with gas turbine will be examined and presented.

GAS TURBINE TOPPING ALTERNATIVES

Gas turbine with waste heat boiler and additional lignite firing

For this case the additional steam required for the production of 300 MW_{el} from the existing steam turbine instead of the actual 250 MW_{el}, is produced in the waste heat boiler for the utilization of the gas turbine flue gas. Due to the thermodynamic characteristics of the high pressure steam for the turbine, (see Fig.2), additional fuel firing is required in the waste heat boiler. Under the assumption that the additional steam is expanded in the existing turbine to produce the total of 250+50 = 300 MW_{el}, the size of the required gas turbine and the total efficiency are calculated.

The steam turbine is now able to produce 300 MW_{el} and an additional 17 MW_{el} are produced from the gas turbine. The total efficiency is increased from today's value of 36% to 38,3%.

The layout for this alternative is shown in Fig.2. In this case the same feedwater heaters can be used for the existing boiler and the new waste heat boiler, but additional fuel preparation equipment and fly ash filters are required for the waste heat boiler due to the simultaneous brown coal combustion. To avoid the additional costs, natural gas can be fired in the waste heat boiler instead of the additional brown coal. It should be mentioned that the high inlet water temperature to the W.H. boiler decreases the efficiency of the W.H. boiler and therefore additional cooling of the flue gas at the W.H. boiler exit is required.

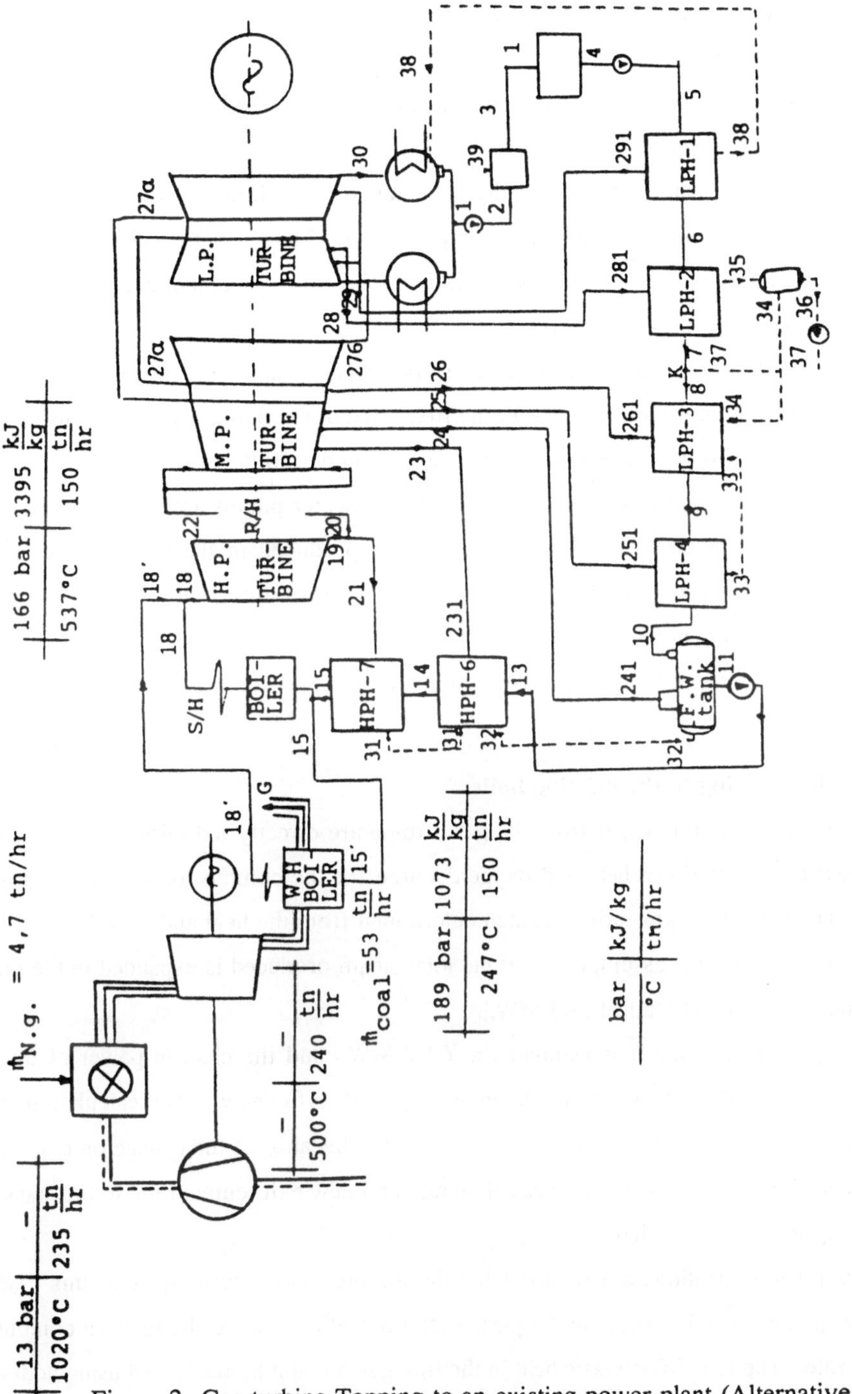

Figure 2. Gas turbine Topping to an existing power plant (Alternative 1).

Gas turbine with unfired waste heat boiler

If the additional firing in the waste heat boiler of the previous test case is neglected, it remains only the possibility of producing steam at an intermediate pressure thus increasing the steam turbine load.

The flue gas characteristics and the heat transfer conditions in the waste heat boiler (pinch-point) determine the additional steam properties. The capacity of the existing turbine under the assumption of no major modifications (no blade replacement at any turbine stage) determines the total steam mass flow rate.

As it is shown in Fig.3 the steam from the waste heat boiler is led to the medium-pressure steam turbine at a location after the first stage of extraction steam. In this case only the first three low pressure feedwater heaters operate under 100% load whereas for the waste heat boiler an additional feedwater tank and new feedwater pumps are required. For steam turbines of this type 61-70% of the total power is produced in the intermediate and low pressure stages according to the operating conditions. Therefore, with the configuration of this alternative a total production of 250+35 MW_{el} from the steam turbine is achieved. The new gas turbine will produce 76,75 MW_{el} and the total efficiency of this process is estimated at 38,7%.

Gas turbine topping to the existing boiler

In this case the flue gases from the gas turbine are directly led into the existing boiler to be used as an additional heat and oxygen source for the combustion of brown coal (see Fig.4). The size of the gas turbine is then determined from the heat and mass balance in the existing boiler under the assumption that the total steam produced is expanded in the turbine to produce the nominal load of 300 MW_{el}.

The gas turbine power is estimated at 60,7 MW_{el} and the electric power of the total process is 360,7 MW_{el} and it is rendered with 40,8% efficiency. The calculation of the layout data of this case includes an iterative procedure because of the change on combustion conditions which results to new flue gas flow rate and new exit temperature to the stack thus influencing the boiler efficiency.

Several modifications are required for the air preheating system, as in this case the rotary air preheater will operate under partial load conditions due to the reduced combustion air flow rate. The remaining waste heat in the flue gases could be recovered using additional

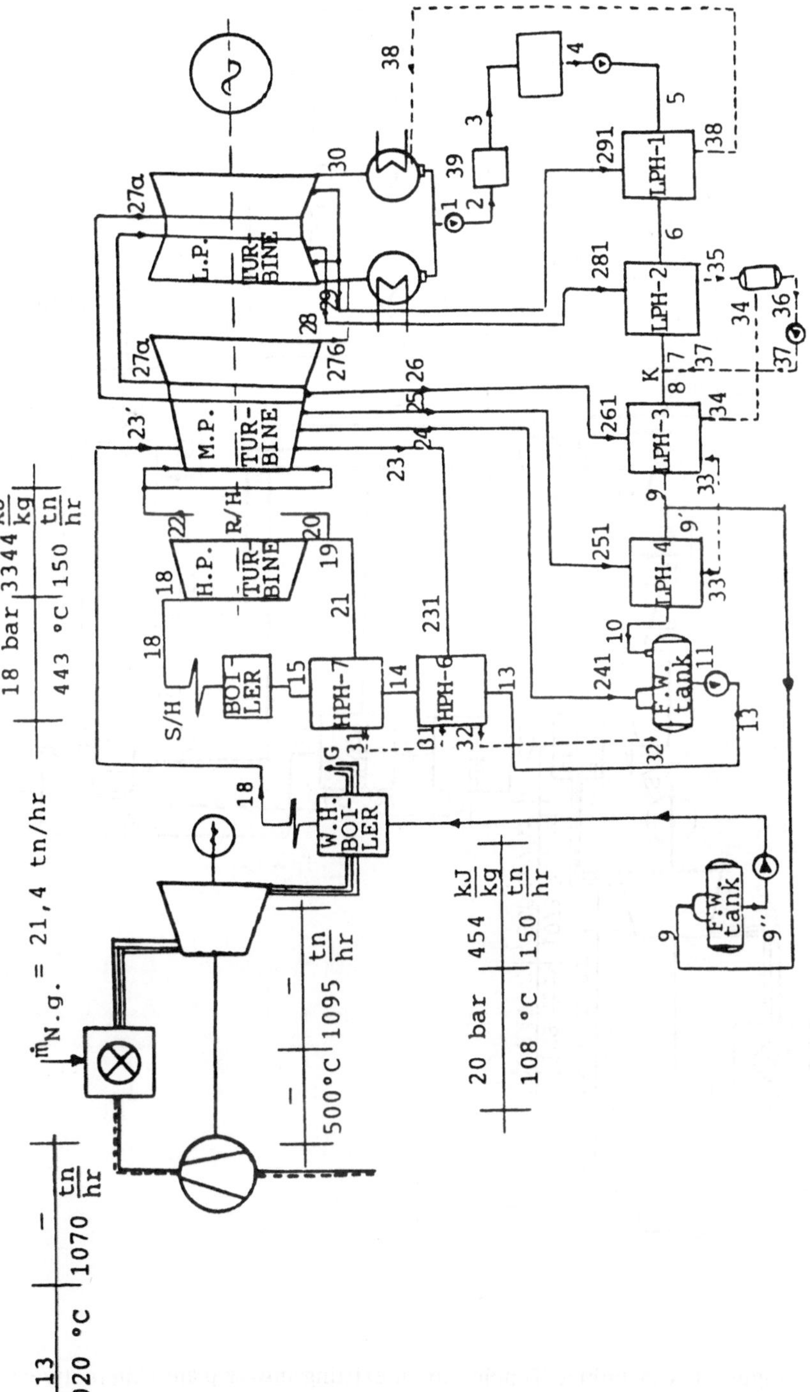

Figure 3. Gas turbine Topping to an existing power plant (Alternative 2).

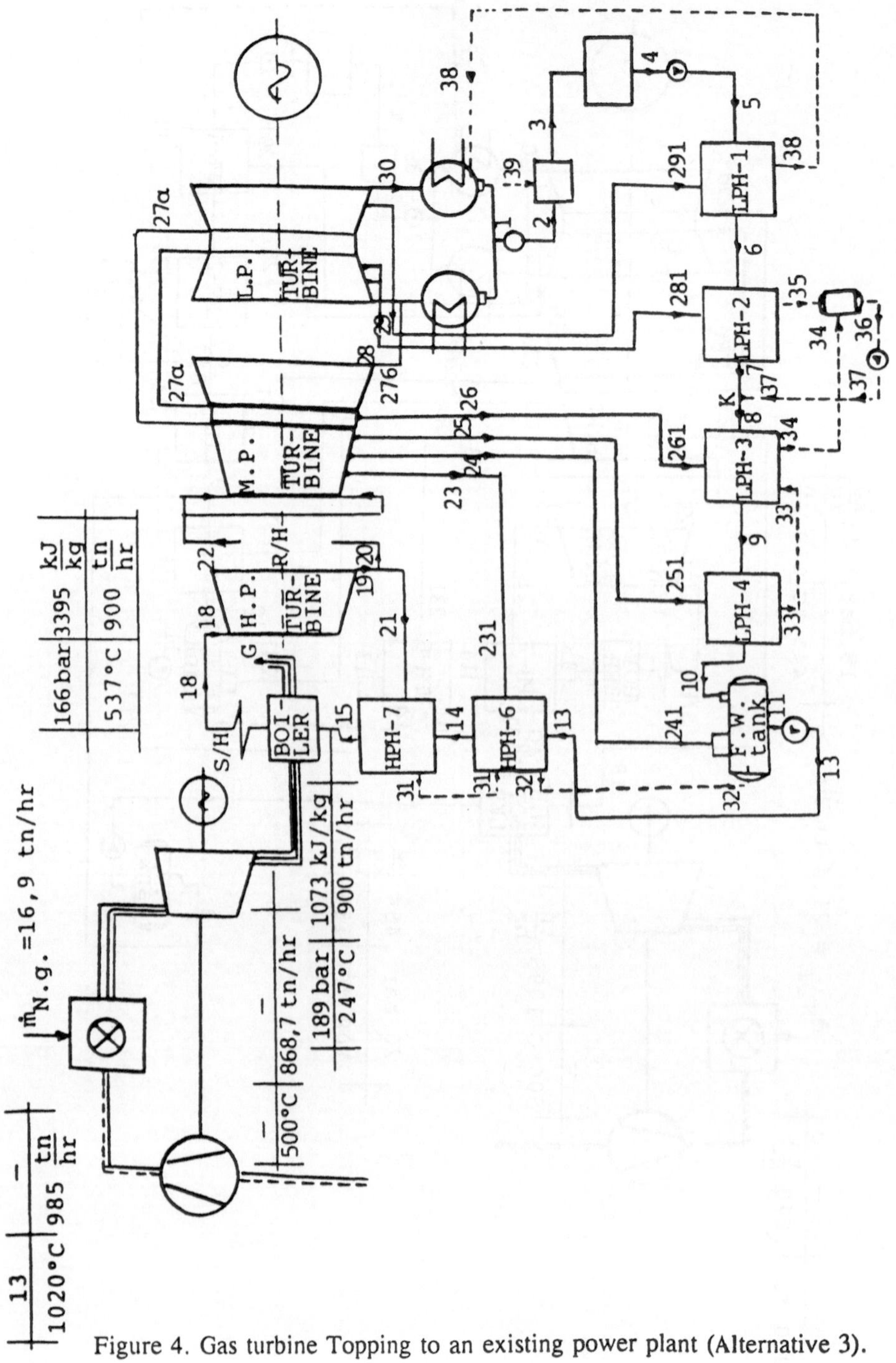

Figure 4. Gas turbine Topping to an existing power plant (Alternative 3).

feedwater economizers which could substitute part of the feedwater heaters operating with extraction steam.

A possible substitution of the Low Pressure Heaters 1-4 by a flue gas heated economizer is shown in Fig.5. This additional economiser will be located in the existing boiler after the last heat exchanger surfaces. In this case the low pressure extraction steam is further expanded in the Low Pressure Turbine. This alternative however is subject to the condition that the existing turbine, condenser and cooling tower could handle the additional steam flow are.

Another alternative could be the substitution of the High Pressure Heaters 6 and 7 by additional flue gas-feedwater heat exchangers or the substitution of L.PH 3 and 4 and H.P.H7. For any of this cases, the flexibility of the turbine, which should operate without the steam extractions, the condenser and the cooling tower is the decisive parameter.

Under the assumption that the additional steam could be expanded in the existing turbine an additional electrical power of approx. 12-15 MW_{el} with an efficiency increase of 2% should be expected.

CONCLUSIONS

The following Table 2 summarises the main results from the examination of the gas turbine topping alternatives presented in this paper.

The third alternative presents the higher efficiency increase potential but it is subject to major modifications of the feedwater and air preheating system.

The alternatives 1 and 2 require the construction of a waste heat boiler. For the alternative 1 the additional fuel preparation equipment and the other supporting devices are required, whereas for the alternative 2 a much simpler waste heat boiler without fans, air preheaters, ESP's etc. is required.

For this alternative (2) a minor modification on the intermediate pressure steam turbine should be also considered.

The reference case for the evaluation of the three alternatives represents the today's situation which is 250 MW_{el} with 36% efficiency. This efficiency is low and immediate heat recovery actions are required.

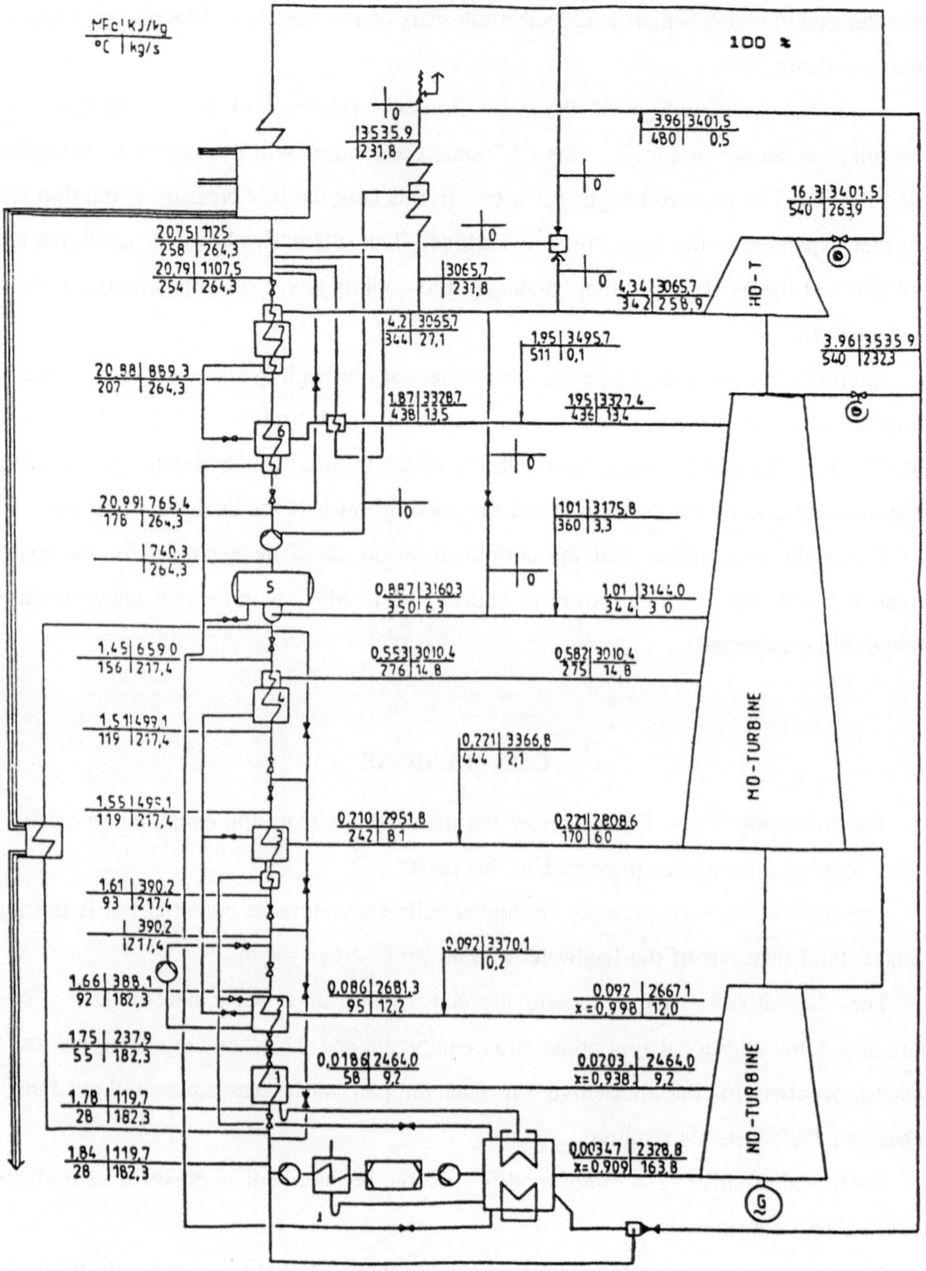

Figure 5. Alternative 3 with external feedwater preheaters.

At a first assesment considering the respective modification costs, the alternative 2 seems more realistic. The final evaluation however should include a detailed cost analysis and the consideration of the unavailability of plant due to the necessary repairs and modifications.

TABLE 2

Comparative results of alternatives 1,2 and 3

	EL. POWER (MW)		CONSUMPTIONS	tn/hr	kJ/kg	TOTAL EFFICIENCY
Reference case	250		Coal : 500	tn/hr	5025	0,36
			Natgas : -		-	
Alternative 1	GT	ST	Coal : 553	tn/hr	5025	0,383
	17	300	Natgas : 4,7	tn/hr	40379	
Alternative 2	GT	ST	Coal : 500	tn/hr	5025	0,387
	76	285	Natgas : 21,4	tn/hr	40379	
Alternative 3	GT	ST	Coal : 500	tn/hr	5025	0,408
	60	300	Natgas : 16,9	tn/hr	40379	

At any case the achieved efficiency increase cannot be compared to the high efficiency of unfired combined cycles which exceeds 50%. Therefore, for the new power stations which will be installed in this area the natural gas combined cycle represents an attractive solution which could also make use of the future technologies concerning the gasification of lignite and the production of SNG available in the next decade.

A BETTER UNDERSTANDING OF WOOD AS A MATERIAL - A WAY TO INCREASED ENERGY EFFICIENCY WHEN MAKING MECHANICAL PULPS?

MYAT HTUN, LENNART SALMÉN and LENNART ERIKSSON
STFI (Swedish Pulp and Paper Research Institute)
Box 5604, S-114 86 Stockholm, Sweden

ABSTRACT

Existing mechanical pulping processes are briefly described in relation to the research approach adopted by STFI to solve the complex problem of reducing electrical energy consumption in refining and upgrading the properties of mechanical pulp fibers.

It is shown that the materials approach involving study of the morphology, physical and chemical properties of wood may be a feasible way to understand the various mechanisms which may be involved when wood is subjected to load.

BACKGROUND

Mechanical pulps are used extensively in many printing papers and carton board grades because of the unique quality of high opacity and high bulk which the mechanical pulps can render the end products. Newsprint and magazine papers for example, are mainly made from mechanical pulp fibers. Today more and more high quality printing papers contain mechanical pulps in the furnish. The reason for this increasing use of mechanical pulps is that the pulp making process is relatively simple compared to that for chemical pulp. The investment costs are lower and the scale factor is less prominent. The pulp yield is very high, i.e. the pulp fibers contain virtually all the wood components which makes the pulp attractive in places where the forest is limited or the price of wood is high. Since the process is mechanical, the environmental load due to chemical treatment, as in the case of chemical pulps, is lower.

The drawbacks with mechanical pulps are that the pulps are weaker than the chemical pulps and their ageing behaviour limits their use. The electrical energy demand to prepare the mechanical pulps is very high, especially in the modern methods that use refiners to disintegrate wood chips. There is also a rule of thumb that more energy is required to produce pulp with better quality. Since quality demands have increased, so also has the energy demand.

The transformation of wood or wood chips by mechanical methods into paper-making pulp fibers, requires a substantial amount of electrical energy. Sweden alone uses approximately 5 TWh of electrical energy annually to produce 3 million tons of mechanical pulps. An energy flow sheet of a modern, integrated printing paper mill in Sweden is shown in figure 1.

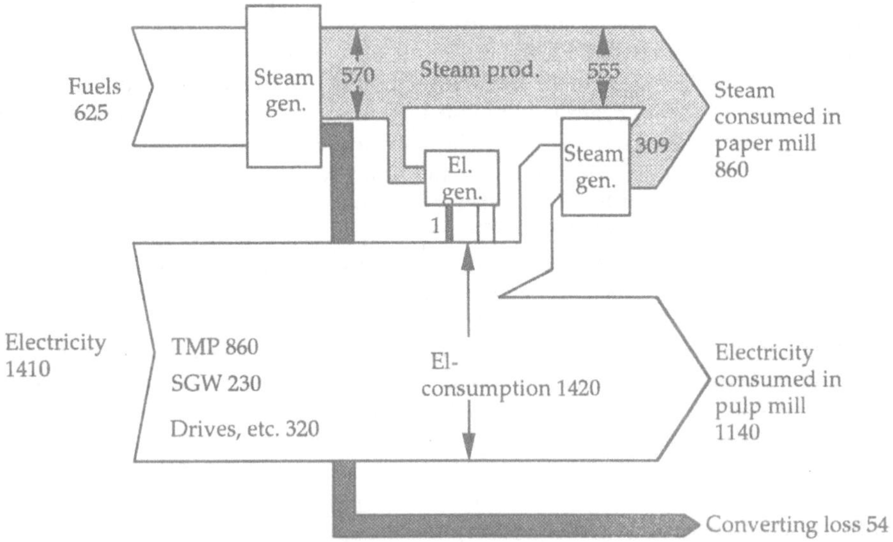

Figure 1. Energy flow sheet in GWh of an integrated printing paper mill.

It has been estimated that during the grinding or refining of wood, less than 5 per cent of the applied electrical energy is being utilized to create well refined pulp fibers. The majority of the energy is transformed into heat without affecting pulp properties. Despite the fact that most of the heat evolved can be recovered and used, e.g. for drying of paper, it has for a long time been a challenge to researchers and equipment suppliers to find a means of lowering the specific energy demand. It is after all cheaper to produce steam from solid/liquid fuels than from refiners consuming high-value electrical energy.

This paper gives a brief description of the existing processes and a discussion of the research approach which STFI has adopted to find means to reduce energy consumption and to produce mechanical pulps with the highest quality.

PULPING PROCESSES

There are several methods available for the manufacture of mechanical pulp. The oldest involves grinding the wood (log) surface with a rotating stone. The pulps produced by this method are called stone groundwood pulps (SGW) and are used mainly for news-print and magazine papers. Although the mechanical properties are poor (it is necessary to reinforce with chemical pulp fibers in the paper production), the optical properties are excellent. The energy required in this process is in the range of 900-1400 kWh per ton of pulp depending on the degree of grinding. A newer development of this method is the application of a pressurized atmospheric environment to the grinding zone as well as to the wood. The pulps produced by this method are called pressurized stone groundwood pulps (PSGW). These are mechanically stronger but have poorer optical properties than SGW. The energy requirement is also higher.

Refiner mechanical pulp (RMP) has been used for some time. In this process, wood chips instead of logs are refined (ground) between two discs, one rotating and one stationary. The pulps produced are stronger than SGW pulp but the optical properties are poorer. The energy consumption is higher, usually around 1800 kWh/ton. The process which was developed in the late sixties and has expanded extensively since then is the so called thermomechanical pulp (TMP) process, see scheme in figure 2.

In this process, the wood chips are preheated with steam prior to refining. The preheated chips are then transported to the refining zone by means of feeding screws. The refining is, as in the case of RMP, done between two discs; one rotating and the other stationary (single disc) or with the two discs rotating in opposite directions (double disc). The refining zone in this case is pressurized, usually around 300 kPa. Refining is performed either in one stage, i.e. using only one refiner to tranform wood chips to fibers, or in two stages. The fibers refined in the primary refiners are screened to separate the untreated fiber bundles. These untreated fiber bundles, so called rejects, are further refined either in a pressurized or in an atmospheric refiner. The steam and the pulp fibers are then separated in pressurized cyclones. The TMP process requires more energy than RMP, usually about 2000 kWh/ton. About 60% of the heat evolved during the refining process is recovered as steam. TMP fibers are significantly stronger than the SGW fibers and can eliminate or reduce the need for chemical pulp fibers as reinforcement fibers in newsprint or other printing papers.

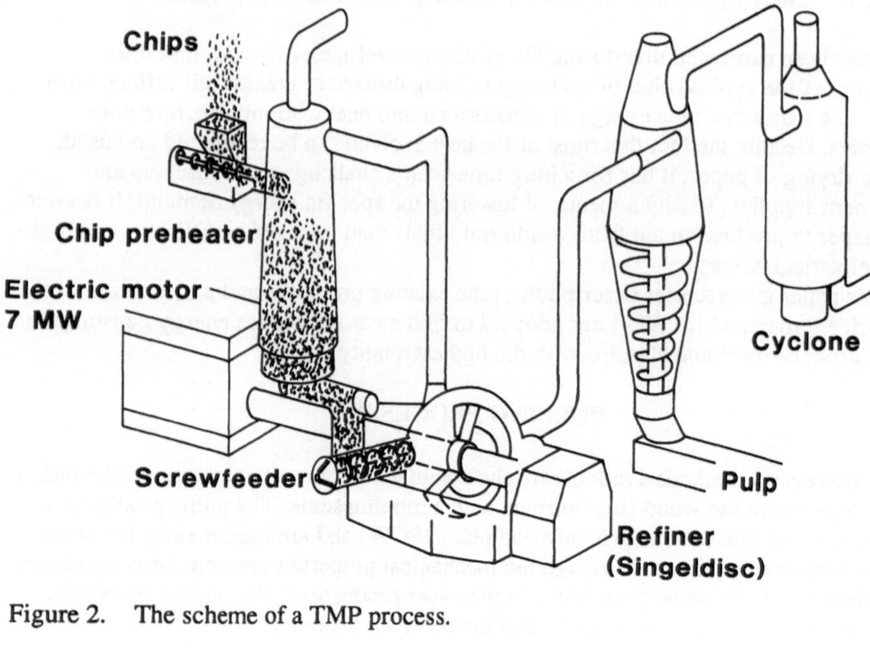

Figure 2. The scheme of a TMP process.

In a modified TMP process, a small amount of chemicals, usually sodium sulphite and alkali, are added to wood chips before refining. This type of pulp - chemical thermomechanical pulp (CTMP) - is somewhat stronger than TMP and is brighter. The drawbacks with these pulps are that energy consumption is higher (2100 kWh/ton) and the opacity is much lower than for TMP. The wood utilization (yield) is also somewhat lower.

STFI's APPROACH

The basic mechanisms behind the mechanical pulping processes are not yet fully understood. The development of the process has been based mainly on extensive "trial and error" studies. In order better to understand the complex problem of refining, STFI has for some years adopted a fundamental approach to wood as a material.

The mechanical pulping process can be divided into several phases. First, a separation of wood into fibers. The success of good separation with minimum fiber damage as well as the creation of "suitable" fiber surfaces is believed to depend on the morphology, i.e. the chemical and physical structure as well as the physical properties of the wood. External parameters such as temperature, loading frequency and loading amplitude are also important for efficient fiber separation. During fiber separation a substantial amount of heat is evolved possibly due to viscoelastic and frictional losses. The separated fibers are not yet ready to make paper with good quality. In order to achieve better papermaking properties, the fibers are further refined. In this phase, fibers are fibrillated as well as producing debris and fibrillar materials, so called fines. Fibrillation of fibers as well as fines contribute to better bonding capability, better optical properties and to better surface characteristics which are requirements for good printing papers. Also in this refining phase, substantial amounts of energy are lost due to increased loss in mechanical energy as heat.

WOOD MORPHOLOGY

In the refiner, the wood material faces loads of various magnitudes which deform the fibers. As wood has a very irregular structure, some fibers will be highly stressed to rupture while others will only dissipate energy as they are deformed visco-elastically. The non uniformity of the material is an essential obstacle towards a more efficient energy utilization in mechanical pulping.

The position at which fracture actually takes place between fibers will depend on the differences in ratio between the elastic properties of the cell wall layers. As illustrated in figure 3, the cell wall of the wood fibers consists of a multilaminar structure, where the chemical composition - and thus for example the elastic properties - vary between layers. The layer connecting the fibers, the so called middle lamella, is for instance rich in lignin, the softening properties of which have a great impact on the fiber separation process (1, 2).

At low temperatures, when the lignin is stiff, the fracture occurs in an uncontrollable manner leading to a greater amount of broken fibers and a high fines content. This is typical of refiner mechanical pulps (RMP). As the temperature is increased, the fracture zone moves outwards rendering a larger long fiber fraction. This is typical of thermomechanical pulps (TMP). With chemical treatments, altering the lignin properties, fibre separation may be further improved, substantially reducing the shives (unseparated fiber bundles) content as for chemimechanical pulps (CTMP). However, with increasing softness of the fiber wall, the energy losses in the subsequent refining also become greater due to an increase in viscosity of the cell wall material.

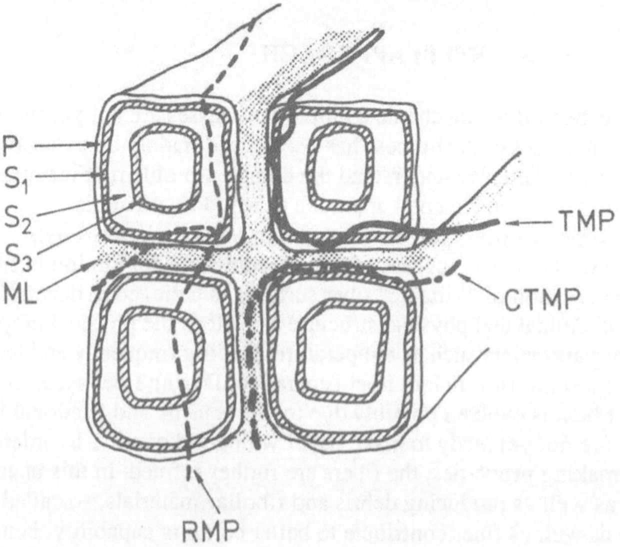

Figure 3. Schematic diagram of fracture zones in soft wood as affected by different mechanical pulping processes, RMP (refiner mechanical pulp), TMP (thermomechanical pulp) and CTMP (chemimechanical pulp) adapted after Franzén (1). The cell wall layers are indicated P (primary wall), S_1, S_2, S_3 (secondary walls) and ML (middle lamella).

THE VISCOELASTIC NATURE OF WOOD

Wood is a composite polymeric material which, due to its amorphous components - the hemicelluloses and the lignin - behaves as a viscoelastic material. This means that the material deforms with time under load and thus that its properties depend on the speed with which the wood is stressed. Especially the softening properties of wood are greatly influenced by the frequency of the load applied to the wood as seen in figure 4 (3). With an increase in frequency, the softening of wet wood, i.e. wet lignin, is shifted to higher temperatures which in turn affects the fiber separation process. The fiber separation may thus to some extent be regulated by the load frequency applied and by the temperature of the fiber separation stage. In the subsequent refining, where the energy demand is substantially greater than in the fiber separation, the way in which the fibers have been separated is also of great importance. A fiber separated at too high a temperature, and thus covered with a layer of lignin, is almost impossible to refine into a suitable papermaking furnish with reasonable energy input.

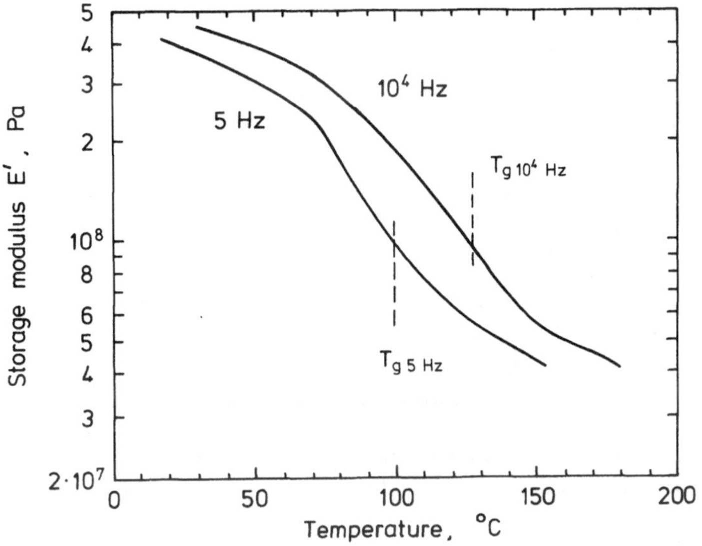

Figure 4. Modulus - temperature curves showing the shift of the softening region of wood to higher temperatures with increasing frequency, adapted after Salmén (3).

THE IMPORTANCE OF THE IONIC STATE

In its native state, the hemicelluloses in the wood contain ionizable groups, carboxylic acid groups, which due to their strong interaction with water influence the properties of the wood (4). It is also possible by chemical treatment to introduce ionic groups into the lignin, carboxylic acid groups by oxidative treatment and sulfonic acid groups by sulfonation. Depending on the environment of the fiber during processing, these ionic groups have different counter-ions attached to them; the most common being sodium, calcium and hydrogen.

The introduction of ionic groups into the lignin, as for CTMP through sulfonation, will somewhat alter the cross-linked structure of the lignin (5) and this leads to a reduction in its softening temperature, figure 5. The softening temperature of the lignin containing ionic groups is always lower when the wood fibers are in the sodium form than when they are in the calcium form, regardless of the nature of the ionic groups. The fiber separation mechanism is thus similar regardless of whether the fibers have been sulfonated or oxidatively treated. It can thus be anticipated that the shives (unseparated fiber bundles) content is reduced as effectively by both chemical treatments.

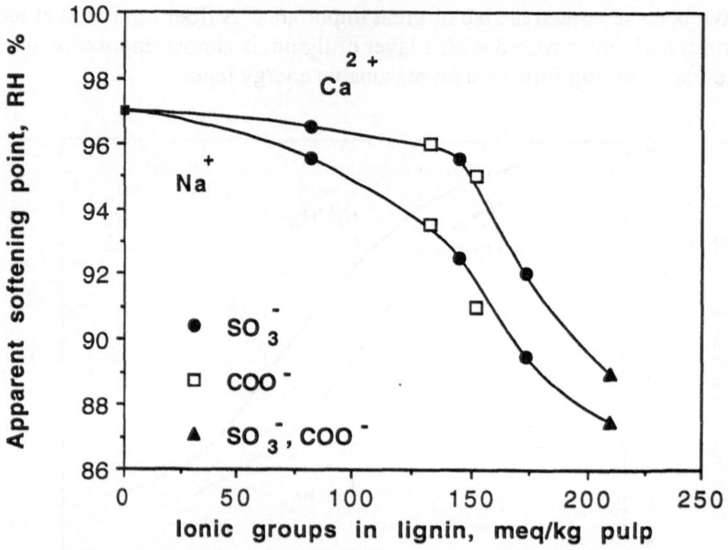

Figure 5. The softening point of wood fibers, i.e. the wet lignin as a function of the amount of ionic groups in the lignin with either calcium or sodium as counter-ion.

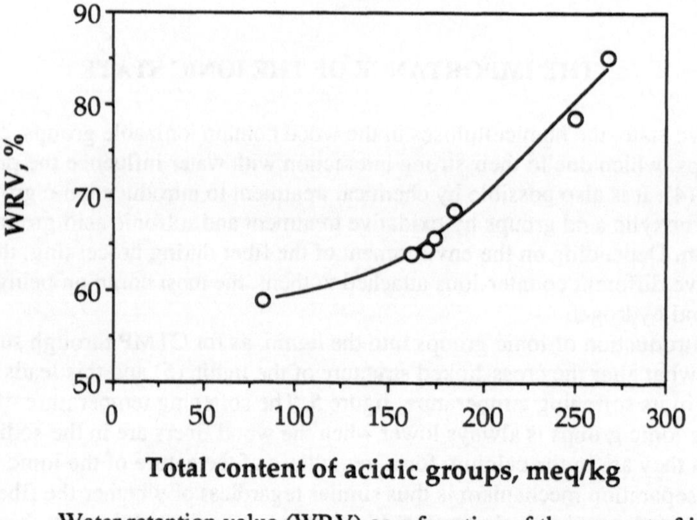

Figure 6. Water retention value (WRV) as a function of the amount of ionizable groups in the wood fibers. The ionic groups have been introduced in the lignin by sulfonation.

The ions have, as pointed out, a strong effect on the water uptake and, as seen in figure 6, the water content measured as water retention value, WRV, increases proportionately with the amount of ionic groups. Thus, the more ionic groups within the wood fiber, the higher is its flexibility and the more easily is it loaded in the refiner without breaking and thus without losing essential properties.

Further, the counter-ion to the ionic groups influences water uptake, as illustrated in figure 7, which shows the water retention value (a measure of fiber swelling) when the counter-ion is changed from the hydrogen form to the sodium form in a thermomechanical pulp. The chemical environment in the refiner accordingly influences the manner in which the wood fiber material being flexibilized is able to sustain the forces applied. This might have an impact on the property - energy consumption relationships.

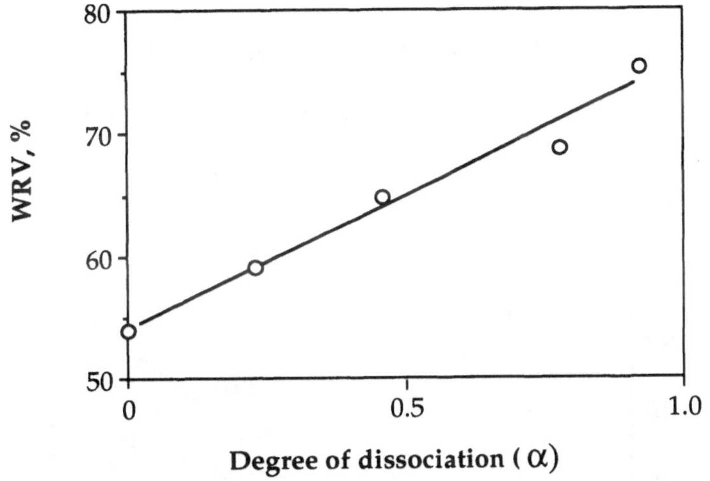

Figure 7. Water retention value as a function of the degree of dissociation from the hydrogen form to the sodium form.

EFFECTS ON PULP PROPERTIES

Since the properties of the wood fibre material are highly affected by its ionic nature, and since this should strongly influence the refiner action, affecting both the fiber separation mechanisms and subsequent refining, it may be concluded that both the mechanical properties of the pulps produced and the amount of refining energy consumed must depend on the refiner environment. Trials in a laboratory refiner also indicate such effects. As seen in figure 8, the tensile index of a pulp - an important strength property - may be substantially increased as a result of the introduction of ionic groups, in this case carboxylic acid groups introduced into the lignin by peroxide treatment. However, the increase in ionic groups may not immediately result in a saving in refining energy, due mainly to an increased energy dissipation within the softer fibers.

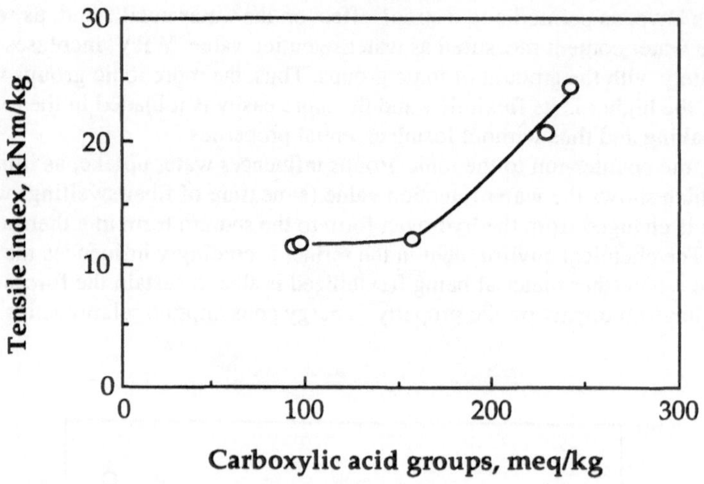

Figure 8. Tensile index as a function of the amount of ionizable groups.

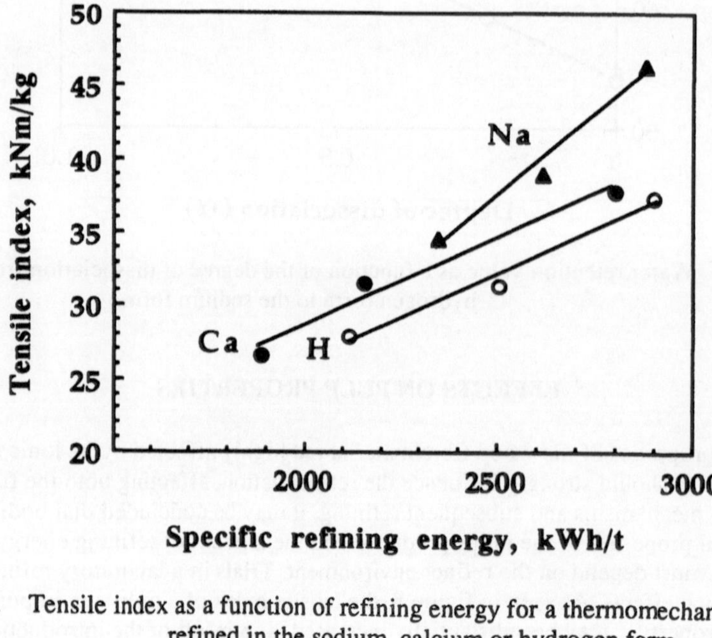

Figure 9. Tensile index as a function of refining energy for a thermomechanical pulp refined in the sodium, calcium or hydrogen form.

By changing the counter-ion, it is however possible to affect the energy demand in refining, as shown in figure 9 for an unmodified softwood pulp refined in hydrogen, sodium and calcium forms. A substantially higher tensile index is reached at the same energy demand with the sodium than with the calcium or hydrogen forms. When compared at the same tensile index, pulping the sodium form requires much less refining energy than does the calcium form. It is thus evident that pulp properties as well as refiner energy demand are highly affected by chemical modifications. This points to a potential saving of refining energy if this concept of controlling the softness of the wood fiber can be utilised properly in full-scale production.

FINAL COMMENTS

The results discussed in this paper suggest that a materials approach to the behaviour of wood under loading may be a feasible way of dealing with the complexity of mechanical refining processes. However, it is too early to say whether the results obtained in the laboratory can be applied industrially.

Before laboratory results can be transferred to industrial application, aspects of process design and the impact on the environment caused by a change in the process have to be considered. The use of oxidative treatments may for instance have an advantage over sulfonation with regard to the environment load.

Before a new process or a process change is proposed on the basis of our fundamental findings, a considerable amount of fundamental work is still needed in order to understand the energy transferring mechanisms which take place during loading of wood, especially the nature of wood fracture and how this fracture is related to the wood properties under different environmental conditions during loading. Furthermore, a deeper understanding is necessary of the refining phase where the fibers are fibrillated and flexibilized. This is the phase where the conversion of electrical energy to heat is greatest.

REFERENCES

1. Franzén, R: Nordic Pulp Paper Res. J., 1(1986):3, 4

2. Yang, J-L, Pettersson, B and Eriksson, K-E: Nordic Pulp Paper Res. J., 3(1988):1, 19

3. Salmén, L: Journal of Material Science, 19(1984):9, 3090

4. Scallan, A.M. and Grignon, J.: Svensk Papperstid., 82(1979):2, 40

5. Atack, D and Heitner, C: Trans Techn Sect., (Can Pulp Pap Assoc), 5(1979):4, TR99

THE INDUCTIVE HEATING OF PACKED BEDS AND ITS APPLICATIONS TO COMPACT FLUID HEATERS AND TO THE REGENERATION OF SPENT ACTIVATED CARBON

P.Duquenne, A.Deltour*, G.Lacoste

ENSIGC, chemin de la Loge 31078 Toulouse Cedex (France)
*ENSEEIHT, 2 rue Camichel 31071 Toulouse Cedex (France)

ABSTRACT

In this paper, a method is presented in order to predict the behaviour of a granular medium submitted to an inductive electromagnetic field. The resulting model is intented to give an easy and reliable characterization of energy transferred to the bed. A series of experiments was carried out and permitted a validation of the hypothesis formulated over a wide domain.

A first application was studied aiming at the sizing of heaters using a stainless-steel balls bed percolated by a cold fluid. Experimental investigation of temperature profiles in both solid and fluid phases and of heat transfer coefficients between them lead to the realization of a compact fluid heater working with small difference of temperature and good efficiencies.

This technique was then applied to the heating of activated carbon beds so as to allow a fast and performing regeneration.

INTRODUCTION

Although induction has proved for long to be an efficient way of bringing huge energy fluxes into electrically conductive materials, its applications remain almost restricted to the metallurgic industries, where it is devoted to the heating of homogeneous pieces.

On the other hand, process and chemical engineering assign a great importance to granular media whenever transfer phenomena occur between a solid and a fluid phase, since they provide considerable contact areas and ensure high values of transfer coefficients.

The matter with inductive heating is that, when applied to a homogeneous material, it generates what is called "skin effect", which means that eddy currents appear in a narrow depth nearby the piece periphery : thus, what is usually seen as an advantage of inductive heating, especially in surface treatment, may bring no betterment to a granular medium, by drastically reducing the active contact area. However, if inductive heating showed to be even in a packed bed, it would be an important improvement compared to all the existing techniques of heating a granular medium.

The aim of this work is then to investigate the behaviour of packed beds submitted to alternative magnetic fields, and especially in two domains basically different, for the chemical engineer as well as for the electrotechnician : in a first case, we will describe the heat transfer between a fluid and a granular bed of highly (electrically) conductive material, and then the mass transfer from a bed of spent activated carbon (comparable to a semi-conductor).

MODEL PRESENTATION

Maxwell's equations

The effects of a magnetic field generating the so-called inductive heating are ruled by Maxwell's equations :

$$\text{rot (H)} = J \quad , \quad \text{div (D)} = \sigma$$
$$\text{rot(E)} = -\frac{\partial B}{\partial t} \quad , \quad \text{div(B)} = 0$$

where $D = \varepsilon E$, $B = \mu H$, and $J = \gamma E$ if electric displacement is neglected.

One understands easily that the resolution of such a system of equations may cause some problems. It is generaly performed through the study of a given domain on the frontiers of which boundary conditions are known - this method requiring the use of a netting of this domain, this netting being hardly suitable with the notion of granular media.

Some simplifications may result of the inductor geometry : provided this inductor is long enough in regard of its diameter (in fact, its length must be at least equal to the diameter),it generates a magnetic field that can be considered as uniform inside the volume it delimits.

Moreover, some shapes allow facilities in the writing of the system above : for example, resolution can be carried out quite easily in the case of thin or infinite plates, cylinders, tubes [1]... Unfortunately, no such simplification may rise from the study of a granular medium. On the other hand, an analogy can be defined that could make this calculation readily feasable : as shown on fig.1, in a first step, randomly packed bed can be considered as a bundle of regular piles, each pile being then supposed to behave like a homogeneous cylinder.

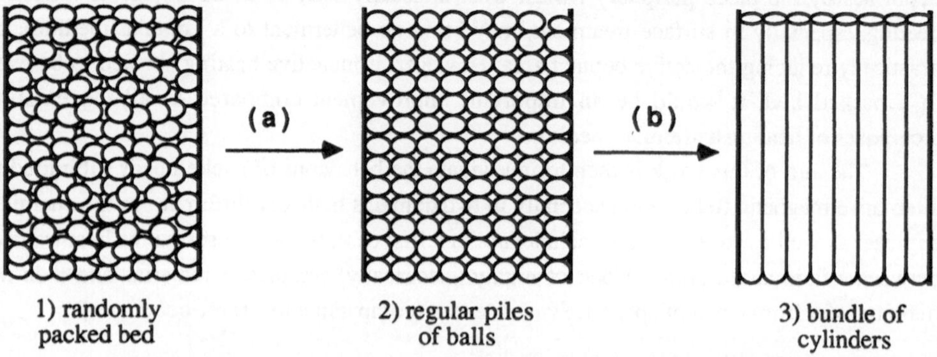

FIGURE 1 : case of a granular bed : two-step analogy

Hypothesis

In order to realize the calculations, we adopt the following hypothesis :

* the inductor is perfect, which means that its length-to-diameter ratio is greater or equal to one; in that case, the inductive field is uniform and parallel to the coil axis.

* the bed is made of spherical and identical particles; bed porosity is uniform and not influenced by reactor walls.

* the cylinders are identical, and their length is equal to the reactor height; they have the same diameter D_b as the balls constituting the bed, and are made with the same material; their total mass is that of the granular bed - so that the calculated behaviour keeps the same meaning in terms of heat generation.

Remembering that power dissipation in a cylinder can be expressed by [1] :

$$P_u = \frac{4\,\pi^2\,R\,N^2}{L}\sqrt{\frac{10^{-7}\,\mu_r\,f}{\gamma}}\;F\,I_e^2$$

The load behaviour can thus be caracterized by an equivalent electric resistance [2] :

$$R_{eq} = \frac{4\,\pi^2\,R\,N^2}{L}\sqrt{\frac{10^{-7}\,\mu_r\,f}{\gamma}}\;F$$

this resistance enabling the calculation of power dissipation in the load from the value of current in the inductor coil.

The conservation of the total mass present in the reactor gives the number of fictitious cylinders :

$$N_c = (1 - \varphi)\left(\frac{D_c}{D_b}\right)^2$$

Assuming these cylinders are electromagnetically independant (due to their high length-to-diameter ratio permitting to consider each of them as a perfect inductor), the total equivalent resistance of a load consisting in a granular bed symbolized by a bundle of Nc cylinders will then by expressed by :

$$R_{eq} = (1 - \varphi)\,\frac{2\,\pi^2 D_c^2 N^2}{L\,D_b}\sqrt{\frac{10^{-7}\,\mu_r f}{\gamma}}\;F$$

EXPERIMENTAL APPARATUS

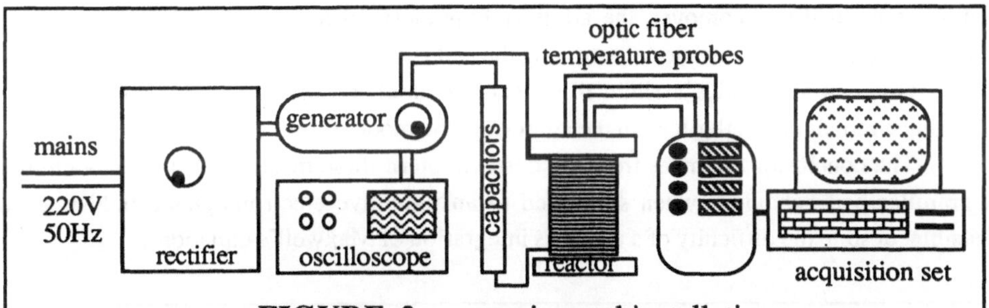

FIGURE 2 : experimental installation

The experimental installation is composed of (fig. 2) :

* a 82 mm - ID, 100 mm - high reactor circled with a 52 turns inductor coil. It is plugged to a generator delivering alternating current with a frequency ranging from 4 to 20 kHz; a set of capacitors enables the circuit to work in resonance conditions.

* an oscilloscope providing the electric characteristics of this circuit; from the values of current, voltage and frequency imposed to the coil one can deduce the load impedance by comparison with the behaviour of the empty reactor and thus the way energy is transferred from the inductor.

* a set of optic fiber temperature probes linked to a computer, allowing the recording of up to four temperatures; the use of optic fiber is made necessary since classic thermocouples would be inductively heated [3]. These probes are intended to make sure temperatures are even in the granular medium ; they also can give a thermodynamic definition

of the electric equivalent resistance :

$$P_u = R_{eq}\, I_e^2 = m\, Cp \frac{\partial T}{\partial t} \rightarrow R_{eq} = \frac{m\, Cp \frac{\partial T}{\partial t}}{I_e^2}$$

* the granular media investigated were lead balls beds with granulometries of 1.8, 3, 3.8, 5, and 7.65 mm. Lead was chosen for its amagnetic properties avoiding the problems linked to the determination of the permeability ($\mu_r = 1$).

MODEL VALIDATION

Figure 3 summarizes the comparisons of experimental equivalent resistances of the different granular beds we tested with the corresponding values predicted by the model.

The figure shows a good agreement between calculations and experimental measurements, which means that the idea of considering a granular bed as a bundle of cylinders is justified; moreover, the arbitrary hypothesis of assigning to these cylinders the same diameter than the particles constituting the bed is validated.

One can also notice a slight divergence for higher values of bed granulometry, the difference remaining negligible : in the aggregate, the model shows to be globally reliable.

The conclusion that rises from the observation of these results is that the modelling of a granular bed behaviour when submitted to an inductive electromagnetic field is easily feasible, despite the difficulty of a rigorous integration of Maxwell's equations.

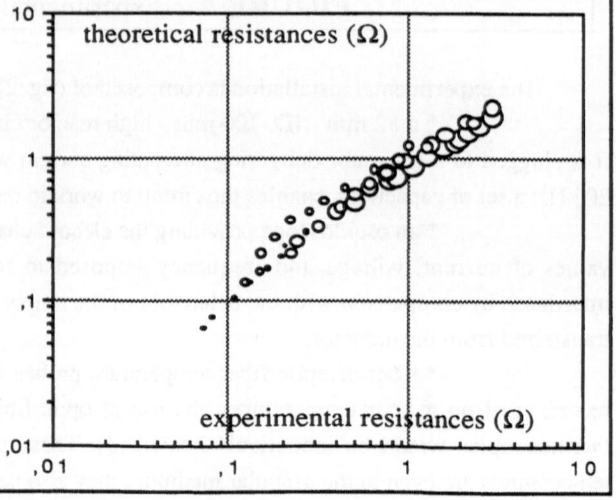

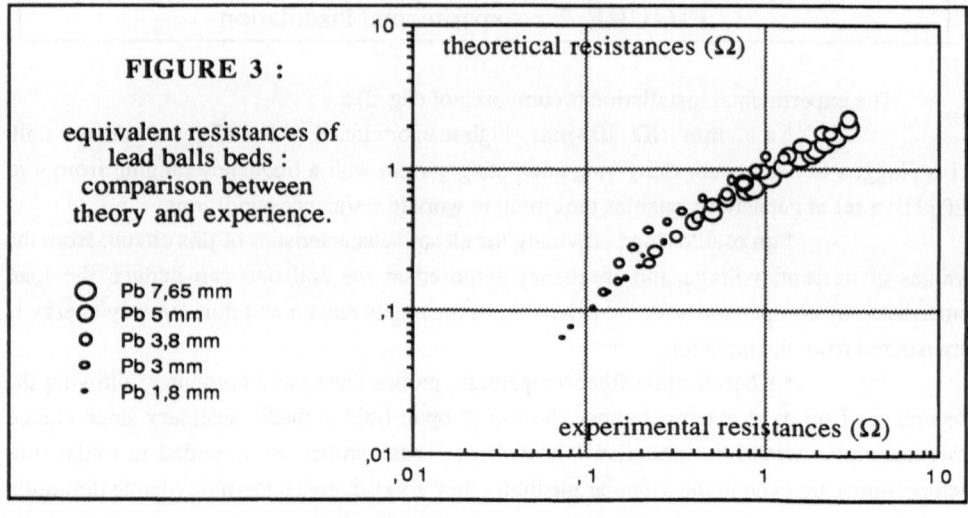

FIGURE 3 : equivalent resistances of lead balls beds : comparison between theory and experience.

APPLICATION TO FLUID HEATERS

Two of the main interests of granular beds are the considerable contact surface existing between them and the surrounding medium, and, when this medium is a percolating fluid, the high values of transfer coefficient resulting of the turbulent flow conditions. That is why their use is widely spread in process engineering each time a transfer (mass - as well as heat - transfer) takes place between a solid and a fluid [4].

Heat transfer coefficients have been investigated [5] : it showed that the traditional good values of coefficients are not affected by the inductive field. Experiments gave the following relation for the calculation of heat transfer coefficient :

$$Nu = 0{,}55\ Re^{0,8}$$

This expression, compared to that of Colburn-Mac Adams, predicts values of transfer coefficients about 20 times higher for a granular heater than for a classical tubular exchanger, with usual values of bed granulometries and tubes diameters. It results in especially compact apparatuses, allowing little pressure drops.Another point is that the high global energetic efficiencies usually encountered with induction are still available [2] : varying with frequency, they can easily exceed 80 % whereas traditional methods harly allow values of 40-50 %.

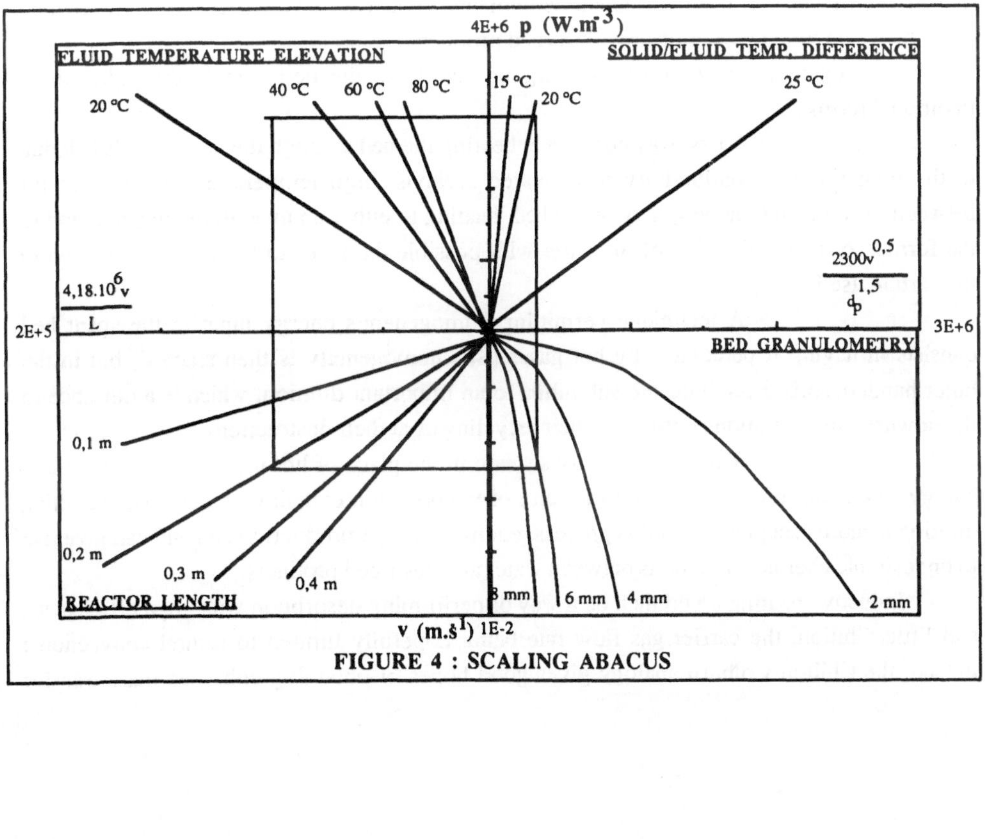

FIGURE 4 : SCALING ABACUS

This experience in both power transmission from inductor coil to granular medium and heat transfer between solid matrix and fluid enabled us to define a method for scaling a heater taking into account such parameters as fluid flow, maximum temperature difference between solid and fluid phases. Figure 4 displays an example of the thermal part of this scaling, adapted to a fluid with physico-chemical properties comparable to that of water. The case represented by a rectangle plotted on the abacus indicates that for a 0.2 m-long heater (bottom left corner) percolated by a fluid having a velocity of 5mm/s (intersection of base with vertical axis), an increase of 30 °C of the fluid temperature (top left corner) with a maximum solid/fluid temperature difference of 20 °C (top right corner) requires the use of 6 mm-diameters balls in the bed (bottom right corner) and a power dissipation of 3 MW/m^3 (intersection of the top with vertical axis). Working frequency and inductor/load energy transmission efficiency will then depend on the material the bed is made with.

APPLICATION TO SPENT ACTIVATED CARBON REGENERATION

Activated carbons have been known for long for their qualities as filtering agents for industrial effluents [6]. Their cost and environmental concerns make it necessary to regenerate spent beds and to recover or destroy desorbed products.

The regeneration requires an energetic supply to the bed, supply that takes three traditional forms :

* A first way consists in heating the bed through the vessel walls [7]: due to the poor thermal conductivity of activated carbons, high temperature gradients result between the core and the periphery of the bed, leading to either an unachieved regeneration in the former, or an overheating of the latter with possible damages to the products or even to the carbon itself.

* A technique permitting homogeneous energy input in the spent bed consists in having it percolated by hot gases [8] : homogeneity is then reached, but in the other hand, desorbed products are submitted to an important dilution, which is a nuisance to the downstream operations (aiming at their recycling or at their destruction).

* A third method uses steam in the place of hot inert gases [9], the main energy flux being brought in under the form of vaporization enthalpy of water.The resulting dilution is more acceptable, but this process requires a steam-production unit and can give rise to undesirable chemical reactions between water and desorbed products.

Inductive heating could then be a way of performing desorbtion with good efficiencies and little dilution, the carrier gas flow rate being hopefully limited to natural convection : indeed, the Chilton-Colburn analogy gives good hopes of obtaining enhanced mass transfer coefficients, as it is the case for heat transfer. Experiments were conducted on activated carbon beds at frequencies of 880 kHz and 3.3 MHz, with results concerning local power dissipation presented on fig. 5 :

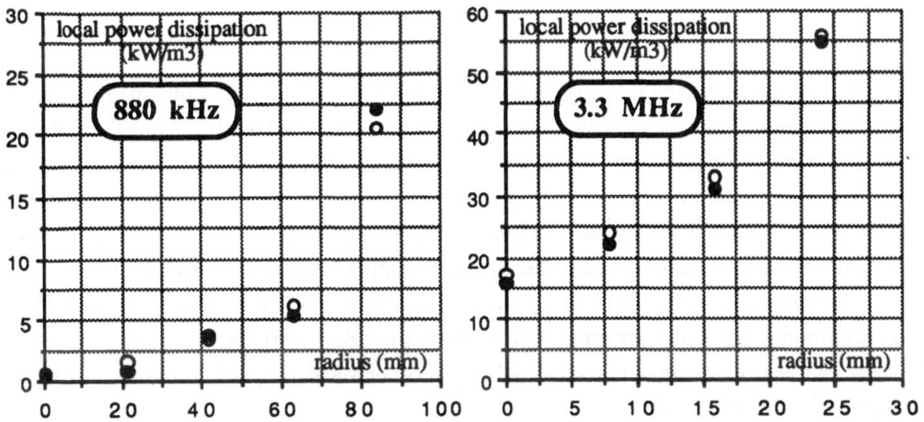

FIGURE 5 : evolution of local power dissipation profiles with frequency two different determinations (see text)

This power dissipation can be determined by two different ways : the first one (black dots) consists in following local temperatures, taking into account radial dispersion; the second (white dots) is focused on moments when this radial dispersion is negligible (beginning of the heating) or deduceable (reaction to the power input cut-off).

Experiments at 880 kHz were made with a 195 mm-wide, 200 mm-high bed. The power dissipation profile shows a behaviour identical to that of a homogeneous piece of electrically conductive material : this profile is characterized by high values at the periphery, rapid decrease linked to the skin effect and a value of zero at the very center of the bed.

In compensation, results obtained at 3.3 MHz (on a 57 mm-wide, 90 mm-high bed) reveal a neat evolution of the behaviour : skin effect is still observable with higher power dissipation in the periphery than in the core, but the energy generated is definitely non-null at r=0, which means that we cannot link this profile to one obtainable in a homogeneous medium.

Comparing these results with the even profile resulting of experiments at 4 to 20 kHz on metal balls beds, we can deduce that the induction effects on granular beds evolve progressively with frequency between two extremes : at "low" frequencies (as it is the case for activated carbon at 880 kHz), eddy currents develop themselves following the reactor periphery. On the opposite, at "high" frequencies (metallic beds), these currents remain confined in each granule independantly from its situation in the bed. And experiments on activated carbon at 3.3 MHz show that an intermediate domain lays where intra-granular and peripheric eddy currents coexist.

CONCLUSION

This study indicated that application of inductive heating to granular media is justified in terms of energetic efficiency and homogeneity. It also pointed out that a quite simple model exists providing good prediction of the behaviour of such an installation.

A consequence of this technique is the opportunity of sizing compact apparatuses allowing high energy transfer rates. Moreover, such equipments are interesting to operate since they are all-electrical : it makes them reliable, safe, clean and easy to automate.

The method investigated here for transfering energy shows qualities that can make it of great importance for many potential applications, provided working frequency is adapted to the load characteristics (granulometry and physico-chemical properties) in order to preserve homogeneity of heating.

NOMENCLATURE

B	magnetic induction	Tesla
Cp	load heat capacity	J/kg°C
D	electric displacement	C/m^2
D_c, D_b	cylinders and inductor diameters	m
E	electric field	V/m
f	frequency	Hz
F	power transmission factor	
H	magnetic field	A/m
I_e	effective current in the inductor	A
J	current density	A/m^2
L	reactor length	m
m	load mass	kg
N	coil's number of turns	-
N_c	number of fictitious cylinders	-
Nu	Nusselt number	-
P_u	power dissipation in the load	W
R	cylinder radius	m
Re	Reynolds number	-
R_{eq}	equivalent electric resistance	Ω
T	temperature	°C
t	time	s

Greek letters :

ε	electric permittivity	F/m
γ	electric conductivity	$(\Omega m)^{-1}$
φ	bed void fraction	-
μ, μ_r	magnetic permeability, relative -	C/m
σ	electric charges density	C/m^3

LITTERATURE CITED

[1] **G. Fournet**, "Electromagnétisme à partir des équations locales", Masson, Paris (1985)

[2] **M. Seghrouchni, G. Ferry, A. Deltour, G. Lacoste**, "Nouvelle conception d'un échangeur ... : caractère volumique du chauffage ", Rev. Gén. Therm. Fr, 337, 28-32 (1990)

[3] **J. C. Barre**, "Mesures de températures sous conditions induites par fibres optiques en verre fluoré de 0 à 800 °C", Congrès Européen sur l'Induction et ses Applications Industrielles, Strasbourg (1991)

[4] **D. Kunii, J. M. Smith**, "Heat transfer characteristics of porous rocks", A.I. Ch.E.J., 6, 1, 71-78 (1960)

[5] **M. Seghrouchni, G. Ferry, A. Deltour, G. Lacoste**, "Nouvelle conception d'un échangeur ... : coefficients d'échange et rendements thermiques", Rev. Gén. Therm. Fr., 349, 17-23 (1991)

[6] **R. W. Soffel**, "Activated carbon", Kirk-Othmer, "Encyclopædia of Chemical Technologies", 3rd ed., vol.4, 561-570, Wiley, New York (1978)

[7] **J. M. Shork, J. R. Fair**, "Parametric analysis of thermal regeneration of adsorption beds", Ind. Eng. Chem. Res., 27, 3, 457-469 (1988)

[8] **A. Jedrzejak**, "Modelling of activated carbon desorption by a circulated inert gas", Chem. Eng. Tech., 11, 352-358 (1983)

[9] **J. M. Shork, J. R. Fair**, "Steaming of activated carbon beds", Ind. Eng. Chem. Res., 27, 8, 1545-1547 (1988)

SESSION 13:

Chemical Reactors

Chairman: Prof. G. Froment

MULTIFUNCTIONAL REACTORS

K.R. Westerterp
Chemical Reaction Engineering Laboratories
Chemical Engineering Department
Twente University
P.O. Box 217, 7500 AE Enschede, the Netherlands

ABSTRACT

Chemical reactors are important energy consumers in case of endothermic processes, whereas energy is produced in exothermic processes. Therefore the efficiency of chemical reactors strongly influences the energy consumption in the world. The principle of the multifunctional reactor promises to improve very much upon the chemical reactors efficiency.
Multifunctional reactors are single pieces of equipment in which, besides the reaction, other functions are carried out simultaneously. The other functions can be a heat, mass or momentum transfer operation and even another reaction. Multifunctional reactors are not new, but they have received much emphasis in research in the last decade. A survey is given of modern developments and the first successful applications on a large scale. It is explained why their application in many instances is still far-away from large scale introduction. The first major successes have been achieved in the reduction of energy losses or increasing specific energy production.

INTRODUCTION

A large research effort is put into the development of chemical reactors. In the last decade the demand for better selectivities and higher conversions has increased more than ever to ensure prevention of the release or removal of products - often in very minor quantities - that would otherwise harm our natural environment. Also, energy saving considerations have strongly influenced the choice of type and design of a reactor and have also lead to new, ingenious configurations and designs.
Also, in the last decade the so-called "multifunctional" reactors have received a strong developmental push, see Westerterp (1992). We owe the word "multifunctional reactor" to Agar and Ruppel (1988), who, in a pioneering paper, have given a survey of the possibilities and some practical examples of the application of the principle of multifunctionality. Its meaning is very simple: in a single piece of apparatus used for reaction, more than one function is executed. The other functions can be: mass, heat and momentum transport operations or even other, independent reactions.
Of course the principle of multifunctionality is not new and has been applied in many instances. Historically the oldest reactors were the stirred pots, and for catalytic processes the adiabatic bed. The retort was a multifunctional reactor combining reaction with mass transfer, distilling off a product. As soon as a cooling jacket or cooling coil was installed in the stirred pot we combined reaction with heat transfer. Also in our combustion engines, reaction and momentum transport are executed in the same cylinder. But today the degree of sophistication in using multifunctionality has increased by an order of magnitude.

We are not going to give a complete survey of the literature on multifunctional reactors nor a full outline of all the possibilities. We will only give a few examples to demonstrate the possibilities and challenges in the use of the principle of multifunctionality. We will discuss practical examples that are still at the research stage, or have already reached commercial application on an industrial scale. We will also, now and then, give a warning not to be over-optimistic; the reality of process economics puts strict limits on new developments. In placing limits we do not want to stop the researcher in his endeavours, but intend rather to stimulate him to make his invention even more efficient so that in the long run it can compete with existing reactors and eventually replace them.

REACTOR CAPACITIES

Reactor capacities, of course, vary widely. However, if one studies reactor capacity in certain branches of the process industries, some approximate ranges can be found. Thus, for the **bulk chemical** and oil industries capacities almost always vary between 4 to 50 kmoles produced per m^3 reactor volume per hour. The lower value relates to an economic boundary where reaction volume becomes too large and therefore uneconomical. The upper boundary refers to a technical limit: it becomes impossible to adequately withdraw the heat produced. For **fine chemicals**, where we mostly use batch processes, capacity levels are much lower and range from 0.1 to 1 kmol/m^3 hr. For **biotechnological processes** capacities are also much lower and range from 1 kmol/m^3 hr for the low cost products to very low values of say $500 * 10^{-6}$ kmol/m^3 hr for high cost chemicals like penicillin. In such processes heat removal rates are no problem, but other operations may be limiting, e.g. oxygen supply may be the dominating factor determining capacity.
In nature, processes are even many decades slower, e.g. in petroleum geology, production rates range from $10 - 200 * 10^{-12}$ kmols converted/m^3 hr.
In our discussion we are primarily interested in reactors for bulk chemicals and will now have a further look at multifunctional reactors, where reaction and heat transfer are combined.

SIMULTANEOUS REACTION AND HEAT TRANSFER

Historically we see that soon the kettle type reactors were equipped with cooling equipment, such as coils and wall jackets, in order to cope with the heat effects of chemical reactions. Temperature rises in adiabatic catalyst beds were kept generally under control by dividing the bed into several parts and installing coolers between them. Then injection of cold feed gases between catalyst beds was practised. For reactions with very high heat effects, the (cooled) tubular reactor was soon developed, despite mechanical difficulties in coping with thermal expansion of the apparatus during operation. Early applications of these reactors were oxidation and hydrogenation reactions, which are highly exothermic. A major progress has been achieved, since for the first time a tubular reactor has been cooled by boiling water, so cooling and steam production in combination. For energy saving this has become almost common practice in modern plants. We should note that in the oil industry, the tubular reactor for endothermic reactions had been introduced long ago: in thermal cracking furnaces the endothermic heat of reaction required was supplied directly in a furnace by combustion flames.
Later the good heat transfer and temperature equilibration properties of the fluid bed reactor were recognised and since then many fluid bed systems have been installed. Agar and Ruppel (1988) describe a clever use of the principle of multifunctionality, in which an inert fluidizable powder is blown through a stationary catalyst bed and there absorbs the reaction heat. After leaving the catalyst bed the heated powder is cooled and returned to the bottom of the bed, where it is again carried through the catalyst bed by the fresh feed gas. With high circulation hot spots can be suppressed effectively and the reactor operates almost isothermally, see also Saatdjian and Large (1985).
The removal of reaction heat by evaporation of a solvent is another technique, which has been applied several times and is used, e.g. in the oxidation of para-xylene using homogeneous catalysts in the well-known Amoco process. The solvent is condensed outside the reactor and recycled. Due

to mixing by the gas flow, these reactors behave like a tank reactor. Van Gelder et al. (1990a, 1990b) have discussed a more sophisticated application of cooling by evaporation, this time in a packed bubble column, which is packed with solid catalyst particles. The gas and liquid flow upward through the column, at the top the liquid and vapour-gas mixtures are separated. The principle of multifunctionality here gives us several significant advantages: 1° No expensive or elaborate cooling systems are required in the catalyst bed, 2° the reactor pressure can be set and hence, the maximum temperature is easily controlled, 3° temperature runaways are impossible because further heat production only increases the evaporation rate, 4° because of the packing, axial dispersion is suppressed, so that plug flow is approached in longer, slender columns. They recommend the method of hydrogen starvation, in cases, where hydrogenation has to be carried out to produce selectively an intermediate product in a consecutive reaction system. This combination of reaction with evaporation also can easily cope with a deactivating catalyst and Westerterp et al. (1988) specially recommend this reactor type with evaporative cooling for application to catalytic hydrogenations in the fine chemicals industries. Compared with a more classical reactor, the partial pressure of hydrogen, and therefore the reaction rate, is reduced due to the vapour pressure of the solvent, but this is more than off set by the ease of heat removal and temperature control.

Adiabatic bed reactor with periodic flow reversal

About 25 years ago in Novosibirsk, Boreskov et al. (1979) started the development of the adiabatic bed reactor with periodic flow reversal (RPFR) and its development was brought to completion by Matros (1985, 1989) during the eighties. The principle was developed aimed at a simplification of chemical process equipment and specially at a reduction of the use of high quality steel in the Sovjet Union, required for industrial scale manufacture of ammonia and sulphuric acid.

As this was not a problem in other parts of the world, it remained a local development. Later, however, it was recognised that the RPFR had great advantages in better pollution control and so Eigenberger and coworkers (1988a, 1988b, 1991) started to develop this multifunctional reactor for the catalytic combustion of volatile organic compounds in exhaust air. We will describe its behaviour on the basis of the oxidation of combustibles present in low concentration in air.

The RPFR works as follows. If the catalyst mass is hot, a cold air stream, containing some combustibles, will be heated up by the hot catalyst. Simultaneously the catalyst mass is cooled down. The air stream being heated up at some place in the catalyst bed will reach the temperature at which the combustibles ignite and the combustion reaction starts. Now, the combustion heat liberated will heat up the solids phase. If in the area of the catalyst bed outlet there is a zone where the catalyst is cold, the hot gas will heat up this cold catalyst. So in this part of the bed a heat wave will travel through the bed towards the outlet. And simultaneously near the inlet, where the hot bed is being cooled down, a cold wave is generated. In the zone between the two waves the bed is hot and here the combustibles are burnt.

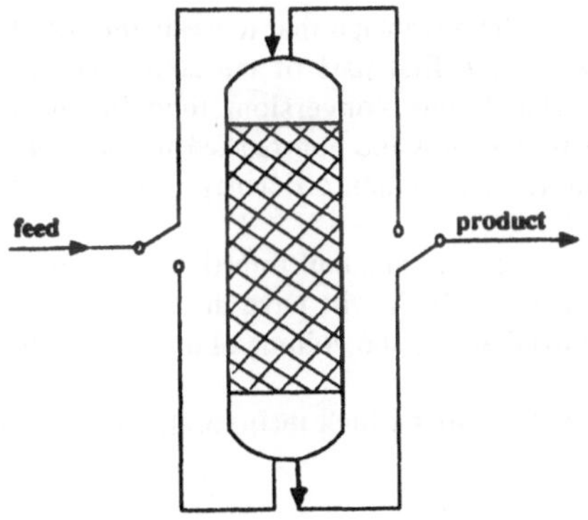

Fig. 1. The adiabatic bed reactor with periodic flow reversal.

By periodically reversing the direction of flow, see Fig. 1, heat can be kept in the reactor and by so doing the temperature in the middle zone will remain high and above the ignition temperatures. To reach a steady state after many flow reversals, the exit gas must be so much higher in temperature than the

feed gas that averaged in the time the liberated reaction heat will be carried away by the exit gases. With mathematical modelling an understanding has been obtained of the operation of the RPFR. In Fig. 2, two calculated temperature profiles are shown at the moment just before flow reversal of the flow direction. The plateau temperature established and the temperature difference between the inlet and outlet streams can be observed. **No energy at all is supplied to the process.** It is kept going by the combustion of the combustibles which are to be removed. The packed bed works partially as a heat exchanger, to heat up and cool down contaminated air, and partially as a combustion reactor. The plateau temperature of the hot zone is influenced by many variables. The concentration of combustibles in the air must never be so high that the oxidation catalysts become overheated and loose their activity. The RPFR has great promises in **the protection of the environment** in the removal of volatile organic compounds from polluted air by catalytic air oxidation **without external energy supply**: only the control problem for variations in composition and concentration of pollutants still has to be solved.
The RPFR therefore operates under dynamic conditions. The hot and cold fronts move back and forth with cycle time an important operating variable in preventing blow out of the reactor. The inlet and outlet zones act as heat exchangers and even at very low adiabatic temperature rises, ΔT_{ad}, due to very low concentrations of combustibles, heat exchange is so effective that the actual temperature rise is many times higher than ΔT_{ad}. Further, the switch valves operate at low temperatures, which increases long service life.
For expensive catalysts the parts of the bed in heat exchange service can be replaced by inert material. We observe that in the RPFR expensive gas-gas heat exchangers are replaced by an additional layer of inert material installed in the same reactor shell.

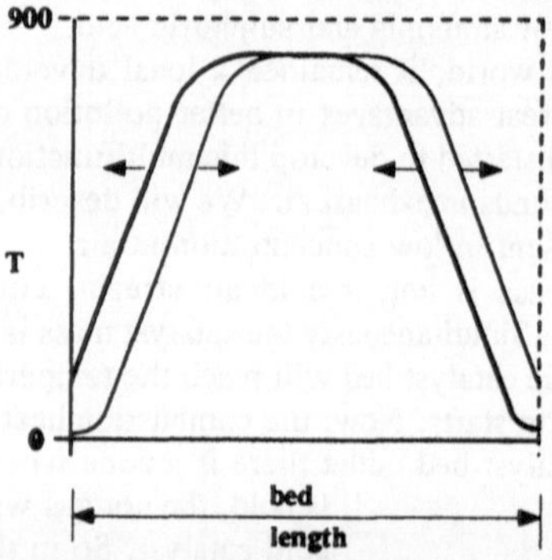

Fig. 2. Temperature profile just before reversal of the flow

For ammonia synthesis and SO_2 oxidation, the RPFR has the advantage that towards the outlet the temperatures drops, meanwhile the reaction continues in the first part of the cold section. For exothermic equilibrium reactions this leads to somewhat higher conversions than in the usual commercial reactors. A disadvantage is that on reversing the flow the non-reacted materials in the bed are now purged into the outlet stream, so reducing the conversion efficiency. This can not be avoided, but for long cycle times the effect is negligible.
In the two examples given, of combined reaction and heat transfer in a multifunctional reactor, with either an evaporating solvent or with periodic flow reversal, both will have the same or higher reactor capacity than their classical counterparts, provided that also the volume of the heat exchange equipment is taken into account.
After these examples we will now refer to multifunctional reactors which incorporate mass transfer in the reaction zone.

SIMULTANEOUS REACTION AND SEPARATION BY MASS TRANSFER

The combination of reaction and separation by mass transfer is not new and many lucid examples can be found in the history of chemical engineering. We do not mean heterogeneous reactions, where mass transfer and reaction are combined in a multiphase reactor system, but refer to reactors in which in the same apparatus one of the reaction components is removed or supplied by a mass transfer separation process. We will consider two older examples of combinations with extraction and distillation.

Schoenemann and Hofmann (1957) elaborated a process to make furfural from xylose in aqueous solution containing HCl. The furfural, which is produced in a combination of parallel and consecutive reactions, can decompose into resins and condensation products, so that yields of only around 10% can be expected. Now, by adding tetraline as an immiscible solvent phase they extracted the furfural as it was produced and obtained yields of over 60% on xylose!

Geelen and Wijfels (1965) studied the equilibrium transesterification of vinyl acetate and stearic acid to vinyl stearate and acetic acid. As a byproduct ethylidine acetate was formed. By executing this reaction in a distillation column they removed acetic acid selectively as soon as it had been formed, and by doing so suppressing the formation of the byproduct almost completely. Moreover, by removing the acetic acid, they could bring the equilibrium reaction to completion with a very high selectivity!

We now will discuss more recent developments in the field of multifunctional reactors, namely the chromatographic reactor, catalytic distillation reactors, membrane reactors and the application of multifunctionality to the methanol synthesis.

<u>The chromatographic reactor</u>

The chromatographic reactor has already received attention for several decades, see Coca and Langer (1983). It is well known, that e.g. in an equilibrium reaction such as A $\leftrightarrows$ B, where the two components are adsorbed with differing strengths, that a pulse of A injected into an inert carrier stream flowing over a catalytically active adsorbent, will result in conversion to B beyond the equilibrium position. This occurs if A is strongly adsorbed and B not, the column is long enough and the reaction is slow. As soon as some adsorbed A is converted, the B produced is carried away by the inert gas, so keeping the concentration of B low in the vicinity of adsorbed A.

Aris and coworkers, see Fisch et al. (1986) and Ray et al. (1990), have adapted this principle very elegantly to a continuous process. Pure A is injected somewhere in the middle of a column, through which a solid adsorbent-catalyst move slowly downwards and an inert carrier gas flows upwards. A (in an equilibrium reaction) is converted into B which is less strongly adsorbed. Now the downward flow rate can be adjusted such that B moves upwards and A downwards. Experimentally they studied the hydrogenation of mesitylene to trimethylcyclohexane TMC. Under their conditions they obtained a top product of 90% TMC, which was above the equilibrium composition.

The operation of a moving bed reactor is delicate and advantages must be sufficiently great to have an incentive for developing the moving bed chromatographic reactor for large scale operation. Agar and Ruppel (1988) in an elegantly novel approach used the principle of the periodic flow reversal to avoid the moving bed operation. They applied the principle to the removal of nitrogen oxides from combustion gases:

$$NH_3 + \text{adsorbent} \leftrightarrow NH_{3\,ads}$$
$$NH_{3\,ads} + NO + 0.25\,O_2 \rightarrow N_2 + 1.5\,H_2O + \text{adsorbent}$$

In commercial steady state processes it is impossible to supply the correct stoichiometric amount of NH_3 to remove the NO. NH_3 is not allowed into the atmosphere and therefore only 80 - 90% of the required amount is fed to the reactor to allow for fluctuations in concentration and flow rates. Agar and Ruppel feed combustion gases to the Denox reactor and periodically give a large shot of NH_3 to the feed stream. The overdosed NH_3 reacts with the NO and the large excess is adsorbed on the Denox catalyst. An adsorption wave thus will travel through the catalyst bed in the direction of flow. After a certain period, the supply of NH_3 is stopped and the reaction continues, consuming the adsorbed NH_3. So a reaction wave also travels through the reactor. Before the NO breaks through, the direction of flow is reversed, and a new pulse of NH_3 is supplied at the inlet, again increasing the amount of NH_3 absorbed. The adsorption front now travels in the other direction, and as soon as the NH_3 pulse is stopped, a second reaction front will also follow. By periodically changing the direction of the flow, the NH_3 is kept within the reactor and break-through is prevented. Furthermore a **complete conversion of NO** is now achieved and fluctuations in the concentrations are now coped with, by the buffering capacity of the catalyst. In the work of Matros periodic flow reversal is applied to a heat transfer situation, whereas in the work of Agar and Ruppel it is a mass transfer application. Their process promises to become the first large scale application of the chromatographic reactor.

The catalytic distillation reactor

The combination of reaction and distillation over a catalyst bed has been introduced to the large scale manufacture of methyl-tertiary butyl ether MTBE, see Smith and Huddleston (1982). To produce MTBE, isobutylene is reacted with methanol, see Fig. 3. The separation is improved by forming a methanol-MTBE azeotrope. Pure MTBE leaves from the bottom of the reaction-distillation column. From the top, a methanol-C_4 azeotrope is distilled. The catalyst used is an ion exchange type resin, which requires low reaction temperatures.

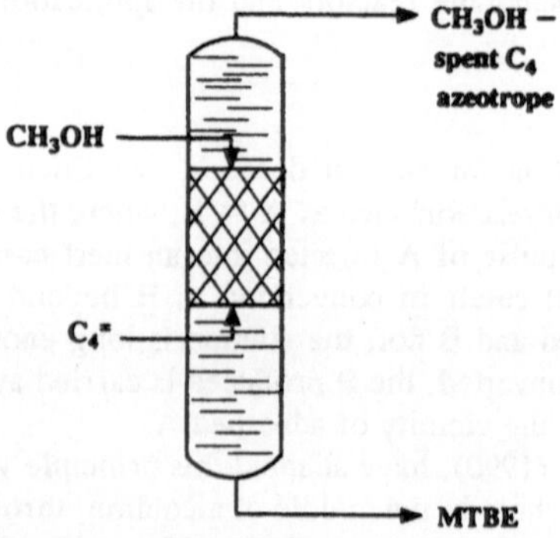

Fig. 3. Catalytic destillation.

Several advantages are apparent. The heat of reaction is used completely in the distillation, so **saving on steam consumption** is achieved. The methanol is fed to the top of the catalyst bed, where the isobutylene concentrations are low, and the high CH_3OH concentration pushes the equilibrium towards MTBE. Further, the MTBE is removed more or less directly after it has been formed. So the MTBE on is kept low and the reverse reaction hardly occurs. Because of the distillation, almost all isobutylene can be converted into MTBE despite the unfavourable equilibrium, provided the catalyst bed is high enough. So by applying the principle of multifunctionality an equilibrium reaction can be forced to near completion in one pass.

Cumene has also been made by catalytic distillation, see Shoemaker and Jones (1987). Many aspects of the relevant vapour-liquid equilibria, the reaction and its kinetics have to be favourable in order that catalytic distillation can be applied successfully. Due to the presence of the vapour phase, the concentrations of some of the reactants in the liquid phase will be lower, thus reducing the reaction rate. For all but expensive catalysts and equipment, this reduction is more than off-set by the deletion of the separate separation step. Some more niches will be found for catalytic distillation, however, many preconditions have to be fulfilled, see also Mayer (1980):

- Temperature ranges for the distillation and the reaction must coincide.
- The reaction has to be fast enough so that the total residence time in the packed catalyst bed is sufficient.
- The reaction can not be so fast that the reaction occurs only in a small part of the bed, otherwise the reaction heat liberated only enhances the distillation locally.

- One of the desired products has to be either the highest or lowest boiling compound in the system, in order that by its separation, the equilibrium shifts in the desired direction, and of course, the reaction must be exothermic. An endothermic reaction would be quenched.

This process has also been recommended for the production of propylene glycol, see Kinoshita (1983) and for the separation of m- and p-xylene, see Saito (1971).

Membrane reactors

Membranes were originally developed for filtration. Dense non-porous membranes for hyperfiltration, also called reversed osmosis, and porous membranes used for ultrafiltration, with pore diameters of 3-200 nm, and for microfiltration, with pores of 0.2-10 microns. Polymeric membranes were almost exclusively limited to biotechnology, see e.g. Belfort (1989). We will restrict ourselves to inorganic membranes.

A distinction can also be made between dense and porous inorganic membranes. Foils of palladium or its alloys are permeable only to hydrogen and silver on dense zirconia only to oxygen. The selectivity of these dense membranes is, of course, very high, but permeation rates are very low. Costs limit the general application of these membranes.

Commercially available porous membranes are mostly based on alumina or zirconia. The membrane is at least 5 micron, but more usually 10-20 microns thick and placed on some intermediate support layers of a few mm thickness. Diffusion through the pores is of the Knudsen type, hence separation factors are related to the permeabilities of the diffusing compounts and to the square root of the ratio of their molar masses. Sometimes for very narrow pores separation factors are much higher, in what usually is called capillary condensation.

In the case of reactions, use is made of the membrane properties in several ways: In the case of equilibrium reactions, by removal of one of the products, so shifting the equilibrium in the desired direction. Side reactions can be suppressed by removal of products subject to decomposition. In cases where we wish to dose reactants slowly, membranes can give more precise control of the reaction. We will give a few examples of applications of inorganic membrane reactors, which, although receiving very much attention, but have not yet passed beyond the research and development stage. For a recent survey e.g. see Hsieh (1991).

In 1966, the first patent was granted for the use of Pd as a **dense membrane** reactor, in which hydrocarbons were dehydrogenated and the reaction rate increased by removal of the hydrogen, since then a great number of applications has been explored. For a recent survey, see Shu et al. (1991). A small scale pharmaceutical application was the synthesis of vitamin K in one step from a mixture of quinone and acetic anhydride in a Pd-Ni membrane reactor with a yield of 95%. This should be compared to the classical multisteps synthesis with a yield of 80%, see Gryaznov (1986). Itoh et al. (1988) dehydrogenated cyclohexane in a Pd membrane reactor at 200°C. At that temperature the equilibrium conversion of cyclohexane would be 19%, but by continuous removal of H_2 conversions above 99% were reached. Many applications have been tried. Selectivity is very high, but success on a large scale is hindered by the low fluxes through the dense membranes.

Porous membranes exhibit low selectivities but have much larger permeabilities. A recent survey was given by Armor (1989). A distriction must made whether the membrane is only a part of the reactor e.g. surrounding a catalyst bed or that the reaction takes place in the membrane itself. In the latter case the membrane is in most cases impregnated with a catalyst. Three examples follow.

Wu and Liu (1992) studied the dehydrogenation of ethylbenzene to styrene over a bed of catalyst pellets, surrounded by a ceramic membrane. H_2 diffused more rapidly through the membrane than the aromatics. Usually this reaction is a.o. accompanied by hydrocracking of ethylbenzene to toluene and methane. By applying the membrane separation this side reaction was strongly suppressed and the styrene yield increased due to reduced H_2 concentration in the reactor. Fluxes were however low.

Champagnie et al. (1990) used Pt impregnated alumina membranes to increase the conversion of ethane to ethylene in the temperature range of 450-600°C. They achieved conversions six times higher than the equilibrium conversion due to the selective removal of hydrogen. In catalyst-impregnated membranes reaction times are very short, hence only reactions with fast kinetics can be executed. Furthermore a typical problem emerges, namely the balance between reaction rates and diffusional permeation rates. The permeation rate of H_2 in the above example must be larger that the dehydrogenation rate of ethane, in order to reduce the local hydrogen concentrations and so force the equilibrium in the desired direction. At the same time the permeation rate of ethane must be lower than the reaction rate. Usually there is a relatively narrow window of operation where conditions are favourable. This dehydrogenation of ethane was taking place with high fluxes.

For **control of a reaction** a membrane can be used to keep reactants separated, e.g. for partial combustion reactions in a porous, ceramic membrane. Reactants are fed to different sides and diffuse from either side into the membrane. A catalytically active material has been deposited in the pores. There is no pressure difference over the membrane. For fast reaction rates the permeation rates are limiting resulting in a small reaction zone somewhere in the membrane. For instantaneous, irreversible reactions the zone reduces to a reaction plane. Product(s) diffuse out of the membrane to both sides. The location of the reaction zone inside the membrane is such that the molar fluxes of the reactants are stoichiometrically equal. After an increase in the concentration of a reactant on either side of the membrane, the reaction zone will shift in such a way that the stoichiometry is again met, see Fig. 4. Sloot et al. (1990) proved this experimentally for the Claus reaction:

$$2H_2S + SO_2 \rightarrow 3/8\ S_8 + 2H_2O$$

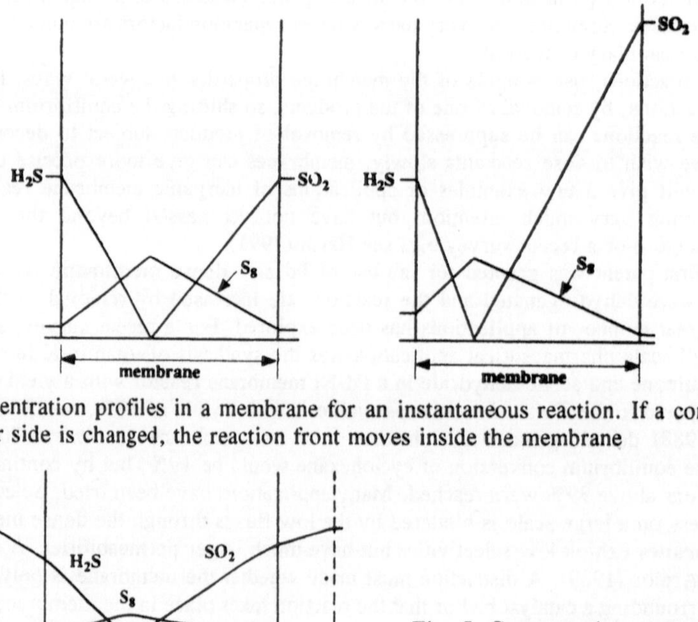

Fig. 4. Concentration profiles in a membrane for an instantaneous reaction. If a concentration at either side is changed, the reaction front moves inside the membrane

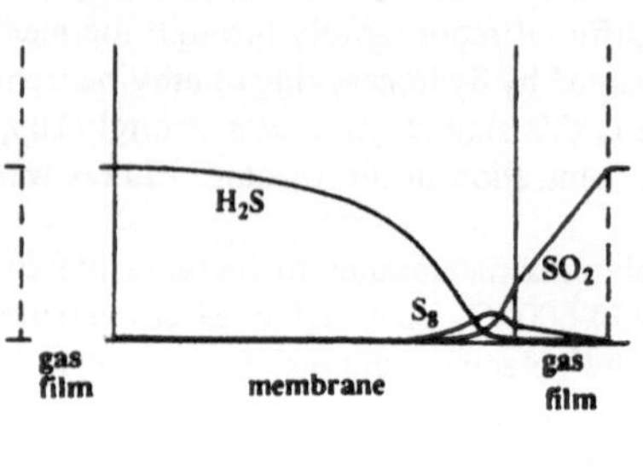

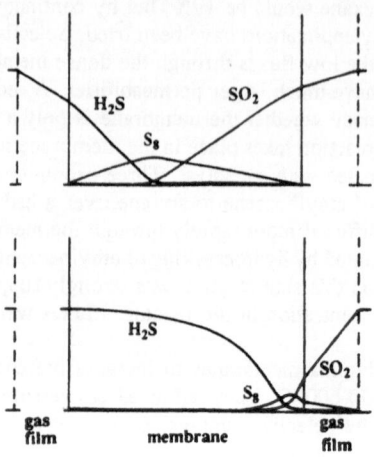

Fig. 5. Concentration profiles in a membrane for a fast reaction without (left) and with (right) a pressure difference over the membrane.

The drift of the sulphur to both sides of the membrane could be prevented by applying a pressure difference, so forcing the product in one direction, see Fig. 5. Fluxes were low.
Much work has also been done in the mathematical description of the behaviour of membrane reactors. Mohan and Govind (1986, 1988a, 1988b) published a set of papers in which, by simulation, they analyzed the influence of the important parameters, including: membrane characteristics and dimensions, reactor geometry and heat transfer rates, the pressures on both sides of the membrane, flow rates of the feed and the sweep gas, temperatures and concentrations in the feed, kinetics, equilibria, and reaction heats. Also permeability variations, recycle streams, and direction of the flow (being co- or countercurrent) were taken into account.
Membrane reactors are still under development and many **problems** still have **to be solved**. These include: the geometry of the reactor, the sealing and packing of the individual units -e.g. in the case of tube shaped membrane in a large array and under greatly varying thermal stresses-, and the fouling of membranes, e.g. by coke, for which steam has to be added. From the reaction engineering point of view major obstacles to their commercial introduction are:

- the low permeation rates. In most experimental studies until now the fluxes are still 10-1000 times too low to give a reactor volume with an acceptable capacity.
- the need for a sweep gas to remove the component(s) at the permeate side in order to keep concentrations low and consequently concentration differences over the membrane high, involves an additional separation step.
- the lack of an efficient method to supply or withdraw heat to or from membrane reactors on a large scale.
- the difficulty of keeping pressure differences across a membrane to almost zero, if desired, in large scale industrial reactors.

Much work still has to be done to tackle all these problems and we still have a long way to go. Some niche applications for membrane reactors surely will be found, the example mentioned for the vitamine K synthesis gives confidence. In view of the low capacity of membranes, more large scale applications should be sought in the fine chemicals industry.

<u>Gas-Solid-Solid Trickle Flow Reactor (GSSTFR) and Reactor Section with Interstage Product Removal</u> (RSIPR)

Equilibrium processes like the syntheses of ammonia and methanol require a cumbersome system to separate the product from the unconverted reactants and to recirculate these reactants to the reactor. **Much energy is wasted** in these separations and recirculations. Moreover, the feed preparation for these plants is complicated because a very pure feed of a stoichiometric compositions is required. Any impurity or reactant in excess, would accumulate in the recycle loop and slow down reaction rates and lower conversions. To overcome these problems Westerterp and Kuczynski (1987a, 1987b) developed two processes to achieve complete conversion in one pass through the reactor. In the gas-solid-solid trickle flow reactor GSSTFR they trickled a solid adsorbent countercurrently to the methanol synthesis gas through a fixed catalyst bed. The adsorbent selectively removed the methanol as it was formed. At high production rates they achieved 100% conversion and they ran a miniplant for 100 hours on a stoichiometric feed with the gas outlet closed and only a stream of solids, loaded with methanol leaving the reactor section.
To realise a process based on the GSSTFR on an industrial scale requires a cumbersome system for methanol desorption and solids recirculation. Therefore Westerterp et al. (1989) also developed a reactor section with interstage product removal RSIPR, based on absorption of methanol in a liquid solvent an the reaction temperature. Also this process was operated successfully on a miniplant scale. In four sets of reactors and absorbers 97-98% conversion can be achieved in a once-through operation without recycle, see Fig. 6. This RSIPR principle is based only in proven technology.
Although much energy is saved by applying the RSIPR process, its main strength lies in the fact it **can** also **handle non-stoichiometric feeds**. Any reactant in excess will pass through the reactor section as an inert material. An excess CO or H_2 leaving the RSIPR can be used as fuel or for

further processing in carbonylation, Fischer-Tropsh or hydroformylation processes and hydrogenations respectively. In Fig. 7 a block diagram shows the plants combined in a large methanol manufacturing unit. In the steam reforming unit, an excess of hydrogen, and in the partial oxidation unit, a feed stream deficient in hydrogen are produced and by careful balancing an exactly stoichiometric feed can be made.

Now in the case, where the stoichiometric requirement can be dropped, as in the RSIPR, we obtain the following simplifications, see Fig. 8. First no recycle is required anymore in the synthesis unit. Secondly the expensive steam reforming can be left out. As the partial oxidation is not an equilibrium process it also can be executed at higher pressures, say at synthesis pressure. Thirdly the expansion turbine and recompressor can be skipped and the natural gas, at well pressure, can be fed straight to the oxidation and synthesis unit. The excess CO can be used for other purposes, e.g. in a Fischer-Tropsch unit. The oxidation unit delivers a stoichiometric ratio of 1.6 - 1.7, for the methanol production we require 2.0 and for Fischer-Tropsch a ratio of 1.0.

We observe that great savings can be obtained with the RSIPR principle applied to methanol synthesis, see also Westerterp (1992). The GSSTFR and RSIPR approaches can also be applied to consecutive reactions, by removing the desired product within the reactor.

COMBINATION OF REACTIONS

The combination of reactions in one single piece of equipment is also not new. In 1746 in England the so called lead chamber process was introduced to make sulphuric acid. In this process the oxidation of sulphur dioxide, according to $2SO_2 + O_2 + 2H_2O \rightarrow 2H_2SO_4$ was carried out simultaneously with the reduction: $2NO_2 \rightarrow 2NO + O_2$ and later in the process the NO was reoxidized to NO_2 by air. We will present two examples of the application of multifuctionality to modern reactors.

Gryaznov et al. (1973) were the first to propose the coupling of two reactions in a **membrane reactor**, in which a hydrogenation and a dehydrogenation are carried out separated by the membrane. The exothermic reaction enthalpy of the hydrogenation reaction is used in the endothermic dehydrogenation. In a simulation study Itoh and Govind (1989) combined the dehydrogenation of butene-1 with the (more obvious) combustion of hydrogen, which has permeated through a dense palladium membrane. In tubes which were coated with a palladium membrane dehydrogenation catalyst pellets were placed, around the tubes air was supplied to burn the hydrogen. They demonstrated that such a reactor can be operated autothermally and that reasonable conversion rates can be obtained. Maybe the combination of endo- and exothermic reactions can solve the problems of heat supply or withdrawal in membrane reactors? At least, in reactors with dense membranes, it must also be possible to balance locally the heat absorption and production.

Recently Blanks et al. (1990) discussed the use of the adiabatic bed **reactor with periodic flow reversal** for a combination of an endo- and an exothermic reaction. Methane combustion with air $CH_4 + 0.5\ O_2 + 1.88\ N_2 \rightarrow 0.75\ CH_4 + 0.25\ CO_2 + 0.5\ H_2O + 1.88\ N_2$ was combined with the endothermic steam and CO_2 reforming process, $0.75\ CH_4 + 0.25\ CO_2 + 0.5\ H_2O \rightarrow CO + 2H_2$, yielding a synthesis gas claimed to be suitable for Fischer-Tropsch syntheses. Of course, the water gas shift reaction $CO + H_2O \rightarrow CO_2 + H_2$ also occurred. They operated a miniplant and a pilot plant with a cylindrical reactor of 4 m length and diameter of 0.57 m. For a catalyst, they employed a nickel reforming catalyst. Zones of inert material at inlet and outlet were used for the heating and cooling. Natural gas conversions from 85-97%, and CO yields of 75-95%, were obtained. Reactor capacities were around 2.5 kmol/m^3 hr and the process could be run autothermally, without external energy supply. Again a very promising result employing the principle of multifunctionality.

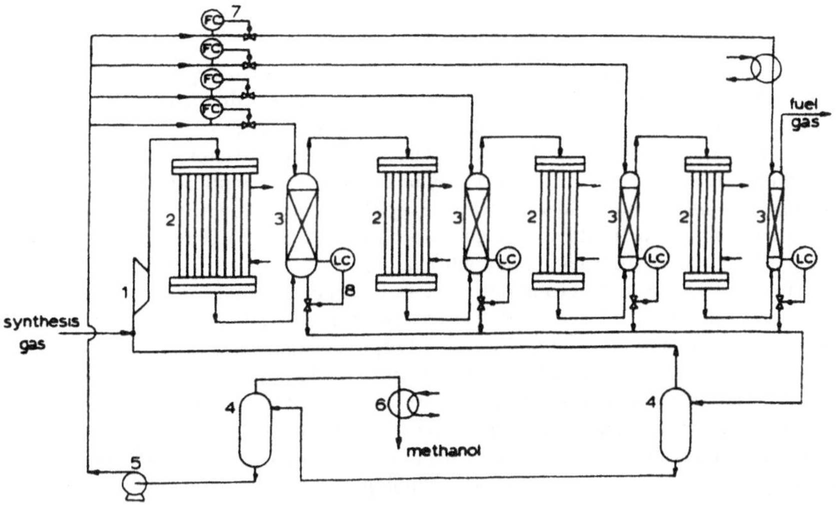

Fig. 6. The reactor section with interstage product removal for the methanol synthesis.

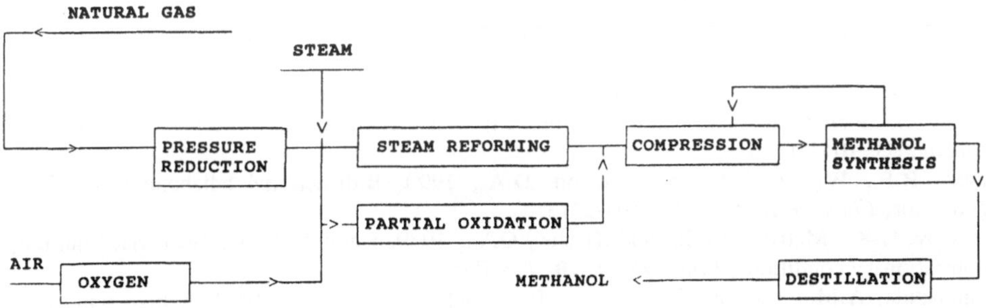

Fig. 7. The line-up of the units in a classical large scale methanol plant.

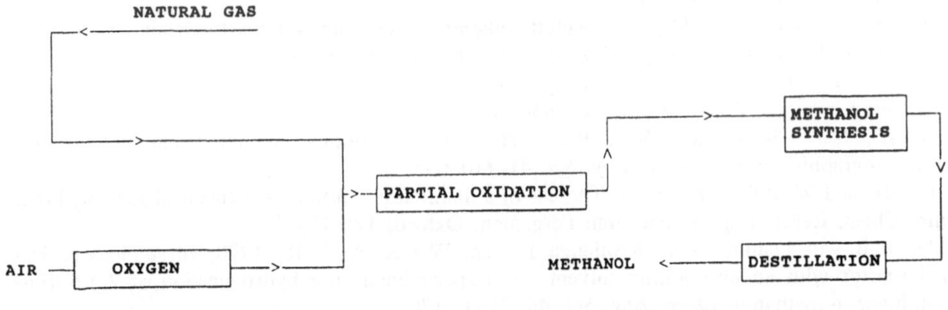

Fig. 8. The line-up of units in a methanol plant with a reactor section with interstage product removal.

CONCLUSIONS

Few large scale applications of multifunctional reactors and their consequences for energy savings could be reported in the outline above. This should not discourage the reaction engineer nor make him pessimistic. Rather it should stimulate him to increase his endeavours to make them a success. The reactor with periodic flow reversal and the catalytic distillation reactor, have already found large scale applications. Dense membrane reactors may be the next. The application of the reactor section with interstage product removal to the methanol synthesis has shown us what great impact the application of the multifunctionality principle can have on energy saving. Other multifunctional reactors are still rather far from large scale introduction.
We should realise that the more exotic our reactors become, the more delicate and difficult their construction and operation. This implies that an operator will not change to more complicated machinery for just minor savings in investments. Large advantages in operating costs and energy savings must be envisaged. This will limit the rapid introduction of multifunctional reactors, despite their scientific and technological challenges. Last but not least we acknowledge the appeal of working on new, alternative and surprising concepts, yet want to stress that for many years much more of our effort in science must still be directed to a better and more profound understanding, and further development, of existing methods and technologies.

REFERENCES

Agar, D.W. and Ruppel, W., 1988, Extended reactor concept for dynamic $DeNO_x$ design, *Chem. Eng. Sci.* **43**, 2073-2078.

Agar, D.W. and Ruppel, W., 1988, Multifunktionale Reaktoren für die heterogene Katalyse, *Chemie-Ing.-Tech.* **60**, 731-741.

Armor, J.N., 1989, Catalysis with permselective inorganic membranes, *App. Catal.* **49**, 1-25

Belfort, G., 1989, Membranes and bioreactors: a technical challenge in biotechnology, *Biotechnol. Bioeng.* **33**, 1047-1066.

Blanks, R.F., Wittrig, T.S. and Peterson, D.A., 1990, Bidirectional adiabatic synthesis gas generator, *Chem. Eng. Sci.* **45**, 2407-2413.

Boreskov, G.K., Matros, Yu.Sh. and Kiselev, O.V., 1979, Catalytic processes carried out under nonstationary conditions, *Kinet. Katal.* **20**, 773-780.

Champagnie, A.M., Tstotsis, T.T., Minet, R.G. and Webster, I.A., 1990, A high temperature catalytic membrane reactor for ethane dehydrogenation, *Chem. Eng. Sci.* **45**, 2423-2429.

Coca, J. and Langer, S.H., 1983, Doing chemistry in the gas chromatograph, *ChemTech.* **13**, 682-689.

Eigenberger, G. and Nieken, U., 1988, Catalytic combustion with periodic flow reversal, *Chem. Eng. Sci.* **42**, 2109-2115.

Eigenberger, G. and Nieken, U., 1988, Abluftoxidation in Monolith-Katalysatoren mit periodischem Wechsel der Strömungsrichtung, *Chem.-Ing.-Techn.* **60**, 1070-1071.

Eigenberger G. and Nieken, U., 1991, Katalytische Abluftreinigung: Verfahrenstechnische Aufgaben und neue Lösungen, *Chem.-Ing.-Techn.* **63**, 781-791.

Fisch, B., Carr, R.W. and Aris, R., 1986, The continuous countercurrent moving bed chromatographic reactor, *Chem. Eng. Sci.* **41**, 661-668.

Geelen, H. and Wijffels, J.B., 1965, The use of a distillation column as a chemical reactor, Proc. third Chem. React. Eng. Symposium, Pergamon, Oxford, 125-134.

Gelder, K.B. van, Damhof, J.K., Kroijenga, P.J. and Westerterp, K.R., 1990, Three-phase packed bed reactor with an evaporating solvent - I. Experimental: the hydrogenation of 2,4,6-trinitrotoluene in methanol, *Chem. Eng. Sci.* **45**, 3159-3170.

Gelder, K.B. van, Borman, P.C., Weenink, R.J. and Westerterp, K.R., 1990, Three-phase packed bed reactor with an evaporating solvent - II. Modelling of the reactor, *Chem. Eng. Sci.* **45**, 3171-3192.

Gryaznov, V.M., 1986, Hydrogen permeable palladium membrane catalysts. An aid to the efficient production of ultra pure chemicals and pharmaceuticals, *Plat. Met Rev.***30**, 68-72.

Gryaznov, V.M., Smirnov, V.S. and Slinko, G., 1973, Heterogeneous catalysis with reagent transfer through the selectively permeable catalyst, in *Catalysis*, J.W. Hightower ed., American Elsevier, New York.

Hsieh, H.P., 1991, Inorganic membrane reactors, *Catal. Rev.-Sci. Eng.* **33 (1 & 2)**, 1-70.

Itoh, N. and Govind, R., 1989, Combined oxidation and dehydrogenation in a palladium membrane reactor, *IEC Res.* **28**, 1554-1557.

Itoh, N., Shindo, Y., Haraya, K. and Hakuta, T., 1988, A membrane using microporous glass for shifting equilibrium of cyclohexane dehydrogenation, *J. Chem. Eng. Japan* **21**, 399-404.

Kuczynski, M., Oyevaar, M.H., Pieters, R.T. and Westerterp, K.R., 1987, Methanol synthesis in a countercurrent gas-solid-solid trickle flow reactor. An experimental study, Chem. Eng. Sci. **42**, 1887-1898.

Matros, Yu.Sh., 1985, *Unsteady state processes in catalytic reactors*, Elsevier, Amsterdam.

Matros, Yu.Sh., 1989, *Catalytic processes under unsteady-state conditions*, Elsevier, Amsterdam.

Mohan, K. and Govind, R., 1986, Analysis of a cocurrent membrane reactor, *AIChE J.* **32**, 2083-2086.

Mohan, K. and Govind, R., 1988a, Analysis of equilibrium shift in isothermal reactors with a permselective wall, *AIChE J.* **34**, 1493-1503.

Mohan, K. and Govind, R., 1988b, Effect of temperature on equilibrium shift reactors with a permselective wall, *IEC Res.* **27**, 2064-2070.

Ray, A., Tonkevich, A.L., Aris, R. and Carr, R.W., 1990, The simulated countercurrent moving bed chromatographic reactor, *Chem. Eng. Sci.* **45**, 2431-2437.

Saatdjian, E. and Large, J.F., 1985, Heat transfer simulation in a raining packed bed exchanger, *Chem. Eng. Sci.* **40**, 693-697.

Schoenemann, K. and Hofmann, H., 1957, Die konsequente Anwendung der chemischen Reaktionstechnik, *Chem.-Ing.-Techn.* **29**, 665-671.

Shoemaker, J.D. and Jones, E.M., 1987, Cumene by catalytic distillation, *Hydroc. Process.*, June, 57-58.

Shu, J., Grandjean, B.P.A., van Neste, A. and Kaliaguine, S., 1991, Catalytic palladium-based membrane reactors: a review, *Can. J. Chem. Engng.* **69**, 1036-1059.

Sloot, H.J., Versteeg, G.F. and van Swaaij, W.P.M., 1990, A non-permselective membrane reactor for chemical processes normally requiring strict stoichiometric feed rates of reactants, *Chem. Eng. Sci.* **45**, 2415-2421.

Smith, L.A. and Huddleston, M.N., 1982, New MTBE design now commercial, *Hydroc. Process.*, March, 121-123.

Westerterp, K.R. and Kuczynski, M., 1987, A model for a countercurrent gas-solid-solid trickle flow reactor for equilibrium reactions. The methanol synthesis, *Chem. Eng. Sci.* **42**, 1871-1885.

Westerterp, K.R., Kuczynski, M. and Kamphuis, C.H.M., 1989, Synthesis of methanol in a reactor system with interstage product removal, *IEC Research,* **28**, 763-771.

Westerterp, K.R., van Gelder, K.B., Janssen, H.J. and Oyevaar, M.H., 1988, Development of catalytic hydrogenation reactors for the fine chemicals industry, *Chem. Eng. Sci.* **43**, 2229-2236.

Westerterp, K.R., 1992, Multifunctional reactors, *Chem. Eng. Sci.*, **47**, 2195-2206.

COMPUTATIONAL FLUID DYNAMICS APPLIED TO CHEMICAL REACTION ENGINEERING

P. TRAMBOUZE
Institut Français du Pétrole
C.E.D.I.- B.P. N° 3
69390- SOLAIZE- FRANCE

ABSTRACT

In the following lecture we try to summarize the conclusions of a survey[1] undertaken at the request of the Commission of European Communities, with the aim of delimiting the area of application of Computational Fluid Dynamics techniques in the broad area of Chemical Reaction Engineering. We first describe, in general terms, Computational Fluid Dynamics techniques and then consider the potential applications to Chemical Reaction Engineering. Finally, we summarize the main conclusions of our survey and add a few recommendations for future action.

COMPUTATIONAL FLUID DYNAMICS (CFD)

CFD refers to the analysis of fluid flow and related phenomena such as heat transfer, mixing and chemical reactions. By using computer techniques, CFD is able to solve the basic equations describing the motion of fluids and the evolution of variables such as pressure, temperature and concentrations.

There are three basic equations governing the behaviour of any fluid element, corresponding to balances for mass, momentum and energy. It is common to call the set of the above equations the Navier-Stokes equations. For solving this set of equations, the space in which the problem is posed is divided into a solution mesh covering the whole space. The above set of partial differential equations is discretized over the mesh and then solved by successive iterations.

---

[1] The complete text of the report should be published shortly

For turbulent systems, as is generally the case for industrial situations, turbulence models must be included in the Navier-Stokes equation. When phenomenon in addition to flow is to be modelled, such as chemical transformations, the appropriate equations must be added, discretized in the same way and solved as part of the equation set. In all cases, sophisticated numerical algorithms are used to obtain the required solution in a reasonable time.

The major difficulty of this approach is linked with the proper choice of the turbulence model. Another difficulty comes from chemical kinetics relationships to be added to the above set of Equations, which are generally nonlinear and introduce further computation complexities. Further difficulty can be found when expressing the necessary boundary conditions.

It is important to realize that the above set of Equations may lead to different types of models (detailed or phenomenological) and that very often additional information coming from empirical models is needed.

For turbulent reactive flow, the adjunction of turbulence models, as already mentioned, leads to a phenomenological model, often called a physical model. Another type of model well suited to the simulation of flows of engineering interest is the large eddy simulation (LES). Using this technique lowpass-filtered Navier-Stokes equations are solved to simulate three-dimensional unsteady motion. The integration of these equations leads to a direct description of grid-scale variables, while the subgrid-scale variables and their influence on the flow field are modelled.

The use of CFD is already widespread in the so-called high technology industries such as nuclear, defence, aerospace. In some instances, CFD calculations may be performed instead of experiments that are dangerous or even impossible to perform. For example, CFD is now being applied to environmental and safety problems. However, it must be clear that CFD can be only **as good as the physical models used**. It is primarily an insight tool, useful for understanding the important features of a system and predicting trends. Therefore CFD may be particularly well adapted for interpolating or even extrapolating our knowledge of the behaviour of a system. It is clear for all researchers involved in the CFD area, that extensive **validation** and **tuning of computer programs** with experimental data are necessary before reliable predictions can be made for a wide range of flow conditions in engineering equipment.

Existing Software

Because of the mathematical and numerical complexity of the method, CFD has tended to evolve to keep the user away from the details of the solution of the resultant equations. Therefore easy-to-use general purpose packages appeared on the market, allowing the non-specialist engineer to set up and solve problems with relative ease. However, this possibility does not hinder the generation of specialized packages, by CFD specialists, to solve a generic problem. A wide range of techniques has been developed to cope with the various types of problems which can be posed. The solution may use finite difference, finite volume or finite element formulations in the discretization of the equations.

There are several aspects to be considered when selecting a CFD software:

1- The system *geometry*: The volume or area in question must be divided into a number of cells. Using the finite volume formulation, a range of geometric representations is possible, with Cartesian or cylindrical coordinates. Body-fitted coordinates (BFC's) may also be used. Finite element methods are able to represent highly complex geometries. Modern codes often provide "grid generators".

2- *Dimensions* of the problem: It is usually wise to build models by starting with a simplified case, in two dimensions rather than three and steady-state rather than transient. Similarly, additional variables such as concentrations, phase fractions, temperatures may be added later. However, it must be clear that the final objective must be to treat the problem as transient and three-dimensional, even if the geometry presents some kind of symmetry.

3- Choice of the *numerical method*: The solutions are generally iterative, and convergence is usually obtained by relaxation techniques. Careful choice of the relaxation parameters makes a major contribution towards the rate of convergence and the obtention of a solution at all. This is often a weakness of the general-purpose codes.

4- The selection of a *turbulence model* is a matter for a specialist and must be looked at with great attention. The sensitivity of the solution to changes in the boundary conditions may be a guide for the quality of the foundations of the model.

5- *Data output*: CFD creates a large quantity of information. It is important of being able to present the salient points in an effective form. Graphical output is a very valuable way of rapidly assimilating a great amount of data.

There is a relatively large number of commercially available software packages (PHOENICS, FLUENT, FLOW3D, ASTEC, FIDAP or EasyFlow).

APPLICATION of CFD to CHEMICAL REACTION ENGINEERING

In most textbooks dealing with Chemical Engineering, it is usual to start with a general presentation of the subject based on the three basic laws of physics: the conservation of mass, the conservation of momentum,the conservation of energy.

Since these general balance equations generally can not be solved for practical problems, it is usual to make simplifying assumptions, leading to standard flow models: the plug flow and the perfectly stirred vessel.

Even if these models are very simplified, they have been extensively used in practice, since they lead to sets of equations that are relatively easy to solve and prove to be useful for practical achievements. However, there are situations where these models are limited and where the use of a complete and detailed description of the physical system seems necessary.

This is the case for fast chemical kinetic reactive flows, where the rate of mixing of reactants and the rate of chemical conversion are of the same order of magnitude.

In a similar manner, when scaling-up a chemical reactor, it is usual to treat separately the chemical aspect of the process and the physical one. By using this simplified procedure, the chemical transformation is studied in pilot-plants and the physical behaviour of the systems is investigated at various sizes in mock-ups.This procedure supposes that the chemical and physical variables can be treated as separable variables. This is often a rough approximation, especially when dealing with multiphase chemical reactors.

As mentioned above, the types of chemical reactors to be considered for potential valuable application of Computational Fluid Dynamics may be divided into two broad categories:

-single-phase systems involving rapid chemical reactions,

-multiphase systems for which the standard simplified description is too approximate, especially for scaling-up purpose.

1- Single-Phase Systems

The case of reactors in which fast reactions occur remains a difficult problem. As we mentioned above, the rate of mixing of the reactants can be of the same order of magnitude as the rate of reaction, and simplified models (plug flow or perfectly stirred vessel) cannot be used. In fact, the difficulty in representing such reactive systems stems from the necessity of characterizing the system at the molecular scale, where mixing and chemical reaction occur simultaneously. At this small scale, mixing is usually called micromixing, with the one occurring at the scale of the reactor vessel being qualified as macromixing. Many developments have occurred during the last forty years tackling this difficult problem of micromixing, and various models have been proposed.

A particularly important process, combustion, having all the characteristics of a very fast chemical reaction performed at the same time as the mixing of the reactants (fuel+oxygen), has been studied for a long time. This difficult problem has been approached by using methods of Fluid Dynamics that Mechanical Engineers have developed for other purposes, such as aerodynamics. Along with the development of ever more powerful computers, progress has been made in this field, and now Computational Fluid Dynamics has become an important subdiscipline, with applications in various practical areas, such as reactive flows research.

It seems logical to try to take profit of all these developments in Combustion Engineering, by applying the same concepts and methods to Chemical Engineering and more specifically to Chemical Reaction Engineering.

Reactive flows are the result of many different mechanisms occurring simultaneously, such as: chemical kinetics, transport of molecular species by diffusion, transport of energy by conduction or radiation, convective flow,thermodynamic equilibrium.

In many instances convective flow is turbulent and the chemical transformation is influenced by the turbulent fluctuations of the flow. The main problem is therefore the coupling between chemical kinetics and fluid dynamics.

During the last decade attempts have been made to apply CFD techniques to chemical reaction engineering problems, showing that progress in this area could lead to a much better understanding of the mechanisms involved. In-line mixing of reactants has been considered by many researchers, especially in the gas phase. On the contrary, few data can be found concerning liquid-phase systems. Relatively, many more papers dealing with liquid-phase stirred tank reactors can be found.

When we look at practical potential applications, the systems of interest are processes involving fast, competitive chemical reactions, where fast mixing is of prime importance in minimizing the by-products. Increasing the selectivity of the transformation means less feedstock consumption, lower separation or purification expenses and fewer waste materials. Therefore improving the mixing rate of reactants could lead to substantial benefits in connection with raw materials, energy and waste disposal.

It seems clear that for such fast reaction systems, the best technology to be selected is in-line mixing, as in a T-tube or a multijet in-line mixer. This very simple technology can be used equally well for gas or liquid-phase systems.

On the other hand, a stirred tank is certainly not very well suited to achieving a very rapid mixing of the injected reactant with the bulk of the tank. In some circumstances, however, its use cannot be avoided. This could be the case, for example, for polymerization reactions involving highly viscous medium. For such applications, where the structure of the polymers or copolymers could depend on the rate of micromixing of the monomers and catalyst in the reaction medium, CFD calculations could be helpful.

2- Multiphase Systems

When dealing with multiphase chemical reactors there is a great variety of cases and equipment to be considered. The simplest way to characterize these systems is to refer to the phases present in each system. Therefore we shall hereunder distinguish the following cases:

- -Gas-Liquid systems,
- -Liquid-Liquid systems,
- -Fluid-Solid systems
- -Gas-Liquid-Solid systems.

When a solid is present, two possibilities occur: the solid can be a reactant or a catalyst. If the solid takes part in the chemical transformation, generally several solid phases will have to be considered.

2.1- Gas-Liquid Systems

Gas-liquid systems are very often encountered in industry. The simplest case is two-phase flow in pipes occurring in many processes and at very large scales. However the detailed modeling of such two-phase flows has not yet been achieved. There are numerous flow regimes, that cannot be described by the same equations, since the interaction terms representing the various transfers between the phases are greatly dependent on the flow regimes. This means that the physical laws governing these systems have not been fully established and that further work is still needed.

When dealing with reactive systems, it is easily understandable that the problem will be even more difficult. For any two-phase system, the first step will be to write for each phase the three fundamental mass, momentum and energy balance equations. The added complexity of the problem will be linked to the writing of the terms expressing the exchanges or transfers between the phases. Gas-liquid chemical reactors generally take the form of vessels or columns, inside of which one phase is continuous and the other is disperse. The two phases can flow either cocurrently, countercurrently or crosscurrently. In order to achieve the chemical transformation it is necessary to have good contact between the disperse phase and the continuous one. This means that the interfacial area and the size of the disperse phase elements (bubble or drop) will be important parameters. As a consequence, it is necessary to express the rate of formation (or of disappearance) of the interfacial area. This last term will generally be related to the exchange of momentum between the phases.

Another complication in relation to the existence of a disperse phase is its description, in terms of size and composition. The elements of the disperse phase (bubbles or drops) are initially generated by a distributor, but later on they can break into smaller elements or, on the contrary, coalesce and produce larger elements. Breakup and coalescence generally interfere with the chemical transformation. Consequently, the description of the disperse phase must be adapted to the case studied and not simply considered as a flowing phase; the concepts of Population Balance or Probability Density Functions (PDF) must then be used to describe the behaviour of the disperse phase. Again the rates of breakage and coalescence must be put in the equation expressing the interfacial area balance. This is a very important physical phenomenon, that is still far from being completely understood.

One can find a very large number of publications dealing with gas-liquid reactors considered from a conventional overall chemical engineering point of view; on the contrary only a few reports have been published treating gas-liquid reactors by using CFD techniques. Up to now, research has been done on bubble columns, which is certainly the simplest geometry used as a gas-liquid reactor. It seems important for this type of system, where several flow regimes are possible, to perform unsteady-state simulation, in order to take into account large fluctuations of flows, that certainly play an important role in backmixing and mass transfer, and consequently influence the chemical conversion. Reactors such as gas-lift systems could certainly be simulated equally well.

Papers dealing with the modelling of flow in gas-liquid stirred tank reactors can also be found. Other types of gas-liquid reactors, such as countercurrent columns (plate or packed tower), could also be considered. However, we don't think this is a first priority, since design methods for such equipment have been relatively well established. This fairly good mastery is certainly linked to the fact that the internals of these columns control the flow behaviour of both phases quite well, and therefore simulation of these flows would lead to only marginal improvement.

On the other hand, equipment of the scrubber type, very often used for gaseous effluents washing and cleaning, could certainly be largely improved by using CFD techniques. As a matter of fact, such equipment is generally designed empirically and must be built on a large scale to treat large volumes of gaseous effluents. The final performance is mainly governed by the flow pattern of the gas inside the equipment and by the more or less strong interaction or momentum transfer between gas and liquid. Even if this phenomenon can be studied on a small scale, using, for example, a single nozzle for liquid dispersion, the design of a large industrial scrubber equipped with several rows of nozzle lines requires a good understanding of the flow pattern in each part of the volume. This could be achieved or at least fairly well approached by CFD computation.

For very rapid reactions, such as acid-base neutralization, in-line mixing could be recommended. In this case, ejectors or venturis are potential pieces of equipment whose design should certainly be improved by using CFD techniques.

Another technique particularly well suited for CFD calculations is the falling film system, where the liquid phase flows in the form of a film along a vertical wall and the gas phase is put into contact with the free surface of the liquid film. This equipment, often used as an evaporator, is also of interest for very rapid and exothermic reaction.

2.2- Liquid-Liquid Systems

There is a great variety of equipment that can be used as a liquid-liquid chemical reactor. Two broad classes can be distinguished: the stirred tank and various countercurrent columns. It is well known, that for liquid-liquid systems, it is generally easier to disperse one phase into the other than to completely separate the phases after their contact. The direct consequence of this, is that the rate determining step to be considered when designing a liquid-liquid contactor is generally the coalescence of the droplets of the disperse phase.

Therefore, in any attempt to simulate the interactions of the two liquid phases, it will be extremely important to express correctly the rates of drop formation, breakage and coalescence. Knowledge on this subject is still very limited, and it is absolutely necessary to get experimental data before trying to perform any fluid flow computations. The concept of population balance or PDF will also be absolutely necessary to represent a set of drops with various sizes and various concentrations of chemical species.

For very rapid reactions, the on-line mixer could also be selected and selectivity improvements be achieved by better local reactant ratio control. Very often, rapid heat transfer will simultaneously be required (for example, nitration of glycerine). In such cases, CFD computation could certainly be of some help.

2.3- Fluid-Solid systems

Two variants must be considered: the solid is a catalyst or the solid is one of the reactants taking part to the reaction.

2.3.1- Catalytic reactors

A solid catalyst is generally used in the form of grains gathered together in large numbers and forming a mass which may be:

- -a fixed bed,
- -a moving bed,
- -a fluidized bed,

-an entrained bed,
-a slurry.

In each case, the interaction between the fluid(s) containing the reactants and the catalyst particles must be sufficiently high to promote mass and heat transfers.

In a fixed bed, the catalyst grains are immobile and the fluid(s) flow through the interstices of the bed. In a moving bed, the catalyst grains are subjected to very slow motion and therefore may be supposed to be immobile when interactions with the fluid(s) are considered. The flow of one fluid phase through such catalytic beds may certainly be analysed using CFD techniques, but, as for packed columns, it does not appear to be the best case of application of these computations. The problem of initial distribution of the feed at the top of the bed, especially if the feed is a gas-liquid mixture, may however be considered for potential improvements.

The realization of strongly exothermic or endothermic reactions in fixed beds necessitates the use of multitubular systems. Examples of processes using this technique are, for example, phthalic anhydride production or steam reforming of light hydrocarbons. The design of such equipment must take into account the heat transfer inside the catalyst bed as well as the reaction kinetics and fluid flow, using a conventional two-dimensional model. Certainly CFD techniques leading to a nonstationary three-dimensional model could provide an improved description of these reactors.

The cases of fluidized or entrained bed seem to be much more interesting for CFD applications. In these cases, the interactions between fluid and solid are stronger, since the particles are sustained or even entrained by the flowing fluid. This will also apply to separation devices required for recovering and recycling the entrained particles (such as cyclones). A relatively large number of papers dealing with fluidized or entrained beds can be found in the recent literature. However, it must be stressed that a better understanding of the interactions between fluid and particles still has to be gained in order to describe correctly fluidized or entrained beds. A good example of the application of CFD techniques to an important industrial process can be found with catalytic cracking technology.

A large number of reactions are also achieved by using three-phase reactors (ebullated bed or a stirred tank with the catalyst under the form of a slurry). These cases could also be studied, using CFD techniques, but it seems that a preliminary good understanding of systems with one fluid phase and one solid is required before dealing with more complex systems.

2.3.2- Reactors for the transformation of solids

A great variety of technologies is used to convert solid reactants. In such equipment it is essential to control the transformation of the solid and generally to achieve complete conversion into the desired product. This means that the residence time distribution of the solid becomes a very important function. The size of the solid grains is also important and will generally change during conversion. Therefore, it seems essential to describe the evolution of the solid phase by using the population balance concept. For each technology, but for fixed bed, a description of the solid flow will be necessary and should result from experimental measurements.

The case of reactions or operations leading to the formation of solids is also interesting to consider. A first example can be found in the Chemical Vapor Deposition (CVD), widely used for the manufacture of microelectronic components. Another example concerns the Spray Dryer used for the manufacture of solid material with a given particle size distribution. A third example is related to very important operations in the chemical industry, i.e. precipitation and crystallization.

Validation- Experimental Techniques

It has been already stressed that, even if CFD techniques offer new and extensive possibilities, detailed validation and tuning of computer programs with experimental data are necessary before reliable predictions can be made for a wide range of flow conditions in engineering equipment.

We have identified two classes of potential applications:

-systems involving **rapid chemical reactions**,

-multiphase systems whose **scaling-up** could be made easier by using CFD techniques.These two classes of problems will be treated in relatively different ways.

In the first case, simulation will be more fundamental, resulting sometimes in Detailed Numerical Simulation or Large Eddy Simulation. Therefore, experimental validation will be used for adjusting the physical laws introduced in the model, especially the turbulence model. The experimental work can generally, in this case, be a full-scale experiment. However, in order to make measurements easier or at least possible, simulation fluids will often be required. For the same reasons, a chemical reaction having the same main characteristics (rate and kinetic scheme) as the chemical system of practical interest, should also be found in order to simulate the interaction between the turbulence and the chemistry. The variables to be measured are the same as the variables in the computer model: local velocities, local concentration of chemical species, pressures, temperatures, etc.. and fluctuations of these variables.

In the second case, simulation will rather be useful for showing the effect of the size of the equipment on its behaviour. Therefore it will be necessary to adjust the model by comparison with experimental data obtained at reduced scales. It will generally not be absolutely essential to perform experiments with a reactive system. Of course, if experimenting with a simulated chemical system is possible, it should be done in order to obtain additional useful information. As a matter of fact, fluid flow simulation may be sufficient, since the chemical reactions involved are generally not fast and therefore there is no direct interaction between turbulence and chemical kinetics. The influence of the fluid flow pattern on chemical conversion should then be calculated by using the CFD code, in which the chemical kinetics will have been incorporated. This complete simulation code must also be checked by comparison with some experimental data obtained in a pilot plant. Finally, we are not far from the concept of mock-ups, for long-time used for scale-up purposes. However, the types of measurements to be performed will be more complete and sophisticated.

MAIN CONCLUSIONS and RECOMMENDATIONS

1- **Computational Fluid Dynamics** (CFD) and its applications to Combustion studies have developed quite rapidly during the last ten years. This fast growing hybrid branch of Mechanics and Mathematics is certainly to be considered as a potentially useful and efficient tool in the field of Chemical Engineering and more specifically in the area of **Chemical Reaction Engineering** (CRE).

2- The difficulties in this new approach stems from the consequence of the complexity of the mechanisms to be simulated simultaneously: fluid dynamics, chemical reactions and physical aspects of each system considered. Another difficulty comes from the numerical treatment of the equations for the final model, resulting in very sophisticated and diversified mathematical treatments. Therefore, it seems essential, for the success of any research aiming to apply present fluid dynamics knowledge to chemical reaction engineering problems, to achieve truly **cooperative and interacting teamwork**, including specialists in various disciplines.

3- The types of chemical reactors to be considered for potential performance improvements when applying CFD as a new tool for their design are numerous; two broad classes of problems have be identified as relevant to this new approach:

-Systems involving **fast chemical reactions**. In-line mixing equipment should preferably be studied for this type of reactions.

-Multiphase systems, whose **scaling-up** still has to be performed with great difficulty based on very simplified models. From looking at the various types of systems found in practice, it appears that gas-liquid and fluid-solid systems should be considered first. However, basic knowledge is still missing concerning the physical behaviour of these systems, especially for the coalescence of bubbles and the momentum transfer between gas and solid. Specific research should be done in order to get this missing information.

4- Presently there are a certain number of **existing CFD software** packages available commercially or developed by various research laboratories. This is certainly an interesting starting point, but we can never be sure that the numerical results given by any of these software packages are applicable to a practical industrial case, **without checking** these results in one way or another.

5- On the **experimental** side, there are a great deal of measurement methods able to give local values of temperature, pressure, velocities, concentrations of chemical species, phases ratio, size of bubbles, drops or particles, as well as to produce an overall visualization of flows. Though these methods are continuously being improved, they are generally limited to transparent continuous phases or to low concentrations of a disperse phase. Consequently, it seems absolutely necessary to improve existing methods or to implement new ways of measurement in order to be able to deal with the current industrial environment.

THE USE OF COMPUTATIONAL FLUID DYNAMICS TO ANALYSE VARIOUS DESIGNS OF CHEMICAL REACTOR

M.J. Tierney, B.A. Splawski, G.L. Quarini, CFDS, Harwell Laboratory, England

ABSTRACT.

Considerable work on Computational Fluid Dynamics (CFD) and its validation by experiment has been sponsored over the last twenty or so years by European governments. In this paper we give our own perception of how CFD is used in practise, for a particular unit operation. The work is similar in flavour to the consultancy services offered to companies, and to projects carried out within industry.

The flow patterns in three stirred systems are presented. In each a different mathematical model of agitation has been set up, to correspond with the availability (or paucity) of measured data. The algorithms are based on: the use of a specified velocity field within the swept volume of the impeller; a time-averaged local drag coefficient between the impeller and fluid; a specified field of applied force within the swept volume. For the last problem, the implications for residence time distribution and chemical reaction (first order) are considered.

NOMENCLATURE.

A_p - Projected cross sectional area of a baffle.
C_d - Drag coefficient.
$\mathbf{F}_v$ - Contribution to Navier Stokes equation.
h - Impeller height.

k - Turbulence kinetic energy.
N - Number of impeller blades.
R - Impeller radius, or contribution to Navier Stokes equation.
r - Radial position, equal to $(x^2 + y^2)^{1/2}$.
S - Stirrer tip speed.
T - Torque.
u,v,w- Velocity components of vector **U**.
α- Angle of blade to vertical.
ε- Rate of dissipation of k.
θ- Cylindrical co-ordinate, equal to arctan (y/x). -
ϕ- species mass fraction.

INTRODUCTION.

In this paper we discuss the application of CFD to batch and continuous stirred tank reactors (CSTRs). Work is similar to industrial projects - carried out for contacts free of charge - although in certain cases key dimensions, structural details and operating conditions are changed to protect commercial secrets.

Tank reactors are used widely in a range of industries, such as brewing, pharmaceuticals and polymers. Continuous units produce the best quality of product only when residence times are consistent. More generally, adequately rapid reactions are maintained by vigorous mixing at the micro- or macro-fluid level [1]; a certain turbulence intensity must be achieved throughout the vessel. In opposition to this requirement, certain materials are damaged by excessive shear.

Middleton [2] has shown that good scale-up predictions can be made using CFD. The quality of estimate is almost surprising, given the assumption of isotropic turbulence within the k-epsilon model used. Middleton's CFD calculations are superior to scaling based on dimensionless groups; here similarity is achieved only for two or three sets of parameters.

We discuss here our own models of stirred tank reactors, listing the appropriate boundary conditions and describing three representations of the impeller region. Our results are in the form of velocity vectors, residence time distributions and contours of species concentration. We discuss the appropriateness of turbulence models, and meshing of geometric details.

A significant amount of future work is identified. This includes experimental validation and the development of algorithms for reaction schemes.

MODELLING.

This includes the system geometry, boundary conditions, momentum sources from the impeller, and chemical kinetics.

We consider the following structural components: the main tank compartment, with dished end; baffles; the impeller; the liquid surface; and (possibly) inlet/outlet pipes. Additional objects such as cooling coils or instrumentation probes may warrant attention in other work. In terms of CFD the required finite volume mesh is complex, and unlikely to be faithfully represented by mono-block codes (that is, programs for which the grid is a structured arrangement of M times N times P cells). Multi-block [3] or unstructured [4] schemes are recommended.

In particular, the use of cylindrical coordinate systems is valid only for axisymmetric flows and geometry. This is certainly not the case for continuous processes, wherein some flow from the inlet traverses the tank centre-line. Multiblock schemes are recommended because they remove the singularities corresponding to r = 0, and permit a better representation of complex shapes (Figure 1).

The inner surface of the tank is represented by the classical log-layer [5]. Here, we recognise that the boundary layers are thinner than even the smallest elements, so that production (or loss) of momentum and turbulence has to be represented by correlations. The free liquid surface is considered as a plane of symmetry: it is impermeable to flow, and the gradients of all quantities are zero. A volume of fluid - or VOF - approach would be more physically correct, but incurs greater computational cost [6].

The CSTR inlets are regions of specified velocity, and the outlets have zero velocity gradient. The k-epsilon model is used to represent turbulence. The impeller is included through alterations to the Navier- Stokes equations for the transport of momentum. Put simply, these are:

$$\rho\frac{\partial}{\partial t}U = (\text{other standard terms}) - RU + F_v \qquad \text{(EQ 1)}$$

R and F_v are additional terms added by the user. Fv has units of pressure gradient or body force per unit volume.

<u>Model 1:</u> We impose velocities in the swept volume by setting R and F_v to large values, far more significant than the standard terms. Turbulence quantities are also specified. Working in a cylindrical frame of reference (directions r,θ,Z):

$$u = -r\sin(\theta)\left(\frac{S}{R}\right);v = r\sin(\theta)\left(\frac{S}{R}\right);w = \sin(\alpha)\,r\left(\frac{S}{R}\right) \qquad \text{(EQ 2)}$$

$$k = 0.002\,(u^2+v^2);\varepsilon = \max\left(0.1, \frac{k^{1.5}}{6R}\right) \qquad \text{(EQ 3)}$$

where R,S are the impeller radius and tip speed.

Alternatively, quantities may be prescribed following laser doppler velocimetry measurements.

<u>Model 2:</u> The stirred velocity, Us, is greater than the liquid's. The momentum input is related to this difference in velocities through a drag coefficient.

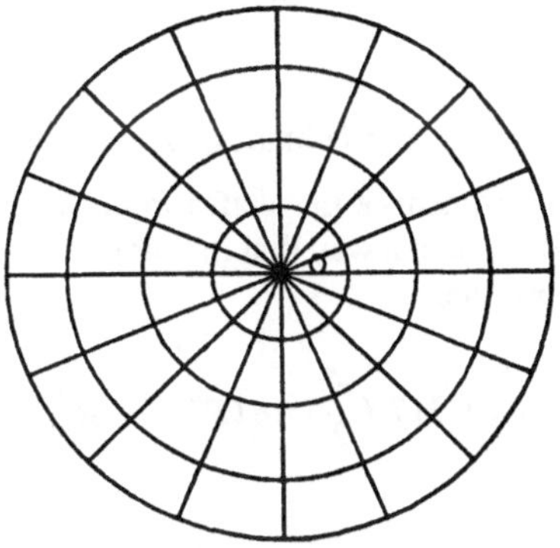

Cylindrical, incuring a singularity at r = 0.

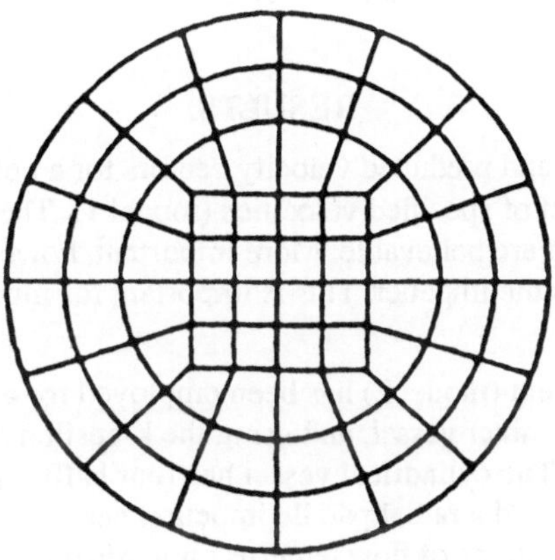

Multi-block scheme.

Figure 1 Cylindrical and multi-block meshes.

$$R = 0.5C_d\rho(|U_s| - |U|)(\frac{NA_p}{V}) \quad \text{(EQ 4)}$$

In addition, the constant term is set as follows:

$$F_v = -RU_s \quad \text{(EQ 5)}$$

A_p is the cross sectional area of each of N blades, whereas V is the swept volume.

Model 3: Suppose a measurement of the shaft torque, T, is available. An appropriate field of Fv can be subscribed, in such a way that its integration over V correctly predicts T. If we set Fv proportional to rn, then:

$$|F_v| = (n+3)T(\frac{r^n}{2})\pi h(R^{n+3} - R_i^{n+3}) \quad \text{(EQ 6)}$$

Here R_i is the inner radius of the swept volume. We set n = 1.

Kinetics are applied only in one study. Material is consumed at a rate given by:

$$r = -K\phi \quad \text{(EQ 7)}$$

This item is added to the equation for mass transport. Its use is - admittedly - naive given the strong dependence of most reactions on temperature.

RESULTS.

Figure 2 shows the mesh and predicted velocity vectors for a polymerisation process. The stirrer is modelled by a set of specified velocities (model 1). The predictions show a set of circulation patterns which are believable. More important, however, is that the bulk liquid is nowhere near as fast as the impeller. This is important for mixing and hence product quality.

The use of a drag coefficient (model 2) has been employed for a demonstration calculation, simulating flow in a mixer vessel, and using the k-epsilon model to represent the effects of turbulence [7]. The cylindrical vessel has four baffles positioned at 90 degree intervals around the wall, and a radial paddle impeller, near the mid-height position. Figure 3 shows ASTEC predictions of flow patterns on vertical plane slices through the model centre, together with turbulence kinetic energy levels on the same plane. Again, the vessel is not expected to provide adequate mixing, because of potential dead regions near the top and bottom. Further analysis with a finer mesh is needed to confirm our opinion.

Similar predictions are made for a CSTR, using specified torque (model 3, see Figure 4). There is a tendency for some flow to by-pass the mixed region, and we believe more thought should be given to the shape and position of the inlet and impeller. To show this more clearly, we have carried out a time dependent simulation of a stepped injection of dye into the vessel entrance. Concentrations in the exit pipe are plotted against time, to

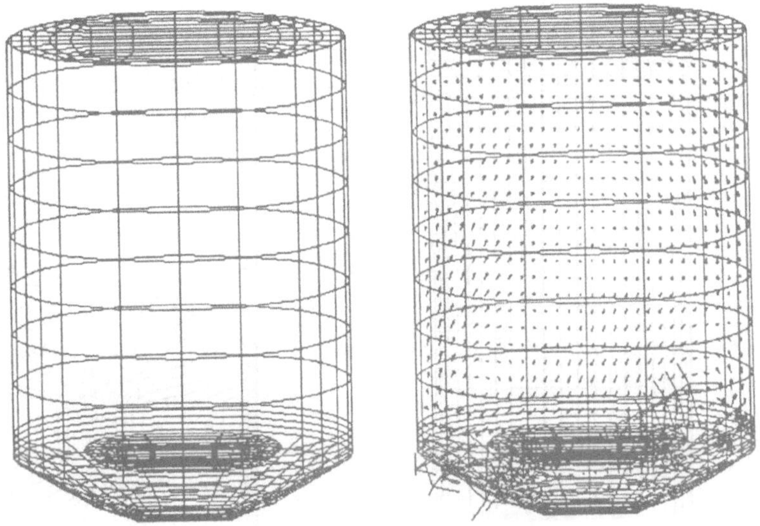

Figure 2 Velocity vectors, indicating poor momentum transfer (model 1).

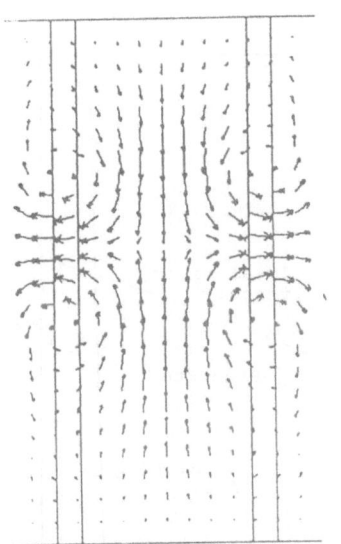

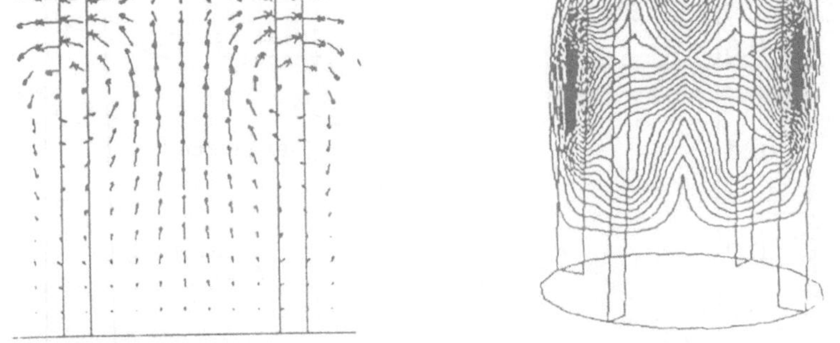

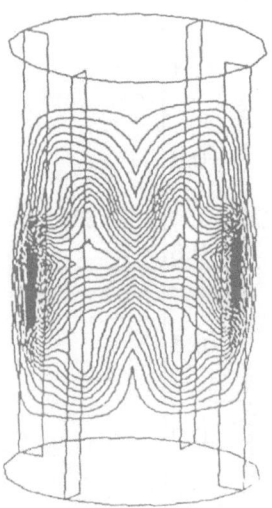

Figure 3 Mixer vessel flow patterns and turbulence levels (model 2).

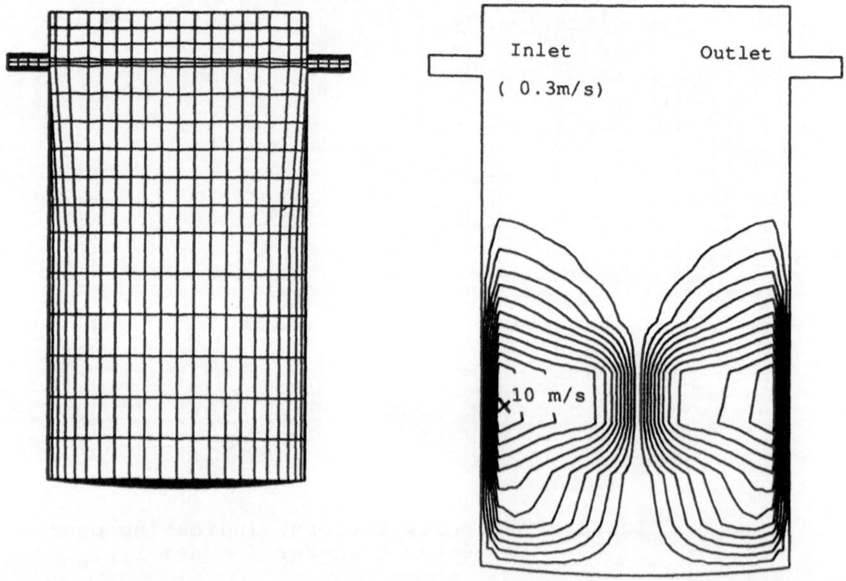

Figure 4 Mesh plus contours of speed in a CSTR (model 3).

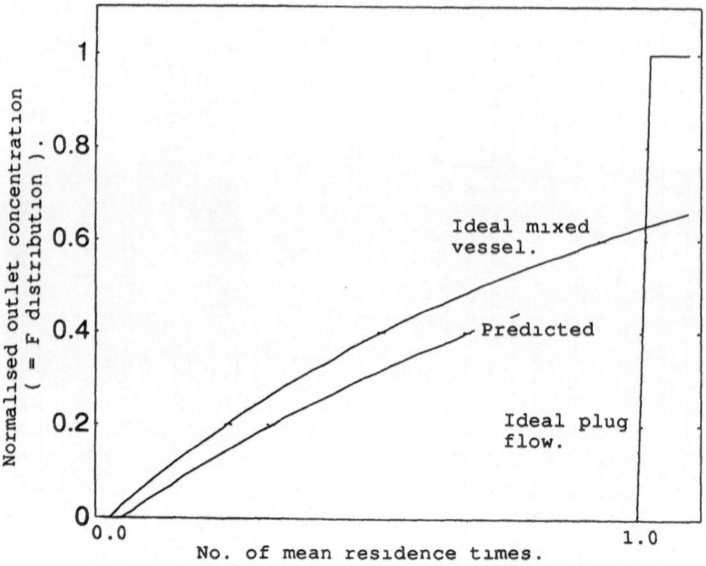

Figure 5 F distribution for a CSTR (see ref. [1]).

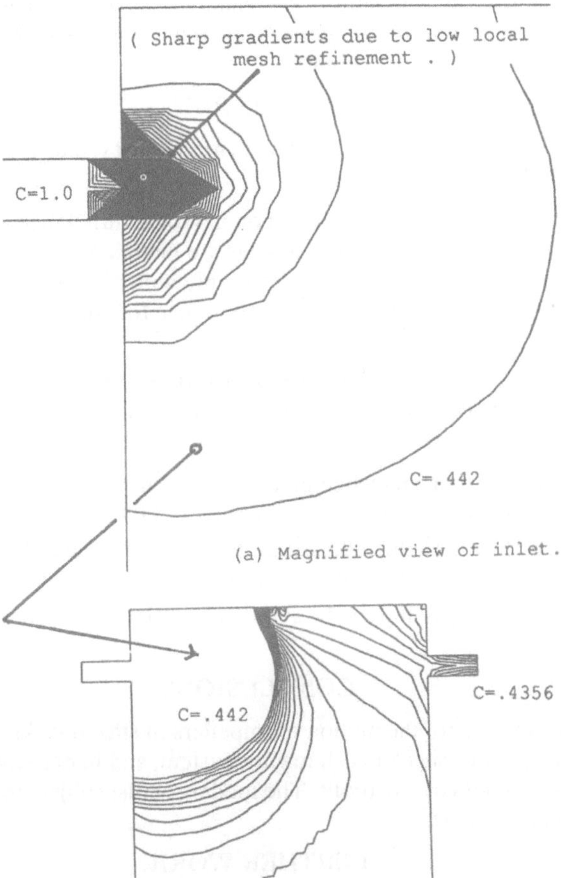

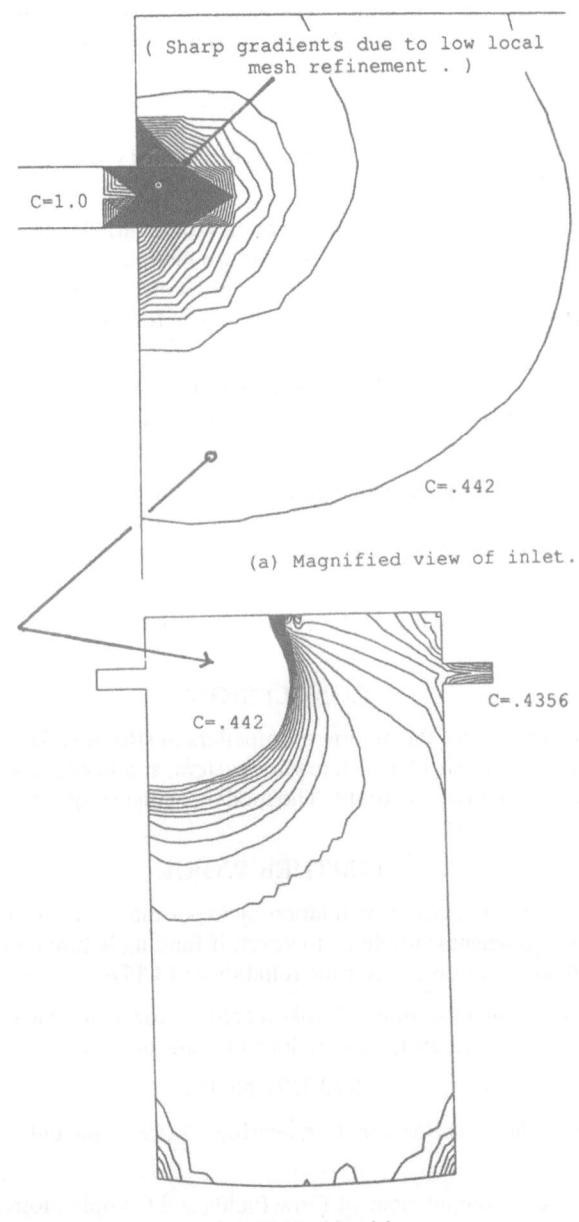

(a) Magnified view of inlet.

(b) Bulk liquid.

Figure 6 Contours of reactant concentration (1st order kinetics assumed).

give a representation of residence time distribution (Figure 5). The resulting 'F' function lies between that of ideal plug flow and mixed vessels [1].

Concentration contours show regions of comparatively high reactant concentration, corresponding to by-passing and hence low residence time. Long residence times within the dead zones result in lower values (Figure 6).

DISCUSSION.

We refer to modelling of turbulence and geometric features, plus the implications of CFD for equipment designers.

The k-epsilon model assumed here is based on isotropic turbulence. Certainly, algebraic and differential second moment closure models are thought superior for flows with significant swirl [8], and where turbulence varies sharply with direction. Such approaches are vital in modelling cyclones [8,9], but their importance for stirred tanks warrants further debate [10].

Generally, our meshes are coarse, with between 1200 and 12000 elements. In particular, higher resolution could be applied at baffles (1 cell wide) or inlets (2 x 4 cells in cross section).

Used as an engineering tool, and when given appropriate thought, CFD makes a valuable diagnostic contribution to design. The bad mixing on Figure 2, typical of some configurations that we have witnessed, undoubtedly results in a poorer product. Current practise in the UK is to write off the profits on inferior grade polymers; typical losses are 90/tonne.

Referring to Figure 4, by-passing can be avoided by pointing the inlet downwards or using baffles.

CONCLUSIONS.

We identify three models for the motion of impellers in stirred tanks. Flow patterns are significant for the performance of each reactor system, and in extreme casees could - if mismanaged - lead to loss of any profit. The models whilst subject to further refinement, are suitable diagnostic tools.

FURTHER WORK.

This is required mostly in terms of validation by experiment. Ideally, expensive laser doppler velocimetry equipment is needed. However, if funding is limited measurements of residence time distributions may infer the reliability of CFD.

Also of interest is (a) the modelling of rapid reactions, for which the equations for scalar transport are said to become stiff, (b) reactions in multi-phase flows.

REFERENCES.

1. O. Levenspiel. 'Chemical Reaction Engineering', 2nd edition, published by John Wiley and Sons, 1972.

2. J. Middleton et al. 'Computations of Flow Fields and Complex Reaction Yield in Turbulent Stirred Reactors, and Comparison with Experimental Data', Chem. Eng. Des., 64, 18-22 (1986).

3. N. Wilkes and A. Burns. 'FLOW3D Manual', available from CFDS., Harwell Laboratory, Didcot, Oxon OX11 ORA. November 1991.

4. R. Lonsdale. 'The ASTEC Code: An Algorithm for Solving Thermal- Hydraulic Equations in Complex Geometries', proceedings of the 5th International Conference on Numerical Methods in Laminar and Turbulent Flow, Swansea, May 1988.

5. W. Rodi. 'Turbulence Models and their Application in Hydraulics', IAHR State of the Art Paper, Delft, Netherlands, 1980.

6. R. Lonsdale, 'ASTEC Manual', available from CFDS.

7. R. Lonsdale et al. 'Process Industry Solutions from CFD Software', proceedings of the World User Association Computational Fluid Dynamics (WUACFD), Basel, Switzerland, May 24-28 1992.

8. B. Launder, 'Current Capabilities for Modelling Turbulence in Industrial Flows', Applied Scientific Research, An International Journal on the Application of Fluid Dynamics. Guest Editor R.V.A.Oliemans, Vol 48, nos 3-4. October 1991. Published by Kluwer Academic Publications, Holland.

9. J. Hargreaves and R. Silvester, 'Computational Fluid Dynamics applied to the analysis of deoiling hydrocyclone peformance', Trans.I.Chem.E., vol. 68, part A, pages 365-386, 1990.

10. A. Brucato, M. Ciofalo, F. Grisafi, L. Rizzuto, 'Computer Simulation of Turbulent Fluid Flow in Baffled and Unbaffled Tanks Stirred by Radial Impellers'. Computer Applications to Batch Processes, Cengio, Italy, March 1990.

NEW METHANOL PROCESSES

K.R. Westerterp
Chemical Reaction Engineering Laboratories
Chemical Engineering Department
Twente University
P.O. Box 217, 7500 AE Enschede, NL

ABSTRACT

The well-established methanol synthesis processes of Lurgi, ICI and Kellogg are discussed first. Then follows a survey of new processes under study or development, in which the methanol synthesis is performed in the Slurry Bubble Column Reactor, the countercurrent Trickle Flow Reactor, the Gas Solid Solid Trickle Flow Reactor or the Reactor Section with Interstage Product Removal.

These new processes are compared with one another and with the existing processes. The process with interstage product removal in the reactor section is highlighted as the most promising. It is based on a "once through" operation and therefore can handle non-stoichiometric syngas feeds. This results in great simplification of the syngas generation scheme and consequently very large savings in investment and operating costs.

INTRODUCTION

Methanol is one of the largest chemicals in tonnes produced per year. It is considered as such to become a motor fuel, whereas it is a base chemical for the production of MTBE, as octane booster in motor fuels. Therefore the energy efficiency of the methanol production is very important and even will grow more so in future. The nowadays commercial processes to make methanol from e.g. methane natural gas have an efficiency of 60-65%, that is of the raw material only 60-65% ends up as chemical CH_3OH, the rest has been used as energy to make the process feasible.

Methanol production nowadays is almost exclusively based on the catalytic conversion of hydrogen and carbon monoxide. To this end modern units use a highly selective copper catalyst, which requires reaction temperatures of at least 210°C for optimal catalyst activities. The methanol yield of the strongly exothermic equilibrium reaction

$$CO + 2H_2 = CH_3OH \quad \text{with} \quad -\Delta H\ (298\ K) = 91\ kJmol^{-1}$$

decreases with increasing temperature. In order to achieve both high reaction rates as well as high conversions per pass, a reactor pressure of at least 4 MPa is required. To use the raw materials economically recycle is necessary. The reactor product is cooled so that methanol condenses out of the un-converted reactant stream. This is then recompressed, heated, and recycled to the reactor feed. In a previous study the possibilities for savings in this classical methanol syntheses were discussed [1, 13]. Really large savings in investment costs, raw materials and energy consumption can only be achieved, if the yield per pass can be improved to the extent that recycle of unconverted synthesis gas is no longer necessary. This requires that we should "avoid" or "break through" the equilibrium.

In this paper two alternative processes will be presented which require no recycle for the production of methanol. They are the gas-solid-solid-trickle flow reactor (GSSTFR) and the reactor system with interstage product removal (RSIPR). For technical and economic comparisons we will refer to the Lurgi low pressure methanol synthesis, which is recognized as an efficient, commercially available process. The ICI low pressure methanol process is also commercially available and has good market acceptance. It uses a cascade of adiabatic packed bed reactors, in which the production per unit of catalyst volume is lower than in the Lurgi process, the inlet temperature to the beds is adjusted by injecting cold feed. Instead of using a "cold shot" converter, Kellogg removes the reaction heat after each bed in coolers, which produces steam improving the energy efficiency.
The most promising new process tried on miniplant scale the socalled Reactor Section with Interstage Product Removal is further elaborated and compared with an entire methanol plant including the syngas generation according to the Kellogg and Lurgi principles. The new process promises large investment savings and considerable cost price reductions for methanol, whereas the efficiency is expected to increase to 80-85%. Large energy savings are expected.

<u>The Lurgi and ICI low pressure processes</u>
Since its introduction in the market in 1971 the Lurgi process [2] has obtained an increasing share of the world market for methanol plants. The converter which incorporates an integrated heat exchanger, that is a cooled tubular reactor, presents an elegant solution to the problem of the removal of reaction heat. The highly active but also very temperature sensitive copper catalyst, is located in a large number of parallel reactor tubes, which are surrounded by a single cooling jacket filled with boiling water. In this way the liberated reaction heat is converted directly into high pressure steam. The reactor tubes have a small diameter, so that the catalyst beds in the tubes are almost isothermal. In this way the formation of dimethylether is suppressed and simultaneously high conversions of 45 to 50% per pass are possible. Figure 1 shows a

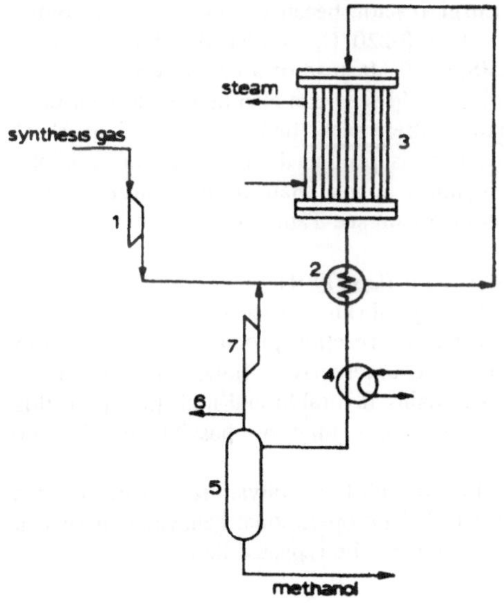

Fig. 1. Simplified flow scheme of a Lurgi methanol synthesis unit. 1. Feed gas compressor, 2. Heat exchanger, 3. Lurgi reactor, 4. Methanol condensors, 5. Gas-liquid separator, 6. Purge and 7. Recycle gas compressor.

simplified flowscheme of the Lurgi process. The fresh synthesis gas is compressed to reactor pressure and mixed with the recycle gas. Before entering the reactor the feed is heated up in a heat exchanger by the hot gases leaving the reactor. The reactor product leaving the heat exchangers enters into a cooler-condensor where condensed methanol is separated and the gases are recycled via the recycle compressor to the reactor.

A part of the recycle gases is purged in order to prevent an accumulation of inert components in the recycle stream. Usual recycle ratios are 4 to 6. In the purge stream part of the synthesis gas is lost, so that the yield on feed stock is less than 100%, usually around 90%. To improve upon this yield the purge ratio has to be diminished, which then requires much higher reactor pressures in order to compensate for the high partial pressures of the inert components.

We will compare two possible new processes with this process, both of which have been tried out on a miniplant scale [3, 4, 5].

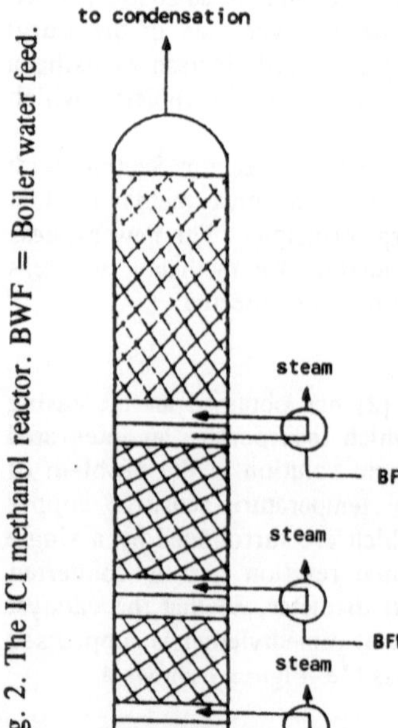

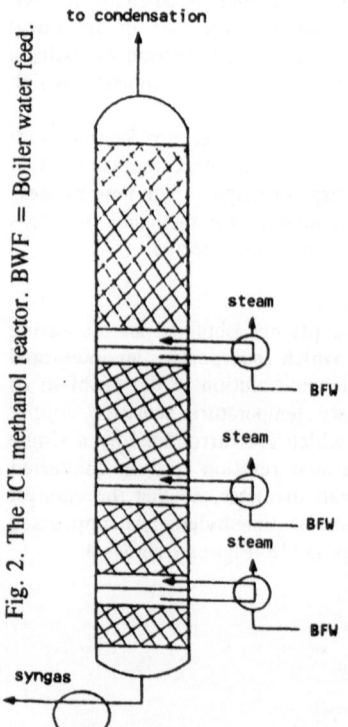

Fig. 2. The ICI methanol reactor. BWF = Boiler water feed.

The ICI reactor, the cold shot converter, consists of a number of adiabatic catalyst beds. The feed temperature and the bed entry temperatures are each individually adjusted carefully in order to obtain the maximum average production rates over each bed. The required temperature adjustments are achieved by injecting cold feed gas into the bed exit gas. The four beds are located within the same shell. The specific amount of catalyst is higher and the heat exchange area of the coolers is also larger than that required for the Lurgi reactor, because heat transfer coefficients in tubes packed with catalyst are higher than in the empty tubes of the coolers. The successive catalyst beds are larger because the increasing CH_3OH content decreases the reaction rates.

Instead of using a cold shot injection and so diluting the feed gas, Kellogg applies external coolers producing steam, see Figure 2, to increase the energy efficiency. The rest of the process flow diagram is the same as for the Lurgi reactor of Figure 1 and not shown. Steam produced in the interstage coolers has a lower pressure than in the Lurgi reactor because they operate at outlet temperatures of 210-220°C, whilst the Lurgi reactor operates at 250-265°C. Instead of a system with interstage coolers, cold feed can also be used to cool the reaction mixture after each bed. In this "cold shot" converter no steam is produced. The investment in the ICI reactor is much lower than in the Lurgi reactor, specially with the cold shot converter. For a further comparison see Table. 1.

<u>A methanol process with a gas-solid-solid-trickle flow reactor (GSSTFR)</u>

The GSSTFR in recent years has been proposed for equilibrium processes and consecutive reactions applications. The principle of this reactor is that the reaction product is removed from the reaction zone using a selective adsorbent [6], in countercurrent flow, so that for an equilibrium reaction the rate in the forward direction remains favorable and high [3, 4]. In this way even complete conversion of reactants in a single pass may be possible despite the thermodynamic limitations.

In Fig. 3 it can be seen how the solid adsorbent trickles through the catalyst bed and in Fig. 4 a GSSTFR is sketched with one single adiabatic catalyst bed. The operational behaviour of such a reactor has been extensively discussed elsewhere [7] and will not be repeated here.

Table 1 REACTORS FOR METHANOL PROCESSES

Name	Methanol synthesis reactor	advantages	disadvantages
Existing reactors			
Lurgi	cooled tubular reactor	good temperature control, no ether, high capacity, lower recycle ratio	expensive reactor, SR required
ICI	multiple adiabatic beds with intermediate cooling with cold feed gas	simple construction, cheap reactor	ether formation lower capacity higher RR's., SR required
Kellogg	multiple adiabatic beds, external coolers producing steam	more expensive than ICI, better energy efficiency	as with ICI reactor
Proposed reactors			
Sherwin and Frank	slurry bubble-column reactor (SBCR)	easy heat removal, good temperature control, no ether, lower RR	very low capacity, inert solvent, SR required
Berty et al.	countercurrent trickle flow-reactor (TFR)	high conversions, even beyond "equilibrium"	SR required, difficult temperature control, difficult heat removal.
Westerterp et al.	Gas-Solid-Solid Trickle-Flow Reactor (GSSTFR)	100% conversion in one pass, no SR required, RR = 0	cumbersome technology, no "solids" pump
Westerterp et al.	Reactor Section with Interstage Product Removal (RSIPR)	100% conversion in one pass, no SR required, proven technology, cheap syngas generation, RR ≈ 0.	high temperature absorbers required

(RR = recycle ratio, DME = dimethylether, SR = stoichiometric ratio)

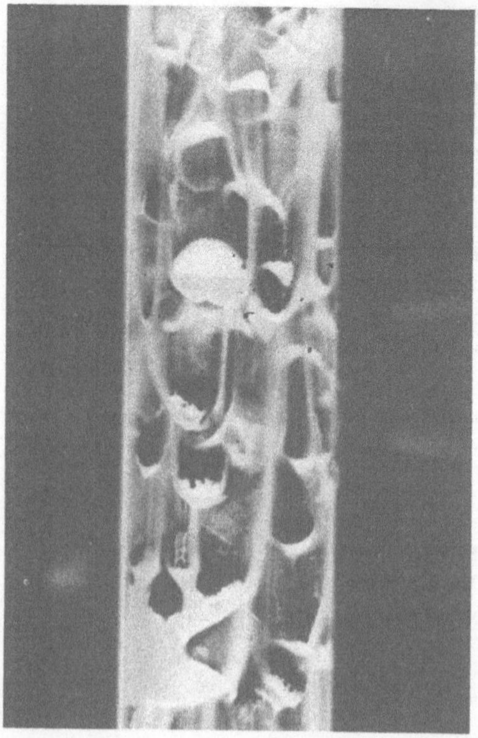

Fig. 3. Trickle flow of a solid through a bed of catalyst particles.

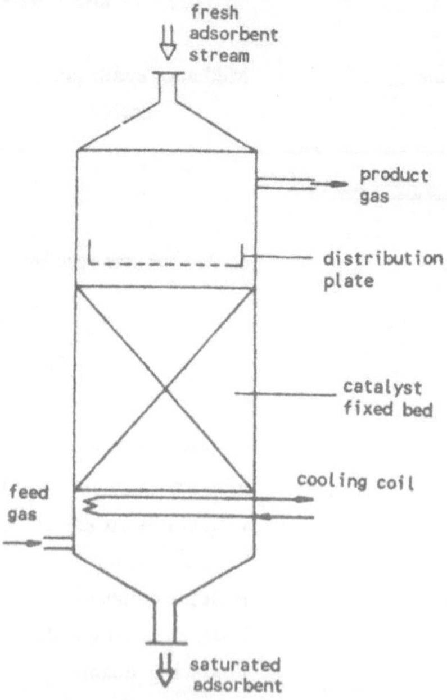

Fig. 4. A catalyst bed for gas-solid-solid trickle flow.

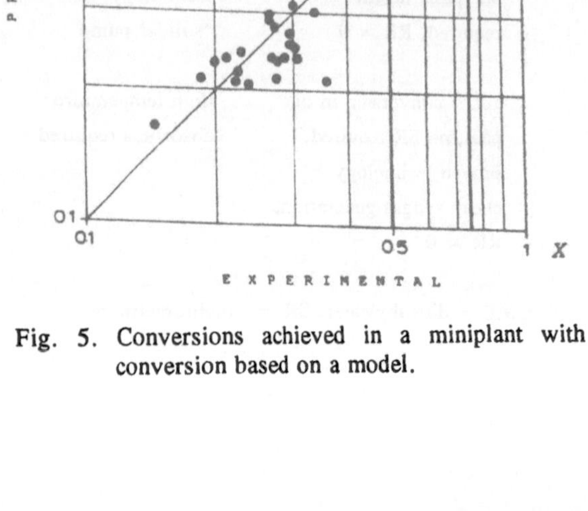

The process has been tested in a miniplant on a production scale of 1 to 1.5 kg CH_3OH/kg catalyst hour. Fig. 5 shows experimental data as predicted by a mathematical model describing the GSSTFR. We can observe that many experimental data points are found in the region beyond the chemical equilibrium.

In total seven runs were done achieving 100% conversion of the key component. Also for a stoichiometric feed the miniplant was run continuously for more than 100 hours with the gas outlet closed and only the stream of adsorbent, saturated with methanol, leaving the GSSTFR. Based on the experience with the miniplant a preliminary design of a possible large scale plant was made.

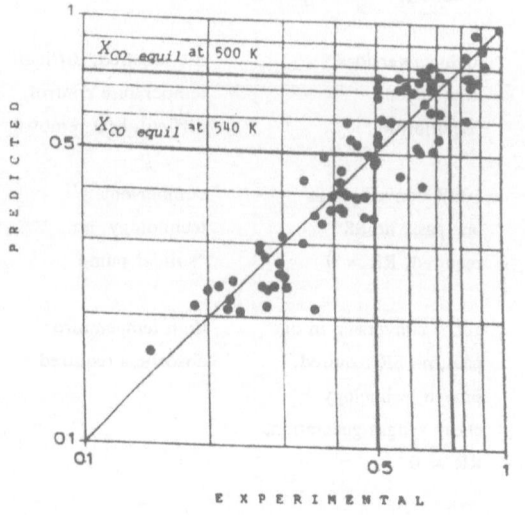

Fig. 5. Conversions achieved in a miniplant with a GSSTFR compared with calculated conversion based on a model.

The proposed reactor section given in Fig. 6 has been amply discussed elsewhere [7]. In Fig. 6 receival and storage vessels for the adsorbent are noticeable. We do not yet have a "solids pump" to bring the solids stream from a low pressure to a high pressure again and therefore vessels in cyclical operation have been used in the design. This method of bringing solids to a high pressure is also used in the Shell and Lurgi coal gasification processes.

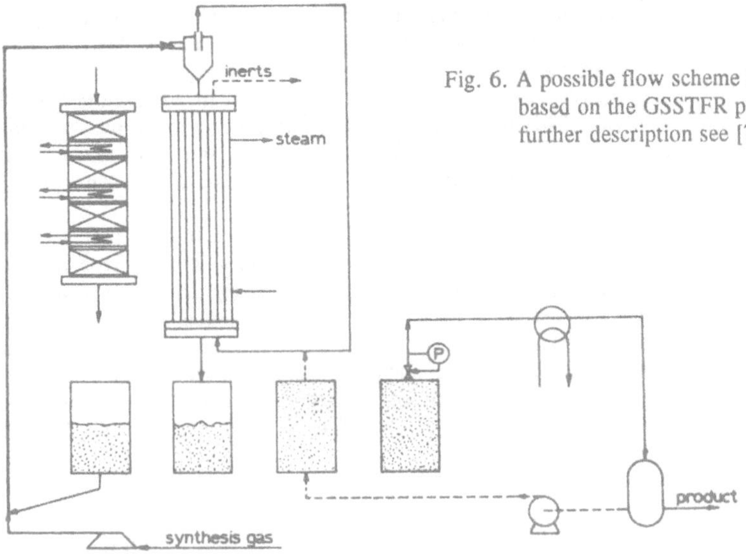

Fig. 6. A possible flow scheme for a plant based on the GSSTFR principle. For a further description see [7a] and [7b].

In the proposed process the yield on raw materials is considerably higher because no gas has to be purged and energy consumption is reduced because no recycle is required. Thus large savings are achieved in raw material and energy consumption, however not so in investments. This is mainly due to the complicated system of solid adsorbent circulation. The large high pressure vessels in cyclical service are very expensive.

As the technology to "pump" solids streams is not yet available, the question arises whether it may be possible to use liquids instead and still obtain full conversion in one pass through the methanol synthesis plant. This led to the concept described in the next section.

A methanol process with a reactor section with interstage product removal (RSIPR)

Instead of adsorbing the methanol on a solid, it is also possible to absorb it selectively into a liquid [8]. This absorption takes place preferably at or near the temperature level of the reaction, in order to obtain high energy efficiency. We could let the liquid trickle countercurrently over a catalyst bed, but this would introduce additional resistances to the transfer of reactant to the catalyst surface and therefore result in larger and more expensive reactors. A realistic alternative, to achieving a high conversion per pass, is to use Lurgi, ICI or Kellogg reactors, with product absorbers after each reactor or catalyst bed.

In the absorbers, methanol is selectively absorbed from the reactor product at reaction temperature. The un-converted reactants pass on to the next reactor. If absorption takes place at close to the reactor temperature, the very expensive heating and cooling of large gas streams can be avoided. To this end we have chosen tetraethyleneglycol-dimethylether (TEGDME) as the solvent. It has a very low vapour pressure, excellent thermal stability and good solubility for methanol at high temperatures.

To test the principle a miniplant was built [5] consisting of a reactor, an absorber and a second reactor. No deactivation of the catalyst in the second bed occurred, and in a two stage flash almost all methanol absorbed could be removed. The miniplant has been in operation satisfactorily for more than a year.
In Fig. 7 a possible design is shown based on the results of experiments on the miniplant, for the synthesis of methanol in a RSIPR. Four sets consisting of a Lurgi reactor and an absorber, are shown. It is a matter of economics how large the number of sets is. The sizes of reactors and absorbers diminish in the direction of the flow, because the gas volumes decrease due to conversion and selective removal of the methanol vapours.
The flow of inert components, or of excess reactant, remains constant in all reactors since they are not absorbed in the TEGDME. This means that the concentration of inerts increases in the direction of the flow and reaction rates decrease somewhat due to dilution of the reactants. In the last absorber the temperature is much lower to prevent any methanol loss. The residual gases can be burnt in the steam reforming furnaces.
In a cascade of four reactor-absorber sets a conversion of over 97% of the key reactant can be easily achieved. The reactor capacity will be considerably higher than in a Lurgi reactor because of the much lower average level of inert concentrations, recycle being absent. In principle all reactors can be run at the same temperature level, so they can be installed in the same pressure vessel with a single boiling water jacket. About 10 tons of TEGDME are circulated per ton of methanol produced. The TEGDME, loaded with methanol, follows a two stage expansion in flash vessels. All methanol and water is flashed off and the solvent is recirculated to the absorbers. Further details are given elsewhere [7]. We further point out that the **RSIPR** is **completely based only on proven technology** !

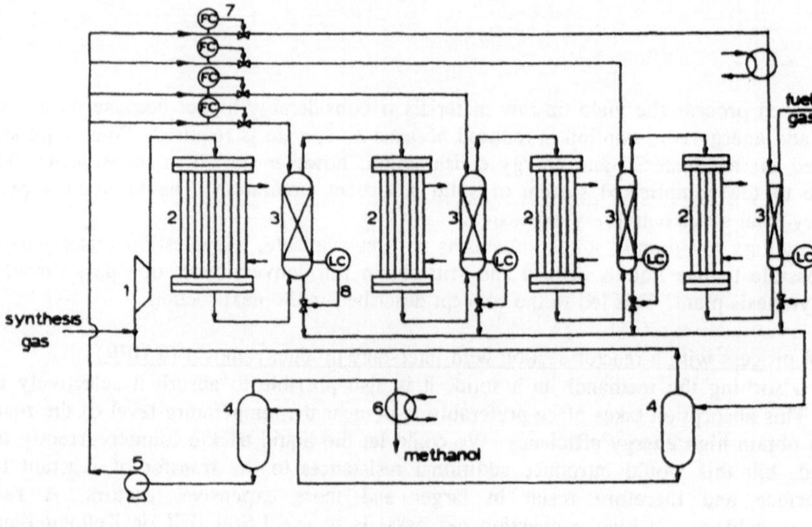

Fig. 7. A possible flow scheme for a plant based on the RSIPR principle. 1. Feed gas compressor, 2. Reactors, 3. Absorbers, 4. Flash vessels, 5. Solvent recirculation pump, 6. Methanol condensor and 7. Solvent distribution over the absorbers.

In Fig. 8 a possible flow diagram of the Kellogg reactor used in a RSIPR is given. As basically all methanol is removed after each bed, the inlet composition to each bed is the same, except for the amount of inert gas. Thus each bed can be operated at the same optimum inlet temperature. This means that the one and same cooler can be used for all streams leaving the

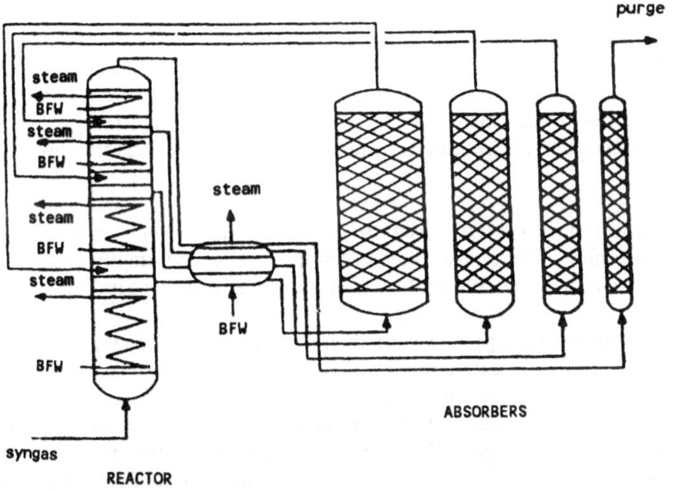

Fig. 8.
A possible reactor-absorber flow scheme for a plant using a Kellogg reactor and the RSIPR principle. Further flow diagram is like in Fig. 7. BFW = Boiler water feed.

catalyst beds. The specific catalyst content of each bed can remain the same, so the actual amounts of catalyst decrease in each successive bed because of the decrease in gas flow rates: the methanol formed has been removed. Further, the flow scheme is the same as sketched in Fig. 7 and therefore not shown.

Other processes proposed

Sherwin and Frank [10] proposed making methanol in a three-phase reaction. Fine catalyst particles are suspended in an inert liquid and syngas bubbled through this slurry in a bubble column. Kodra and Levec [11] have studied this process extensively and compared it to a trickle flow reactor. It is claimed that heat removal in this Bubble Slurry Column Reactor System (BCSRS) is much easier. A pilot plant is operating in the USA.

Berty, Krishnan and Elliot [12a, 12b] proposed using a countercurrent trickle flow reactor system with TEGDME as the third, liquid phase. This is in principle the same as in the RSIPR system, except that absorbtion is not separated from the reaction anymore. They also found conversions, beyond the equilibrium composition at the reactor conditions, because the methanol is countercurrently removed from the reaction zone.

Comparison of processes

Processes based on either the GSSTFR or the RSIPR give a (almost) complete conversion in one pass of the raw materials, circumventing the recycle necessary in existing processes. For a non-stoichiometric feed the key reactant will be completely converted and the excess of the other reactant will pass through the reaction section as an inert. In the case of stoichiometric feed, at complete conversion no gas at all will leave the reactor section.

The GSSTFR is a very elegant process, but, as already mentioned, we regretfully we do not possess a solids pump. The use of vessels in cyclical operation demands heavy investment in large process vessels, so that no investment savings can be expected at the present status of the technology. Therefore we will not discuss the GSSTFR process further.

The BCSRS requires a much larger reactor than the conventional processes. For a catalyst weight fraction of say 20% it can easily be shown that the specific reactor volume will increase about five fold. In view of pore diffusion limitations the catalyst particles must be very small in order to avoid low effectiveness factors. As no higher conversions can be expected than in the Lurgi, ICI or Kellogg reactors, this process will also not be discussed in detail.

The TFRS concept also requires much larger reactor volumes. For catalyst pellets of 4.8 x 4.8 mm Berty and coworkers [12b] reported production rates of 0.09-0.27 kg methanol/kgh at conversions of 40-78% at 100 bars pressure in a small laboratory reactor. This indicates that the catalyst effectiveness factors are in the order of 5-15%. This is due to diffusion limitations in the pores, which are now filled with liquid absorbent. As complete conversion is not reached with this principle, and in view of the much larger reactors required, we will not discuss this process further. See again Table 1.

In comparing alternative processes, it is first required to determine the essential parameters which govern the process economics. These are: the investment, the consumption of raw materials and the specific energy consumption. We will restrict ourselves to comparing the Lurgi and the RSIPR processes:
For investment costs the most important savings in the methanol synthesis are obtained through the absence of a recycle system. There is no condensor train required to condense methanol out of un-converted reactants. The part with the recycle compressor is absent, and reactor volumes, at the same pressure and temperatures, are 40-60% lower. Heat exchange areas for heating and cooling of the streams are much lower or absent. The packed bed absorbers, two flash vessels and the solvent pump will increase the investment costs, but overall investments in the reaction section are significantly lower in a RSIPR compared to the Lurgi process.
With respect to raw material and energy costs the specific raw materials consumption in the methanol synthesis is equal to or better than the conventional processes; this depends on the number of reactor-absorber sets. The conversion of CO_2 is slower than that of CO and H_2, but CO_2 is also coabsorbed in the TEGDME. Not so the H_2, CO and CH_4. It has been estimated that under normal conditions only 1 to 3% of the synthesis gas, except CO_2 coabsorbs in the TEGDME. So by purging part of the small amount of gas flashed off in the first flash vessel, the CO_2 content in the syngas can be controlled, so keeping the CO_2 at the level required for maximum reaction rates. By not having a recycle system, the net steam production in the reactors will be considerably increased. Overall, the raw material and energy costs are much lower in the RSIPR than in the conventional processes.

The entire methanol plant, the influence of the RSIPR principle on the synthesis gas generation

The ***most important improvement*** resulting from the RSIPR system, is that ***it can handle non-stoichiometric syngas feeds***. This ***influences*** not only the reactor section itself, but ***the entire methanol plant***, especially the syngas generation. The fact that the un-converted excess reactant is removed directly with the off-gases, improves raw material and energy consumption of the entire plant. We will elaborate on this for both small and large methanol plants, as well as considering the generation of synthesis gas.
For syngas generation many different raw materials can be used. For simplicity we will consider only natural gas consisting of pure methane. Syngas can be produced by the well-known steam reforming process:

$CH_4 + H_2O \rightarrow CO + 3H_2$

or with a partial oxidation:

$2CH_4 + O_2 \rightarrow 2CO + 4H_2$

For this second process an oxygen plant is required. In the steam reforming process the shift reaction is also important:

$CO + H_2O \rightarrow CO_2 + H_2$

The steam reforming reaction is endothermic and the reaction heat is supplied by burning part of the natural gas feed. In the partial oxidation process steam is also added to the reactor feed to prevent carbon deposits on the catalyst.

Small methanol plants
In small methanol plants no partial oxidation is used and therefore also no air splitter is included. We need a hydrogen consumer, because more hydrogen is produced than is required for the methanol synthesis. Therefore large purge streams are generated. The purge gas is used to heat the steam reformer furnaces. In cases where the RSIPR process is used, a much cheaper syngas generator can be built. That part of the methane that is not converted in the reactor, now no longer accumulates in the recycle loop, but passes "once through" the reactor section and can be used for heating the steam reforming furnaces. A much higher level of CH_4 can now be accepted in the synthesis gas, so that the reformers can be operated either at lower temperatures or at higher pressures. This leads to lower investment, a lower specific energy consumption and lower natural gas consumption compared to the syngas generation and methanol synthesis using the Lurgi and ICI processes.

Large methanol plants
For large methanol plants, see Figure 9, the two synthesis gas processes are used simultaneously, see [9]. The steam reformer supplies excess hydrogen, whereas the partial oxidation supplies insufficient hydrogen for methanol formation. By careful balancing and control of the two units an exactly stoichiometric feed, still containing inerts, can be produced.

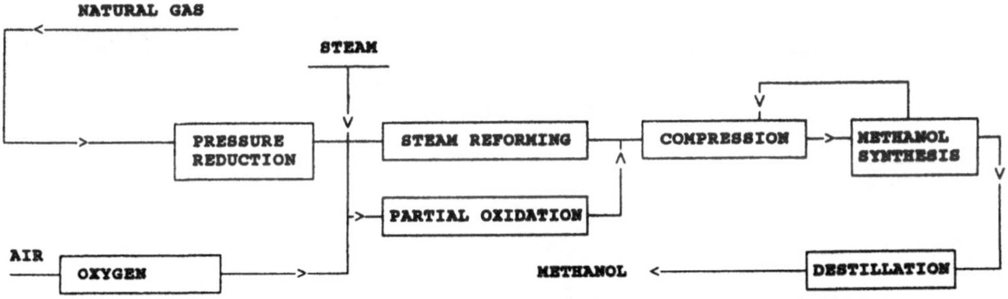

Fig. 9. A block scheme for a large methanol plant, according to a conventional process.

At the expense of higher investment in an air splitter, and a partial oxidation unit the purge flow and hence the consumption of natural gas can be reduced.
Very large savings can now be achieved by using a RSIPR, which also can handle a non-stoichiometric syngas. First, the recycle loop can be deleted, see Fig. 10a. Also now that we can handle non-stoichiometric mixtures, we don't need the two processes to generate the syngas anymore and can choose the cheapest generator. So the very expensive steam reforming unit can be deleted, see Fig. 10b. The partial oxidation is not an equilibrium reaction and therefore can also be operated at much higher pressures, say at the pressure of the methanol synthesis, see Fig. 10c. In this case we do not need to compress the syngas anymore and further we do not need to expand the natural gas, but instead use it at the pressure it comes from the well. In the end all these savings amount to about 50% lower investment costs compared to the conventional processes. Raw material and energy savings also lead to much lower cost prices of say 20-30% lower. More precise figures can only be given, when exact data are available on pressure and composition of the natural gas.

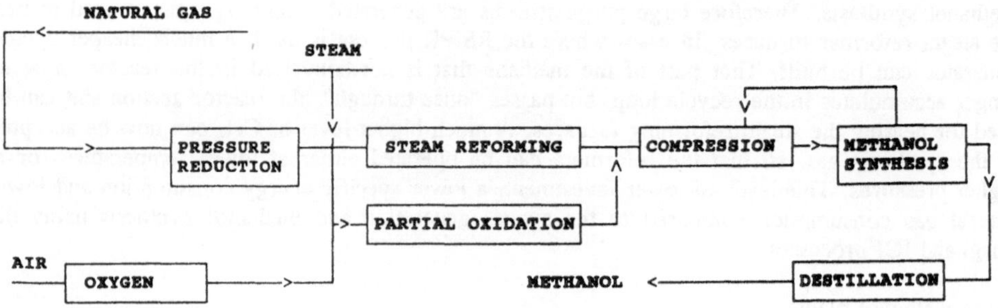

Fig. 10a.

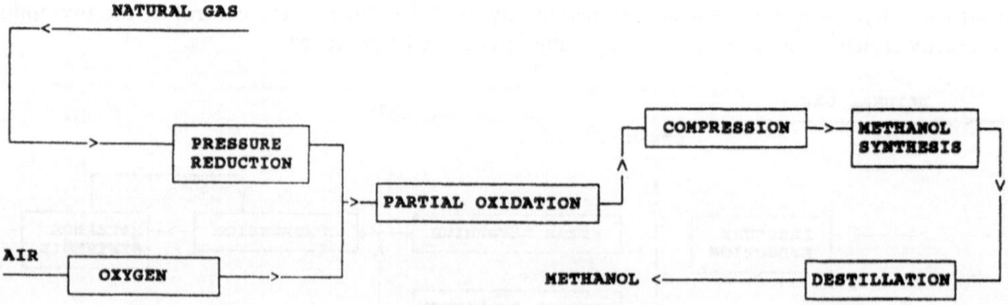

Fig. 10b.

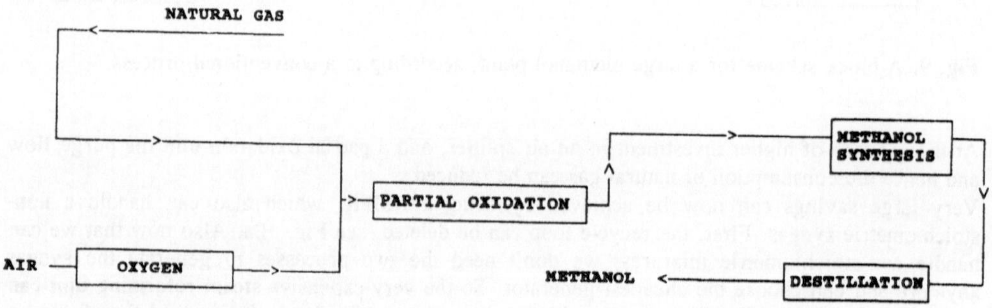

Fig. 10c.

Fig. 10. Investment savings in a large methanol plant by applying the RSIPR principle.

a: no recycle.

b: no steam reformer.

c: raising the pressure of the partial oxidation, no expansion turbine, no syngas compressor.

CONCLUSIONS

Of course, a small methanol plant can also be equiped with a partial oxidation unit, air splitter and RSIPR unit. All units will also have a product distillation unit.
We observe that we have developed two new processes for the production of methanol and have proven them on a miniplant scale. Both have the advantage they can be run with a non-stoichiometric feed to the methanol synthesis unit. The first process based on the Gas Solid Solid Trickle Flow Reactor is very elegant, but its economics are maybe not so attractive and further, part of the technology of repressurizing flows of solids is not yet available.

The second process, based on a Reactor Section with Interstage Product Removal, is perhaps not so elegant, but promises strongly to have very favourable economics. Most of the investment and natural gas savings are obtained in the synthesis gas generation. We therefore feel that a decision is needed to construct a demonstration unit for a methanol process using a reactor section with interstage product removal.

ACKNOWLEDGEMENT

We recognize the many efforts of those young chemical engineers, who, under the supervision of Professor Van den Berg, and on the basis of the ASPEN programs, carried out the process designs and investment calculations for the many alternative methanol processes mentioned in this paper, and also Dr. Vrijland for the economical evaluations. Further the author wishes to acknowledge the fine cooperation of Michal Kuczynski, who in his Ph.D. project, operated the miniplants and contributed strongly to an understanding of the processes in the early part of this project.

REFERENCES

[1] Westerterp, K.R. and Kuczynski, M. (1986) Hydroc. Proc., 11, 80-83.
[2] Supp, E., (1973), Chem. Technol., 3 (7), 430-435.
[3] Westerterp, K.R. and Kuczynski, M. (1987) Chem. Eng. Sci., 42, 1871-1885.
[4] Kuczynski, M., Oyevaar, M.H., Pieters R.T. and Westerterp, K.R. (1987) Chem. Eng. Sci., 42, 1887-1898.
[5] Westerterp, K.R., Kuczynski, M. and Kamphuis, C.H.M. (1989) Ind. and Eng. Chem. Res., 28, 763-771.
[6] Kuczynski, M., van Ooteghem, A. and Westerterp, K.R. (1986) Colloid & Polymer Sci., 264.
[7a] Westerterp, K.R., Bodewes, T.N., Vrijland, M.S.A. and Kuczynski, M. (1988) Hydroc. Proc., 11, 69-73.
and
[7b] Westerterp, K.R., Kuczynski, M., Bodewes, T.N. and M.S.A. Vrijland (1989) Chem.-Ing.-Techn., 61, 193-99.
[8] Kuczynski, M., 't Hart, W. and Westerterp, K.R. (1986) Chem. Eng. Process., 20, 53-58.
[9] Supp, E. (1984) Hydroc. Proc., July, 7, 34C-34J.
[10] Sherwin, M.B. and Frank, M.E. 1976, Hydroc. Proc., 11, 122-124.
[11] Kodra, D. and Levec, J. (1991) Chem. Eng. Sci. 46, 2339-2350.
[12a] Berty, J.M., Krishnan, C. and Elliot, J.R. (1990), October, Chemtech, 624-629.
and
[12b] Same authors:
1991, IEC Research, 30, 1413-1418.
[13] Westerterp, K.R., Proceedings Eurogas '92, Trondheim, Norway, June 1-3, Norwegian Petroleum Society.

REVERSED FLOW METHANOL SYNTHESIS

K.M. Vanden Bussche and G.F. Froment,
Laboratorium voor Petrochemische Techniek, Universiteit Gent,
Krijgslaan 281, B9000 Gent, Belgium.

and

W. Glasz, H. Bosch and L.L. Vandierendonck,
DSM Research, P.O. Box 18, 6160 MD Geleen, The Netherlands.

ABSTRACT

Reversed flow operation of a fixed bed reactor for methanol synthesis was studied in a bench scale reactor. Typical experimental results are presented. Capacitive effects, both on the catalyst surface and in the gas phase, are illustrated.
A reactor simulation model, accounting for transients in the bulk gas phase and in the gas phase inside the catalyst, was developed and combined with the existing TISFLO flowsheeting package, for the simulation of the complete synthesis loop. An internal heat exchanger was added in the centre part of the reactor, to remove the reaction heat. Based on the modeling, an economic evaluation is presented.

TABLE OF SYMBOLS

a_v	:	external particle surface area per unit of reactor volume. $[m^2{}_s/m^3{}_r]$
A_b	:	heat exchange surface at the bed side. $[m^2]$
c_j	:	concentration of component j. $[mole/m^3{}_g]$
c_p	:	specific heat of the gas at constant pressure. [J/mole/K]
c_{ps}	:	specific heat of the solid at constant pressure. [J/kg/K]
c_t	:	total concentration. $[mole/m^3{}_g]$
d_p	:	pellet diameter. $[m_s]$
F_t	:	total molar flux. $[mole/m^2{}_r/s]$
F_c	:	molar flux of carbon. $[mole/m^2{}_r/s]$
F'	:	total volumetric flow rate. $[m^3{}_g/s]$
h_f	:	gas solid heat transfer coefficient. $[W/m^2{}_s/K]$
k_{gj}	:	mass transfer coefficient for component j. $[mole/m^2{}_s/s]$
$\dot{m}$	:	total mass flow rate. [kg/s]

M_m	:	average molecular weight.[g/mole]
N_j	:	molar flux for component j in the pellet.[mole/$m^2{}_s$/s]
p_t	:	total pressure.[Pa]
R	:	universal gas constant.[J/mole/K]
r_j	:	reaction rate of component j.[mole/kg_s/s]
t	:	clock time.[s]
T	:	gas phase temperature.[K]
T_c	:	temperature of the cooling medium.[K]
T_s	:	solid temperature.[K]
U_f	:	overall heat transfer coefficient between gas and cooling medium.[W/m^2/K]
U_s	:	overall heat transfer coefficient between solid and cooling medium.[W/m^2/K]
W	:	catalyst weight.[g]
y_j	:	gas phase mole fraction of component j.[-]
$y_{js}{}^s$	:	mole fraction of j at the solid surface.[-]
z	:	axial reactor coordinate.[m_r]
Z	:	total reactor length.[m_r]
ϵ	:	void fraction of the catalyst bed.[$m^3{}_g$/ $m^3{}_r$]
ϵ_s	:	pellet porosity [$m^3{}_g$/$m^3{}_s$]
μ	:	gas phase viscosity.[Pa s]
ξ	:	coordinate in the pellet.[m_s]
ρ_B	:	bulk density.[kg_s/$m^3{}_r$]
ρ_s	:	solid density.[kg_s/$m^3{}_s$]
λ_{ea}	:	effective axial conductivity of the solid phase.[W/m_r/K]
$\Delta H_{f,j}$	:	enthalpy of formation of component j.[J/mole]
Ω	:	cross section of the reactor.[$m^2{}_r$]

Superscript o indicates inlet conditions, subscript o initial conditions.

INTRODUCTION

Traditionally, methanol is produced in multitubular or quenched multibed reactors from a rich synthesis gas, fed at 220 to 250 °C. The conversion to methanol in these types of reactor is equilibrium limited. In the C.I.S., reversed flow technology is applied for SO_3 production from metallurgic off gas containing SO_2 and for the total oxidation of hydrocarbons in effluent gases. Thereby the flow direction is periodically reversed. This type of operation is explored here for methanol synthesis. Preheating the feed may prove superfluous and even lean syngas can be processed autothermally. The temperature profile in the reactor may favour the carbon conversion to methanol, reducing the carbon loss in the purge. A simulation model accounting for all experimentally observed gas phase transients is used. It is linked to a flowsheeting package for an integrated simulation of the complete synthesis loop. This work, performed at the 'Laboratorium voor Petrochemische Techniek' of the Ghent University (Belgium) and at DSM in Geleen (The Netherlands), was funded in part by the Commission of the European Communities in the framework of the Joule programme, sub-programme Rational Use of Energy.

REVERSED FLOW OPERATION OF A FIXED BED CATALYTIC REACTOR.

Consider a catalyst bed with an initial uniform temperature of 300 °C. Feeding a mixture of CO, CO_2 and H_2 to the reactor at a low temperature, e.g. 100 °C, causes progressive cooling of the solid. This is accompanied by the exothermic synthesis reactions on the catalyst. The cocurrent heat and concentration waves, thus generated, move along the reactor as shown in Figures 1 and 2. The catalyst evolves from an ignited to a non ignited steady state as the waves travel beyond the corresponding reactor section.

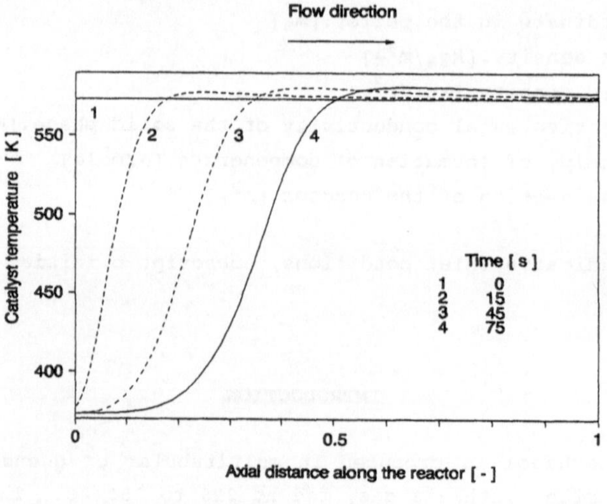

Figure 1. Temperature profiles at the startup of the reversed flow reactor.

This will obviously result in the extinction of the reactor if the waves are allowed to reach the exit. This can be prevented if the flow is reversed. In this way, part of the heat is trapped inside the reactor and starts to move in the opposite direction.

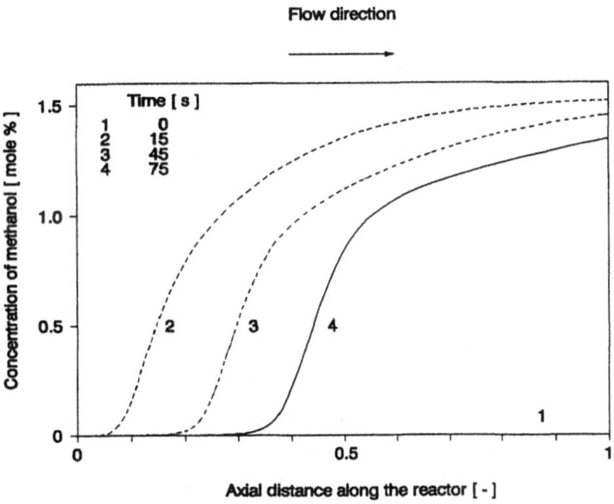

Figure 2. Methanol concentration profiles at the startup of the reversed flow reactor.

From a certain number of reversals onwards, a stationary cyclic regime is achieved, whereby temperature and concentration profiles at a given moment in the semi-cycle are exactly reproducible from one semi-cycle to another. Figure 3 shows the temperature and corresponding methanol concentration profile at a given moment in time for such a 'fully converged' semi-cycle.

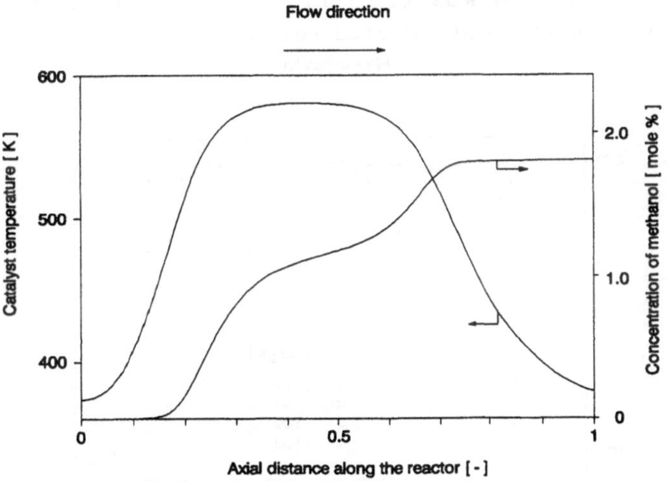

Figure 3. Temperature and concentration profile for a fully converged semi-cycle.

The shape of the heat wave moving through the reactor is beneficial for the conversion to methanol. As the catalyst reaches a temperature of 180 °C, the reactions take off and the conversion quickly reaches the equilibrium value corresponding to the maximum catalyst temperature. Then, as the temperature decreases again, the equilibrium value is shifted towards higher conversions and more methanol is formed. Whether or not the equilibrium conversion is reached when the temperature drops below 180 °C again, depends on the slope of the upstream end of the wave. The reversed flow reactor may therefore lead to higher conversions than the conventional MBAR, with its continuously rising temperature profile.

The temperature rise is significantly larger than the adiabatic temperature rise. Indeed, the gas is preheated by the first encountered layers of catalyst and then reacts on layers further downstream, thereby heating the solid. The catalyst bed itself acts as a regenerative heat exchanger and allows for autothermic operation, even for lean feeds.

Possible drawbacks of the reversed flow operation are the variation of exit concentration and temperature with time and the occurrence of pressure waves, caused by the flow reversal itself. Both are easily circumvented by installing an extra vessel acting as a buffer.[1-3]

BENCH SCALE EXPERIMENTS FOR REVERSED FLOW METHANOL SYNTHESIS.

The equipment has been described elsewhere [4]. Figure 4 shows the experimental temperature profiles at various moments during one semi-cycle, for the conditions summarised in Table 1. The feed composition is typical for a conventional multibed reactor.

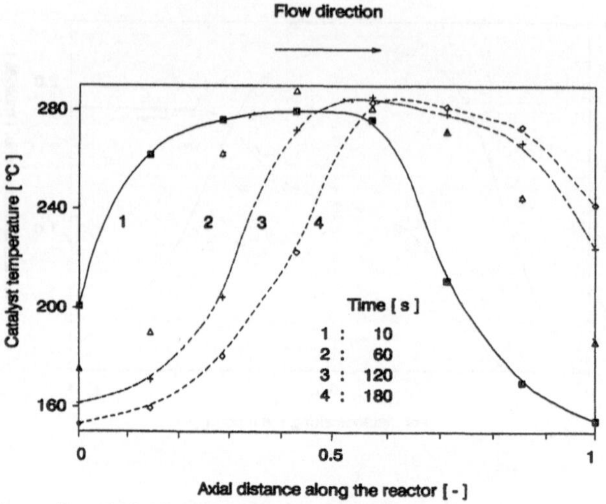

Figure 4. Experimental temperature profiles in the bench scale reactor for the conditions shown in Table 1.

TABLE 1.
Experimental conditions for the reversed flow experiments.

Operating conditions		
T°	K	426
Pt°	bar	50
F'°	l(NTP)/min	5.06
W	g	30
cycle time	s	360
Feed composition		
CO	mole %	4.1
H_2	''	81.8
CO_2	''	3.0
inert	''	11.1

Figure 5 shows the corresponding exit concentration profiles. During the first 30 seconds of the semi-cycle, the profiles are disturbed by the flow reversal. Upon reversal, process gas, that had not yet reached the hot section of the catalyst bed, is swept back out of the low

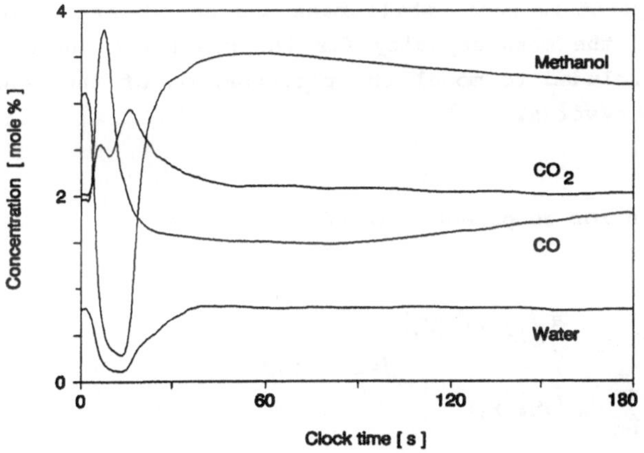

Figure 5. Evolution of the exit concentration of water, methanol, CO and CO_2 for the experimental conditions shown in Table 1.

temperature zone and creates a dip in the concentration profiles of water and methanol, and a maximum for CO, CO_2 and H_2 (not shown). Water and methanol adsorb in the cold section of the reactor and desorb as the temperature wave passes by, resulting in a further broadening of the dips of the reaction products. The second peak occurring in the carbon dioxide concentration profile is ascribed to formate desorption resulting from the temperature rise in the bed [4]. After this initial perturbation, the temperature and concentration waves move concurrently through the bed without distortion, yielding a nearly constant exit concentration of the reaction products. Towards the end of the semi-cycle the conversion gradually decreases because the amount of catalyst which is at reaction temperature decreases and because of the rising temperature at the reactor exit.

The stability of the $Cu/ZnO/Al_2O_3$catalyst was tested under reversed flow conditions. The periodical temperature fluctuations didn't have any negative effect on the physical strength, the activity or the selectivity of the catalyst.

REACTOR SIMULATIONS.

Based upon the experimental observations, a simulation model was developed, consisting of equations (1) to (6), to be integrated under the conditions (7) and (8). The model is of the heterogeneous one dimensional type, accounting for intraparticle and interfacial gradients [5-6]. The catalyst pellets were assumed to be spherical and isothermal.
The transient term in the solid energy equation (6) enables the modeling of the moving temperature wave, while mass and energy capacities for the bulk gas phase and the mass capacity for the gas phase inside the catalyst pores were included to model the perturbation of the concentration profiles upon flow reversal.

Continuity equation for component j in the gas phase

$$\varepsilon C_t \frac{\partial y_j}{\partial t} = -F_t \frac{\partial y_j}{\partial z} + a_v k_{gj} (y_{js}^s - y_j) - a_v y_j \sum_{i=1}^{N} k_{gi} (y_{is}^s - y_i) \qquad j=1,\ldots,N-1 \tag{1}$$

Change in the total molar flow rate

$$\varepsilon \frac{\partial c_t}{\partial t} = -\frac{\partial F_t}{\partial z} + a_v \sum_{i=1}^{N} k_{gi}(y_{is}^{s}-y_i) \tag{2}$$

Gas phase energy equation

$$c_p \varepsilon c_t \frac{\partial T}{\partial t} = -F_t c_p \frac{\partial T}{\partial z} + a_v h_f\ (T_s-T) + \frac{A_b}{\Omega Z} U_f\ (T_c-T) \tag{3}$$

Pressure drop equation

$$-\frac{dp_t}{dz} = \frac{F_t T}{p_t} \frac{R(1-\varepsilon)}{d_p \varepsilon^3} \left(\frac{150(1-\varepsilon)\mu}{d_p} + 1.75 M_m F_t\right) \tag{4}$$

Continuity equation for component j in the solid phase

$$\varepsilon_s \frac{\partial c_j}{\partial t} = \frac{1}{\xi^2} \frac{\partial(\xi^2 N_j)}{\partial \xi} - r_j \rho_s \tag{5}$$

Energy equation for the solid phase

$$\begin{aligned} \rho_B c_{ps} \frac{\partial T_s}{\partial t} &= \lambda_{ea} \frac{\partial^2 T_s}{\partial z^2} + h_f a_v\ (T-T_s) \\ &+ \frac{A_b}{\Omega Z} U_s\ (T_c-T_s) + \rho_B \sum_{i=1}^{N} r_i(-\Delta H_{f,i}) \end{aligned} \tag{6}$$

Boundary Conditions

$$\begin{aligned} y_j(0,t) &= y_j^o \\ p_t(0,t) &= p_t^o \\ F_t(0,t) &= F_t^o \\ T(0,t) &= T^o \\ T_s(0,t) &= T_s^o \\ \frac{\partial T_s}{\partial z}(z,t) &= 0 \qquad , z=0,\ Z \\ N_j(\xi=0,z,t) &= 0 \\ N_j(\xi=\frac{d_p}{2},z,t) &= k_{gj}(y_{js}^{s}-y_j) \end{aligned} \tag{7}$$

Initial Conditions

$$\begin{aligned} y_j(z,0) &= y_{jo} \\ T(z,0) &= T_o(z) \\ T_s(z,0) &= T_{so}(z) \end{aligned} \tag{8}$$

This set of equations was combined with the kinetics of Graaf et al. [7]. Non ideality of the gas phase was accounted for using the Soave Redlich Kwong equation of state [8]. The adequacy of the kinetics was tested by simulation of a classical MBAR, consisting of 4 quenched beds, using the industrial feed conditions shown in Table 2.

TABLE 2.
Process Conditions used in the simulations of the MBAR and the reversed flow reactor.

Catalyst		MBAR	RFR I	RFR II
Density	$kg/m^3{}_s$	1775	1775	1775
Porosity	$m^3{}_g/m^3{}_s$	0.5	0.5	0.5
Amount	tons	122	122	183
Pellet diameter	m_s	.0054	.0054	.0054
Reactor				
Diameter	m_r	4	4	4
Length	m_r	8.3	8.3	12.43
Operating Conditions				
T°	K	513.2	398.2	398.2
Pt°	bar	98	98	98
$\dot{m}$	kg/s	59.9	59.9	59.9
cycle time	s	–	60	60
Feed Composition				
CO	mole %	4.10	4.10	4.10
H_2O	''	0.10	0.10	0.10
MeOH	''	0.60	0.60	0.60
H_2	''	81.10	81.10	81.10
CO_2	''	3.00	3.00	3.00
inert	''	11.10	11.10	11.10

Half of the feed entered the reactor as cold shot. The calculated exit concentration of methanol was 5.0 mole %, providing excellent agreement with the industrially observed value of 5.1 mole %. The former value was subsequently used as a reference in the calculation of the performance of the reversed flow reactor.

For a first reversed flow reactor simulation, the same conditions and dimensions were used as for the MBAR, except for the lower inlet temperature of 125 °C, leaving a margin of 5 °C with respect to the condensation temperature of water and methanol at their expected concentrations in the high pressure effluent.

The calculations yielded a maximum bed temperature of 320 °C, well above the thermal stability limit of the catalyst. Furthermore, the high temperature had a negative influence on the equilibrium of the methanol synthesis reactions, yielding a time averaged concentration of methanol of only 3.1 mole %, compared to the 5.0 mole % obtained in the MBAR. An internal heat exchanger was therefore required in the reversed flow reactor. A boiler, producing steam at 40 bars, was chosen. The simulation model of the reactor containing the internal heat exchanger was then linked to the TISFLO flowsheeting package. In this way, an adequate simulation of the complete synthesis loop, presented in Figure 6 and containing the reversed flow reactor, was possible.

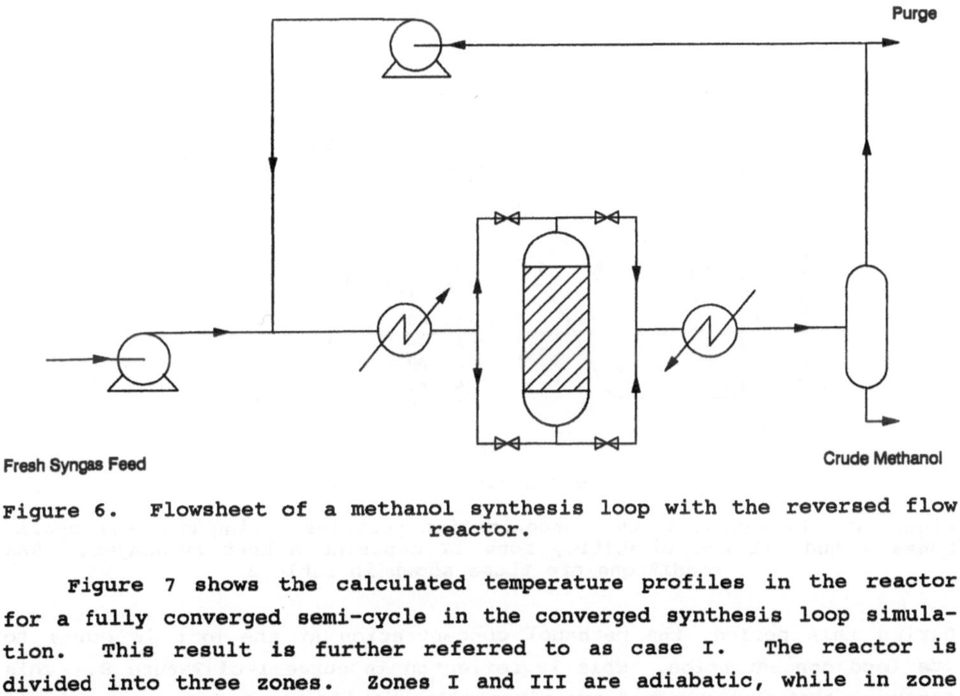

Figure 6. Flowsheet of a methanol synthesis loop with the reversed flow reactor.

Figure 7 shows the calculated temperature profiles in the reactor for a fully converged semi-cycle in the converged synthesis loop simulation. This result is further referred to as case I. The reactor is divided into three zones. Zones I and III are adiabatic, while in zone II (the middle 60 % of the bed), the cooling medium at 525 K removes heat from the catalyst bed. There is no temperature plateau anymore and this is beneficial for the conversion to methanol, as can be seen from Figure 8, showing a practically monotonous rise in the corresponding concentration profiles. For a while after flow reversal, a certain volume of unreacted gas is swept out of the reactor.

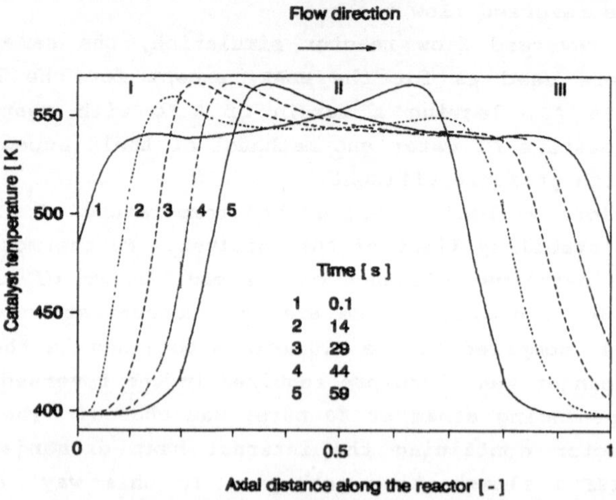

Figure 7. Evolution of the temperature profiles during one semi-cycle. Zones I and III are adiabatic, zone II contains a heat exchanger. The conditions are those shown in Table 2.

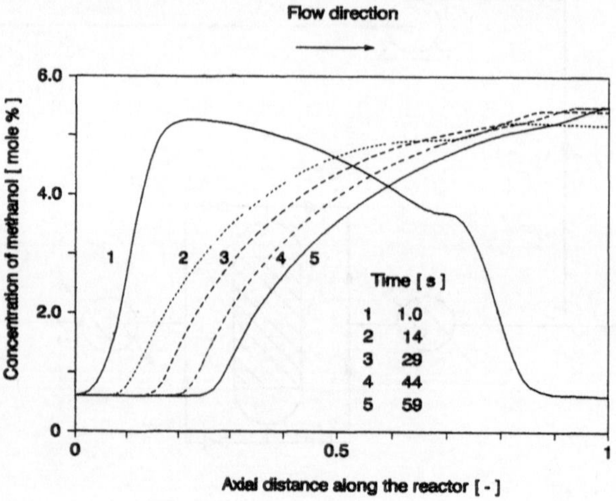

Figure 8. Evolution of the concentration profiles during one semi-cycle. Zones I and III are adiabatic, zone II contains a heat exchanger. The conditions are those shown in Table 2.

During this period, the methanol concentration at the exit is equal to its feed concentration. This is reflected in curve 1 of Figure 8. This feature, that was observed experimentally, is illustrated more clearly in Figure 9, showing the evolution in a semi-cycle of the exit concentration of methanol. The time averaged exit concentration of methanol is 5.04 mole %, slightly higher than the 5.0 mole % obtained in the MBAR.

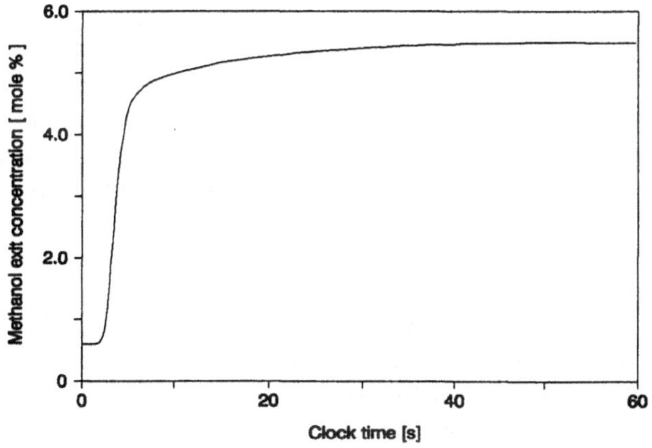

Figure 9. Simulated evolution of the exit concentration of methanol during one semi-cycle.

For a second case, the amount of catalyst used was increased, since only part of the catalyst is at high temperature at each moment in time under reversed flow operation (see e.g. Figure 4). The temperature of the cooling medium was decreased to 516 K, so that steam at 35 bars was produced. In order to avoid hot spots in the bed in the first seconds of the semi-cycle, the heat exchanger now extended over the middle 80 % of the bed, keeping the total heat exchanging area constant at 5500 m^2. This case is further referred to as case II. The combination of more efficient cooling and a larger amount of catalyst led to a time averaged methanol exit concentration of 5.3 mole %.

As summarised in Table 3, both reversed flow cases perform slightly better than the MBAR in terms of conversion. Furthermore, over 20 MeW of power is produced in the form of low pressure steam, in both cases.

TABLE 3.
Survey of the simulation results for the MBAR and the RFR cases. (1) Axial distance, in percentage of the bed, over which the heat exchanger extends. The heat exchanging area is taken to be 5500 m^2 in both cases. (2) Time averaged concentration at the exit.

		MBAR	RFR 1	RFR2
Psteam	bar	-	40	35
% cooled(1)	-	-	60	80
Q	MeW	-	22	21
y_{MeOH}(2)	mole%	5.0	5.04	5.26

ECONOMIC EVALUATION.

In Table 4 the estimated investment and variable costs are summarised for the MBAR and the two RFR cases, for a 350000 ton/year methanol plant. Only the costs that differ from one case to another have been entered explicitly. The item, marked as 'General Costs' accounts for all other costs common for the three cases. The main additional investment involved in the reversed flow reactor is the internal heat exchanger. Besides this, the feed/effluent heat exchanger has to be somewhat larger than in the MBAR case, because of the lower driving force. Due to the higher conversion, a number of secondary effects slightly decrease the investment cost of the synthesis loop.

The variable costs are entered in the Table, taking the MBAR as a reference. Since the smaller pressure drop inside the RFR has not been accounted for in the economic evaluation, the decrease in the compression energy is only due to the higher conversion to methanol.

TABLE 4.
Estimated investment and operating costs for the MBAR and the two RFR reactors. (H.E. = heat exchanger)

Investment cost [thousand US $]	MBAR	RFR 1	RFR 2
Reactor Vessel	2050	2050	2600
Internal H.E.	0	1450	1450
Product/feed H.E.	1450	2050	2050
Synthesis loop	58000	57600	55000
Valves	0	230	230
Subtotal	61500	63380	61330
General costs	54500	54500	54500
Total investment	116000	117880	115830
Increase in variable costs [thousand US $ / year]			
Steam raised	0	-3170	-2850
Compression Energy	0	-7	-43
Increased MeOH production	0	-19	-122
Total increase in variable costs	0	-3196	-3015
R.O.I. [%]	25	27.3	27.6

Assuming a total investment of 116 million US $ for a MBAR plant, with an R.O.I. of 25 %, the RFR cases yield slightly better values of 27.3 and 27.6 %, due to the much lower operating cost of the RFR.

The difference between the MBAR and the RFR case I mainly lies in the investment for the internal heat exchanger and in the value of the produced steam. It is therefore essential that the steam is used efficiently somewhere in the plant. However, it is beyond the scope of the present paper to determine the optimal integration of the newly acquired energy.

The difference between RFR case I and II lies mainly in the lower investment for the loop itself. Because this reduction was calculated using a fairly general approach, a more detailed evaluation of the complete loop is needed for an optimisation of the reactor size.

CONCLUSIONS.

A bench scale reactor proved the feasibility of reversed flow operation of a fixed bed reactor for methanol synthesis. Based on the experimental observations, a computer simulation model was developed, accounting for the transient effects occurring in both the gas and the solid phase. The reactor model was subsequently coupled to the TISFLO flowsheeting package and used for the simulation of an industrial methanol synthesis loop. Provided that an internal heat exchanger removes part of the reaction heat, thereby producing medium pressure steam, the reversed flow reactor yields higher conversions than the conventional MBAR technology. Though further optimisation is still possible, the economic evaluation already presents favourable perspectives for the process.

ACKNOWLEDGEMENT

This work, performed at the 'Laboratorium voor Petrochemische Techniek' of the Ghent University (Belgium) and at DSM in Geleen (The Netherlands), was funded in part by the Commission of the European Communities in the framework of the Joule programme, sub-programme Rational Use of Energy.

REFERENCES

1. Froment G.F., Reversed flow operations of fixed bed catalytic reactors. In '<u>Unsteady state processes in catalysis.</u>', ed. Matros Y., VPS B.V., Utrecht, 1990.

2. Matros Yu., Unsteady processes in catalytic reactors. Studies in Surface Science and Catalysis 22, Elsevier, Amsterdam, 1985.

3. Matros Yu., Catalytic processes under unsteady state conditions. Studies in Surface Science and Catalysis 43, Elsevier, Amsterdam, 1989.

4. Neophytides S. and Froment G. A bench scale study of reversed flow methanol synthesis. Ind. Eng. Chem. Res., 31, 1583-1589, 1992.

5. Neophytides S.N., Vanden Bussche K.M., Zolotarskii I.A. and Froment G.F., to be published

6. Froment G. and Bischoff K., Chemical reactor analysis and design. Second edition, John Wiley & Sons, N.Y., 1990.

7. Graaf G.H., Stamhuis E.J. and Beenhakkers A.A.C.M., Kinetics of low pressure methanol synthesis. Chem. Eng. Sci., 43 (12), 3185-3195, 1988.

8. Soave G., Equilibrium constants from a modified Redlich-Kwong equation of state. Chem. Eng. Sci., 27, 1197-1203, 1972.

ACHIEVEMENTS AND TRENDS IN REFORMING SYSTEM DESIGN

PAN ORPHANIDES
Consultant
13 Spetson street, 15342 Ag. Paraskevi Greece

ABSTRACT

Natural Gas and other light hydrocarbons, reformed to Synthesis Gas are the main feedstocks for the production of Ammonia, Methanol, Hydrogen and for many other catalytic synthesis in the chemical industry, requiring syngas of various H2/CO ratios. Conventional and state of the art Steam Reforming Processes (chemistry, reaction kinetics, flow diagrams, industrial applications, efficiencies, disadvantages and limitations), are described. New developments and trends to reduce investment cost, to improve efficiencies, to increase operational safety and to reduce environmental impact are discussed in detail.

INTRODUCTION

Hydrogen and CO are two of the most important building blocks of the chemical industry. Hydrogen is mainly used in ammonia and methanol synthesis and petroleum refinery. Mixtures of Hydrogen and CO are used in the OXO synthesis for the production of higher alcohols and of synthetic fuels. CO is a major component in the production of paints, plastics, foams,pesticides and insecticides. The production of hydrogen and CO , called usally Synthesis Gas or **Syngas**, is carried out mainly by the following processes, when natural gas or other light hydrocarbons are used as feedstock:

* Steam reforming (primary or primary/secondary)
* Autothermal oxygen-enhansed reforming, Partial oxydation

Steam reforming is the process most widely applied for the generation of syngas and hydrogen. Utilizing conventional supported nickel catalyst, the highly endothermic reaction between natural gas and steam is usually carried out in a direct fired reformer. Secondary or autothermal reforming is a type of steam reforming that utilize the heat of partial combustion, by air or oxygen of feedstock to supply the heat required to sustain the endothermic steam reforming reaction on an adiabatic catalyst bed. The feedstock to the secondary or autothermal reformer is either natural gas or partly steam reformed natural gas or a mixture of both. Partial oxidation (POX) do not utilize catalyst and depend on partial combustion, usually by oxygen of the feedstock to internally supply the heat of reaction.

Although heavy oils and coal are less expensive feedstocks, the capital cost for heavy fuel oil POX units, or for coal gasification can be two to three times more expensive than for a natural gas reforming plant. Coal is used in countries without gas utilization or for political reasons. From the other side it shall be said that a dozen of advanced coal gasification processes are under development around the world and that this development stands in sharp contrast to practically no further development in the field of partial oxydation. Reforming of natural gas is expected to remain the most cost effective way to generate syngas for at least the next decade and probably beyond year 2000.

Beyond this time frame, new advanced technology reforming processes will have an increasing share in the syngas production. Technologies using regenerative or nuclear energy sources will become also available mainly for hydrogen production. These technologies will all involve water splitting by one or the other energy source. Nuclear energy may be used for water electrolysis as sto rage mean of electricity produced during off-peak hours. Thermochemical and hybrid processes will be technologically proven and solar-, geothermal-, biomass-, wind-, sea water thermal gradient- hydrogen production will still be in development, whereas fusion energy the last one to become available, could be used for direct thermal decomposition of steam.[1]

In this report we will focus on the key factors (efficiency, operational safety, cost and environmental impact) affecting the today state of the art and the new developments in the natural gas reforming systems design only. within the time frame considered, this feedstock and this technology, will continue to supply more than 80% of syngas generated world wide.[2]

In the Diagram 1 below the economically feasible routes to transform natural gas to chemical products and to synthetic fuels, as well as to reducing gas are shown

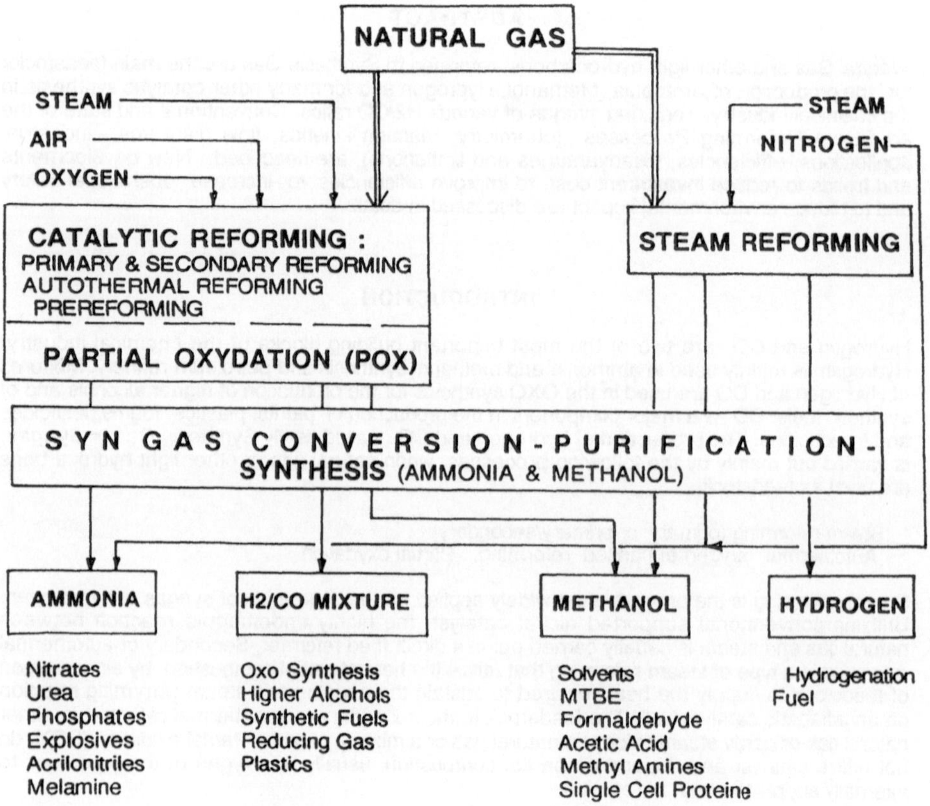

Fig. 1: Block Diagram of Natural Gas Reforming (Conventional and State of the Art Technologies, Products and Derivates)

SYNGAS GENERATION BY CONVENTIONAL DESIGN REFORMERS

Steam and oxygen reforming technologies for natural gas and light hydrocarbons are reviewed along with their respective advantages and disadvantages

Conventional and state of the art steam reforming [Primary Reforming]

In a conventional steam reforming process hydrogen and CO/CO2 are produced by reacting methane with steam over a nickel catalyst at high temperatures. The catalyst is contained in tubes located in a box type fired furnace that provides the large endothermic heat of reaction. The basic reactions of the steam reforming of methane are expressed by the following equations:

$$CH_4 + H_2O = CO + 3H_2 \quad (1)$$

$$CO + H_2O = CO_2 + H_2 \quad (2)$$

$$2CO = C + CO_2 \quad (3)$$

$$CH_4 = C + 2H_2 \quad (4)$$

$$CO + H_2 = C + H_2O \quad (5)$$

$$CH_4 + CO_2 = 2CO + 2H_2 \quad (6)$$

Reaction (1) is known as **steam - methane reforming,** while reaction (2) is refered to as the **water gas shift reaction.** Both reactions are reversible and approach equilibrium. In addition to the desired reactions (1) and (2), other side reactions (3, 4, and 5) are under certain conditions also possible. Selection of suitable catalyst and appropriate operating conditions will promote reactions (1) and (2) and suppress reactions 3 to 5. Reaction (6) is taking place in steam reformers where CO2, available downstream the CO2 removal section is recycled back to the reformer in order to increase CO formation in the syngas{ methanol , oxo syngas production}. The conversion is favoured by high steam to carbon ratio (in excess of the stochiometric quantity), high outlet temperature, low pressure and high catalyst activity. There are restrictions and limitations for a maximum conversion:

Too high steam to carbon ratio will make the process inefficient energy wise , large quantities of excess steam have to be condensed and high steam flow will increase the equipment cost. Low steam to carbon ratio will improve the energy efficiency, but can form carbon in the pores of the catalyst and other undesirable byproducts.[3]

Too high operating temperature pose problems of heat transfer and lowers the mechanical strength of the tube material (temperatures up to 960°C under pressure of about 35 bar).. Low operating temperature reduce the conversion, but some times this offers an attractive possibility for conversion completion under optimum conditions downstream the primary reformer.

Too low operating pressure increase the cost of equipment and makes the reforming process inefficient, as there is always necessity to compress the produced syngas at higher pressures for further processing (purification and synthesis). There are mechanical limits for higher reforming pressures, but in general higher reforming pressure reduce the overall operating and investment cost.

Catalyst with too high activity could at first stage improve the conversion rate but because of its sensitivity, it may lose easily activity due to impurities in the feedstock, or because of maloperation. Small size catalyst in a given total reformer tube volume will present a higher catalyst mass and thus a higher conversion rate and a better approach to equilibrium, but also gives a higher pressure drop. Loss of activity may lead to tube overheating, as reduced conversion means less heat absorption for the endothermic reforming reaction.[4]

Several configurations of reforming furnaces are in use today, characterized mainly by the disposition and position of the burners. As a fired reformer is heat transfer limited, it is a main concern to design the reforming furnace in such a way, that the burner flamme do not reach the hot tube wall and that the highest heat flux can be given only to zones of low process temperature and high feedstock partial pressure. The most common types of fired reformer configurations are : the **Top Fired** and the **Side Fired** Reformer.

Top fired reformer use multiple rows of tubes with burners located in the arch on each side of the tubes. The heat to the tubes is supplied by the radiating products of combustion. Main advantage of this configuration is the few burners relative to the tubes, the higher radiant efficiency, the presence of the high heat flux zone in the "cold" inlet of the feedstock and the very large tube number which can be accommodated in one radiant box. Main disadvantages is the limitation in the heat input control and the hot operating level at the top.

The Side fired reformer has multiple radiant wall burners along both side walls and one row of tubes in the middle of the box. The heat to the tubes is emitted from the radiant walls. The main advantage is the uniform heat distribution and the very good heat input control. Disadvantages are the required large number of burners, the lower radiant efficiency and the size limitation of the fired box (single box for 100 to 150 tubes).
For critical reformer applications, i.e. CO2 recycle a side fired reformer is prefered, due to the higher risk of carbon formation on the catalyst.

Other type of fired reformer are the **Bottom fired** and the **Terraced wall** Reformer.

The heat flux and the tube skin temperature at the upper part of the reformer are high in the case of the top fired reformer. In the side fired and the terraced reformer the heat flux along the reformer tube and the conversion are more uniform.

<u>Reformer efficiency and Environmental impact</u>

In a conventional reforming process only 40% of furnace duty is absorbed by the endothermic heat of reaction. About 35% are recovered in the form of waste heat export steam , by utilizing part of the latent heat of the syngas for CO2 removal and by preheating feedstock-combustion air or BFW in the convection section of the reformer, where flue gas is cooled from 900 to 950 ^{o}C to about 125 to 150 ^{o}C. 20 to 25% of the heat applied are lost in the stack, in cooling water and in heat losses.

Depending on the burner configuration (low NOx burners) and the combustion air preheatingtemperature the NOx content in the flue gas can be kept below 200 mg/Nm3, but many high efficiency reforming plants with combustion air preheating temperatures above 450 ^{o}C have higher NOx emmission levels, despite the use of low NOx burners. High NOx values occur in some reforming plants for ammonia production, when non scrubbed purge gas from the synthesis section containing NH3 are used as fuel in the reformer. Depending on the content of sulfur in the fuel gas , there is a small SO2 emmission from the stack. Even 800 ppm of sulfur in the fuel gas, which is for natural gas or LPG a very high value, the SO2 emmission is relatively low with regard to emmission regulations in USA or Germany. It is evident that large quantities of CO2 are released in the atmosphere from the reformer stack . Noise with the today forced draft burners is not any more a major problem, as it was with the old generation induced draft burners.

<u>Reformer Tubes</u>

The most critical item in a fired reformer are the tubes in which the catalyst is placed and constitute up to the 30% of the total reformer cost. The most common materials are :

25 Cr / 20 Ni (HK40)

25 Cr / 35 Ni, Nb (HP with Nb)

For mechanical and process reasons , a typical reformer tube is 11 to 14 m long, with ID of 100 to 130 mm and wall thickness between 8 and 15 mm. Tubes are usually centrifugally casted and machined in the internal borehole.

The well known HK 40 material is suitable for steam reformers operating at low pressure. For higher reformer pressures, application of more expensive materials like HP with Nb prove to be more cost effective. Despite these improvements reformer tube materials work inevitably under creep conditions. Consequently their service life is limited. A temperature increase of only 20 °C above design temperature can reduce service life by 50% (from 10 to 5 years)[5]. Several methods are tried to estimate residual tube life. Non distructive tests are not very reliable. Destructive tests on a regular basis in combination with dimensional checks provide better basis for tube replacement decision,while minimizing the risk of unexpected tube failures.[6]

The major reasons of Reformer tube failures are:[7]

- Poor catalyst performance, uneven flow in tubes
- Thermal and pressure cycling due to many shut downs and start ups
- Overfiring during start up, or loss of steam supply during shut down
- Poor burner operation, uneven heating of tube
- Condensation at the bottom of the tube or thermal shock because of water carry over

Conventional and state of the art of combined reformer, oxygen enhanced reformer - oxygen/autothermal reformer - gas heated reformer(GHR)

Ammonia production is the most important reforming operation. There are more than 400 reforming plants in operation around the globe producing syngas for ammonia synthesis. The nitrogen required for the ammonia synthesis is usually supplied in the form of process air injected together with steam reformed gas in an other reformer called **Secondary** or **Autothermal Reformer**.The secondary reformer consists of a refractory lined vessel housing in the upper part a mixing and burner assembly, a combustion zone in the middle and a catalyst bed in the lower part.
In Fig. 2a and 2b sections of typical secondary reformers are shown

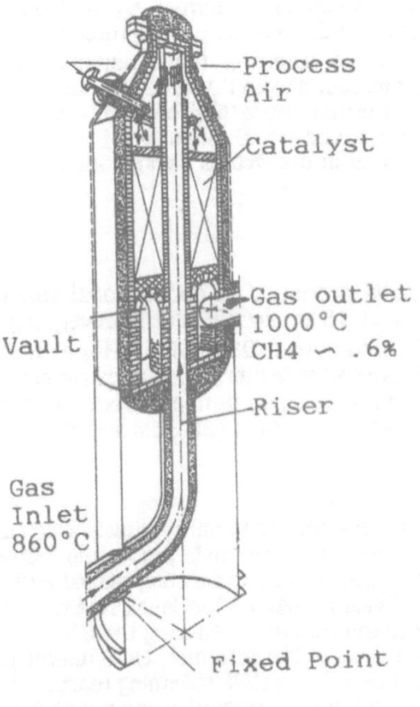

Fig. 2a UHDE design secondary reformer

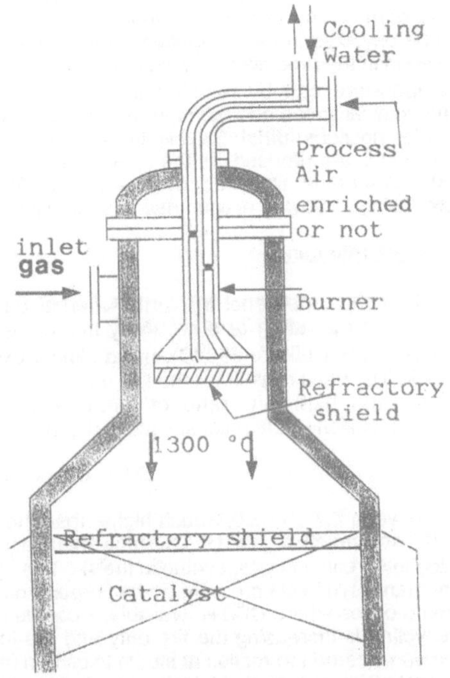

Fig. 2b Kellogg design secondary reformer

The catalyst is operating under high temperatures in the upper part of the bed and must be able to withstand temperatures up to 1370°C and abrasion resulting from the highly turbulent gas flow. Usually a shield layer from refractory material is laid at the top of the catalyst bed. Oxygen is partly converting methane, hydrogen, and carbon monoxide of the steam reformed gas in the upper part of the secondary reformer. Principally hydrogen is consumed in the combustion zone, together with partly converted methane and CO,

$$CH_4 + 1/2O_2 = CO + H_2 \quad (7)$$

$$H_2 + 1/2\,O_2 = H_2O \quad (8)$$

$$CO + 1/2O_2 = CO_2 \quad (9)$$

supplying the necessary heat for the almost complete conversion of the remaining methane and CO in the catalytic steam reforming zone at the lower part of the secondary reformer in conformity with reactions (1) and (2)

To be able to give at least the stochiometrically required quantiny of nitrogen for the ammonia synthesis, a minimum amount of residual methane shall be left at the exit of the primary steam reformer. From the other side the heat supplied from the partial conversion (oxydation) of methane, hydrogen and CO shall be sufficient to complete the catalytic steam reforming reaction till a very small fraction of residual methane (methane slip) of about 0.6%. To achieve that, the gas is heated in the partial oxydation zone above 1300°C and leaves the reformer at about 980 to1040°C. The gas is cooled in special design waste heat boiler followed usually by a steam superheater.

<u>Oxygen - enhanced reforming</u>

A consequent development of the combined reforming for ammonia production, is the use of oxygen enriched air as oxydant in the autothermal secondary reformer . In this mode of operation more load is shifted from the primary to the secondary reformer reducing the required size of the primery and the severity under which it operates, but more steam reformed gas has to be oxydized in the combustion zone of the autothermal secondary reformer. By injection of oxygen in the process air oxygan content can be increased to 32 to 35%, primary reformer gas temperature can by reduced from 860 to730°C. To have reformer tubes under similar mechanical stress conditions, reformer pressure can be increased from 20 bar to 36,7 bar, thus achieving substantial saving in compression power. Further more fuel consumption and emissions are reduced, as the conversion efficiency of the autothermal reformer is 65 to 70 % against 40 to 45 in a fired reformer. It is selfunderstood that in the overall savings the cost of producing or buying oxygen has to be considered.

<u>Oxygen reforming</u>

92% of the methanol in North America and Europe is produced in conventional steam reforming operating at a relatively moderate pressure of 15 to 24 bar and relatively high temperature (860 to 890°C), using almost exclusively natural gas (93 to 97% CH_4) as feed stock. The main reason for the low pressure is a compromise for higher methane conversion.. Given the high H/C ratio of methane and the additional hydrogen produced by the decomposition of the process steam, the so called stochiometric number (SN), defined as

$$SN = (H2 - CO2)/(CO + CO2) \quad (10)$$

is between 2.7 and 3.0, much higher than the optimum value for methanol synthesis which is 2.05 and much higher H2/CO ratio than required in many other synthsis processes (Oxo alcohols, reducing gas, synthetic fuels) . This hydrogen in excess has to be compressed at the methanol synthesis pressure, to be purged from the synthesis for use as fuel in the reformer. In some cases where CO2 is available, it can be recycled to the reformer, reducing the SN . CO2 recycling is increasing the fire duty and the investment cost of the reformer. CO2 reforming prevents carbon formation at steam to carbon ratio as low as 1.5, as CO2 reforming reaction (5) is faster than the carbon forming reaction (3). The advantage of CO2 reforming can not be used in conventional steam reformers for methanol, as they operate usually at S/C ratios of 2.5 and above.

By combining steam reforming and **oxygen reforming,** as shown in the arrangement of Fig. 3 it is possible to generate a syngas with a stochiometric number of 2.05. The reformed gas leaves the steam reformer with a high methane slip and it is further reformed with oxygen in the autothermal reformer with the rest of natural gas which is bypassing the steam reformer. The specific oxygen consumption is about 0.4 t per ton of produced methanol. The fired reformer duty is reduced by 45 to 50% and the makeup gas compression requirements by more than 50%.

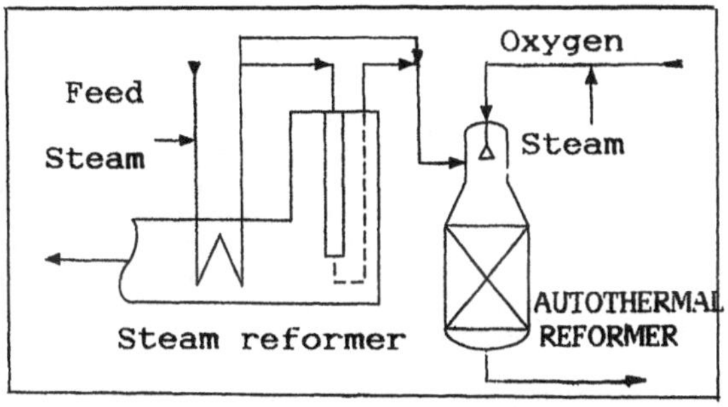

Fig.3 Combined Reforming System

The advantages/disadvantages of combined reformig for methanol production against conventional steam reforming are the following:

low residual methane in the syngas, optimum SN , higher reforming pressure = low compression energy, almost offsetting energy for air and oxygen compression.

less natural gas consumption per ton of methanol because of better thermal efficiency of autothermal reforming

less NOx and CO2 emission. For 50% of methane steam reformed and oxygen reformed together with bypass methane directly oxygen reformed, the emmissions are reduced by about 75 and 35% respectively.[8]

single train for capacities up to 5000 tpd ,because of reduced equipment size.

for plant size of about 2500 tpd capacity total invesment cost including air separation unit are up to 15% higher according to some estimates[9] or less by some others[10], than the case of conventional single train steam reforming

the combined reforming process is viable and commercially proven, but only limited references are available and the operation of these plants with oxygen burners is still less reliable than the well proven steam reforming

high natural gas prices and environmental restrictions could make combined reforming a viable solution for very large scale single train methanol plants

reduction of S/C ratio to increase thermal efficiency in connection with high CO partial pressure may result in severe corrosion phenomena of enhanced carburisation of high austenitic metallic surfaces, phenomena known as metal dusting.

Autothermal oxygen reforming alone can also be used to produce stochiometric makeup gas suitable for methanol synthesis . To avoid carbon formation a special design for mixing oxygen with the feedgas is required. Lower temperatures at the inlet have to be used so that only a small fraction of the reaction can start at the mixing point, the bulk being initiated on a special ignition catalyst placed at the upper part of the bed,where normally the combustion zone of a conventional autothermal reformer is located.Some special design burners are operating successfuly in few small size oxygen reformers. Instead an adiabatic reactor can be placed in front of the oxygen reformer filled with a special catalyst. This reactor is an adiabatic steam reformer, called **Prereformer** and operates at temperatures of about 550 ^{O}C.[11] Prereformer is also used in connection with conventional steam reformers in order to reduce the firing duty of the reformer and to hydrogenate and convert heavier hydrocarbons. As the reaction in the prereformer is endothermic too, reheating of the prereformed gas is required to take full advantage of the operation, but this is complicating the whole prereformer setup.

Assessment of the today state of the art in natural gas reforming technology

A drastic development has taken place in the reforming system design in the last ten years. This development is achieved in a series of process improvements and by the introduction of new materials, including mechanical and manufacturing innovations in the area of burners, reformer tubes , convection systems, catalyst, instrumentation and automatic process control design, insulation materials. In the area of energy conservation the concepts of combustion air preheating, lower S/C ratio, improved energy cycles, i.e. the use of gas turbines, high efficiency steam turbines and compressors, the use of low level heat recovery systems, vapour recompression systems, PSA and membrane separation technology , have led to an increase in the energy efficiency, the plant safety and reliability, as well as environmental improvements to levels considered highly uneconomical in the past.

Despite all the above improvements, many weak points are still inherent to the today state of the art reforming technology: Improved efficiency created a considerable amount surplus waste energy which had to be exported outside the reforming plant and this is not always feasible or desirable. Preheating the combustion air to temperatures as high as 505 to 550 ^{o}C increase the preheater cost significally and also the NOx content in the flue gas to levels requiring denitrification.Furthermore high combustion air temperatures in connection with high feedstock- steam mixture temperatures and lower steam to carbon ratios increase the heat flux at the upper part of the tubes and the risk of "hot bands" and carbon formation there[12]. This in connection with higher reformer pressure brings reformer tubes near to the creep limits. Special techniques were successfuly applied to shift reforming load and severity away from fired reformers (prereformer, combined reforming, overstochiometric secondary reforming [13]), but still today fire reformers are somewhat complicated and require close attention to operate safely. Heavy investments are needed to efficiently recover the large amount of produced waste heat.

An interesting development to reduce load and reforming severity in fired reformers is the use of the **adiabatic prereformer** [14]and the **exchanger type reformer** (ETR) [15]. The ETR is a presurized refractory lined vessel containing free standing open end reformer tubes supported from a tube plate located at the bottom cooler side of the reformer. About 25% of the total feedstock mixed with steam and preheated to about 380 ^{o}C is entering the ETR from the bottom and flows up through the tubes and it is steam reformed by the convection heat supplied from the rest 75% of the feedstock already steam reformed in a conventional fired reformer and entering at the top of the ETR, where mixed with the already reformed 25% portion of the syngas is flowing downwards along the tubes. This last development is the breakthrough point, which opened the way to the complete replacement of the fired steam reformer by more efficient advanced reformer designs, which we will discuss in the following paragraphs.

ADVANCED DESIGN REFORMING SYSTEMS

In the Fig. 4 the schematic diagram of a reformer configuration (LCA) is shown developed by ICI and in operation at Severside UK since April 1988 at a 450 tpd ammonia plant. In this type of reformer configuration the mixed feedgas with steam pass through bayonet type tubes filled in the annulus space with catalyst, while the reformed gas from the secondary reformer , at about 960 to 1000° C flows on the shell side providing the heat for the primary steam reforming.

To complete reforming to acceptable methane slip levels, when air is used as oxydant, an overstochiometric gas is produced, requiring the elimination of the excess nitrogen from the syngas befor the final synthesis to ammonia[16]

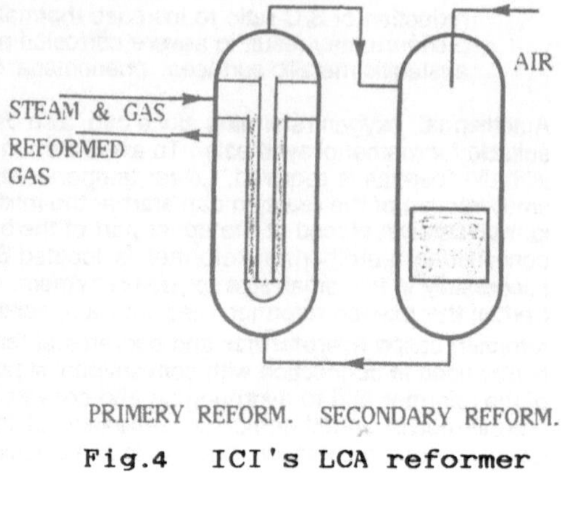

Fig.4 ICI's LCA reformer

The basic chemistry and the underlying principles of steam reforming are unchanged from that in a conventional steam reforming for ammonia production. The secondary reformer catalyst was in the original version a monolithic type of extremely high activity having a volume of only 1/8 of a conventional secondary reformer catalyst, but because of the rapid pressure drop build up due to blockage of the gas passage, it has been replaced in early 1992 by a conventional type secondary reformer. Increased pressure drop has been also observed in the primary, **gas heated reformer** (GHR) due to breakage of the lower part of the catalyst and was directly related to the large number of thermal cycles (stops, restarts) during the early operation period. With an improved bayonet tube design and with less shut downs it is hoped that this problem is resolved now. In Fig. 5 below a section through GHR is shown.

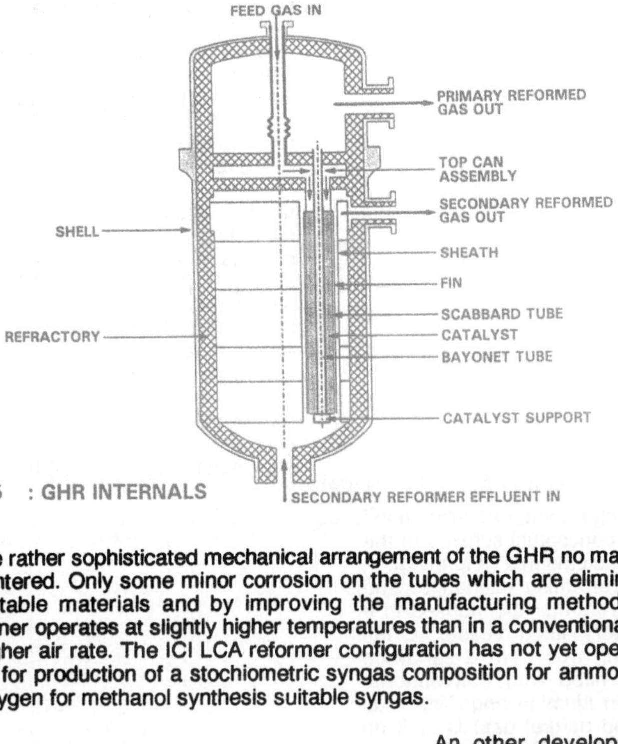

FIG. 5 : GHR INTERNALS

Inspite of the rather sophisticated mechanical arrangement of the GHR no major problems have been encountered. Only some minor corrosion on the tubes which are eliminated by selection of more suitable materials and by improving the manufacturing methods.The secondary reformer burner operates at slightly higher temperatures than in a conventional plant, due to the relatively higher air rate. The ICI LCA reformer configuration has not yet operated with oxygen enriched air for production of a stochiometric syngas composition for ammonia synthesis, nor with pure oxygen for methanol synthesis suitable syngas.

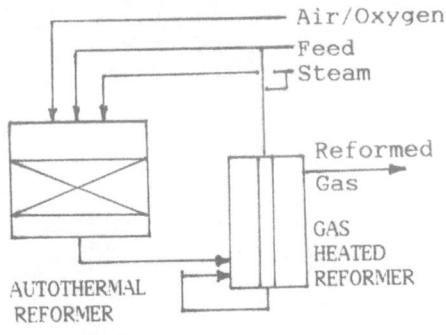

Fig. 6 MW Kellogg KRES reformer

An other development announced by MW Kellogg is schematically shown in Fig 6.[17] The effluent of an autothermal reformer is supplying the heat for completion of the conversion in a open tube reforming exchanger. The autothermal reformer is designed to operate either with enriched air or with oxygen. Kellogg reforming exchanger system (KRES) is not yet in operation, but MWK is claiming that the concept is based in proven components and in their patented open tube reforming exchanger having a mechanicaly simple design. In Fig 7 the reforming exchanger is shown.

The flow split between autothermal reformer and reforming exchanger is 75/25 %.

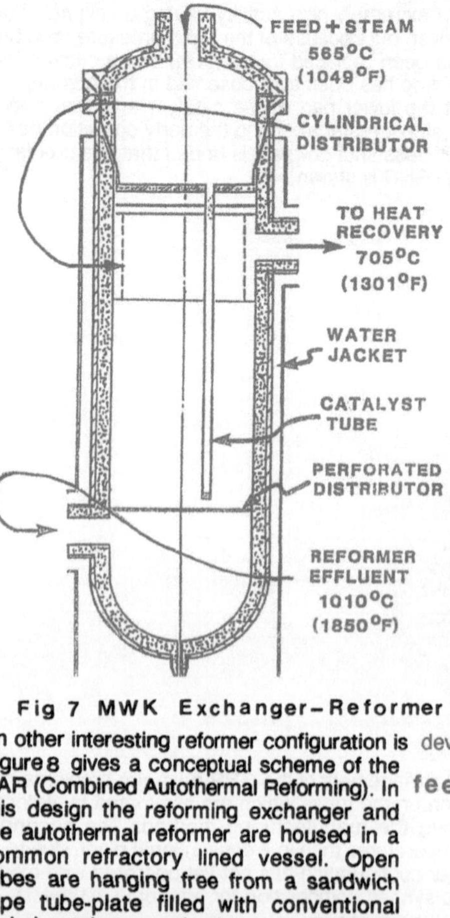

Fig 7 MWK Exchanger–Reformer

In case enriched air or pure oxygen will be used,Kellogg's proprietaty water cooled burner will be applied, shown in Fig 4b, which has already had commercial reference in operation with pure oxygen and steam. The enriched air stream enters the top of the reformer through a refractory protected injector tube and flows out at the periphery of the lower part of the injector. Mixed feed enters tangentially to the reformer getting a vortex like flow. Because of the lack of hydrogen in the feed to the autothermal reformer the design of the burner is very critical in order to ensure a conversion free from carbon formation. Steam to carbon ratio is 3.3 to3.8. This is a high S/C ratio selected to minimize carbon formation risk and carburization corrosion(metal dusting).

In this design, as well as in all other reforming systems without fired reformer there is no waste heat available to produce steam, to preheat fuel , BFW and process air. This is usually supplied by a separate heater fueled with natural gas, or from the exhaust of the turbine driving the air/oxygen compressor. Feed and steam is heated befor entering the reforming exchanger by the effluent of this exchanger. The reforming exchanger is designed to operate at 38 bar pressure.

An other interesting reformer configuration is developed bv UHDE[18] Figure 8 gives a conceptual scheme of the CAR (Combined Autothermal Reforming). In this design the reforming exchanger and the autothermal reformer are housed in a common refractory lined vessel. Open tubes are hanging free from a sandwich type tube-plate filled with conventional nickel catalyst on alumina rings.The total feed (desulfurized natural gas) is split up into two parts, a primery and a secondary. The primery feed is mixed with steam and it is steam reformed flowing down in the catalyst tubes. These reformer tubes are externally heated by the hot partly reformed gas, returning up from the partial oxydation chamber below. In this chamber secondary feed and partly steam reformed gases are partly oxydized with oxygen or oxygen enriched air, supplied through peripherical water cooled injectors. Secondary feed and oxydant injectors are creating a strong vortex movement resulting in a high turbulent flow pattern with the gas blown out of the catalyst tubes. This is ensuring a temperature and reaction homogenisation. Unlike ICI's LCA system CAR is designed to use oxygen, or oxygen enriched air.

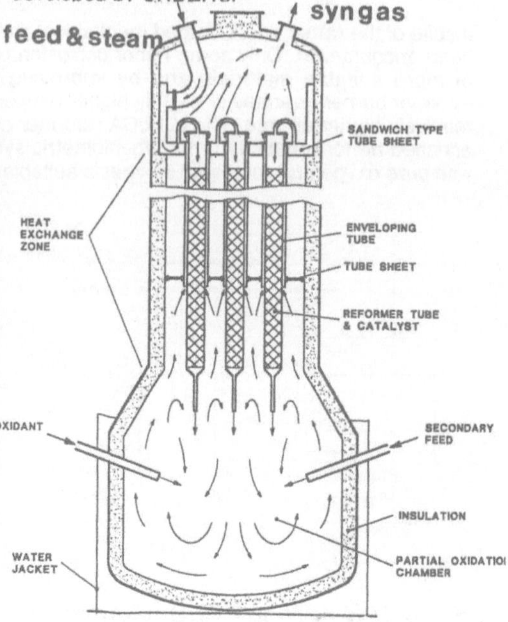

The fact that part of the feed gas is only partially oxydized, without undergoing a subsequent secondary steam reforming, deteriorates slightly the efficiency, as POX is a reaction, against catalytic reforming taking place away from equilibrium. From the other side the arrangement of the oxydation chamber at the bottom, eliminates some risks for catalyst deterioration inherent to autothermal oxygen, or even air secondary reformers.

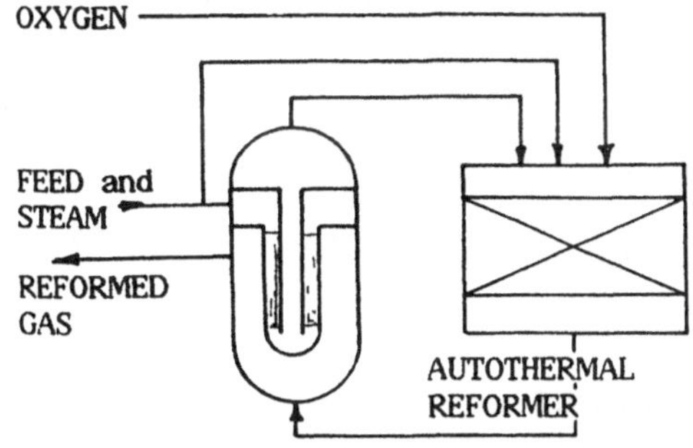

Fig.9 Oxygen reformer and GHR

UHDE is designing the CAR also as a simple reforming exchanger without the partial oxydation chamber. The purpose is to be able to operate it in connection with an authothermal or conventional reformer. ICI , MWKellogg and others are following the same route. [19]
An interesting possibility could be the reverse of that followed by Kellogg shown in the configuration Fig. 6. The autothermal reformer is receiving a partly steam reformed gas-stream together with a mix of feed and steam, as shown in Fig. 9. ICI's GHR would be a suitable reforming exchanger type. Suitable Exchanger type reformer is also the **Tandem Reformer** developed and designed by GIAP[20].
The scheme of Fig. 9 facilitates the operation and reduces the reforming severity in the autothermal reformer. Ignition catalyst or special design burners are not required in this configuration.

A demonstration CAR reformer is in operation in Strazske Czechoslovakia since December 1990. The CAR is producing a gas containing 67% H2, 24% CO, 7.6% CO2 . Oxygen consumption is 0.5 mol/mol of C. Steam to carbon ratio is 1.6 total and 2.5 in the primary part. Operating pressure 17 bar. The reformer is fed with 2600 Nm3/ h natural gas in the primary and with 1508 Nm3/h in the secondary. 2340 Nm3/h of 95% oxygen are consumed. After 9 months of troublefree operation enhanced carburization was found on HK and Incoloy 800 material. New more appropriate materials will be tested during a modification which will be implemented in September this year. In the mean time the CAR is operating with slightly higher S/C ratio [21].

DISCUSSIONS AND CONCLUSION

Reforming technology of natural has progressed in the last ten years impressively. Autothermal reformers in connection with gas heated reformers is certainly a proven process scheme with already good industrial-commercial references. Specific energy consumption against state of the art conventional reformers is only slightly better, when oxygen has to be produced in cryogenic air separation plant and no credit is given for the coproduced nitrogen. Plant complexity will decrease,due to the elimination of the fired reformer with all its large waste heat recovery equipment, inspite of the added air separation unit and the oxygen compressor, but oxygen reformer operation will be a source of concern until the operation reliability of the various designs is proven.

Large scale single train oxygen reforming plants for methanol production will have certainly an economic advantage against conventional plants, especially when natural gas prices increase above 4 US$/MMBTU, and if there are not any hidden, unknown yet factors, which may hamper the scale up from the existing small size demonstration units to large plants. For instance the conditions and mechanism under which metal dusting is occuring are still not very well known, and the construction of oxygen reformers operating at methanol synthesis pressure (elimination of syngas compression) may be not without problems.

In environmentally sensitive areas the operation of non fired reformers has certainly big advantages against conventional reformers.

Even today medium size oxygen reformers fed with low cost surplus oxygen to enrich air, or the Tandem reformer with an air operated Secondary Reformer, in connection with the C.F. Braun **Purifier** process[22] is an attractive alternative for revamp of old, less efficient ammonia plant steam reformers, or for new ammonia plants for up to 1100 MTPD capacity.

The selection of the most suitable syngas reforming system design depends from several key factors, which must be carefuly considered. Either steam reforming or oxygen reforming or combination of both could be the optimum depending on the syngas end use requirements, feedstock - utilities (oxygen, CO2, power) availability and cost. Process optimisation requires consideration of all main process parameters such as reforming pressure, temperature, steam to carbon ratio, CO2 to carbon ratio, as well as efficient waste heat requirements and integration.

Mechanical design considerations are equally important to achieving a reliable and economic advanced design of a reforming plant. Some of the more critical areas are the oxygen reformer configuration, the burner design, and the reforming exchanger mechanical design.

An energy price increase will certainly accelerate worldwide the development and multiply the applications of the Advanced Reforming System Design.

REFERENCES

[1] W. Balthasar "Hydrogen production and Technology: today, tomorrow and beyond" . Int. J. Hydrogen Energy Vol. 9, No 8, pp 649-668, 1984

[2] K.S. Raghuraman "Steam Hydrocarbon reforming technology - A review", pp 5, KTI Netherlands Synposium , Jan 1984 Zoetermeer NL

[3] L. Storgard "Evaluation of Catalyst Performance in Natural Gas Steam Reforming" Nitrogen 91 Conference Copenhagen June 91.

[4] "Reforming catalysts for the production of ammonia",Nitrogen No174, pp 23-24, Jul-Aug 88

[5] "High-temperature High pressure service", UHDE GmbH publication Nov. 91.

[6] "Reforming the front end" , Nitrogen No 195, pp 22 - 31, Jan-Feb 92.

[7] "Ammonia plant failure statistics", AIChE Ammonia Plant Safety Symposium, Panel discussion 1986.

[8] D. Kitchen "Energy Efficient Small Ammonia Plants", AIChE Ammonia Plant Safety Symposium, San Diego, Aug 1990.

[9] R.V. Schneider , and J.R. LeBlanc "Choose optimal syngas route", Hydrocarbon Processing, March 1992.

[10] G.L.Farina and E. Supp "Produce syngas for methanol" ,Hydrocarbon Processing March 92.

[11] R. Vannby and W. Madsen "Adiabatic pre-reforming" , Ammonia Plant Safety Symposium Los Angeles , Nov 1991.

[12] J. Rostrup-Nielsen "Catalytic Steam Reforming", Cata lysis Science and Technology, Vol 5, Springer Verlag 1984.

[13] K.G. Christensen, J.H. Gosnell and B.J. Grotz "Flexible design provides economical ammonia plant", Nitrogen (191) , May-June 1991.

[14] D.N. Clark, and W.G.S. Henson, "Opportunities for savings with pre-reformers, AIChE Ammonia Plant Safety Symposium , Minneapolis August 1987. "C-I-L a successful debut for the AMV process", Nitrogen 162 , Jul-Aug 1986.

[15] S.I Wang and N.M. Pattel, "Hydrogen production by Enhanced Heat Transfer Reformer", AIChE Summer national meeting, Session I , Denver Colorado, 22 Aug 1988.

[16] K.J.Elkins, I.C.Jeffery, D. Kitchen and D. Pinto " The ICI Gas-Heated reformer (GHR) system", Nitrogen 91 conference, Copenhagen June 91.

[17] R.V. Schneider III, "Advances in Reforming System Design", AIChE Ammonia Plant Safety Symposium, San Diego August 1990.

[18] N. Tiagaradjan , "CAR - Future Prospects", UHDE Ammonia Symposium , Dortmund June 1992.

[19] R.N. Udengaard , L.J.Christiansen, " A Heat Exchanger Reformer for Hydrogen Production", AIChE Meeting, Orlando, March 1990.

[20] S. Sergeev " Low Energy Ammonia Process (LEAP) " REVAMP Frtilizer Plant Retrofitting Conference , Budapest Hungary September 1992. Organizer British Sulfur .

[21] H. Marsch " CAR operation experience ", Ammonia Safety Symposium, San Antonio USA September 1992.

[22] B. J. Grotz , L. Grisolia " The Braun Purifier Process" , Nitrogen No 199, Sept-Oct 92, p. 39 and Press release of Brown & Root Braun in the 92 AIChE Ammonia Safety Sypmosium, Sept. 1992 in San Antonio USA

ENERGY SAVINGS IN ALUMINIUM ELECTROLYSIS BY CONTINUAL MONITORING AND CONTROL OF THE AlF_3 CONTENT OF THE CRYOLITIC MELT

V. VASSILIADOU and I. PASPALIARIS
Laboratory of Metallurgy
National Technical University of Athens
GR-157 80 Zografou, Athens

D. STEFANIDIS and D. GEORGANTONIS
Aluminium de Grèce
GR-320 03, Paralia Distomou

ABSTRACT

AlF_3 content in the cryolite bath is an important cell parameter, strongly related to cell temperature and heat balance, therefore affecting current efficiency and energy consumption. Traditional control of AlF_3 content in the bath consists in periodic bath sampling and analysis- a procedure having a long time lag. In the present paper a new method for the AlF_3 determination and control based on the relation between the cell resistance and bath temperature and composition, is proposed. Extensive experimental results in the industrial cells of "Aluminium de Grèce" have shown that the method can be successfully used for the description of the cell thermal state and prediction of the appropriate AlF_3 additions by using a suitable formula.

INTRODUCTION

The industrial production of primary aluminium by the Hall-Héroult process is based on the electrolytic reduction of Al_2O_3 dissolved in molten cryolite ($3NaF.AlF_3$). The overall reaction can be written as follows :

$$2Al_2O_3 + 2C \Rightarrow 4Al + 3CO_2 \tag{1}$$

The metal is deposited at the cathode, consisting of a liquid aluminium pool over the cathode electrode, and the oxygen is dıscharged at the carbon anode and reacts with it forming CO_2. Reduction takes place in the temperature range 940 - 960 ^{0}C in a 4-5 cm thick layer of molten bath between the anode carbon and the liquid metal, called interpolar distance.

The electrolytic cells used in "Aluminium de Grèce", consist of rectangular steel shells, lined with thermal insulation that surrounds an inner lining of carbon which contains the highly corrosive fluoride electrolyte and molten aluminium. Thermal insulation is adjusted to provide sufficient heat loss to form a ledge of frozen electrolyte on the inner walls (to protect them from erosion) but not at the bottom of the cell which must remain bare under the prebaked anodes to provide electrical contact with aluminium. Alumina is fed into the cells by a point feeding system consisting of a mechanical feeder located in the centre of the cell between the two anode rows. A crust of frozen electrolyte and alumina covers the top of the cell around the anodes. The cells operate with a current of 70000 A and an average cell voltage of 4 volts.

BATH ADDITIVES

Due to its unique capacity for dissolving alumina, molten cryolite ($3NaF.AlF_3$) remains the major component of the electrolyte; certain additives are also introduced to the bath in order to improve its physico-chemical properties. The ideal additive should, first of all, decrease the solubility of aluminium in the melt and lower the liquidus temperature. The additive should also increase or at least not decrease alumina solubility (for easy operation), increase the electrical conductivity and decrease the density (for phase separation from aluminium). Aluminium fluoride is the most common additive and his addition is referred to as varying the "ratio" of NaF to AlF_3. Variations are considered with respect to the stoichiometric cryolite composition which is 40 wt % AlF_3; In the industrial practice three alternative terminologies are used to characterise the melt: the wt % AlF_3 in excess of the stoichiometric cryolite, the cryolite ratio (CR) expressing the molar ratio NaF / AlF_3 and the weight ratio of those two salts called bath ratio (BR).

In the industrial cells AlF_3 is the electrolyte component that is consumed at the highest rare (20 kg / ton Al) and is usually present in a 5-15 wt % excess over cryolite stoichiometry. The addition of AlF_3 decreases the density, the liquidus temperature, especially at high concentration, and reduces the co-deposition of sodium. These advantages are partly counteracted by an ıncrease ın vapour pressure and a lowering in the electrical conductivity of the melt.

Alumina is fed into the bath with the "point-feeding" system, batchwise in small amounts. The point feeding is controlled so as to achieve a low alumina concentration allowing the optimum operating conditions. Usually alumina concentration ranges among 2-8%.

CURRENT EFFICIENCY

The total energy consumption E (kWh/kg Al production) is related to current efficiency CE (ratio between the metal produced to theoretical production base on the Faraday's law) and cell voltage according to the following formula [1]:

$$E = \frac{2.980V}{CE} \qquad (2)$$

where C_E is the current efficiency and V is the cell voltage.

Low current efficiency is primarily related to re-oxidation of Al dissolved in the electrolyte, by the anode gas. The degree of dissolution is effected by a number of variables including temperature, current density, interpolar distance, composition of the electrolyte and cell design. Temperature has been proved to be, in the industrial cells, the most important parameter affecting the current efficiency. Therefore, great improvement in current efficiency can be achieved by lowering the cell temperature and this can be achieved by certain additives such as AlF_3. Addition of AlF_3 decrease the liquidus temperature allowing cell operation at lower temperatures, where aluminium solubility and the consequent oxidation losses are lower.

Reported data [3,4] indicates that increasing the temperature by 1^oC decreases current efficiency by about 0.1- 0.18 %. The effect of temperature in current efficiency is not depending on the current density or the interpolar distance. Also, from a study carried out in a 135 kA cell a formula was deduced calculating the current efficiency as a function of temperature, % AlF_3 excess, metal height and cell age [2]. It is therefore usual practice for the cells to operate with an excess of AlF_3 over the stoichiometric cryolite and at a limited temperature range 940-960 ^oC.

HEAT BALANCE

During normal cell operation, at the desired temperature a heat balance is established, the total energy input is equal to the energy needed for electroreduction plus the heat losses. The heat balance of the cell is, in short term, disturbed by numerous operations as alumina feeding, anode changes, metal tapping, anode effects. If energy input is less than the energy required, cell operation changes: temperature is decreasing, AlF_3 concentration is increasing and more electrolyte is freezing. This is the case of the "cold" cell operation. On the other hand if the energy input is higher than required , the cell presents the opposite changes and its operation is characterised as " hot". The usual industrial practice is for the cell to operate close to the "cold" conditions. Generally, due to the thermal balance between the molten and the frozen bath any change of the bath composition alters the liquidus and therefore the operating temperature.

CELL CONTROL

High current efficiency and therefore low energy consumption is based, to a large extent, on the application of a control system. Cell control is achieved by maintaining, close to the optimum conditions, a set of variables such as current, cell voltage, interpolar distance, cell temperature, bath composition and others. Difficulties in controlling the above parameters are associated with the fact that most of them are not independent and their continuous monitoring are not always easy. Presently, the only continuously measured parameters are the cell voltage and potline current. Cell control is achieved as follows:

The cell resistance or also called pseudo-resistance derived by the formula

$$R = \frac{V - 1.65}{I} \tag{3}$$

where V is the cell voltage , I the cell current is measured and compared to a predetermined set-point. If the deviation is larger than a certain limit the primary reaction will be to adjust the interpolar distance by raising or lowering the anodes level. The cell pseudo-resistance is also strongly depended on the alumina concentration.

The resistance vs alumina concentration curve can be used as a knowledge base for the condition of the cell. This is the basis of the control system, which contains a rather sophisticated logic operation with additional inputs and safeguards.

Alumina feeding in the cells of "Aluminium de Grèce" is, therefore, based on the resistance vs. alumina concentration curve and the continuous measurement of the cell resistance. Alumina is fed into the bath in small amounts with a predetermined rate: The cell is initially fed with a quantity of alumina less than the one required theoretically (underfeeding). As electrolysis proceeds the alumina concentration decreases up to a certain point at which alumina feeding rate increases significantly and becomes greater than the theoretical (overfeeding). At that moment that alumina reaches the initial value the above procedure is repeated. For the determination of the cell feeding rate the gradient of cell resistance vs time curve, is calculated and compared to a predetermined set-point. If the gradient is higher than the set point, alumina feeding rate is increased. In this way is possible to control indirectly, one of the two most important bath variables, the alumina content and to keep a nearly constant interpolar distance.

CONTROL OF AlF_3 CONTENT IN THE BATH

AlF_3 concentration, the other most important bath variable, is strongly coupled to other parameters, heat balance and current efficiency. Therefore, continuous determination and control of AlF_3 content allow the cell to operate close to the optimum conditions, with high current efficiency and low energy consumption. Under the current industrial practice, AlF_3 content is controlled by periodic bath sampling, analysis and corrective additions, a time-consuming procedure which is usually practised once per week per cell. Due to the large time lag, changes may appear at bath conditions between the time of the sampling and the time of the corrective action.

CORRELATION BETWEEN BATH TEMPERATURE AND AlF_3 CONTENT

In order to achieve an effective cell control, which will reduce the delay between measurements and corrective actions, a technique for the determination of AlF_3 additions, based on bath temperature measurements has been proposed [6]. A linear model between bath ratio NaF / AlF_3 and temperature has been obtained and was used to calculate the daily AlF_3 additions to the cell, according to the equation:

$$A_i = A_o + 5\,(T_i - T_t) + 2\,(T_i - T_{i-1}) \qquad (4)$$

Where

A_o is the daily AlF_3 addition which depend on the cell age only
T_i is the actual bath temperature
T_t is the target bath temperature
T_{i-1} is the actual bath temperature the day before

Physical explanation of the relation between temperature and AlF_3 concentration has been adequately discussed [6,8].

In order to verify this technique and derive a suitable mathematical model relating bath temperature and AlF_3 concentration for the electrolytic cells of "Aluminium de

Grèce", a series of experiments has been carried out. Bath temperature and AlF_3 content was monitored in 80 different cells and the results are presented in figure 1. Regression analysis performed on these data gave the following linear equation:

$$\% \, AlF_3 = 197.4 - 0.199 \, T \qquad R^2 = 0.78 \tag{5}$$

where, T °C is the bath temperature and % AlF_3 is the AlF_3 content in the bath.

A series of experiments was carried out in 80 cells, whereby cell temperature was measured and AlF_3 content of the bath was analysed and compared to that predicted by equation (5). The comparison is shown in figure 2.

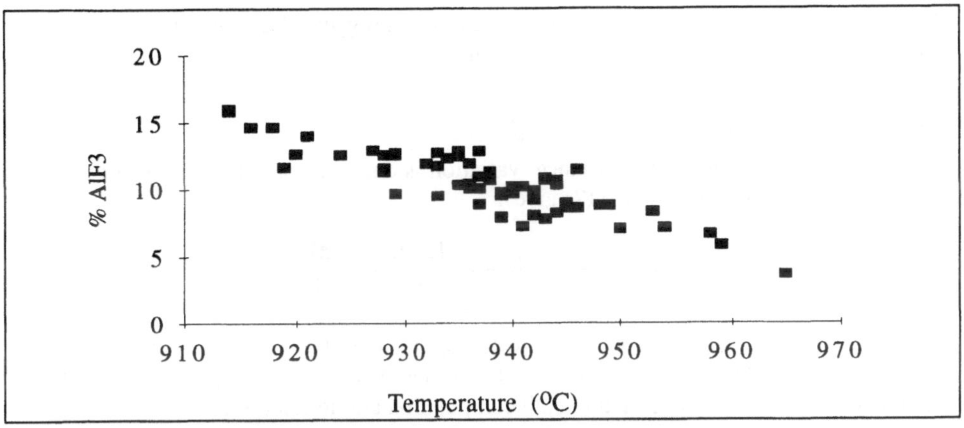

Figure 1. Relation between AlF_3 concentration and temperature

It is obvious that a good prediction can be made; however, it is considered that continuous temperature measurement is not practical. Therefore the technique described in the next paragraph was developed.

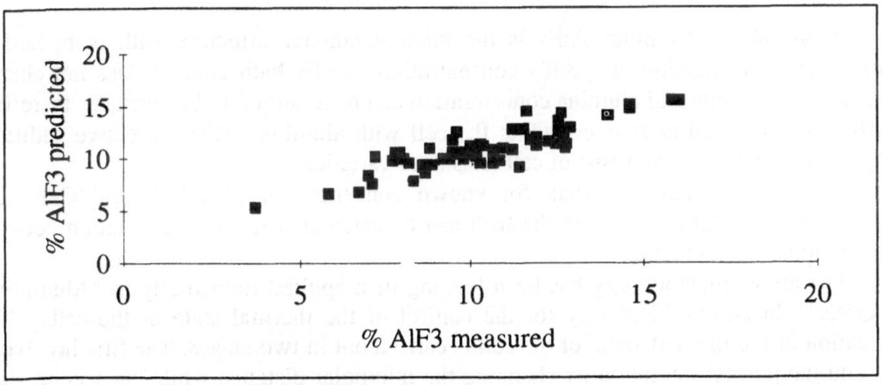

Figure 2. Relation between the measured % AlF_3 and the predicted from the model

AlF_3 DETERMINATION BASED ON BATH RESISTANCE VARIATION MEASUREMENTS

In the present paper, a new method for the indirect determination of the AlF_3 content in the bath, based on the relation between the electrical resistance, temperature and the composition of the bath is described.

Under specific conditions of the interpolar distance l_1, bath resistivity ϱ and anodic surface S, bath resistance is given by the formula:

$$R_1 = \varrho \frac{l_1}{S} \tag{6}$$

By changing the interpolar distance to l_2 and assuming that bath resistivity and anodic surface remain constant, bath resistance becomes

$$R_2 = \varrho \frac{l_2}{S} \tag{7}$$

Therefore, assuming that cell resistance variation is due only to bath resistance variation, the subtraction of the formula (6),(7) gives:

$$\Delta R_{cell} = \Delta R_{bath} = R_2 - R_1 = \varrho \frac{l_2 - l_1}{S} = \varrho \frac{\Delta l}{S} \tag{8}$$

The above formula indicates that by measuring R_1 and R_2 at the specific interpolar distance l_1 and l_2 and assuming that S is known, the bath resistivity ϱ can be calculated. Bath resistivity is strongly related to bath composition and temperature:

$$\varrho = f(Al_2O_3,\ CaF_2,\ AlF_3,\ T) \tag{9}$$

Comparing equation (9) and (8) we have :

$$\Delta R_{cell} = F(Al_2O_3,\ CaF_2,\ AlF_3,\ T) \tag{10}$$

In the above formula, AlF_3 is the main parameter affecting bath composition; temperature is a function of AlF_3 concentration, CaF_2 bath content does not change very much with time and alumina concentration can be assumed to be constant; therefore, if ΔR_{cell} is measured after overfeeding the cell with alumina, AlF_3 corrective additions can be estimated from the value of cell resistance variation.

Formula (10) indicates also that, for known concentration of Al_2O_3 and CaF_2 and temperature, the AlF_3 content in the bath can be calculated if an exact relation between bath composition and ΔR_{cell}.

The above methodology has been feeding of a applied industrially in "Aluminium de Grèce" in an empirical way for the control of the thermal state of the cells . The regulation of the thermal state of the cell is carried out in two stages. The first involves a fast and temporary regulation by changing the interpolar distance while the second stage tends to maintain the already modified operating conditions by adding the appropriate amount of AlF_3. More specifically, after overfeeding the cell with Al_2O_3, the interpolar distance is increased by Δl_1, the R_{1cell} is measured, and the interpolar distance is increased again by Δl_2, and R_{2cell} is measured again (Figure 3). ΔR_{cell} is calculated as

$$\Delta R_{cell} = R_{2cell} - R_{1cell} \tag{11}$$

The cell resistance variation value is at first used for the immediate regulation of the cell thermal state according to the following cases:
- If $\Delta R > \Delta R_{max.}$: the cell operation is characterised as " cold" ; The situation is combated by a temporary increase in interpolar distance, with results in an increase in cell resistance and therefore in heating of the cell.
- If $\Delta R < \Delta R_{min.}$: the cell operation is characterised as " hot" and the reverse action is taken.
- If $\Delta R_{min} < \Delta R < \Delta R_{max.}$: the cell operation is characterised as " normal" and therefore no action is taken.

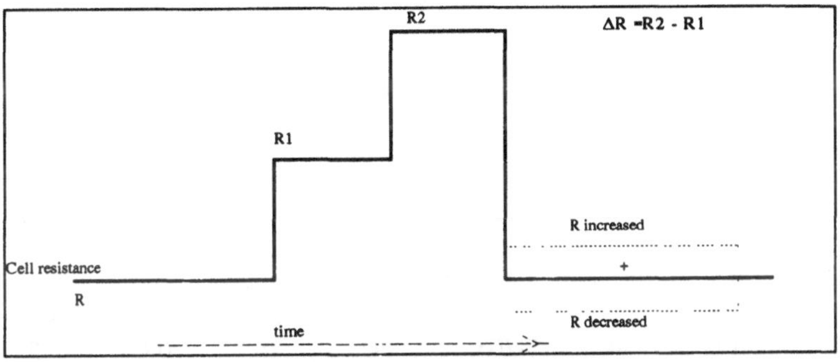

Figure 3. The methodology of the cell resistance variation

The above mentioned regulation is temporary and in order to attain a long term control of the thermal state of the cells an appropriate amount of AlF_3 is added into the bath. The amount of AlF_3 corrective addition is calculated by an empirical formula which is based on the average values of cell resistance variation. AlF_3 is introduced into the bath with point feeding in a way similar to alumina feeding.

In order to check the validity of the assumptions made for the application of the proposed technique a series of experiments, described in the following paragraphs has been carried out in the industrial cells of "Aluminium de Grèce ". An attempt has also been made to deduce from the experimental data an exact relation between the cell resistance variation and the bath composition.

In this series of experiments a set of cell operating parameters, as bath content in Al_2O_3, AlF_3, CaF_2, LiF,MgF_2, KF, bath and metal height, operating temperature were measured during the method application in a number of cells in order to determine their operating range, their correlation and to evaluate their influence in cell resistance variation.

The experimental procedure includes:
1. sampling of the bath always at the same position and always extremely carefully to avoid contamination with alumina. The samples were taken with a cold iron bar which

was immersed vertically into the bath and then quickly pulled out in order to achieve a fast solidification.
2. measurement of the bath temperature with a thermocouple immersed in the same hole that the sampling was carried out.
3. measurement of the bath height simultaneously with the sampling. The bath height was measured at the iron bar after the solidification of the sample.
4. measurement of the metal height in the same way with the bath height measurement.
5. chemical analysis of the bath sample for ingredients: Al_2O_3 with the method of free alumina [7], AlF_3 with the condumetric method [7] and CaF_2, LiF, MgF_2, KF with atomic absorption[7].

Experimental results

During this series of experiments, the above parameters were measured in 80 cells. The experimental results were statistically elaborated. In table 1 the summary statistics and in table 2 the correlation coefficients among the variables are presented.

TABLE 1

Summary statistics of the experimental results

	ΔR	%Al_2O_3	%AlF_3	%CAF_2	T(oC)	%LiF	Hmetal	Hbain
Mean	2.05	3.97	9.78	6.64	940.4	1.54	19.73	17.76
St. Deviation	0.26	1.49	2.34	0.85	10.6	0.21	1.21	2.71
Range	1.15	6.63	12.00	4.80	66	0.82	5	13
Minimum	1.51	1.79	6.00	5.17	919	1.16	17	12
Maximum	2.66	6.00	14.00	7.97	985	1.98	22	25

From the table 1 we can draw the following observations :

- Cell resistance variation ΔR, measured in a large number of cells, ranges between 1.5 and 2.5 μOhm which are the conventional limits proposed for the application of the technique. Low ΔR values are due to the incomplete change of the interpolar distance. High ΔR values appeared in a small number of cells are due to the cell instability.
- The concentration of LiF, MgF_2 and KF in the bath remains, in all cells, almost constant at the level of 1.5 % , 0.75 % and 1.4 % respectively;
- CaF_2 has a slight variation (6- 7.5 %), but can be assumed to be constant.
- Due to the different thermal state of each cell, there is a variation in bath temperature (920-980 ^{o}C), bath height (12-25 cm), metal height (17-22 cm) and AlF_3 content in the bath (5-16 %) from cell to cell.
- Although the Al_2O_3 content in the bath at the time of the AlF_3 determination procedure was expected to be about 3.5 % in all cells, experimental results showed that it varies between 2 and 6 %.

From the table 2 we can make the following observations :

- There is a strong relationship between temperature and AlF_3 concentration described by a linear model :

$$\% \ AlF_3 = 197.4 - 0.1994 \ T \qquad R^2 = 0.78 \qquad (12)$$

- Cell resistance variation is primarily affected by the AlF_3 concentration but also by the Al_2O_3, CaF_2 concentration and temperature. An attempt has been also made to correlate cell resistance variation as a function of bath temperature and composition. From the experimental results a poorly correlated model (R^2=0.30) has been derived.

The poor correlation coefficient of the model is due:

1. to the deviation of interpolar distance variation Δl
2. to the deviation of the alumina concentration values generated partly from the alumina analysis error and partly from the cell operation.

TABLE 2
Correlation coefficients

Parameter	ΔR	% Al2O3	% AlF3	% CaF2	% LiF	T (oC)	Hbath	Hmetal
Hmetal	-0.34	-0.12	0.07	0.07	0.05	-0.03	-0.2	1
Hbath	-0.06	0.09	-0.18	-0.4	0.07	-0.065	1	
T (oC)	-0.53	0.3	-0.82	0.06	0.17	1		
% LiF	-0.17	-0.27	-0.33	0.08	1			
% CaF2	-0.2	-0.17	-0.15	1				
% AlF3	0.56	-0.38	1					
% Al2O3	0.11	1						
ΔR	1							

CONCLUSIONS

High current efficiency and low energy consumption can be achieved by continual monitoring and control of the concentration of AlF_3 in the aluminium electrolytic cells.

Experimental results in the industrial cells of " Aluminium de Grèce " have shown that there is a strong correlation between bath temperature and AlF_3 bath content. The concentration of AlF_3 in the bath can be predicted from bath temperature using model (equation 5). This simple way of controlling AlF_3 concentration is not applied in the current industrial practice because continuous measurement of temperature is not considered to be technically feasible.

A new method for the determination of AlF_3 corrective additions based on the measurement of cell resistance variation is proposed. The value of ΔR_{cell} has been successfully used for the qualitative description of the cell thermal state and for the prediction of the appropriate AlF_3 additions by using an empirical formula developed in the " Aluminium de Grèce ". This method, permits the maintenance of the AlF_3 content in a short range around the optimum value.

The necessary assumptions for the application of the proposed method are fulfilled, except that of the constant alumina concentration after overfeeding the cell with alumina, which has not been verified yet because of analytical errors in the alumina determination.

A first attempt has been made to deduce a precise relation between bath composition and ΔR_{cell}, but due to the fluctuation of the operating conditions in the current industrial practice, a poorly correlated model has been derived.

In order to overcome these difficulties and to extract an adequate model for the prediction of AlF_3 content in the bath as a function of ΔR, a new research programme is already carried out in the Laboratory of Metallurgy(NTUA). This programme concerns the application and study of the method in a laboratory cell with the same electrode configurations as the industrial cell. The laboratory study permits the close monitoring and the exact measurement of bath composition, temperature, cell resistance variation ΔR_{cell} and the interpolar distance variation Δl.

REFERENCES

1. Grjotheim, K., and Welch, B.J., Aluminium Smelter Technology, Aluminium Verlag, Dusseldorf, 1986.
2. Grjotheim, K., and Kvande, H., Understanding the Hall-Héroult Process, Aluminium Verlag, Dusseldorf, 1982.
3. Grjotheim, K., Krohn, C., Malinovsky, M., Matiasovsky, K., and Thonstand, J., Aluminium Electrolysis, 2nd edition, Aluminium Verlag, Dusseldorf, 1986.
4. Haupin, W., "The influence of additives on Hall-Héroult bath properties", J. Metals, November 1991, p.28-32.
5. Welch, B.J., "Aluminium reduction technology-Entering the second century", J. Metals, November 1988, p.19-25.
6. Desclaux, P., " AlF_3 additions based on temperature measurements", Light Metals 1987, Warrendale, PA: TMS 1987 p. 310-313.
7. Vassiliadou, V., "Study of an automatic cell control procedure (test) at the industrial cells of " Aluminium de Grèce ", Athens, 1989.
8. Thonstand, J. and Roselth, S., " Equilibrium between bath and side ledge in aluminium cells- Basic principles", Light Metals 1983, Warrendale, PA: TMS 1983 p. 414-424

ACKNOWLEDGEMENTS

The financial assistance of the General Secretariat for Research and Technology of Greece and the "Aluminium de Grèce" is gratefully acknowledged. Acknowledgements are also extended to the management of "Aluminium de Grèce" for the granting permission to publish the paper.

CHEMICAL FIXED BED REACTOR WITH INTEGRATED HEAT EXCHANGER WITH SINTER METALS

Ton J.M. van Wingerden, Jan der Kinderen,
GASTEC N.V., P.O.Box 137,3700 AC APELDOORN
Prof. John Geus, P.O.Box 80083, 3508 TB Utrecht
Peter Neumann, P.O.Box 2155, 5608 Radevormwald

ABSTRACT

From metal powders porous bodies can be formed. The void spaces can be used to put a catalyst in and gas can flow through the structure. Thus a chemical reactor is constructed. Sintered metal bodies conduct any produced or demanded reaction heat very well. A reactor based on this principle can be used to solve the problems of run-aways in chemical rectors and selectivity.
Based on this principle the relevant sinter metal properties have been investigated:

1. A novel sinter metal production process was used to raise the porosity.
2. The heat conductivity has been determined.
3. A reactor model was calculated.
4. Two model reactions (natural gas combustion and desulphurization) have been investigated.

The results of the measurements, calculations and tests are given. They are very promising: conventional reactors can be simplified and new reactor types are possible in industry.

INTRODUCTION

There is a great number of gas reactions which take place in fixed bed catalytic reactors. These reactions produce or demand heat. Several solutions can be chosen for heat transport, for instance:

1. A series of alternately adiabatic reactors and coolers (or heaters) is constructed in which the heat, produced in the adiabatic reactor is removed in the successive cooler. The number of steps is such, that there is sufficient conversion of the input gases without having a temperature rise, which can cause an undesired shift in the equilibrium or the formation of undesired products.
   An example of this is given in figure 1. where a conventional Claus plant is shown [1].

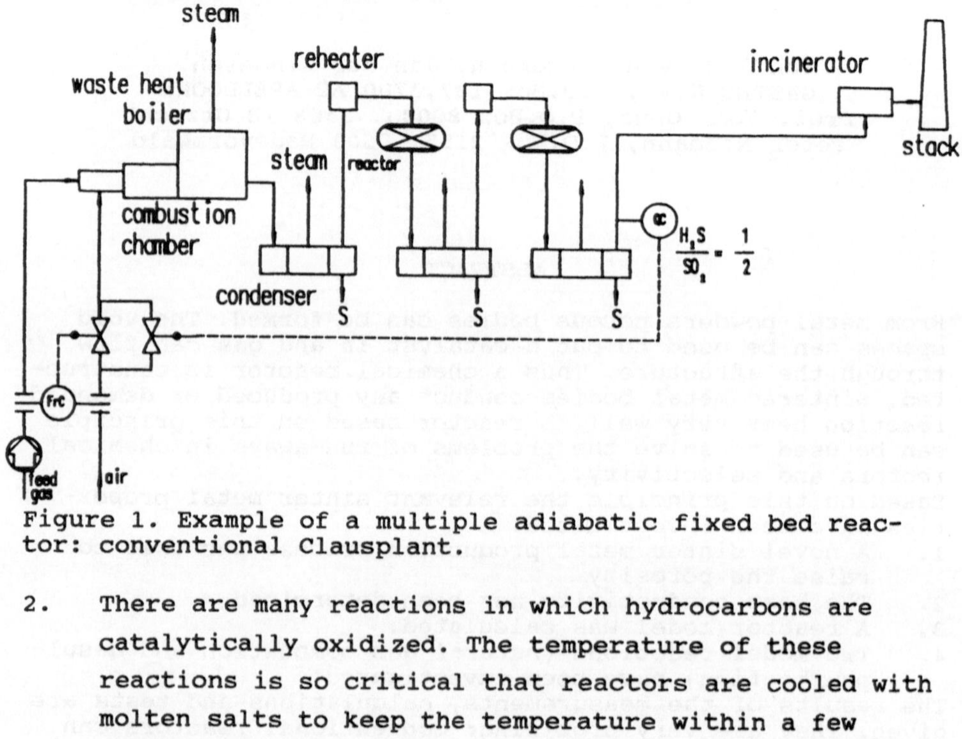

Figure 1. Example of a multiple adiabatic fixed bed reactor: conventional Clausplant.

2. There are many reactions in which hydrocarbons are catalytically oxidized. The temperature of these reactions is so critical that reactors are cooled with molten salts to keep the temperature within a few degrees and the reactor dimensions are only a two or three catalyst particles.
3. Fluidized bed operation will give an isothermal bed. However this option has several disadvantages as attrition, pressure drop, dust emissions etc.

These solutions have to be chosen because it is very difficult to get heat directly out of a packed catalyst bed.

This is due to two reasons:

1. The fixed bed catalyst packing has a very poor heat conductance as a result of the many transition resistances from one particle to another.
2. As the catalyst particles have only very little contact with the reactor wall, there is a considerable heat transfer resistance.

Sinter metal matrices don't have these two problems. Because of the sintering of the metal powder particles are joined together and to the wall (see figure 2.).

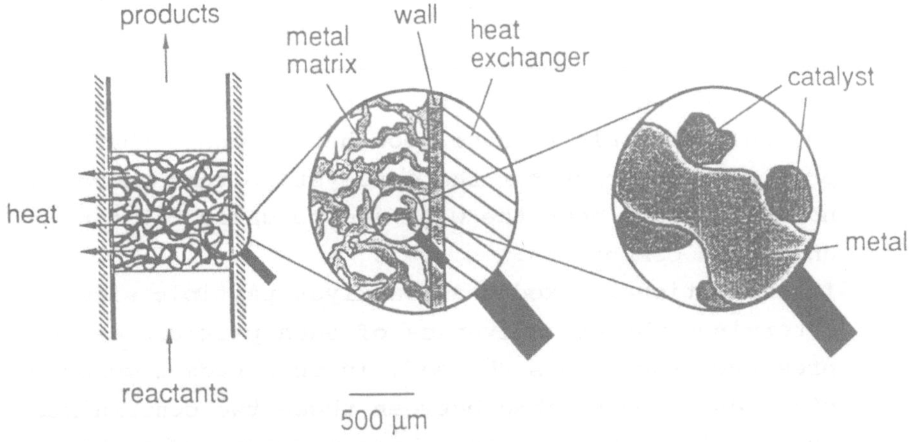

Figure 2. Sintermetal matrix.

This research was done in a cooperation between Krebsöge, a sinter metal producer from Radevormwald in Germany, The Institut für Kernenergie und Energieumwandlung of the University of Stuttgart in Germany, the State University of Utrecht and Gastec, both in the Netherlands.

Krebsöge produced sinter metal samples and parts and determined the flow characteristics. The University of Stuttgart determined the heat conductivities. The University of Utrecht deposited the catalyst on the sintered metal and determined the kinetics of the reactions. This was tested and evaluated in a model by Gastec.

The project was partially funded by the EC-program JOULE.

SINTERED METALS

The advantages of the use of sinter metals are:

1. Although the construction of a reactor with sinter metals in it, is an elaborate and therefore expensive process, the possibility of integrating reaction with heat transfer is making the complete construction cheap.
2. The sintering of the sinter metal to the wall makes it not only possible to have good heat transfer but also provides a carrier for the catalyst so that there is no channelling near the wall and no short cut for unreacted components.
3. In conventional fixed beds catalyst particle size determines the effectiveness of each particle and the pressure drop across the bed. In such beds a compromise has to be reached between these two constraints. In a sinter metal reactor these parameters can be chosen independently because the pressure drop is fixed by the sinter metal matrix while the effectiveness of the catalyst is determined by the catalyst particle size.
4. If fouling occurs in such a conventional reactor it can only be removed by opening the reactor and removing the catalyst and cleaning it outside the reactor. A sinter metal matrix may be cleaned by giving a gas pulse in the opposite direction of the normal flow,

thus blowing out the dust particles which have settled down on the catalyst. In conventional beds the catalyst would have blown out as well.

As an example of an application of sinter metals, the SUPERCLAUS process was chosen:
Conventionally, in a Claus plant H_2S is oxidized to sulphur in two steps (see figure 1):
Firstly, the H_2S is for a third part burned to SO_2 :

$$H_2S + 1.5\ O_2 \Rightarrow SO_2 + H_2O$$

Secondly the non-burnt H_2S reacts with SO_2 to sulphur in the Claus reaction:

$$2\ H_2S + SO_2 \Leftrightarrow 1.5\ S_2 + H_2O$$

This reaction is an equilibrium reaction so that a part of the H_2S will not react. About 3 to 5% will slip through, depending on the number of stages (after each reaction stage sulphur is removed) and the inlet concentration.
SUPERCLAUS is the name of a new process in which a newly developed catalyst is used, which oxidizes H_2S to sulphur in one step:

$$H_2S + .5\ O_2 \Rightarrow 0.5\ S_2 + H_2O$$

Due to the heat effect, this last reaction can only be applied to the last 10 to 20 % of the total inlet H_2S (actual concentration: 1 to 3%).
The SUPERCLAUS reaction was chosen as an example because:

1. The yield of the process and the inlet concentration can be increased by using a isothermal reactor.
2. The process is well known, because it was developed by the University of Utrecht, Gastec and the dutch en-

gineering company Comprimo. It is an industrial application.

3. As explained in figure 3. inside the catalyst particle a side reaction occurs in which sulphur is further oxidized to the undesired SO_2 and therefore the catalyst particle should be as small as possible. In order to avoid problems with pressure drop the application of a sinter metal carrier is advantageous.

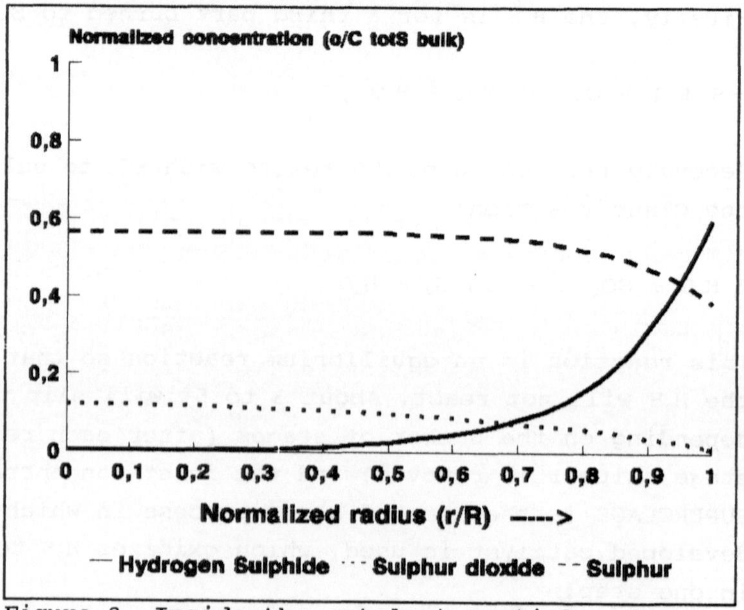

Figure 3. Inside the catalyst particle, S reacts to SO_2.

REACTIONS

The main reaction is to oxidize H_2S to sulphur. However, investigations into the system reveals that there are two other reactions present as well. H_2S is also oxidized to SO_2 directly and the formed sulphur is oxidized to SO_2 in a follow up reaction.
In figure 4 the results are summarized.

POROSITY

The porosity of a sintered metal body which is formed by letting the metal powder flow freely onto a shape after which it is sintered, is low. Too low to, optimize a reactor in which the heat production and heat removal are to be balanced. Moreover the porosity is not the same throughout the body [3].

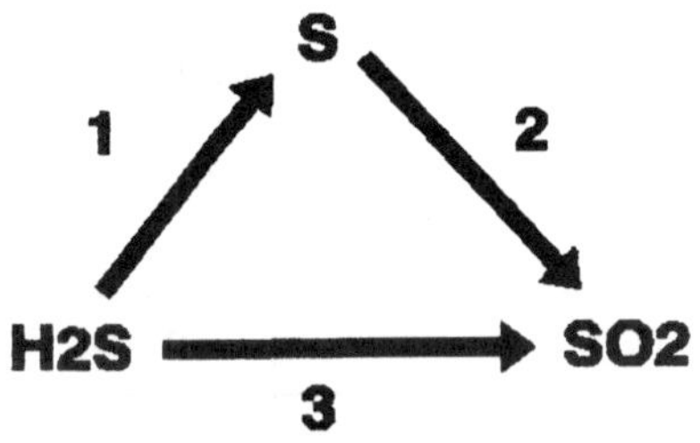

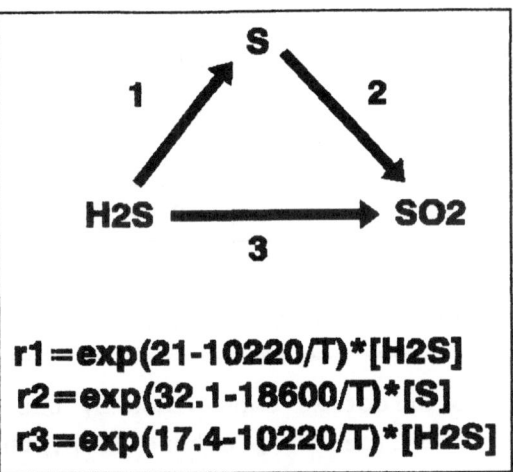

$$r1=\exp(21-10220/T)*[H2S]$$
$$r2=\exp(32.1-18600/T)*[S]$$
$$r3=\exp(17.4-10220/T)*[H2S]$$

Figure 4. Reaction paths.

A new method was tried by Krebsöge (Z/S process) to come to a uniform distribution but variable porosity. This method consists 1
of shooting the metal powder with a resin into its shape. In this way a uniform porosity which varies between 40 and 70% is reached.

HEAT CONDUCTIVITY

The heat conductivity is much better than conventional catalyst carriers. In table 1. this is summarized.
As can be seen sintering of metal particles to a matrix and to the wall improves the heat conductivity significantly. If no sintering is done, the particles do not fit the wall and the porosity near the wall is equal to unity. Therefore there is no heat conductivity and there is a considerable slip of unreacted gases. This can only be solved by lengthening the reactor tubes, thus introducing extra pressure drop.

TABLE 1.

Heat conductivity of sintered metals.

Material	Conductivity W/mK	Contact resistance K/W
Al_2O_3	0.32	1.8
316L, powder	0.55	2.5
316L, sintered	3-12	0

This effect is emphasized by the relative higher velocity to equalize the pressure drop, the absence of catalyst and the higher surface of the walls in (small) tubes.
In sintered metal tubes the slippage is 10 to 15 times less.

SIMULATION MODEL

To come to a design procedure the simple Krischer model was adopted [4]. See figure 5. The model combines two simplified structures of porous materials. The porous material is seen as parallel plates. In one part the heat flow is parallel to the plates (A) and in the other structure the heat flow is perpendicular to the plates (B). The fitting parameter a, which determines which part of the system is parallel, can be derived from measured data.

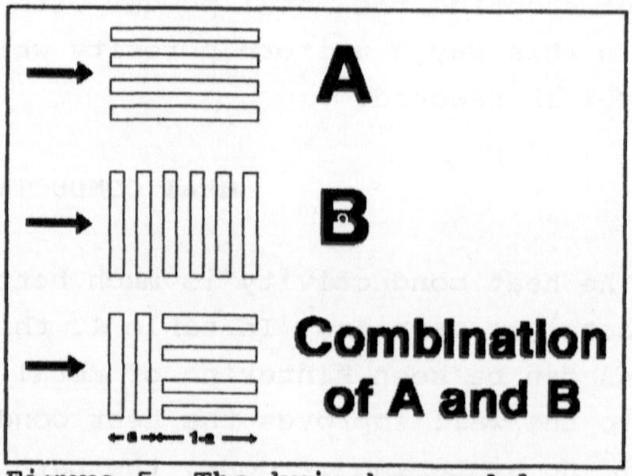

Figure 5. The krischer model.

Both the heat production and removal are functions of the porosity: when the porosity is low there are high conductivities possible but little void space available for catalysts (see figure 6).

To remove the heat the reactor tubes can be put into a boiler of about 85 bars to hold a constant temperature of 573 K (see figure 7). The results of the calculations are summarized in tables 2, 3 and 4.

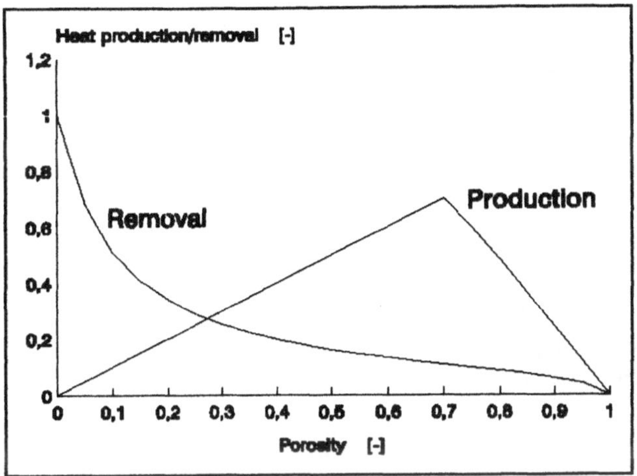

Figure 6. Heat production and removal as function of porosity.

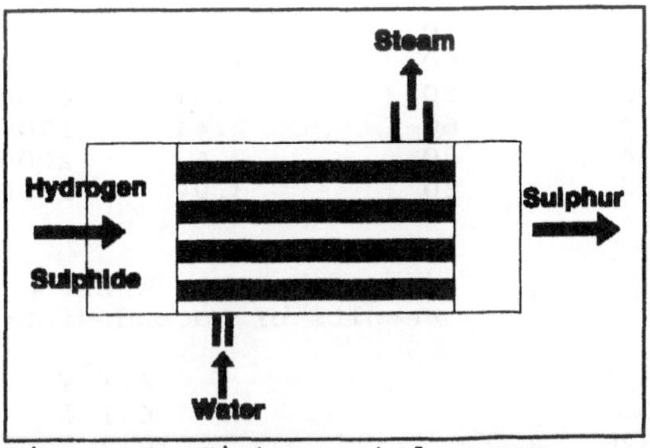

Figure 7. Sinter metal superclaus plant boiler.

TABLE 2.

Data of basic calculations.

Temperature steam	300 °C.
Velocity	2 m/s
Porosity	60%
d_p	10 mm
P_{steam}	85 bar.
Length	1.3 m
Tube diameter	25.4 mm
H_2S_{out}	2.4 % of inlet
S_{out} (yield)	87%
SO_{2out}	10.6%
Number of tubes	3200
Amount Sinter metal	6800 kg
Pressure loss	.03 bar
Residence time	.6 s

TABLE 3.
Results of various velocities and porosities.

Velocity m/s	Porosity %	Length m	Nr.tubes	Diameter mm	Sinter metal kg
2	40	1.83	1200	41	14200
2	60	1.31	2700	28	6800
2	80	1.31	4000	23	3400
4	60	2.44	1300	28	6300
4	80	2.63	2000	23	3400
6	80	3.94	1300	23	3400

TABLE 4.
Results of second sinter metal stage.

H_2S_{in}	4.1 %
O_{2in}	6.1 %
$H2O_{in}$	30.7 %
N_{2in}	59.1 %
Pressure	1.2 bar
Temperature steam	300 °C.
Velocity	2 m/s
Porosity	60%
d_p	10 mm
P_{steam}	85 bar.
Length	1.2 m
Tube diameter	88.4 mm
H_2S_{out}	.045 % of inlet
S_{out}(yield) sec. stage	89% of inlet H2S

Total yield	98.9 %

Continuation of table 4.
Results of second sinter metal stage.

SO_{2out}	0.4 % of inlet H_2S
Number of tubes	275
Amount of Sinter metal	6500 kg
Pressure loss	.02 bar
Residence time	.6 s

It can also be estimated that energy savings of up to 25 % are possible: in a conventional plant the gas has to be reheated after each stage and the waste heat boiler. This is avoided in the new concept.

CATALYTIC COMBUSTION OF NATURAL GAS.

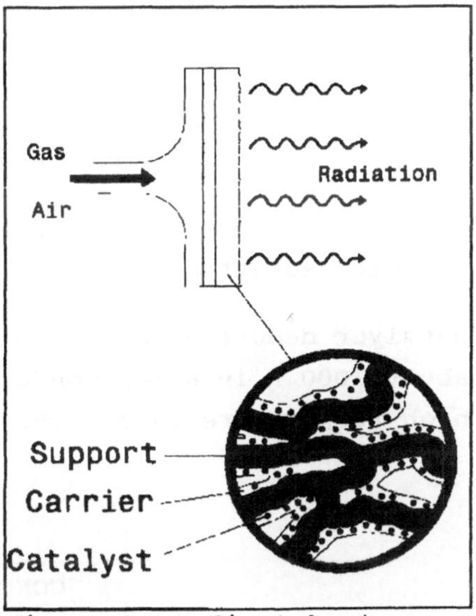

Figure 8. Radiant heater.

In radiant heater (see figure 8) gas and air are flowing through the catalyst. This results in a temperature distribution in the panel. If the temperature is too high the catalyst will be destroyed. The hot spots in this panel are determined by the heat conductivity of the carrier material: if the heat conductivity is low, higher temperature differences are needed to remove the heat away. As the capacity determines the surface temperature, the heat conductivity determines the hot spot temperature. In figure 9 the temperature distribution is more even in the sinter metal case when compared to the ceramic case.
This means that significantly higher capacities can be obtained. If the maximum temperature is 1100 K to prevent

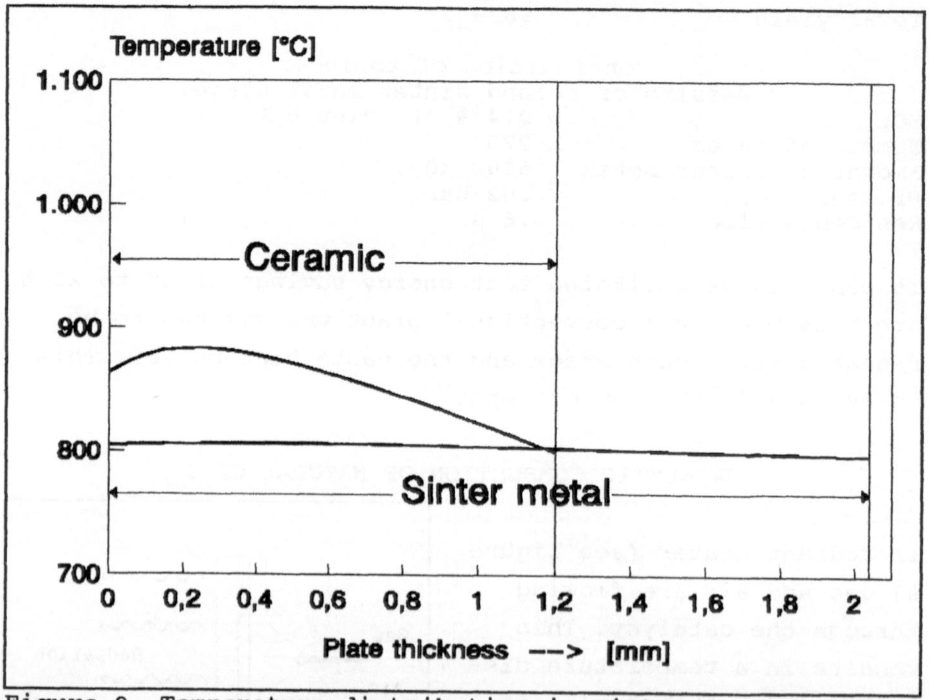

Figure 9. Temperature distribution in sintered metal and ceramic radiant heater.

catalyst deactivation, then the surface temperature can be about 1000 K in a low conductivity panel and 1100 K in a high temperature panel. The capacity can be 50% higher.

CONCLUSIONS

Sinter metal carriers have high potentials for catalyst reactors.

Based on measurements and calculations it can be seen that industrial reactors and simple applications can benefit from the advantage of its properties.

Future R&D

further steps to be taken are:

a. Further development of the sintered metal optimization method (Z/S process).

b. Development of a better model than the Krischer model for the heat removal for which a considerable amount of sintered metal samples are to be measured for porosity and heat conductivity.

c. Development of a general computer model for all kinds of reactions and shapes.

d. Construction of a pilot plant for H_2S oxidation.

REFERENCES

[1] Küpers, G.R., Flow and Temperature Behaviour in a Catalyst bed, Comprimo 22nd ACHEMA Exhibition Congress, 1988.

[2] Suter,D., Radial flow reactor optimization for highly exothermic selective oxidation reactions. Chemical Engineering Science, Presented at ISCRE 11, Toronto.

[3] Neumann, P., R. Sicken, V. Arnhold, New manufacturing process for high porous sinter elements. Presented at the 1992 Powder Metallurgy World Congress in San Francisco.

[4] Schlünder,E.U.,E. Tsotsas, Wärmeübertragung in Festbetten, durchmischten Schüttungen und Wirbelschichten, Georg Thieme Verlag Stuttgart, 1988

SESSION 13:

Chemical Reactors
Continued

Chairman: Prof. K.R. Westerterp

INCREASED YIELD AND ENERGY EFFICIENCY THROUGH PROCESS INTENSIFICATION

J. HANNON & R. KING
BHR Group Limited, Cranfield, Bedford, MK43 0AJ England

ABSTRACT

The concept of Process Intensification is described with particular emphasis on chemical reactors. Examples of applications are given. Experimental reaction data from a high-intensity loop reactor are compared with simulations of a stirred tank under identical chemical conditions. Depending on reaction chemistry, a loop system can produce a higher yield with reduced energy consumption.

INTRODUCTION

Process Intensification was a term introduced during the 1970s to describe order of magnitude reductions in plant size or increases in processing rates by selection of alternative process designs. The term has come to mean more generally any improvement in plant performance. Intensification can be achieved in four ways:

Equipment Intensification

whereby a stirred tank might be replaced with a high-intensity mixer

Chemical Intensification

where a change in feedstock might lead to higher yield

Mechanical Intensification

where one piece of equipment accomplishes several tasks (e.g. pump as mixer)

Layout Intensification

where judicious plant layout saves energy (e.g. PINCH Technology)

Potential benefits include

- increased process yield and product quality
- reduced raw materials consumption
- reduced environmental impact
- improved safety
- reduced energy consumption

In this article, methods are described for intensifying chemical reactors in particular, which has knock-on effects both up- and downstream Attention is further restricted to intensification by changes in reactor geometries

REACTOR INTENSIFICATION

The chemical reactor is the nucleus of the process plant This is where products and waste are initially formed Reactor intensification can lead to

- higher process yields
- reduced separation requirements
- reduced overall plant size
- reduced overall energy consumption

In a typical chemical production facility, energy consumption by the reactor forms a small fraction of the total plant load Units such as distillation columns, heat exchangers and compressors are both capitally more expensive and use more energy than the reactor By employing reactors of different geometry, the energy cost of the reactor can be reduced, but this may in itself have little impact on the overall energy use However if the reactor can be made to produce a higher quality product which requires less treatment by auxiliary plant, the overall plant energy use can be substantially reduced This is so even if to achieve higher quality product, the intensified reactor is energy intensive

MIXING EFFECTS ON REACTIONS

A broad class of important reactions have a high intrinsic speed Examples include organic oxidations, polymerisation, neutralisation, some precipitations, chlorination, hydrogenation and azo-coupling When these are carried out in conventional reactors (e g stirred tanks, bubble columns), reaction starts before mixing of reagents is complete This can lead to reduced reaction rates (limited by slow mixing) or more seriously, by product formation Mixing patterns therefore strongly affect yield Improved reactor designs would match mixing rates, residence times and their distributions with the specific requirements of the reaction

Evidence for these effects is widespread in the chemical engineering literature Paul & Treybal (ref 1) found that yield from a competitive consecutive reaction correlated with impeller tip speed Bourne (ref 2) found that moving a feed point from the liquid surface into the impeller discharge significantly reduced by product formation for an azo coupling Meyer et al (ref 3) found similar sensitivities with the precipitation of barium sulphate Oxidation of organics is an important multi-phase example (ref 4) Polymerisation, such as in reaction injection moulding, is another (ref 5)

Many of the reactions described, particularly those susceptible to by-product formation, would be best carried out in high intensity plug flow reactors with short residence times Examples of commercial designs include Sulzer and Kenics motionless mixers (Figure 1) These produce a small high-intensity

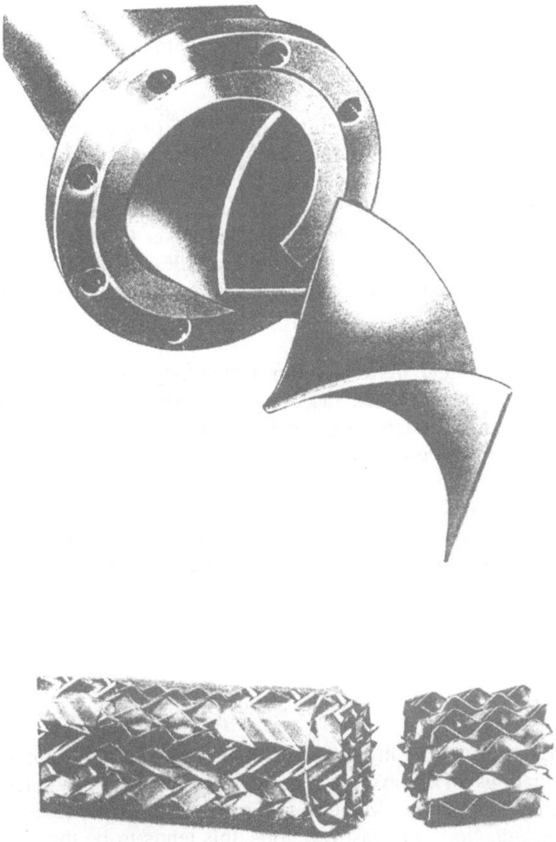

FIGURE 1: Kenics and Sulzer Motionless Mixers

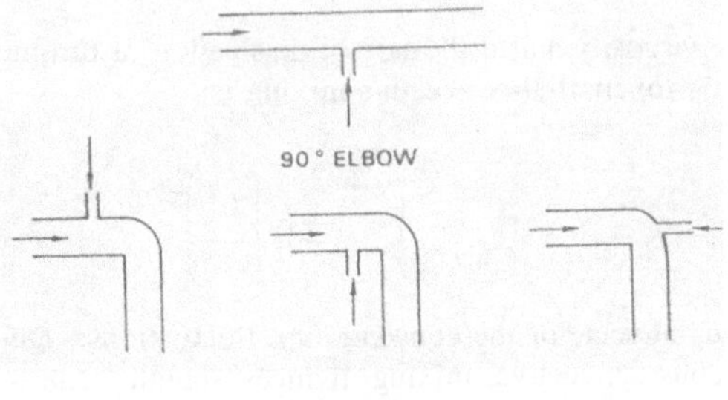

FIGURE 2: Pipeline Tee Mixers

mixing zone which allows the reactions to proceed at or near their intrinsic speed, and limit by-product formation by preventing products from coming into contact with reactants and maintaining thermal homogeneity.

A US based pharmaceuticals manufacturer recently upgraded a process involving precipitation of a pharmaceutical intermediate by intensified mixing (ref. 6). Use of a semi-batch reactor resulted in long batch times and a product with a wide particle size distribution. Milling was required downstream to achieve a uniform particle size, but this caused undesirable exposure of personnel to dust. The stirred tank was replaced by a simple pipeline tee mixer (Figure 2). By varying flowrates and concentrations, it was possible to obtain the desired particle size and shape without the milling stage. This is a concrete example of how better reaction conditions have benefits both up- and downstream.

Two main barriers exist to the wider uptake of this technology by industry. One is the natural conservatism of chemical engineers. When faced with a choice between conventional and new designs, the conventional is almost certain to be selected. Only when this results in conditions of severe adversity will alternatives be considered.

The second is the lack of reliable design data for high-intensity reactors. Most devices in this category were developed for specific applications, such as viscous polymer processing (some static mixers) and milling and grinding (rotor-stator mixers). A substantial research programme is now underway to provide design data for fast chemical reactions (ref. 7) and the comparison presented below draws on some of this work.

THEORETICAL BASIS

There are two important types of mixing in liquids: macromixing and micromixing. Macromixing is the equalisation of concentrations on a coarse scale in the reactor volume. It determines features such as the residence time distribution (RTD). Micromixing refers to the local reduction of concentration fluctuations and scale of segregation. For fast reactions, this tends to be the more critical stage.

Micromixing in turbulent liquids takes place in three stages: inertial-convective, viscous-convective and viscous-diffusive. These correspond to the three subranges of the turbulent concentration fluctuation spectrum (ref. 8). Inertial-convective mixing describes the reduction in size of blobs of reagents from a coarse to a fine scale, whose lower bound is defined roughly by the Kolmogoroff microscale:

$$\lambda_K = \left(\frac{\nu^3}{\varepsilon}\right)^{\frac{1}{4}} \tag{1}$$

where ν is the kinematic viscosity and ε the rate of dissipation of turbulent kinetic energy. A characteristic time-scale for inertial-convective mixing is:

$$t_{ms} = 2.0\left(\frac{L_s^2}{\varepsilon}\right)^{\frac{1}{3}} \tag{2}$$

where L_s denotes the integral scale of the concentration fluctuations. During and following inertial-convective mixing, viscous-convective mixing reduces length scales further, to the Batchelor microscale:

$$\lambda_B = \frac{\lambda_K}{\sqrt{Sc}} \tag{3}$$

where Sc denotes the Schmidt number A characteristic time-scale is

$$\tau_\omega = 12.7 \left(\frac{\nu}{\varepsilon}\right)^{\frac{1}{2}} \tag{4}$$

Viscous-convective mixing increases concentration gradients in fine-scale fluid elements so that molecular diffusion can act rapidly to remove inhomogeneity A characteristic time-scale for viscous-diffusive mixing within a small deforming eddy is

$$t_{DM} = 2 \left(\frac{\nu}{\varepsilon}\right)^{\frac{1}{2}} \text{arc sinh } 0.05\ Sc \tag{5}$$

Certain operating conditions favour certain mixing stages For example, an increase in fluid viscosity under turbulent flow conditions causes viscous-convective and viscous-diffusive mixing to slow down, whilst not affecting inertial-convective mixing It is possible, by minimising the initial length scale of the reactant blobs, to reduce the inertial-convective timescale, without affecting the other two Broadly speaking, this requires low feedrates of additive Under these conditions, inertial-convective mixing can be ignored Baldyga & Bourne (ref 9) have presented a mathematical model for this simplified problem which has been validated in several different types of reactor

$$\frac{dV}{dt} = E\ V \tag{6}$$

$$\frac{dC_i}{dt} = E\ (C_e - C_i) + R_i \tag{7}$$

$$E = \frac{\ln 2}{\tau_\omega} \tag{8}$$

Equation (6) describes the growth of the reaction zone volume which is initially equal to the volume of the additive stream Equation (7) describes the rate of change of each component concentration in the reaction zone, by mixing with the local environment and consumption or production by chemical reaction Equation (8) defines the mixing parameter in terms of kinematic viscosity and turbulent energy dissipation rate

This model is valid for liquids of relatively low viscosity (i e $Sc<4000$) It is applied below to a stirred tank reactor

EXPERIMENTS WITH A COMPLEX REACTION

Bourne & co workers (ref 10) have accurately measured the kinetics of the competitive consecutive and parallel reactions between 1 and 2 naphthol and diazotised sulfanilic acid which form several dyestuffs

A + B → R	k_1=12238 m^3/mol s
A + B → T	k_2=921 m^3/mol s
R + B → S	k_3=1 835 m^3/mol s
T + B → S	k_4=22 25 m^3/mol s
C + B → Q	k_5=124 5 m^3/mol s

On the basis of the model presented above, yield is a function of the following dimensionless groups.

$$X_{R+T} = f\left(\frac{k_i}{k_4}\ ,\ \frac{N_{AO}+N_{CO}}{N_{BO}}\ ,\ \frac{C_{AO}}{C_{CO}}\ ,\ \frac{V_{A+C}}{V_B}\ ,\ \frac{k_4 C_{BO}}{E}\right) \tag{9}$$

where k_i denotes the kinetics of reaction step i

When B is added to A and C with rapid mixing, R & T are preferentially formed If mixing is slower, Q is favoured and if it is very slow, S is favoured The amount of each dyestuff in the product mixture is thus related to the rate of mixing experienced in the reaction zone This scheme (or a subset) is typical of many simplified industrial reactions

We carried out this reaction in aqueous solution at 298 K in a loop reactor (Figure 3) containing a Kenics Static Mixer of 39 mm diameter Under these conditions, the kinematic viscosity (ν) of water is 0 89 x 10^{6} m^2/s Mixing energy is derived from passage of fluid over the blades of the mixer and thus varies with throughput

2500 litres of a fresh solution of A and C, buffered to pH 10 with a concentration ratio of 1.4 was pumped once through the mixer and collected in a receiving tank The mixer was therefore operated in continuous plug flow mode, while the overall mode of operation was batch A small stream of highly-concentrated B (50 mol/m^3) was added directly into the mixer A and C were in stoichiometric excess of 500% The volume ratio was 3000 Nominal liquid throughput varied from 0 5 to 1 25 l/s (1 l/s = 3 6 m^3/hr)

The intrinsic reaction time of the slowest reaction in the scheme is of order 0 01 s under these experimental conditions Reaction was completed within a few centimetres of the feed tubes Samples were taken downstream of the reaction zone, immediately diluted with buffer solution and then analyzed spectrophotometrically This yielded the quantity of each dyestuff present (or yield) A check on the validity could be made by comparing these concentrations with the known volume and concentration of B added Mass balances were within 5% of closure

SIMULATIONS

To compare results from the loop reactor with conventional equipment, equations (6) to (8) were used to predict the performance of a semi-batch stirred tank reactor (Figure 4) of 2 0 m diameter, agitated with a pitched blade turbine (0 6 m diameter, 4 blades at 45 degrees) pumping downwards, under the same chemical conditions (volume ratios, concentrations, etc) and liquid physical properties as the loop reactor The impeller rotational speed was 80 rpm

Turbulent energy dissipation rates were estimated from Laser Doppler Anemometry data (ref 11) The mean value was 0 1 W/kg and values of 0 01 and 2 W/kg were used for the liquid surface and impeller regions, respectively In order for viscous convective mixing to control (and the model to be applicable), the batch (or feed) time was taken to be 100 times the circulation time Simulated results are compared below with measurements in the loop reactor

RESULTS & DISCUSSION

Tables 1 & 2 (columns 1 and 2) compare experimental data for yield of desired products R & T in the loop reactor with model calculations for the stirred reactor Yield increased with increasing throughput in the loop, and varied with addition point in the stirred tank, being lower near the liquid surface than in the impeller region Surface addition is more common in practise, being more

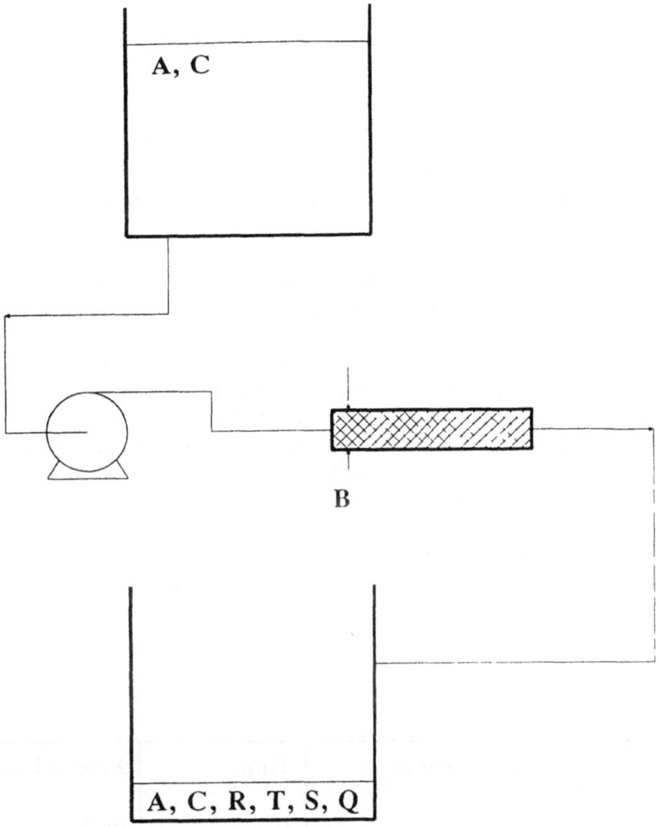

FIGURE 3: Loop Reactor Used for Reaction Measurements

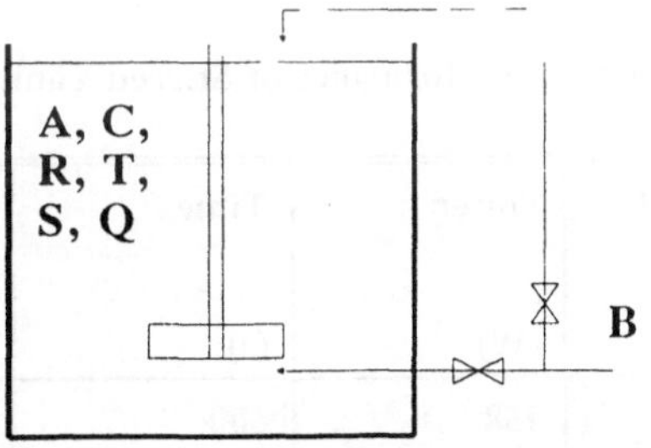

FIGURE 4: Stirred Tank Reactor Used in Simulations

convenient. Yield in the loop was generally higher than in the tank.

The total energy consumed in mixing is the product of power consumption and batch time. These data are tabulated in columns 3,4 and 5. The loop reactor required as little as 4% of the energy consumed by the tank and still produced a higher yield than surface addition. This energy saving is significant, even assuming very moderate pump efficiency. Batch times are longer with the loop reactor, but can be shortened by increasing the throughput. At high flowrates and short batch times, the energy consumed by the loop would approach and exceed that of the tank system, but the loop would maintain a higher yield.

Energy consumed per mole of R & T produced can be readily calculated and is given in column 6. The most favourable value in the tank is still 5 times higher than the least favourable in the loop.

Superior performance of the loop stems from high intensity mixing in the reaction zone and the near-plug flow pattern. These ensure that rapid reaction takes place as soon as B enters and that R & T are not recirculated back into the reaction zone, where degradation to S can occur. Similar results have recently been obtained with Sulzer mixers (ref. 12). The loop has the added advantage that it can be operated in fully-continuous mode without loss of yield. Yield from the stirred tank would decrease further if operated in fully-continuous mode.

These results apply for the particular reaction and conditions of experiments and simulations. Suitability of loop reactors for improved mixing depends on the mixing-sensitivity of the reaction considered.

TABLE 1: Performance of Loop Reactor

Loop Reactor Flowrate (l/s)	Yield of R & T (%)	Power (W)	Time (s)	Energy Used (kJ)	Energy per unit product (kJ/mol)
0.5	65	1.06	5000	5.3	0.2
0.75	68	3.6	3333	12.0	0.42
1.0	72	8.5	2500	21.3	0.71
1.25	74	16.6	2000	33.2	1.08

TABLE 2: Performance of Stirred Tank Reactor

Stirred Tank Addition Point	Yield of R & T (%)	Power (W)	Time (s)	Energy (kJ)	Energy per unit product (kJ/mol)
Surface	37	258	600	154.8	10.04
Impeller	70	258	600	154.8	5.31

The above calculations are restricted to the reaction unit. Higher yield also means reduced need for separation and purification, representing a further substantial cut in energy use as well as capital and running costs.

CONCLUSIONS

For certain rapid reactions with complex pathways, in-line high-intensity reactors potentially offer higher yield and greater energy efficiency than stirred tank reactors. Design data for these systems is now becoming available, which will encourage wider use of this technology.

NOMENCLATURE

A	species A	-
B	species B	-
C	concentration of species	mol/m^3
E	engulfment rate	1/s
k	reaction velocity constant	$m^3/mol.s$
L	length	m
N	number of moles	moles
Q	species Q	-
R	species R	-
R	rate of reaction	mol/s
S	species S	-
Sc	Schmidt number	-
T	species T	-
t	time	s
V	volume	m^3
X	yield	-
ε	rate of turbulent energy dissipation	W/kg
λ	length scale	m
ν	kinematic viscosity	m^2/s
τ	time scale	s

Subscripts:

0	initial
A+C	of species A and C
B	Batchelor
	of species B
DM	diffusive mixing
e	environment
i	species i
	reaction i
K	Kolmogoroff
ms	inertial-convective
R+T	of species R & T
s	segregation
ω	vortex

REFERENCES

1. Paul, E. & R. E. Treybal, AIChEJ, Vol. 17, No. 3, p. 718 (1971)
2. Baldyga, J. & J. R. Bourne, Chem. Eng. Comm., Vol. 28, p. 231 (1984)
3. Meyer, T. et al., Chem. Eng. Sci., Vol. 43, No. 8, p. 1955 (1988)
4. Ramshaw, C., Proc. Int. Chem. React. Eng. Conf., p. 685, John Wiley & Sons, New York (1984)
5. Tucker, C. L. & N. P. Suh, Polym. Eng. Sci., Vol. 20, No. 13, p. 875 (1980)
6. Liu, P. et al, AIChE Annual Meeting, paper 66b, Chicago (1990)
7. Zhu, M., Design Guide for Motionless Mixers as Gas-Liquid Reactors, BHR Group Report HDG02, July 1992.
8. Tennekes, H. & J. L. Lumley, A First Course in Turbulence, MIT Press (1972)
9. Baldyga J., & J. R. Bourne, Chem. Eng. Journal, Vol. 42, p. 83 (1989)
10. Bourne, J. R. et al, Ind. Eng. Chem. Res., Vol. 31, 1992
11. Hannon, J., PhD Thesis, Cranfield Institute of Technology, in preparation.
12. Bourne, J. R. et al, Ind. Eng. Chem. Res., Vol. 31, 1992

DEVELOPMENT OF AN INDUCTIVELY HEATED REACTOR

DAVID GARDNER
EA Technology
Capenhurst, Chester, Cheshire, CH1 6ES, UK

ABSTRACT

A new concept in electrically heated reactors is being developed that will increase the selectivity and conversion of certain catalytic reactions. Most types of catalysts can be heated, which eliminates any requirement for high cost high frequency generating equipment. The method is based on the idea of induction heating which has already been successfully applied to traditional engineering operations. The method is particularly suitable for petrochemical applications where the main reaction is endothermic. Heat can be added at any or all points of the catalyst bed so that the ideal temperature profile can be obtained. Hydrodynamic tests show that the new design of reactor performs as a standard packed bed reactor. The whole range of temperature profiles have been obtained, from decreasing temperature with bed depth through essentially isothermal operation to increasing temperatures, all for the same reaction.

INTRODUCTION

Chemical plants can be split into a number of generic units which form the building blocks of every process. These are commonly referred to as the unit operations and comprise such functions as separation, heat transfer, transport of materials etc. Perhaps the most crucial of all operations is the reactor as, by definition, this is where an incoming material will be chemically altered from one state to another.

Quite often reactions are represented in very simple terms, for example the well known but very rarely encountered reaction:

$$A \dashrightarrow B$$

For this reaction it is the rate at which B is formed that is of importance. This rate can often be expressed as:

Rate = k (A) where (A) is the concentration of A at any time and k is the rate constant.

For simple systems the rate constant is given by:

$k = A \exp(-E/RT)$ where A is the Arrhenius factor, E is the activation energy, R the gas constant, and T the temperature.

This is a very simplistic approach but the fact that the rate is highly dependent on temperature always holds. As the temperature appears in the form of an exponential term it can be appreciated how sensitive reaction rates are to temperature.

Much time and effort is spent in trying to find the exact relationship between the rate constant and temperature and concentrations of reactants. Even for relatively simple reactions the kinetics are not always found with ease. However, reactions are always sensitive to temperature variations and so if the correct temperature can be held within a reactor it is possible to run a whole plant very close to its optimum efficiency.

It is on this premise that the work at EA Technology has progressed. The electromagnetic field produced is not expected to alter chemical kinetics in any way, but merely to provide a method of heating the reactor and its internals to obtain ideal temperatures. These temperatures can be obtained on the small scale reactor used for calculating chemical kinetics e.g. differential reactors, but cannot be obtained on large scale equipment due to problems of heat transfer. The new design of reactor proposed is able to heat large vessels uniformly, with very little limitation on the power density.

Induction Heating

Induction heating is achieved by the passage of current through a material when the current is induced from a separate source. Whatever the configuration, induction heating always involves a varying magnetic field. Figure 1 shows the two methods by which eddy currents can be induced into a material.

The frequency of this field can vary by many orders of magnitude, from a few hertz to hundreds of kilohertz. Through heating of a workpiece is achieved at lower frequencies, whilst at the higher frequencies only the surface of the material will be heated by the magnetic field. At all these frequencies the method of heating is that of the Joule effect (I^2R losses). If the frequency is increased still further, into the radio frequency range, the dielectric properties of the material become important and determine the heating effect.

One of the most important factors in induction heating is the concept of skin depth. If a magnetic field is being generated near a metal surface the strength of that field will drop as it is used to generate a current in the metal. Although the magnetic field will extend indefinitely into the material a point is reached where most of the available energy will be dissipated. The definition of skin depth is given as the distance at which the amplitude of the field falls to 1/e of the incident amplitude. As the power associated with a wave is proportional to the square of the amplitude this means that at one skin depth into the material 86% of the energy has been dissipated. More energy is available, but compared to the surface the power density is relatively small.

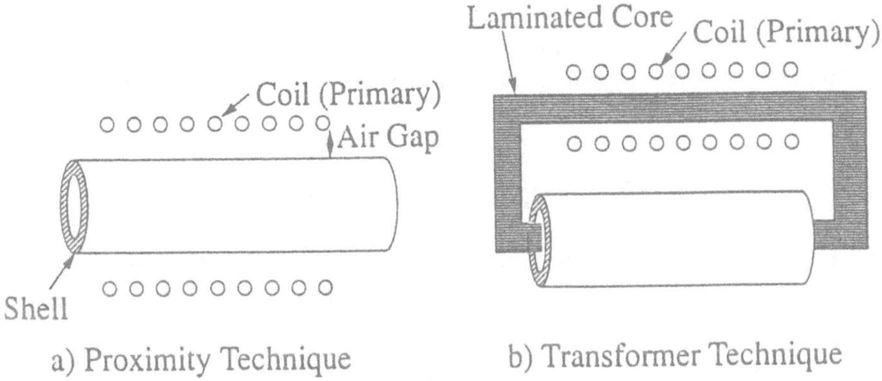

Figure 1 Methods of Induction Heating

The value for the skin depth is dependent on the resistivity; permeability; and the frequency in the following manner:

$$d = (2\rho/\mu\omega)^{1/2} \quad (1)$$

where ρ = resistivity

μ = permeability ($=\mu_r\mu_o$)

ω = angular frequency ($=2\Pi f$)

Typical values at 50 Hz and ambient temperature:

copper	= 9.3 mm
mild steel	= 9.0 mm
stainless steel	= 70 mm

When induction heating is taking place the workpiece forms a secondary winding to a transformer. In some instances this may not be obvious but it is always true. The calculation of voltage and current flowing in an ideal situation is quite straightforward:

$$V_p/N_p = V_s/N_s \quad (2)$$

where V is the voltage and N is the number of turns (the subscripts p and s denote the primary and secondary windings respectively). Therefore, for a single secondary turn the voltage will be equal to the voltage in the primary divided by the number of turns on the primary winding. As the power must be equal (ideally) then the current must increase in the secondary compared to the primary by the same factor.

Although no system can be ideal the performance of 50 Hz equipment is very efficient. The delivered power at the workpiece will be over 90% and so the overall efficiency (which takes into account heat loss) will approach 90%.

EXPERIMENTAL METHOD

Three pieces of equipment were designed and constructed to assess the impact the new reactor design would have on the operation of a fixed bed catalytic reactor. A plastic model was built to test the fluid dynamics of the new internal design; a section through the reactor was used to monitor the operation of the induction heating method; and two reactors were built to accommodate a catalyst packed bed of 13 litres, and assess the effect on the chosen test reaction.

The dehydrogenation reaction of ethylbenzene to produce styrene was chosen as a test case for this study. The catalyst used is typically made from iron oxide with various inhibitors and promoters included. Reaction temperatures vary between 550°C and 650°C. Steam is used to reduce the partial pressure of the reaction mixture and also to de-coke the catalyst via the water-gas shift reaction. Commercially this process is now carried out under a slight vacuum using two radial flow reactors in series. The pressure is kept low so that the equilibrium conditions can be improved towards the production of styrene. Typical conversions through a single adiabatic reactor may be as high as 45% with a selectivity of over 90%. The temperature drop over the catalyst bed will approach 100°C. This temperature drop is due to the high heat demand of the reaction to convert the ethylbenzene into the styrene monomer. In the second reactor, after some interstage heating, the overall conversion can be increased to 70%. The temperature drop is not as great, but side reactions will start to play a more important role in the process.

An ideal temperature profile for the reactor can be found by calculating the change in reaction rate with temperature for the styrene producing reactions and the change in the rate of production of all by-products. When this is done a resultant temperature profile is found that increases with conversion and hence increases with bed length. This is shown in figure 2, which includes the temperature profiles obtained when using the split reactor system and an isothermal reactor. By comparing these three sets of curves it can be seen that neither the isothermal reactor nor the split reactor approach the ideal temperature profile. The split reactor system could approach the ideal if many more reactors were used with each subsequent reactor inlet temperature being slightly higher than that preceding it. However, economic and practical considerations rule out this approach. In an attempt to achieve the ideal temperature profile a new design of reactor has been proposed that effectively provides a large number of reactors with interstage heating, but all within the same reactor.

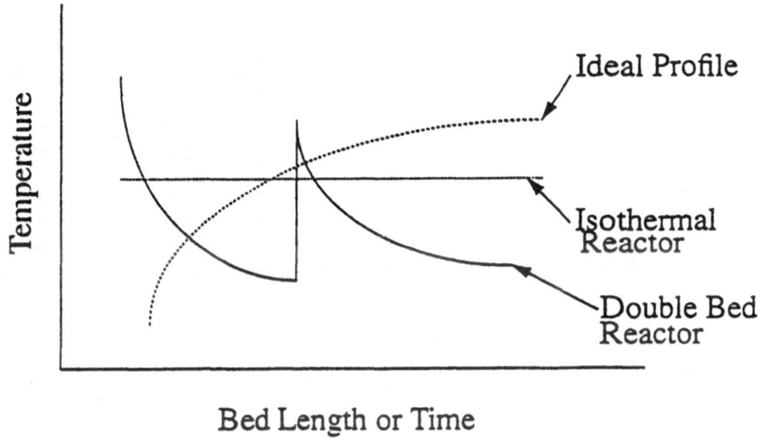

Figure 2 Temperature Profiles for Different Reactor Configurations

Operation of the Styrene Test Rig

A small scale rig was built for experimental verification of the original theoretical work. This consisted of the equipment necessary to raise steam to 1000°C and ethylbenzene to over 700°C. The maximum flowrate of ethylbenzene was just over 5 kg/h and that of steam was over 10 kg/h. Two reactors were built with both able to contain a catalyst bed of 30 cm diameter and 35 cm deep. The first reactor operated as the first stage of a standard adiabatic system. This was used as a test case for comparison with the inductively heated reactor, which could operate using both methods of induction heating (refer to figure 1). The main aim of the experiments was to manipulate the temperature profile within the reactor, but also to compare the performance of the inductively heated reactor compared with the adiabatic reactor.

RESULTS

The experimental equipment with the adiabatic reactor in place was run for more than 300 hours with over fifty experiments being performed. Table 1 gives a selected number of results for the adiabatic reactor. The temperature given is that of the steam/ethylbenzene as it leaves the superheater unit.

TABLE 1

Conversion and selectivity using the adiabatic reactor

Experiment	Flowrates EB (l/hr)	Flowrates Steam (kg/hr)	Temperature (°C)	Conversion	Selectivity
1	2.5	8.0	681	44.6	90.4
2	3.75	8.0	680	35.6	91.6
3	1.5	9.0	688	60.9	90.3
4	4.0	5.9	691	32.0	89.4
5	4.11	6.31	704	37.0	86.8
6	4.6	8.4	717	41.9	87.4

Figure 3 shows the temperature with time of the twelve thermocouples for the final experiment, which is the one that gave the best results for conversion and selectivity. This shows the approach to steady state as well as the temperature drop through the reactor once there.

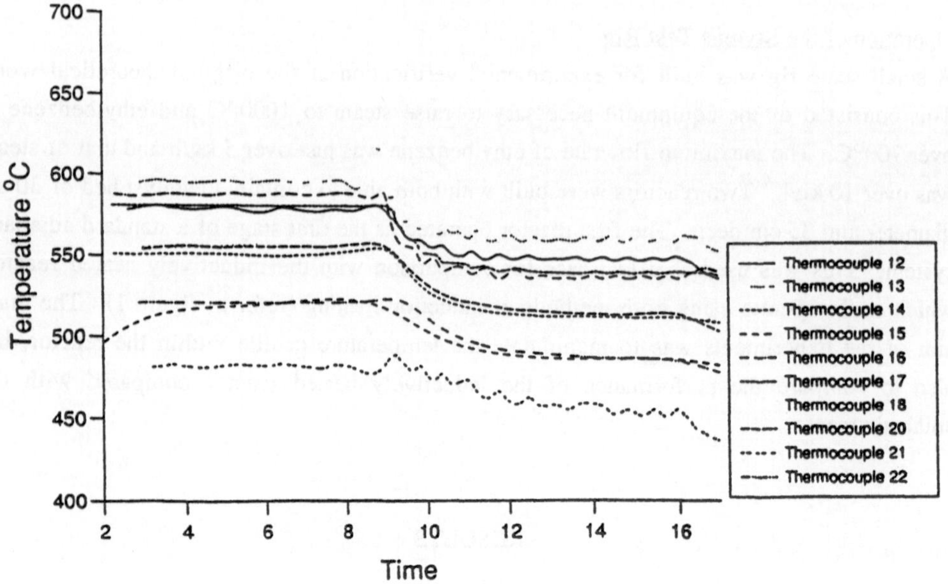

Figure 3 Temperature vs time for experiment 6

Table 2 shows the conversion and selectivity for some of the experiments carried out on the inductively heated reactor. The power to the transformer and coils was adjusted so that varying temperature profiles within the catalyst bed could be achieved. Experiments 8 and 12 can be compared to experiment 6. Although the conversion is lower the selectivity is much greater, even though the ratio of steam to ethylbenzene in experiment 12 should produce a lower selectivity. The improvement is due to the ability to lower the incoming reactant temperature as heat is now being added exactly where it is needed, i.e. inside the reactor.

TABLE 2

Conversion and selectivity for the inductively heated reactor

Experiment	Flowrates		Temperature (°C)	Conversion	Selectivity
	EB (l/hr)	Steam (kg/hr)			
7	2.5	8.0	648	48.1	90.0
8	3.81	8.0	621	39.1	89.9
9	3.81	10.5	654	36.8	90.5
10	3.81	10.5	644	42.4	90.5
11	3.81	10.5	656	55.4	87.8
12	5.1	7.5	608	36.6	89.4

Figure 4 shows the temperature variation with time for the experiment that most closely attained the isothermal catalyst bed condition being sought in the first series of experiments.

After these experiments were completed the power into the reactor was increased further and at the same time the inlet temperature was dropped by lowering the degree of superheat in the ethylbenzene and steam lines. Figure 5 shows one experiment in which the outlet temperature exceeded the inlet temperature by more than 100°C. The effect that the induction heating had on the process is seen after the power was turned on at 07:00. At this point the temperatures at the base of the reactor rose very sharply whilst those at the top increased only slightly.

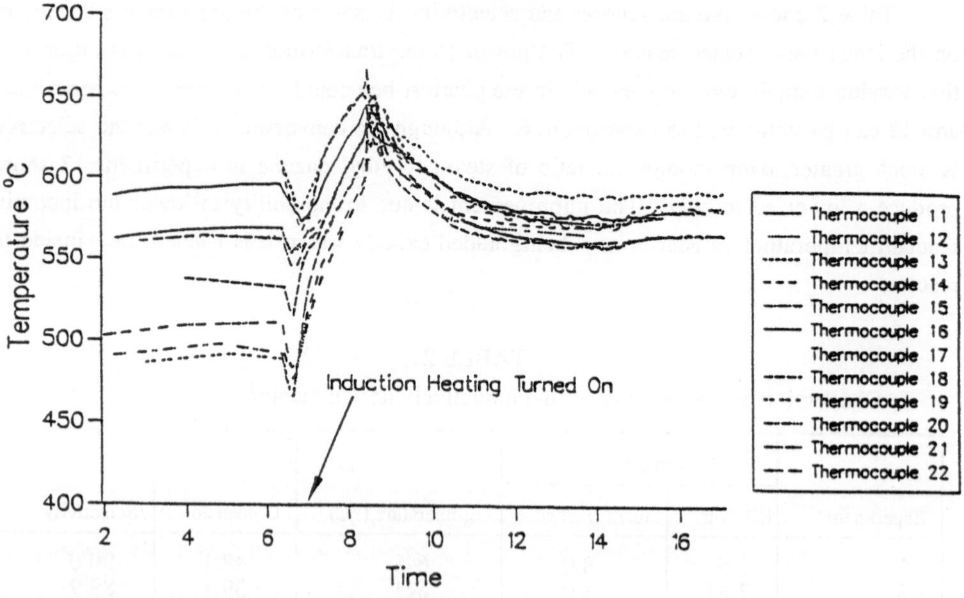

Figure 4 Temperatures in a near isothermal reactor

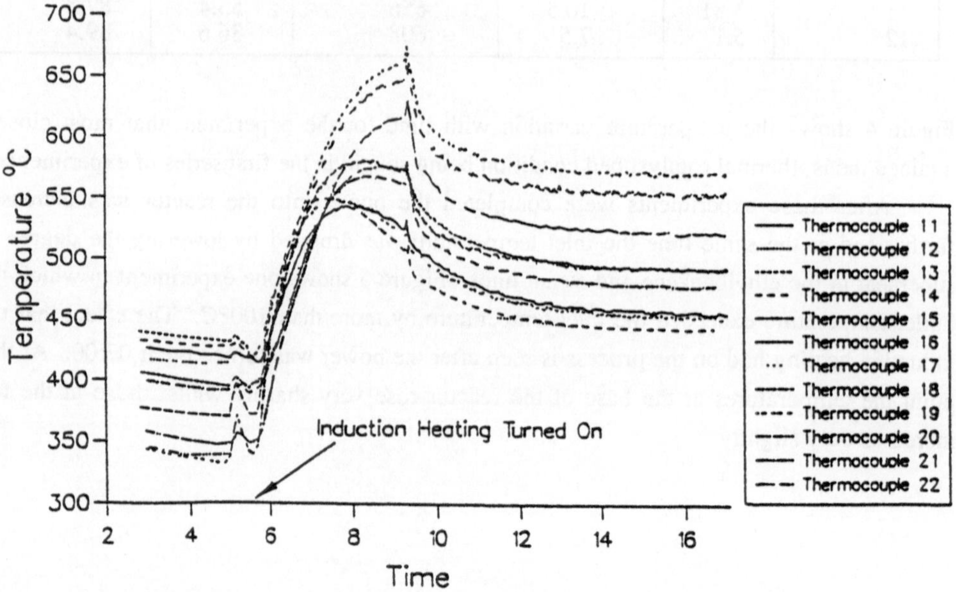

Figure 5 Temperatures for a reactor with increasing temperature with bed depth

DISCUSSION

A large scale experimental test rig has been built and successfully operated for the reaction to produce styrene from ethylbenzene. Two reactors have been used, the first was operated as the first stage of a typical industrial double bed adiabatic installation, and the second reactor used the new technology to heat the catalyst bed. The first reactor was fully optimised to obtain the best conversion and selectivity possible from the equipment. Lower selectivity for conversion was obtained than would normally be expected due to the heat loss from the reactor, but the results did approach those expected from the catalyst used.

The new reactor design was tested by three rigs, one to test the fluid dynamics, one to optimise the induction heating, and finally the large rig to produce styrene. The fluid dynamics equipment verified that the new reactor insert performed in the same manner as a packed bed, and the same correlations could be used for predicting dispersion coefficients. The induction heating test rig was used to assess the theoretical model developed for the heating of the reactor inserts. Results from these experiments showed that the temperature profile within the reactor can be manipulated quite easily by appropriate design.

The large scale experimental rig performed exceptionally well. The temperature was easy to control and could be altered during an experiment. Sufficient power was used to replace all heat loss and heat of reaction. Later experiments increased the power still further and the temperature in the axial direction increased with bed depth, a complete reversal of the normal situation. This temperature inversion improves the equilibrium position of the reaction.

It was not possible to fully optimise the second reactor, but the experiments that were carried out did show that significant improvements can be obtained using the new induction heating method.

Economic Assessment

A very simple approach can be used to give a preliminary assessment as to the potential viability of using an electrically heated reactor as compared with using standard equipment. The capital costs comprise the cost of the electrical equipment and the added cost of material used as plates inside the reactor. The running costs will reflect the higher price of electrical heating as compared with more conventional methods of increasing temperatures (e.g. oil fired heat exchangers).

A calculation for a plant producing 20,000 tonnes of styrene per year will be used to illustrate the procedure for producing an initial comparison of costs. For this flow the heat of reaction is 1MW, which needs to be supplied by the magnetic field. If the electrical efficiency is 90% 1.12 MW is required. If any sensible heat is added to the reactor the power required would be another 1 MW (to raise the reactant temperature from 400°C to 600°C). Assuming a capital investment of £50/kW this capital cost would be £112,000. The cost of the new plates would be approximately £20,000.

The running costs are calculated on the tariff of 3.83p/kWh, which is the current rate in the UK for most industrial companies. For an isothermal reactor (i.e. only the heat of reaction being supplied) the cost per year is £343,000. Increasing the reactant temperature by 200°C results in a running cost of double this amount. The cost of fuel oil firing (assuming 100% efficiency) gives a running cost for the adiabatic reactor of £55,000 and £110,000 for the extra 200°C temperature rise.

Looking at only the running costs the isothermal reactor would have to run with an increase in yield of 2.3% to recover the cost of the electrical heating, whilst for the extra heating case the increase in yield to recover the heating cost would be 4.6%. Using the kinetics available a single bed isothermal reactor increases the yield by over 50% compared to a single bed adiabatic reactor. Compared to a double bed adiabatic reactor (and hence a large increase in the capital cost) the single bed isothermal reactor increases the yield by 3.4% (which gives a payback period on the capital of less than one year if styrene sells at 36.9p/kg). Again using the kinetics available in the literature the potential increase in yield using the "ideal" temperature profile is 16%. This reduces the payback time to less than two months.

This is obviously a very simple approach as the integration of the reactor with the rest of the plant has not been taken into account. However, it does show that the cost of electricity is not prohibitive for these given conditions.

CONCLUSIONS

An inductively heated reactor has been designed, built and tested on the reaction to produce styrene from ethylbenzene. The new reactor design was able to heat the catalyst bed to replace the heat loss and the endothermic heat of reaction. This resulted in a temperature profile that approached isothermal conditions.

Increasing the power to the reactor resulted in an increasing temperature with bed depth. This compares very favourably with the theoretically ideal profile predicted from the reaction kinetics. It was not possible to optimise the inductively heated reactor, but improvements over the fully optimised adiabatic reactor were realised. Further improvement would be quite easy to achieve.

Although the reactor design has been successful, several areas for new research have been identified. Some of these are being covered under the current programme of work at EA Technology and specifically address the modelling and control of the magnetic field inside the reactor for one specific design. Other areas that need to be tackled include the mass and heat transfer through the new reactor, catalyst design to take full advantage of the new system, and tests on new reactor configurations.

The manufacture of styrene has been used as a test case but other similar reactions can also benefit from this new technology. The process is very flexible and should be used whenever temperature control is important.

SESSION 14:

New Process Routes

Chairman: Prof. K.R. Westerterp

A MODIFIED CAPROLACTAM PROCESS:
ENVIRONMENTAL AND TECHNICAL OPTIMIZATION OF THE RASCHIG SECTION

C.GOATIN, S.TONTI, R.STROZZI, F.CAINELLI,
P.FURLAN, F.PIGNATARO.
Research Centre EniChem Anic
Via della Chimica 5, 30175 Porto Marghera (VE), Italy

ABSTRACT

More than 50 % of the installed caprolactam (CPL) world capacity makes use of the process that produces hydroxylamine sulphate, via the Raschig reactions, cyclohexanone-oxime from cyclohexanone plus hydroxylamine and caprolactam via Beckmann rearrangement.
EniChem Anic has developed a new process for hydroxylamine sulphate production which results competitive with the widely applied Raschig process. The new technology is called "Direct Raschig Process" (DRP). With respect to the usual process via ammonium nitrite, its main strengths are:

- Elimination of the nitrite stage production with a reduction of total energy consumption of 40 %.

- Very high efficiency in raw material with a saving of 60 Kg NH_3/ton CPL (equivalent to 1.8 GJ/ton CPL).

- Very high degree of NOx conversion does not require a specific treatment in order to comply with the new EC pollution limits (500 mg/Nm^3 for each pollutant).

The Direct Raschig Process can make use of NOx gas (1.5 - 10 % vol) from nitric acid plants and ammonium bisulphite solution, obtained from tail gases containing SO_2.

INTRODUCTION

For the production of caprolactam (raw material for nylon 6) many processes have been developed, but very few have reached the industrial stage.
The conventional route that produces 90 % of caprolactam is the conversion cyclohexanone to cyclohexanone-oxime by hydroxylamine salt, followed by Beckmann rearrangement to give caprolactam.

In the conventional Raschig Process, which is the most widely used, NOx at a given molar ratio (obtained from NH_3 oxidation) reacts with ammonium carbonate.
The resulting ammonia and ammonium nitrite solution reacts with SO_2 to hydroxylamine disulphonate which is hydrolyzed to hydroxylamine sulphate:

$$NH_4NO_{2(aq)} + NH_4OH_{(aq)} + 2SO_{2(g)} \xrightarrow{0°C} HON(SO_3NH_4)_{2(aq)} \quad -74 \text{ Kcal (a)}$$

$$HON(SO_3NH_4)_2 + 2\ H_2O \xrightarrow{20 \div 100°C} (NH_2OH)\cdot H_2SO_4 + (NH_4)_2SO_4$$

The hydroxylamine sulphate reacts with cyclohexanone and ammonia, to form cyclohexanone-oxime and ammonium sulphate as a co-product:

$$(NH_2OH)\cdot H_2SO_4 + C_6H_{10}O + 2\ NH_3 \longrightarrow C_6H_{10}NOH + (NH_4)_2SO_4 + H_2O$$

Finally cyclohexanone-oxime undergoes a Beckmann rearrangement by means of oleum and the resulting intermediate is hydrolyzed, by neutralization with aqueous ammonia to crude caprolactam:

$$C_6H_{10}NOH \xrightarrow{oleum} C_6H_{10}NOH\cdot H_2SO_4 \xrightarrow{NH3(aq)} C_6H_{11}ON + (NH_4)_2SO_4$$

The hydroxylamine sulphate synthesis via NH_4NO_2 presents some problems:

- High NOx concentration in the effluent process gas, which requires a specific abatement.
- Low NH_3 to hydroxylamine yield.
- Ammonium bicarbonate supplied for ammonium nitrite preparation.
- High energy consumption.

In this paper, all the work which has carried out on the caprolactam plant of Enichem Anic (Porto Marghera site), in order to have an improvement in the Raschig section, has reported.

OPTIMIZATION OF THE RASCHIG PROCESS

An environmental and technical optimization of the nitrite and hydroxylamine sulphate synthesis on the caprolactam plant of Enichem Anic has carried out in subsequent steps:

I) A drastic NOx reduction in effluent gases and yield optimization, with use of sodium nitrite as an intermediate instead of ammonium nitrite.

II) Reduction of energy consumptions by partial substitution of SO_2 gas and NH_4OH with an ammonium bisulphite solution.

III) Further optimization of hydroxylamine disulphonate synthesis applying a new technology which avoids the nitrite production (DRP) (1-3).

Raschig via Sodium nitrite (step I and II).

Since the first start up of Porto Marghera caprolactam plant, the problems connected with the use of ammonium nitrite for hydroxylamine disulphonate production have been overcome using $NaNO_2$ obtained by NOx reaction with NaOH solution:

$$NO + NO_2 + 2\ NaOH \longrightarrow 2\ NaNO_2 + H_2O$$

$$NaNO_{2(aq)} + 2SO_{2(g)} + NH_4OH_{(aq)} \xrightarrow{0°C} HON(SO_3)_2NaNH_{4(aq)} \quad -74\ Kcal \quad (b)$$

$$HON(SO_3)_2NaNH_4 + 2\ H_2O \longrightarrow (NH_2OH)\cdot H_2SO_4 + \tfrac{1}{2}\ Na_2SO_4 + \tfrac{1}{2}\ (NH_4)_2SO_4$$

The major advantages of this modification are:

- A higher NH_3 yield
- Lower emissions

TABLE 1
Hydroxylamine disulphonate processes from nitrite
yield, by-product, selectivity and emission comparison.

Raschig process	via NH_4NO_2	via $NaNO_2$
yield and by-product selectivity		
% NOx to disulphonate	71	90
% NOx to nitrate	8	3
% NOx to nitrilo-trisulphonate	3	3
% NOx to nitrogen	12	3
ammonia related yield		
% NH_3 to Hydroxylamine Sulphate	63	82
pollutant in total effluent gas		
NOx as NO_2 mg/Nm^3	3500 (*)	500
SOx as SO_2 mg/Nm^3	500	500
salt coproduction (Kg salt/Kg cyclohexanone-oxime)		
Sodium Sulphate		0.75
Ammonium Sulphate	2.80	2.05

* gas which must undergo specific treatment to comply with the anti-pollution law requirement

As shown in Table 1, the process coproduces sodium sulphate (SS) and ammonium sulphate (AS) at the same time.

In order to reach a commercial grade of SS and AS a proprietary process is applied for crystallization, recovery and purification of those two salts. These operations bring about a higher energy consumption than the usual route which produces only AS.

A further improvement in energy saving, has been obtained by feeding the hydroxylamine disulphonate synthesis, an ammonium bisulphite aqueous solution instead of SO_2 and aqueous NH_3 (4).

This concerns to the use of $NaNO_2$ and NH_4HSO_3 solution and allows to remove with industrial water the heat generated in reaction.

$$NH_4OH_{(aq)} + SO_{2(g)} \xrightarrow{30^\circ C} NH_4HSO_{3(aq)} \quad -24\ Kcal \quad (c)$$

$$NaNO_{2(aq)} + NH_4HSO_{3(aq)} + SO_{2(g)} \xrightarrow{0^\circ C} HON(SO_3)_2NaNH_{4(aq)} \quad -50\ Kcal \quad (d)$$

The resulting energy saving is about 0.7 GJ/ton CPL.

The comparison between the two processes, based on the energy budgets for different sections (excluding ammonia, caustic soda and SO_2 relative to the salts coproductions), is shown in Table 2 where the ammonia consumption for hydroxylamine sulphate synthesis is calculated considering that 30 GJ are necessary to produce 1 ton of NH_3 (5).

TABLE 2
Energy budgets in GJ/ton CPL
(for different Process sections)

Raschig process	via NH_4NO_2	via $NaNO_2$
Ammonia to burner for NOx	7.4	5.6
steam recovery	(2.2)	(1.7)
Salt co-production - steam, oil	6.1	6.7
- e. power	0.4	0.6
Additional energy consumption	+ 0.4 *	- 0.7 **
Energy budget	12.1	10.5

* due to NH_3 consumption in catalytic DeNOx treatment
** due to use of NH_4HSO_3 solution.

Considering that the hydroxylamine consumption to produce cyclohexanone oxime, is nearly stoichiometric for both the processes, the advantage, from the energetic point of view, is attributed to the Raschig via $NaNO_2$ is about 1,6 GJ/ton CPL.

Direct Raschig Process (step III).

This new proprietary process relates the hydroxylamine disulphonate synthesis by using NOx, at a given molar ratio NO/NOx, and ammonium bisulphite solution.
In this case the intermediate nitrite is formed "in situ" and reacts with the ammonium bisulphite to give hydroxylamine disulphonate according to the following reaction:

$$NO_{(g)} + NO_{2(g)} + 4NH_4HSO_{3(aq)} \xrightarrow{0^\circ C} 2HON(SO_3NH_4)_2 + H_2O \quad -90\ Kcal \quad (e)$$

The DRP has been studied in the EniChem Anic Research Centre of Porto Marghera by means of a bench scale pilot plant.
The first laboratory experiments were carried out by some gas-liquid adsorbers with recycling of the solution through a heat exchanger and showed the necessity to carry out the reaction (e) in two steps:

- in a main reactor (3 adsorbers) at pH 5.8
- in terminating reactor where pH is lowered to 3.5.

In order to scale up the process, the main requirements as a result from pilot experiments, are:

- Very low pressure drop to avoid gas compression.
- High gas-liquid contact to improve productivity and conversion.
- Low liquid hold-up to minimize by-product (nitrilo-trisulphonate).
- Good heat removal to obtain a reaction temperature as costant as possible and to improve selectivity.
- The main reactor must have more reaction steps and the possibility of adding NO_2 in order to adjust the oxidation degree of NOx at a proper level in each zone.

Beside these requirements, two other considerations were useful to define the best asset of the industrial reactor:

- Heat removing during the reaction and where it is generated.
- Co-current gas-liquid directions are preferable for selectivity and flow dynamics reasons.

The main results by optimization of DRP process (3) are reported in table 3:

TABLE 3
Direct Raschig Process
yield, by-product, selectivity and emission

Raschig process	directly from NOx
yield and by-product selectivity	
% NOx to disulphonate	90
% NOx to nitrate	4
% NOx to nitrilo-trisulphonate	3
% NOx to nitrogen	3
ammonia related yield	
% NH_3 to Hydroxylamine Sulphate	82
pollutants in total effluent gas	
NOx as NO_2 mg/Nm^3	250
SOx as SO_2 mg/Nm^3	250
salt coproduction	
(Kg salt/Kg cyclohexanonoxime)	
Ammonium Sulphate	2.80

The new DRP in comparison with Raschig via $NaNO_2$ has:

- Further lowering the emissions which complies with the new EC limits (less than 500 mg/Nm^3 for each pollutant) with an energy saving of 0.1 GJ/ton CPL (related to the different exotherm of reactions d and e).
- An additional advantage derives from the coproduction of only one type of salt (AS) with an further energy saving of 0.8 GJ/ton CPL .

The energy budget for caprolactam production for different process sections is 9.6 GJ/ton CPL via DRP in comparison with 10.5 GJ/ton CPL via $NaNO_2$ and 12.1 GJ/ton CPL via NH_4NO_2.
EniChem Anic has recently started up a unit of 25 kton/year (Fig. 1) above the caprolactam capacity of 110 kton/year via $NaNO_2$. It can make use of concentrated and diluited NOx gas, from nitric acid plant at low pressure.

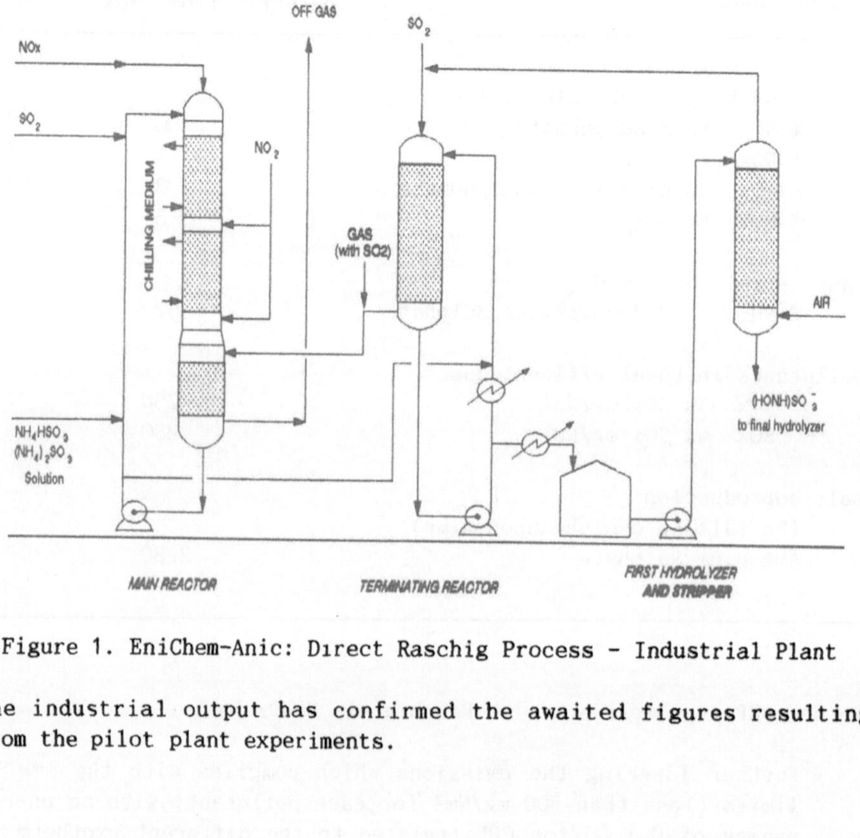

Figure 1. EniChem-Anic: Direct Raschig Process - Industrial Plant

The industrial output has confirmed the awaited figures resulting from the pilot plant experiments.

CONCLUSIONS

As a result of this work we can conclude that EniChem Anic has developed a new process for hydroxylamine sulphate production, which results competitive with the widely applied Raschig process. The new technology is called "Direct Raschig Process".
The main strenghts of the new technology are:

- Elimination of the nitrite stage production.
- Very high efficiency in raw materials and energy consumption.
- Very high degree of NOx conversion which avoids a specific treatment on the off-gas.

The new process which makes use of NOx gas (1.5 - 10 % vol) from nitric acid plants and ammonium bisulphite solution obtained from tail gases containing SO_2, shows a clear advantage with respect the usual Raschig Process for that the energy budget and environmental aspect are concerned.

REFERENCES

1. S.Tonti, C.Goatin, R.Cecchin, G.Talamini : Italian Patent 1.152.229

2. C.Goatin, S.Tonti, G.Talamini, R.Cecchin : Italian Patent 1.170.253

3. C.Goatin, S.Tonti, G.Olivieri, R.Cecchin, R.Strozzi : La Chimica e l'Industria 1, 1990, 13-7

4. F.Sormani, F.Luraschi : ICP January 1988, 35-41

5. Low-Energy Ammonia Processes : Fertilizer International 293, January 1991, 41-6

SESSION 15:

Exergy Analysis

Chairman: Prof. E.P. Gyftopoulos

EXERGY LOSS: A BASIS FOR ENERGY TAXING

GERARD HIRS
Professor of Energy Technology
University of Twente
Postbus 217, 7500 AE Enschede, The Netherlands

ABSTRACT

The paper introduces exergy loss or entropy added as a basis for energy taxing. Exergy loss will be shown to account objectively for all energy input and output in its different forms and qualities: electricity, fuel, feedstock, product and heat of high as well as low temperature.
Analogously to Value Added Tax (VAT), the tax is termed Entropy Added Tax (EAT). The paper lists and discusses advantages and disadvantages of EAT. The main advantage of EAT is the strong incentive for investors to conserve energy in combination with the absence of interference with free market operation. The main disadvantage might be the temptation to governments to use EAT to increase the overall tax burden.

INTRODUCTION

Coase (1) is one of the first economic scientists having considered how to compensate for environmental damage in an economic system. In his book "The firm, the market and the law" important conclusions are that it is not important for an economy which person or institution compensates but that transaction cost should be low. It is implied that free market operation, but not without clear regulations and its enforcement, will be beneficial to solving environmental problems.

Georgescu-Roegen (2) considers classical economic theory too mechanistic and has introduced an economic understanding of irreversibility and entropy. In his book "The Entropy Law and the Economic Process", he concludes that all economic activity is to a great extent wasteful and goes hand in hand with entropy creation. Although entropy creation is in itself harmless, it means a variety of physical and chemical processes that might cause environmental damage. Of course, damage is not a necessity. Damage will occur if such processes are excessive in quantity or in intensity. Entropy creation is a useful common yardstick for this.

In this paper entropy creation, as a term, is considered to be too general. In fact entropy creation is a process natural to nature and not exclusive to human beings and their economic activity. The paper will concentrate on the part of entropy creation attributable to human, economic activity. For that purpose, the term entropy added will replace entropy created. It should make clear that all human, economic activity inevitably adds entropy. The economic objective is to add value. Adding value without adding entropy is physically impossible. As value and entropy are interlingued, both should be utilized in economic thinking. In economic terms value added indicates what is considered desirable. Entropy added indicates what is undesirable or, within limitations, tolerable. To minimize the ratio of entropy added and value added becomes the new economic objective.

It is clear that the introduction by Georgescu-Roegen of entropy in economic thinking is important. At the same time the pragmatic approach of Coase is attractive. This paper is the first step in a further development consistent with the work of both authors. The basic idea is to replace Value Added Tax partially or completely by Entropy Added Tax.

ENVIRONMENTAL TAX CONCEPTS

Environmental taxes are abundant. There are two types. They have in common that citizens and firms have to pay. They have different objectives and differ in the utilization of the revenue. Taxes of the first type are identical with service taxes. Waste collection and incineration can be financed in this manner. Such a tax is attractive because its revenues can be compared directly with the quality and the quantity of the service provided. In principle the service does not have to be provided by local government and the financing does not have to be a tax. Organisational constructions including local government and private enterprise are also possible. Transaction cost should be low.

Recently, several environmental taxes of the second type have been introduced. Such taxes are not directly coupled to a service. Two predecessors outside the environmental field clearly illustrate this. The tax on alcohol is not used to cure alcoholics and to compensate damage done to their families, their employers or society as a whole. The tax on fuel for cars far exceeds what is necessary for traffic infrastructure and maintenance. Most likely these taxes were meant to change human behaviour but it is well known that alcohol consumption does not decrease under the influence of tax, neither does the number of cars and the mileage. However, tax on fuel for cars did result in a tendential preference for smaller cars with low fuel consumption.

Apparently, environmental taxes of the second type will not change human behaviour much but they do encourage investment in energy conservation and other environmental measures.

Several new environmental taxes are now being discussed within the European Community. On the national level in The Netherlands a fuel tax is being discussed. This tax has met with opposition from energy intensive and other industry. The most severe threat that industry can use, "we will leave the country" has been used. For a country like The Netherlands with an internationally oriented industry including many multi-nationals this would be a disaster. Government is now trying to save this tax proposal by introducing tax exemptions for parts of industry. The discussion continues.

On the Community level a CO2-tax has been discussed but the member countries are still very divided on the subject.

Such taxes are harmful in economic terms because they create a higher fuel price in general and in addition an artificial difference in fuel price between parts of the world with and without the tax and between parts of industry/society with or without tax exemption. Governments introducing such taxes act like OPEC in disguise. When confronted with the consequences of an increase for all, these governments will try to find a compromise. The transaction cost will be enormous and the danger, an economic crisis as a result of the price increase, will not be prevented.

In summary an environmental tax should be so designed that environmental objectives, such as energy conservation, are achieved. An environmental tax should stimulate the economic process in such a manner that investment in these objectives is attractive. By no means should an environmental tax hamper the economic process, interfere with free market operation, create an economic slow down, disturb free competition, lead to high transaction cost, etc.

An example in present day taxing where the influence on the economic process is benign and effective is Value Added Tax.
For designing an environmental tax, VAT should be used as a model

ENTROPY ADDED TAX

An entropy added tax relies completely on the ability to determine entropy added of individual economic activities. In Fig. 1, within system boundary no. 1 the economic activity of an individual firm or factory is shown. The economic activity is called industrial factory. The system boundary represents the fence around the factory, the economic entity, etc. This definition is crucial. Entropy added by economic activity can be determined by looking at the flows of material and energy that pass.

On the input side flows of energy and material are shown. More specifically, electricity, fuel, intermediate products and feedstock are entering the firm. It is clear that in normal practice all these flows are identified and quantified. As far as quantification is concerned one additional measure for quantification is necessary: exergy. Kotas (3) and several others have explained the exergy concept in detail. For this paper it is sufficient to note that exergy is a measure for energy quality. The definition of exergy is: the maximum quantity of work obtainable from flows of material and energy by bringing these into a state of equilibrium with the environment. In Fig. 1, on the input side, all flows should be translated into exergy. A similar translation should take place on the output side where product flows are leaving the firm. The exergy content of these product flows should also be quantified and translated into exergy. By definition, waste heat and by-products mixing with the environment have zero exergy.
By subtracting the exergies entering and leaving the firm an exergy loss can be determined, see Eq. 1

$$\Sigma E_{in} - \Sigma E_{out} = E_{loss} = T_o * \Delta S \qquad (1)$$

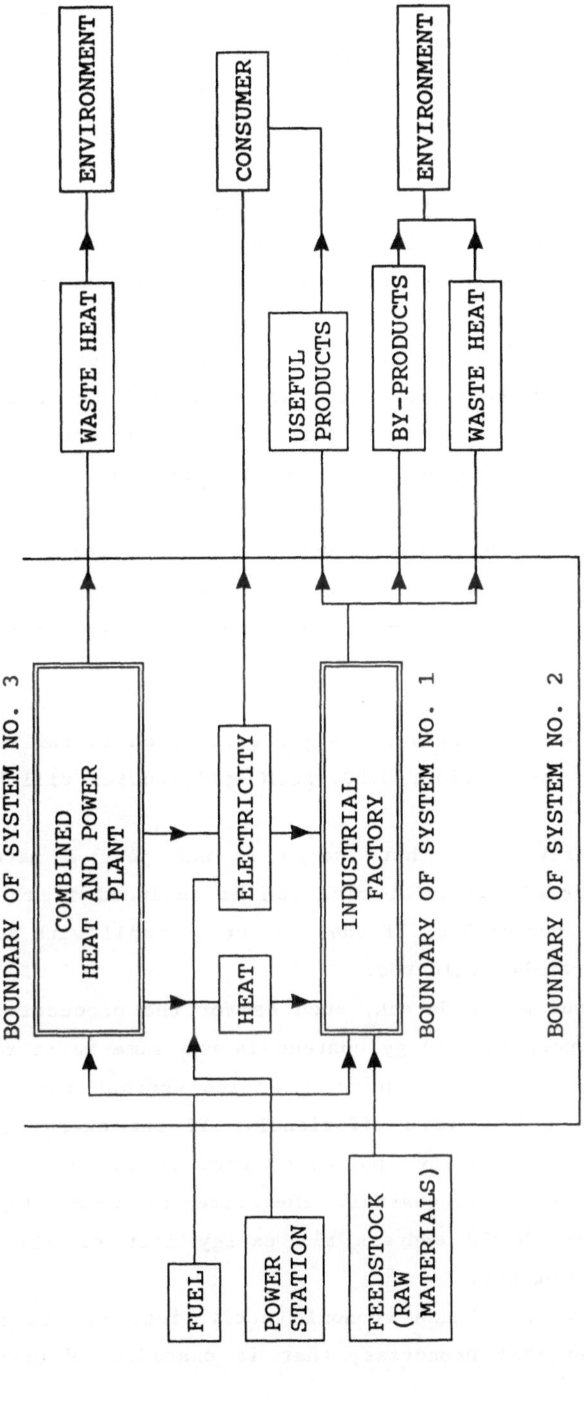

FIG. 1. SYSTEM BOUNDARIES.

Eq. 1 also shows that the exergy lost is by definition equal to the ambient temperature (T_o) times entropy added (ΔS).

Within system boundary no. 2 the same industrial factory is shown and in addition a combined heat and power plant. If this plant is out of operation (case 1) the consequences are:

- the fuel flow to the factory is appreciable as the process heat ahs to be supplied by boilers within system boundary no. 1
- the electricity has to be supplied by the grid outside system boundary no. 2.

If this plant is in operation (case 2) the consequences are:

- the fuel flow to the factory is smaller and can be assumed to tend to zero if all heat is supplied by combined heat and power
- the fuel flow to the power plant can be expected to be higher than in case 1 (to the factory) in practical situations
- the electricity production compensates the electricity from the grid in case 1, in practical situations surplus electricity can be produced.

By subtracting case 1 from case 2, additional fuel flow enters the system within boundary no. 2 and additional electricity exits the system within boundary no. 2. The additional fuel appears to produce additional electricity, which has an efficiency of, e.g., 0.75 which is far above that of a combined cycle power station, 0.50. Both efficiencies will be used subsequently.

Figure 1 clearly shows that exergy is not just a matter of fuels. Exergy content of all material flow can be included if the magnitude of this content is appreciable. If the content is small with respect to that of the fuel, it can be neglected.

If fuel is used as feedstock, such as for the production of polymers, or as energy source, the exergy content is the same basis for quantification. It will be decided at the output side whether the exergy entering the firm will have been used efficiently. If the exergy content of the products such as electricity, polymers, etc. is as high as possible the conversion process is successful. Therefore it seems logical that an environmental tax should address all exergy lost or all entropy added between input and output.

More generally, all human economic activities, can be represented as cascades of industrial processes, that is cascades of exergy converters

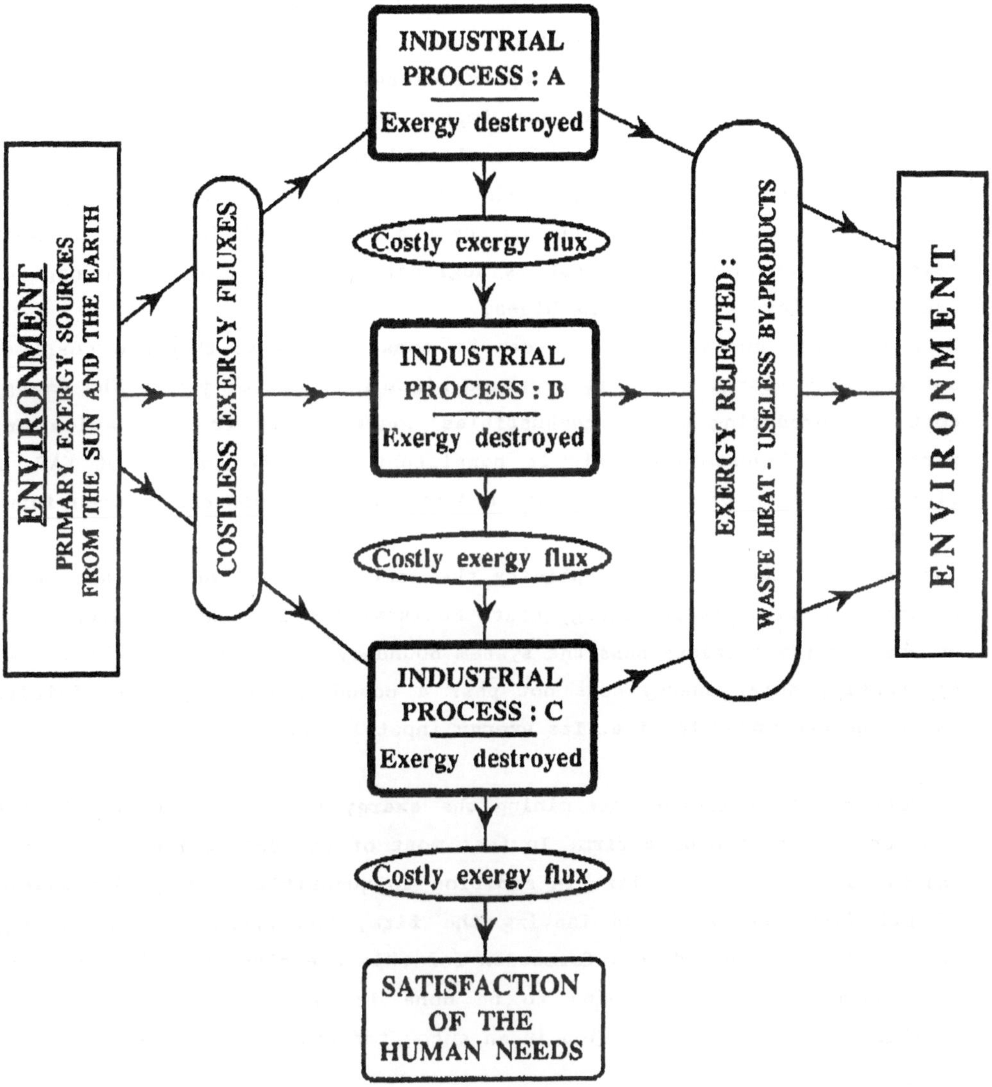

Fig.2 : A cascade of industrial processes presented as a cascade of exergy converters.

(see Figure 2). Each process is a consumer of costly exergy supplied by the preceding process and a producer of costly exergy delivered to the following process.

At each level, the cost of exergy is a function of both the exergy loss during the operation, and the exergy loss for constructing the equipment, that is the amortization of the capital exergy invested in the equipment.

Furthermore, several processes import some costless exergy directly issued from the primary sources (see also Figure 2).

For example the kinetic energy of the wind for windmills and the height of water in the mountain lakes, aboven the sea level for hydropower . In such cases the operating exergy cost is negligible as compared to the capital exergy cost invested in the machines.

Even for coal and fuel oil, it can be considered that they are costless solar exergy stored in the underground since millions of years. The exergy cost for producing usable combustibles comes partly from operation and partly from investments. This is consistent with the concept of EAT. A mining operation of coal will show a jump in exergy context between input and output.

However, for the sake of simplicity the remaining part of the paper will concentrate on operating exergy cost. Decisive for the exergy content will be whether its carriers pass the system boundary.

Apparently, solar energy does not pass a boundary and has by definition not to be accounted for i.e. its exergy input is zero.

Most of the data for determining the exergy loss are available in the book keeping system of a firm. In fact most of the data are necessary for calculating Value Added Tax. In addition to quantities and specifications of all flows entering and leaving the firm, the translation to exergy content is required. Much information for this exercise is available, e.g. from Kotas (3). What remains to be done is creating a data base for carrying out such a translation in an automated book keeping system.

ENTROPY ADDED TAX CALCULATIONS

In Table 1 several cases are listed for which an EAT has been calculated.

The first column gives different cases for energy, or rather, exergy conversion. The next column gives the energy input in terms of exergy. If the input is a fuel the chemical exergy content is close, but not equal, to the lower heating value. In the case of electrolysis the input is electricity and the exergy input is equal to the input in electrical units. In the case of windmills the input is not attributable to human activity.

TABLE 1

EXERGETIC EAT FOR ENERGY CONVERSION

AT DIFFERENT EFFICIENCIES

ENERGY CONVERSION	INPUT (I)	PRODUCT (P)	LOSS (L)	EAT L/P	EAT L/I
COAL	1	0.40	0.60	1.50	0.60
GAS (COMBINED CY)	1	0.50	0.50	1.00	0.50
CHP	1 [1)]	0.75 [2)]	0.25	0.33	0.25
ELECTROLYSIS	1	0.70	0.30	0.43	0.30
WIND MILL	(>)0 [3)]	1.00	0.00	0.00	--
CONSUMER	1	(>) 0.00	1.00	∞ [4)]	1.00

1) For CHP the fuel input is equal to the difference in fuel inputs without combined heat and power and with combined heat and power as in Fig. 1.

2) Likewise for CHP the product in the form of electricity is equal to the difference in electricity without combined heat and power and with combined heat and power as in Fig. 1.

3) No operational input due to human, economic activity, as shown in Fig. 1. The input comes from within the system boundaries which is typical on a site with renewable energy potential.

4) A finite value will be obtained if consumers produce recyclable waste.

Therefore, it is consistent with the definition of entropy added as being due to human, economic activity that the input is zero. The same would apply if a firm would have a hydropower station within its fences. It is evident that the investment and the erection of windmill and hydropower station are attributable to human, economic activity. However, how to deal with investment will not be discussed in this paper. Therefore, Table 1 only presents calculations for the exergy input and output during operation and not for the exergy loss associated with the investment.

The product in column 3 is electricity for all items mentioned with the exception of electrolysis where an input of almost zero exergy is transformed into an output with high chemical exergy. In the case of combined heat and power the fuel input and the electricity output is defined as in Fig. 1 where identical processes without and with CHP are considered. They can be seen to result in an additional fuel input leading to electricity production.

The output of the consumer has been assumed to be zero in column 3. All exergy input is lost in a pure consumer situation. At first sight this might seem meaningless. However, the exergy loss associated with the consumer should be accounted for in an economic process. This situation will only change if the consumer would be a producer, e.g. of waste with an economic value and a recyclable waste content. Alternatively, human labour can be considered to be a product which can be translated into an exergy output.

The fourth column presents the exergy loss: column 2 minus column 3. The EAT for these cases is expressed in EAT units and presented in column 5. The definition of an EAT-unit is the ratio of loss and product in terms of exergy for the case of a power station with an exergetic efficiency of 0.5, see the third horizontal line in Table 1 and column 5.

Column 5 clearly shows that EAT for fuel fired power stations varies from 0.33 to 1.5 per unit electricity. In economic terms this tax is extremely elastic and will strongly stimulate investment in energy saving. When contrasted with a tax based on energy or exergy input it would appear that the elasticity of such a tax is much lower and that the incentive to invest would almost disappear.

The EAT on the electrolysis appears moderate but can be further improved by process development yielding more product and by combined heat and power, see Fig. 1.

EAT favours windmills as no tax is due. This picture might change by accounting for EAT on investment and maintenance.
The lowest line gives the fate of the consumer who has to pay the accumulated EAT and does not have the opportunity to pass on EAT down the line.

By buying a product the consumer will be faced with the entire exergy loss. The only way of reducing EAT is by producing recyclable waste.

Not reflected in Table 1 is the fact that the consumer will have the opportunity to compare the exergy losses of his suppliers. By buying electricity from a power station operating at high efficiency, he would save on the amount of EAT in his electricity bill. On electricity from a windmill he would pay no EAT at all. This is were the analogy between EAT and VAT stops, EAT is a tax reflecting the exergy efficiency of the industry supplying the product and VAT is a fixed percentage on the industry's price.

ENTROPY ADDED TAX COLLECTION

Part of the transaction cost associated with EAT is tax collection. It seems evident that these should be as low as possible and not exceed those of other taxes for the same amount of tax collected. An important cost element is the fact that EAT, like VAT, should be paid in the market where the product is purchased. This means that the tax is returned to exporters at national borders or Common Market. Tax collection cost reduction will be achieved for both kinds of taxes for markets as large as possible. Therefore,it is not recommended to introduce EAT at the national level.

For industry it is essential that EAT is corrected at a border. Otherwise the competitiveness will be influenced negatively or positively by EAT. It is also essential to collect EAT on imports. This will prove to be difficult with imports from countries where an EAT system does not exist and the exergy input is unknown. The best solution in such a case is to base an EAT on experience with local firms. The importer or the foreign firm should be given the opportunity to prove that the actual exergy lost in the production process is smaller than the exergy lost by local firms. By doing so the import of products with minimal exergy lost is encouraged.

It is well known that for collecting VAT it is essential to look twice into the book keeping system. In principle the same should be done for

EAT. However, in many firms the input is more important than the output in terms of exergy. In those cases, the exergy contents of the material input and the product output are small and/or roughly equal. Decisive is the exergy loss specified in electricity and fuel bills. A typical example is the services industry such as engineering firms. Passing on the tax on the exergy loss of such firms to their customers is still a possibility but should be carried out only if the cost of such a transaction is acceptably low.

EAT calculations and collection require a greater effort in chemical industry, steel works, etc. Exergy input and output should be quantified and the exergy contents of the materials and products should be determined. For these calculations the chemical compositions should be known. In the case of steel the exergy contents of coke at the input and of iron or steel at the output will be of importance.

SUMMARY AND CONCLUSIONS

- An environmental tax should stimulate the economic process in such a manner that investment in environmental objectives becomes more attractive.
- An environmental tax should not disturb the economic process and, possibly, improve free market operation.
- Entropy added should be considered as a common measure of the undesired, environmental by-effects of an economic activity; value added is the desired effect.
- Minimizing entropy added in combination with maximizing value added is the new, environmentally inspired, economic objective.
- Entropy Added Tax (EAT) can be modelled after Value Added Tax (VAT).
- EAT should be considered to partially or completely replace VAT
- This paper shows that EAT could work in practice and that EAT-collection cost could be acceptable but that a more detailed study is necessary.
- This paper shows that EAT will be a very powerful stimulus to invest in environmental objectives such as energy conservation.
- By introducing EAT, consumers will be able to compare the magnitude of

EAT for similar products and so form an opinion of the environmental impact of its production.

ACKNOWLEDGEMENT

The author thanks Professor P. Le Goff (Nancy, France) for making valuable suggestions with regard to text and figures.

REFERENCES

1. Coase, R.H., The Firm, the Market and the Law, The University of Chicago Press, Chicago and London, 1985.

2. Georgescu-Roegen, N., The Entropy Law and the Economic Process, Harvard University Press, Cambridge Mass., 4th pr., 1981.

3. Kotas, T.J., The Exergy Method of Thermal Plant Analysis, Butterworths, London, 1985.

EXERGY ANALYSIS OF ADIABATIC AND DIABATIC DISTILLATION COLUMNS : AN EXPERIMENTAL STUDY

R. RIVERO, T. CACHOT, P. LE GOFF

Laboratoire des Sciences du Génie Chimique,

CNRS-ENSIC-INPL, B.P. 451, 1 rue Grandville, 54001 Nancy, France

ABSTRACT

Separation processes involving conventional fractional distillation columns consume large amounts of exergy. A new distillation process, called diabatic distillation, presented in a previous paper [8], consists in replacing the reboiler and condenser, which are normally located at the top and the bottom of the column, by two heat exchangers integral with the column itself. One of the exchangers is incorporated in the lower part of the column below the feed point to supply heat and the other, located above the feed point is used to extract heat. This arrangement minimizes entropy production in the system and thus maximizes the exergy effectiveness. In recent papers we have presented numerical studies of examples of diabatic distillation based on the theoretical performance. This paper presents experimental results which demonstrate the validity of the theory.

INTRODUCTION

The use of several small heat exchangers, either integral with or independent of distillation columns, has been proposed by other authors to give a better distribution of input and output heat flows in separation processes [1-4]. Furthermore, the concept, and even the term "diabatic distillation" and "quasi-reversible distillation", have already been used in published papers notably by another worker in the Laboratoire des Sciences du Génie Chimique" [5-8]. In this investigation we propose to give experimental proof that diabatic distillation columns degrade less energy than adiabatic columns.

THEORETICAL BACKGROUND

Conventional fractional Distillation

Figure 1 shows a typical distillation system for a binary mixture for which the following standard mass, enthalpy, entropy, and exergy balances can be written :

$$F = D + B \quad (1)$$
$$X_f F = X_d D + X_b B \quad (2)$$
$$Q_B + H_f = H_d + H_b + Q_c \quad (3)$$
$$Ex_B + Ex_f = Ex_d + Ex_b + Ex_c + Irr \quad (4)$$

In this Irr is the irreversible exergy loss in the process due to entropy production.

The diagram shown in Figure 2 can be used to obtain the values of specific exergy for each thermodynamic state of a mixture. If such a diagram is not available these values can be obtained from enthalpy and entropy data for the three states of the mixture (for example by using diagrams of enthalpy-concentration and entropy-concentration).

To simplify the interpretation of these various terms we may consider that the mixture is separated by an isothermal process working at the feed temperature. The variation of the so-called "chemical enthalpy" is then :

$$\Delta H_{xs}]T = H_d]T + H_b]T - H_f]T \quad (5)$$

and the corresponding variation of exergy is :

$$\Delta Ex_{xs}]T = Ex_{xd}]T + Ex_{xb}]T - Ex_{xf}]T \quad (6)$$

with : $$Ex_{xj}]T = RT_o \, S \, N_j \, X_j \ln g_j X_j \quad (7)$$

In the case of an ideal mixture, the variation of enthalpy $\Delta H_{xs}]_T$ would be zero and the activity coefficients γ_j would be equal to 1.

In reality the separation process is not carried out isothermally but the "chemical" term can be separated from the "thermal" term in each input and output of heat, so that corresponding variations in the "thermal" exergy can be calculated as :

$$\Delta Ex_t = \Delta H.q = \Delta H\,(1 - T_o/T) \quad (8)$$

The enthalpy and exergy balances can then be written in the following form :

$$Q_r - Q_c = \Delta H_s = H_d + H_b - H_f \quad (9)$$

$$Q_r\theta_r - Q_c\theta_c = \Delta Ex_s + Irr \quad (10)$$

where θ_r and θ_c are calculated using the logarithmic mean temperatures of the heating and cooling media respectively.

The exergy associated with the difference between the heat flows which are supplied to and removed from the system represents the net thermal exergy supplied to the column. A fraction of this thermal exergy is converted into chemical exergy, and the remaining fraction is destroyed[1]. It is important to note that the "chemical" exergy difference, as defined by equation (7) is a portion of the total exergy difference of the separation (eq. 10) ($\Delta Ex_S = \Delta Ex_x + \Delta Ex_t$) and it provides definition of the "chemical" effectiveness of a separation process[2] :

$$\varepsilon_S = \Delta Ex_S / (Q_r q_r - Q_c q_c) \qquad (11)$$

and

$$\varepsilon_x = \frac{\Delta Ex_x I_T}{Q_r q_r - Q_c q_c} \qquad (12)$$

DIABATIC DISTILLATION

The optimisation of a distillation system consists in reducing the irreversible exergy losses and increasing the effectiveness of separation. This can be done by integrating the boiler and the condenser, and their associated heat fluxes, into the column itself. Figure 3 shows the operating principle of such a "diabatic" distillation process. Figure 4 shows the temperature profiles in the heating and cooling media which are practically parallel to and not very distant from, the isobar line. The temperature differences between the heating medium and the fluid being heated are quite small and uniform all along the column.

The liquid is vaporised progressively and uniformly all along the stripping section instead of only in the reboiler. **The reboiler can therefore be eliminated.** In the same way the vapour is condensed progressively and uniformly all along the length of the rectifying section. **The condenser and the reflux head can therefore also be eliminated.**

Figure 5 shows the classic McCabe-Thiele stepwise construction method for a standard distillation column. In such an adiabatic column there are two straight operating lines which have slopes given by the ratio of the molar flow rates of liquid and vapour flowing counter-currently.The steps which are marked off between the operating lines and the equilibrium isobar represent the series of theoretical plates which are equivalent to the distillation column.

---

[1] The exergy associated to the difference in the chemical potential, which should truly be called "excess exergy" or "compositional exergy" [6, 7]. The authors prefer to follow the tradition of most of the thermodynamicists who use the adjective "chemical" to deal with the energy associated to any change of composition of a molecular mixture, with or without chemical reaction.

[2] When applying equations (11) and (12) to experimental conditions it is necessary to substract from the denominators, the exergy losses through the walls of the column to the surroundings.

Figure 6 shows that in the new distillation process proposed here these two straight operating lines are replaced by an **operating curve** which is roughly parallel to the equilibrium isobar. A measure of the irreversibility of the separation process is given by the distance between the operating curve and the equilibrium curve. If, in the limit, these two curves were infinitely close to each other then the exergy losses would be zero and the distillation process would be perfectly reversible. However, it should be pointed out, that bringing these two curves closer together whilst it would reduce the exergy losses would also increase the number of theoretical plates required to perform the separation. It is therefore necessary to look for an economic optimum between the first factor, which is the operating cost, and the second factor which is a capital cost (see later).

EXPERIMENTAL EQUIPMENT

The objective of the "Experiments with a laboratory version of a complete column" is to make an experimental investigation of the influence of certain parameters (flow rates, heat fluxes,...). For this we have constructed a 15 cm diameter column with a power input of some fifteen thermal kilowatts. The experimental column has been designed to allow a separate and successive investigation of the two sections composing each fractional distillation column ; that is the rectifying section and the stripping section. In addition, it can be used as a standard adiabatic column as well as a diabatic column.

The general experimental programme has three main parts :

I Preliminary experiments.
II Rectifying section.
III Stripping section.

This paper deals with the second part of the investigation.

Figure 7 shows a standard distillation column. The part of the column located below the feed point is the STRIPPING SECTION. Depending on how we divide up the different sections of the column the stripping section may or may not include the feed point. Furthermore, in the diabatic case the condenser may condense only the distillate.

Consequently we may distinguish the various types of operating conditions shown in Figure 8 :

The different operating modes

Case IA	Adiabatic rectifying column (mixture reboiler and reflux condenser).
Case IB	Diabatic rectifying column (mixture reboiler and reflux condenser).
Case IC	Diabatic rectifying column with no reflux (mixture reboiler and distillate condenser).
Case IVA	Adiabatic stripping column with reflux (reboiler and reflux condenser).
Case IVB	Diabatic stripping column with reflux (reboiler and reflux condenser).

Case IVC Diabatic stripping column with reflux but with no reboiler (reflux condenser)

Case IVD Adiabatic stripping column with no reflux (reboiler and distillate condenser).

Case IVE Diabatic stripping column with no reflux (reboiler and distillate condenser).

Case IVF Diabatic exhausting column with no reflux and no reboiler (distillate condenser).

EXPERIMENTAL METHOD

Figure 9 is a schematic diagram of the experimental set up and shows the various measurements which can be made.

- The measurements are essentially :

. Temperatures (of the heating fluid, the cooling fluid, and the process fluid).
. Flow rates.
. Compositions (of the liquid on each plate).
. Pressures.

- The parameters which can be controlled are :

For the rectifying column :

. The composition of the mixture xr.
. The amount of heat input to the reboiler Qb.
. The reflux ratio.

For the stripping column :

. The composition of the mixture xr.
. The amount of heat input to the reboiler Qb.
. The feed rate F.
. The amount of heat removed by the first condenser for the reflux Qc1.

In addition must be added to the parameters mentioned above the internal structure (arrangement of the heating and cooling coils) in the diabatic column. However, it is not possible to make an inventory of all possible structures and furthermore, the experimental equipment used at the present time does not allow us to put the coils in parallel. We may therefore distinguish the following cases :

1. Distribution in series throughout the column.
2. Distribution in series for the 5 coils in the lower part.
3. Distribution in series for the 5 coils in the upper part.
4. Distribution in series for the 4 or 6 coils in the centre.
5. Distribution in "alternating" series for the three plates.
6. Distribution in series for the 5 coils at the ends of the column (3 plates above and below).

EXPERIMENTAL RESULTS

The full experimental results for the 68 experiments performed will be presented in a further paper. However, to show the general form we present results for two cases.

- A standard distillation operation (case IA)
- A diabatic distillation with the 10 cooling coils working in series (case IC).

Figure 10 shows the operating conditions when the experimental apparatus is working adiabatically and diabatically in rectifying a water-ethanol mixture. The adiabatic column uses 200 kg/h of cooling water whereas the diabatic column requires 135 kg/h. The heat duty of the condenser is reduced from 12.2 kW to 1.3 kW and the temperature of the water leaving the system is 67°C in the adiabatic case and 97°C in the diabatic case.

Figure 11 gives the experimental results on a McCabe-Thiele diagram and Table I gives the distribution of exergy losses in the two operating modes.

The Grassman diagrams for the adiabatic operating conditions are shown on Figures 12 and 13 respectively. The values given on the diagram are presented as percentages of the total input exergy (to the reboiler and in the feed mixture). The total exergy losses in the adiabatic column represent 94.2% of the input exergy and only 56.5% in the diabatic case. In the adiabatic case the exergy extracted by the cooling water and rejected to the surroundings represents 23.7% of the input exergy. In the diabatic column this exergy is 37.3% but this is not a reject as it can be used as a heat source for other equipment else where in a plant.

Even if the exergy extracted by the cooling water in the adiabatic case was not rejected to the surroundings (its temperature is only 67°C), the exergy losses of the system would still account for 70.5% of the input exergy. The effectiveness of the adiabatic column is 1.32% whilst that of the diabatic column is 7.45%.

CONCLUSION

For the same amount of heat input and the same number of plates, the diabatic distillation column destroys only about half as much exergy as does the adiabatic distillation column (8312 kJ/h vs 14442 kJ/h). At the same time : it consumes less cooling water (135 kg vs 200 kg) ; it has a higher effectiveness (7.45% vs 1.32%) ; and it produces an effluent water at a higher temperature (97°C ve 67°C).

The reduction in the amount of cooling water required is important as far as cooling water must be bought at a certain price per cubic meter independant of its temperature level. Moreover, the effluent water from the diabatic column is at 97°C and could be used for heating other equipment in the plant. This means that in certain conditions it could have an economic value which should be born in mind in the overall economics of the process.

Finally it should be remembered that a stripping column rather than a rectifying column would show an advantage in requiring less heating fluid.

NOMENCLATURE

B	bottom product	γ	activity coefficient
C	condenser	ε	effectiveness
D	top product flowrate	θ	Carnot factor
Efl	effluent exergy losses		
Ex	exergy		
F	feed mixture flowrate		<u>Subscripts</u>
H	enthalpy		
Irr	irreversible exergy losses	b	bottom product
M	mass flowrate	c	condenser
Pex	total exergy losses	d	top product
Q	heat	f	feed
R	reflux	lm	logarithmic mean
S	entropy	r	reboiler
T	temperature	s	separation
x, X	fraction of ethanol in liquid	t	thermal
y, Y	fraction of ethanol in vapour	x	excess (called chemical, compositional).
		0	dead state

REFERENCES

[1] King, C.J., Separation Processes, 2nd Ed., Mc Graw Hill, New-York, 1981.

[2] Naka, Y., et al., An Intermediate Heating and Cooling Method for a Distillation Column, Journal Chem. Eng. Japan, Vol. 13, 1980, pp. 123-129.

[3] Nakaiwa, M., et al., Minimum Reflux Ratio and Possibility of Energy Saving on a Plate-to-Plate Heat Integrated Distillation Column, Sekigu Gakkaishi, Vol. 31, 1980, pp. 81-86.

[4] Kaiser, V., Gourlia, J.P., The Ideal Column Concept : Applying Exergy to Distillation, Chemical Engineering, August 19, 1985, pp. 45-53.

[5] Tondeur, D., Kvaalen, E., Equipartition of Entropy Production : An Optimality Criterion for Transfer and Separation Processes, Ind. Eng. Chem. Res., Vol. 26, 1987, pp. 50.

[6] Rivero, R., Anaya, A., Exergy Analysis of a Distillation Tower for Crude Oil Fractionation, in "Computer Aided Energy Systems Analysis". G. Tsatsaronis et al., Editors, ASME, New-York, 1990, pp. 55-62.

[7] Le Goff, P., Rivero, R., de Oliveira Jr., S., and Cachot, T., Application of the Enthalpy-Carnot Factor Diagram to the Exergy Analysis of Distillation Processes", in "Fundamentals of Thermodynamics and Exergy Analysis", G. Tsatsaronis et al., Editors, ASME, New-York, 1990, pp. 21-28.

[8] Rivero, R., Cachot, T., Ramadane, A., Le Goff, P., Diabatic or Quasi-Reversible Distillation :

Exergy Analysis, Industrial Applications", D. Kouremenos et al., editors, Greg Foundas technical and Scientific Editions, Athens 1991, pp. 129-140.

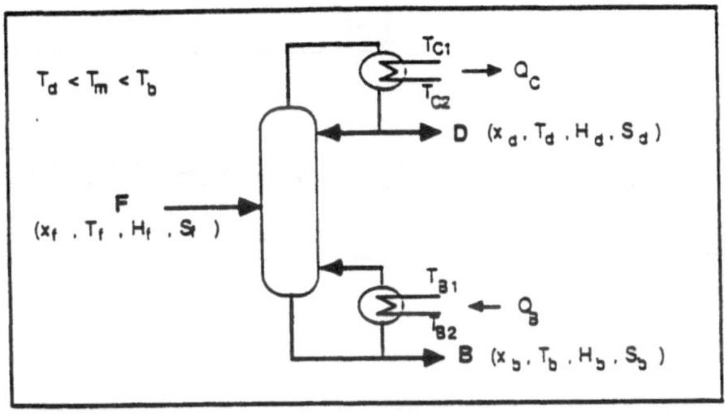

Fig. 1 : Typical distillation system for a binary mixture

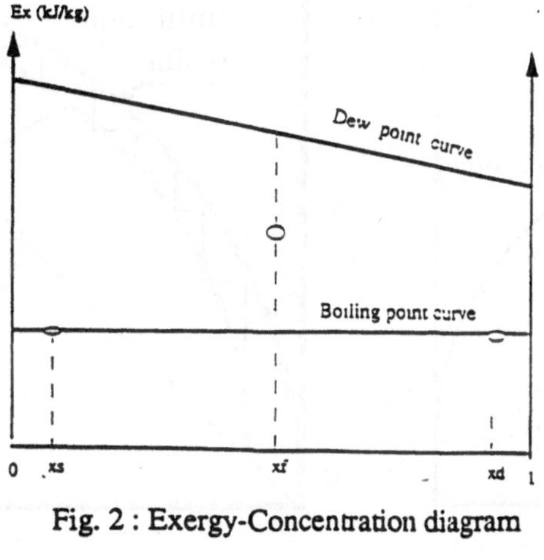

Fig. 2 : Exergy-Concentration diagram

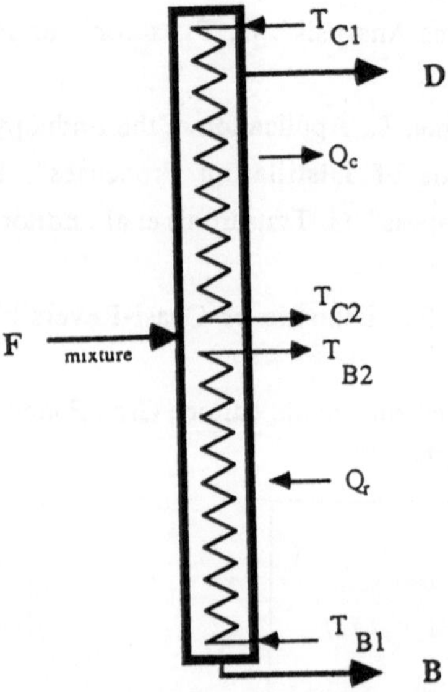

Fig. 3 : Diabatic distillation column

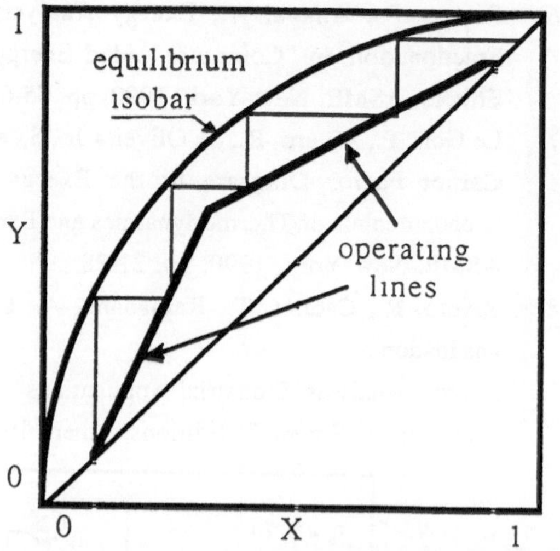

Fig. 5 : Mc CABE-THIELE diagram (adiabatic)

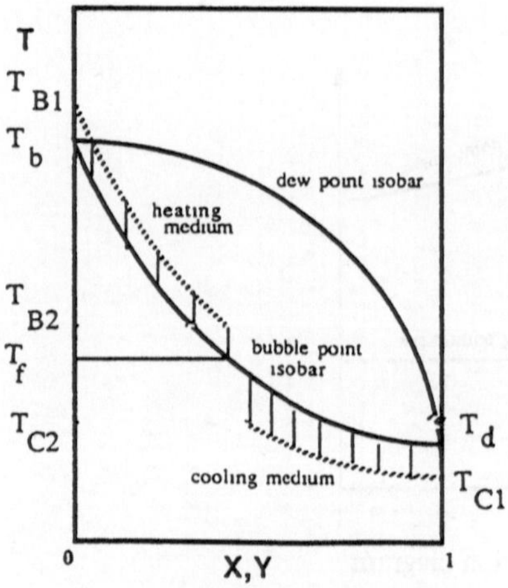

Fig. 4 : Temperature-Composition diagram for distillation

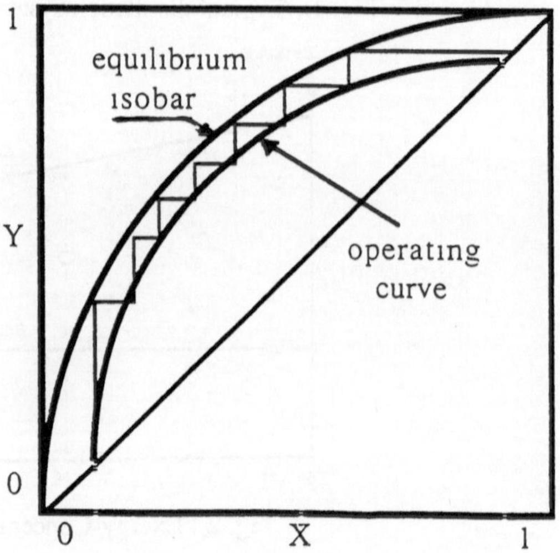

Fig. 6 : Mc Cabe-Thiele diagram (diabatic)

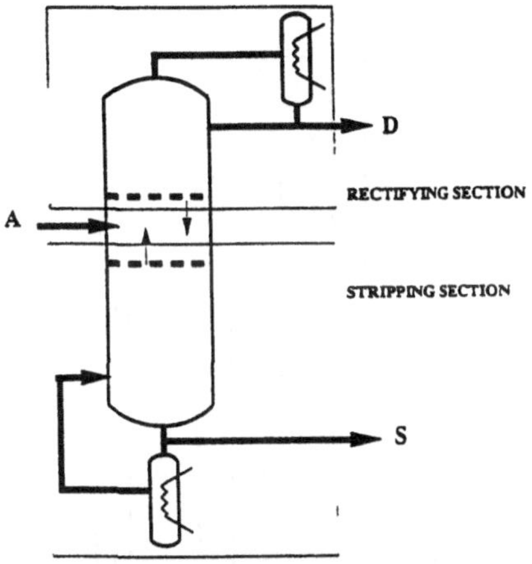

Fig. 7 : The two sections of a distillation column

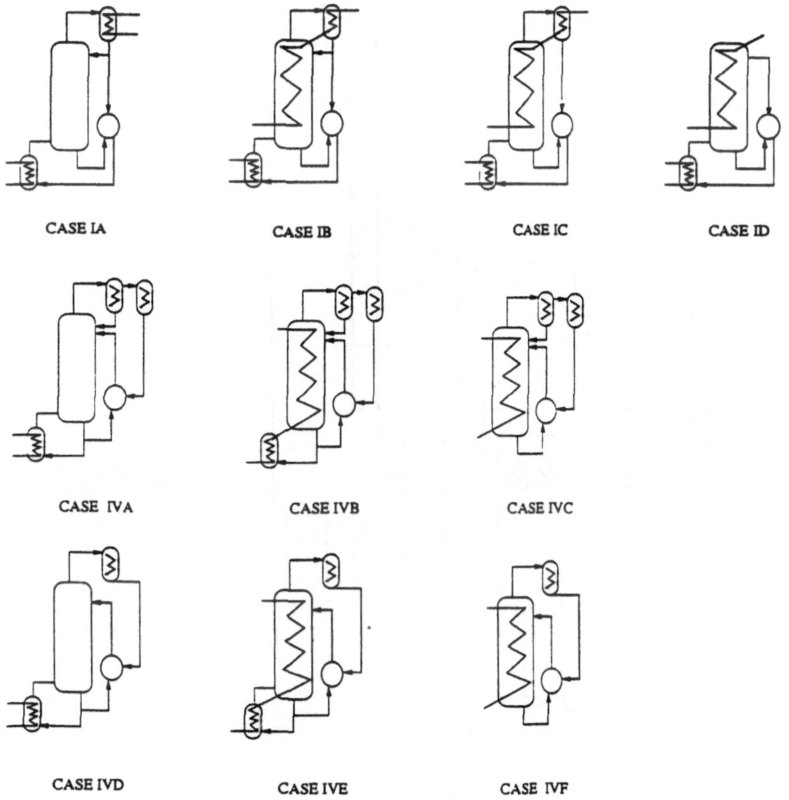

Fig. 8 : Schematic representation of the operation of rectifying and stripping columns.

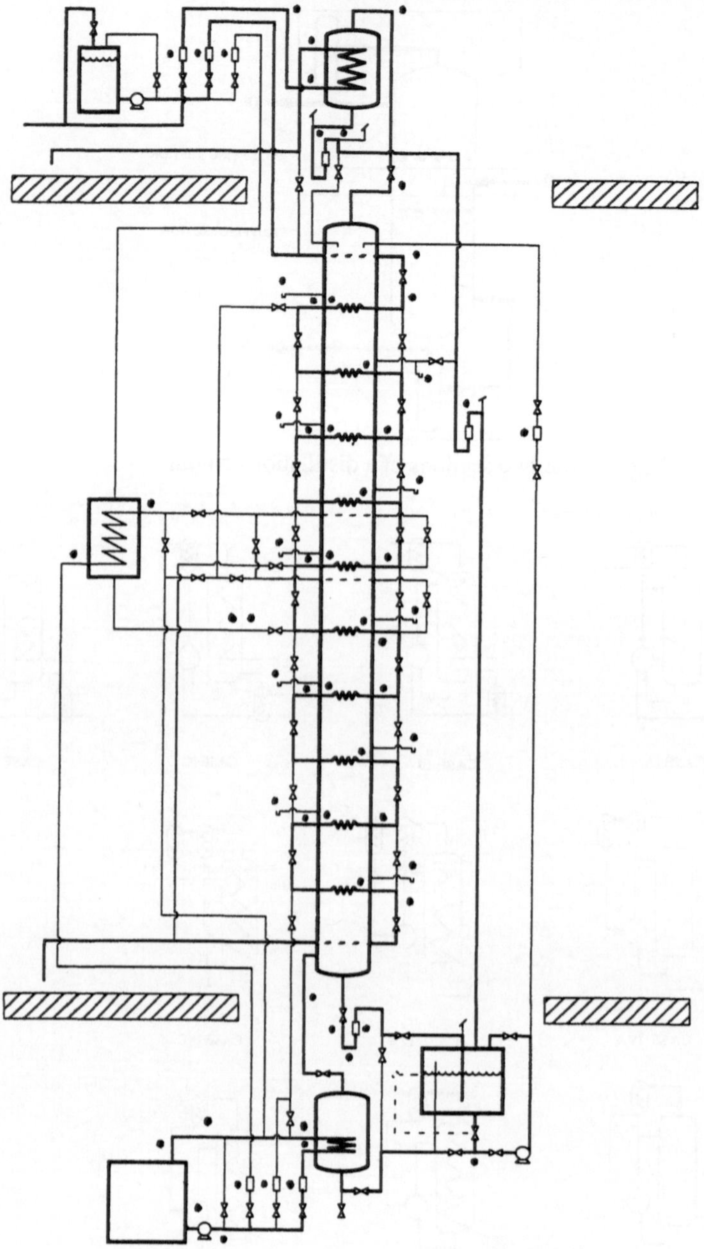

Fig. 9 : Schematic diagram of the experimental equipment showing the different points where measurements can be made.

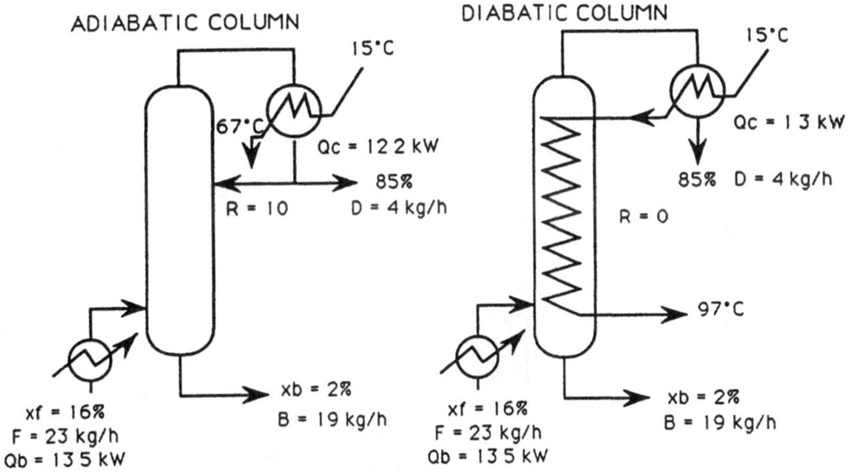

Fig. 10 : Operating conditions of the experimental adiabatic and diabatic distillation.

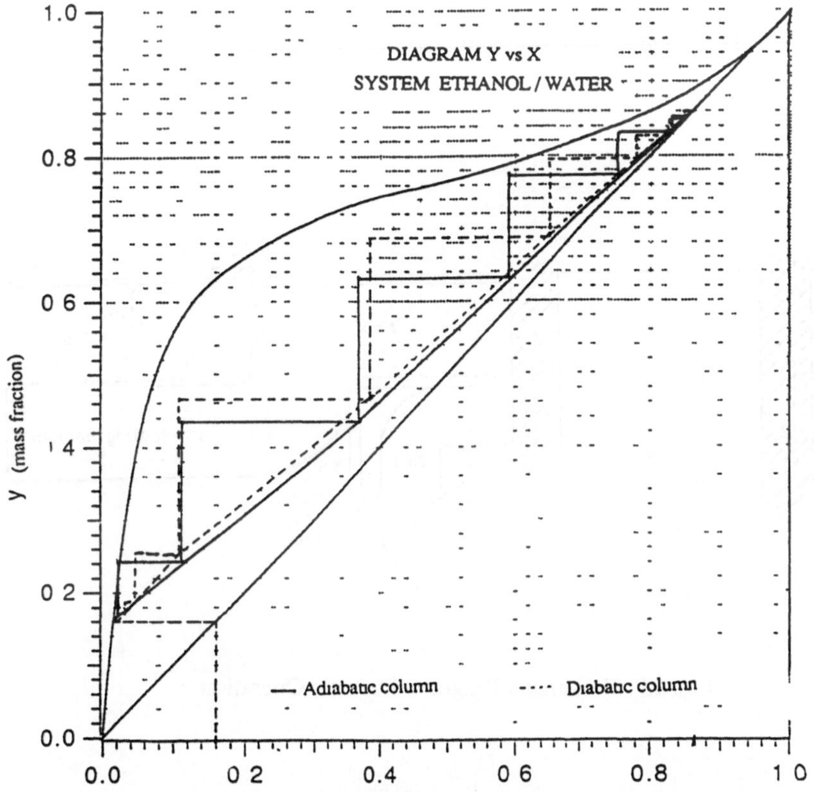

Fig. 11 : Experimental points for the Mc Cabe-Thiele construction

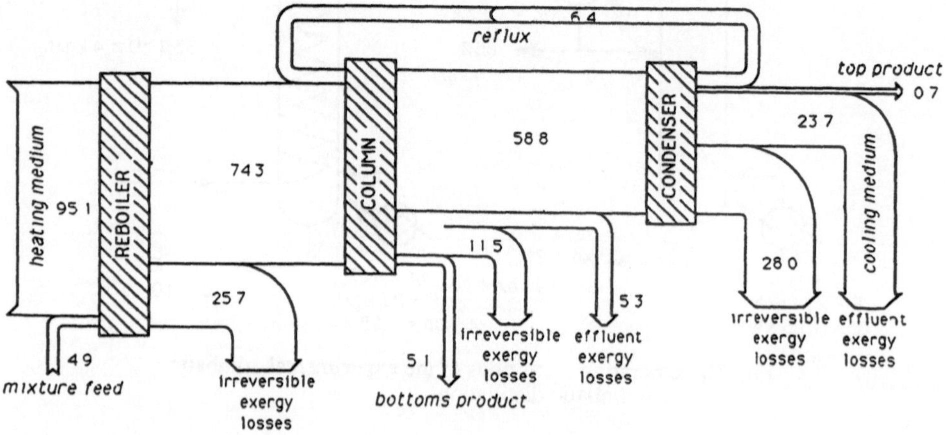

Fig. 12 : Grassmann Diagram. Adiabatic Operation

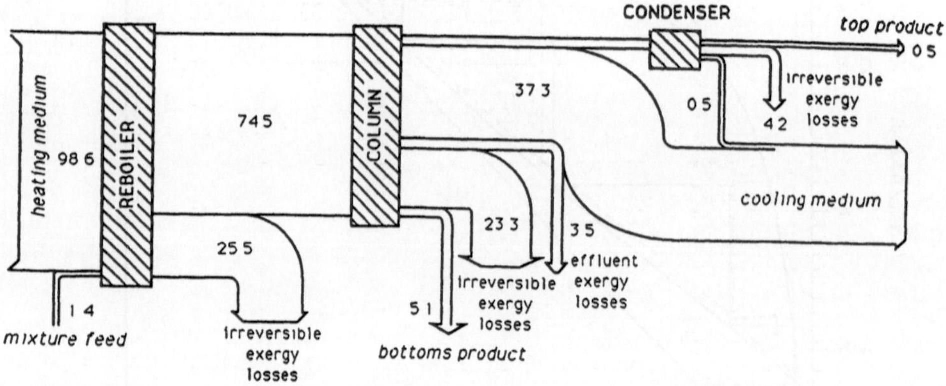

Fig. 13 : Grassmann Diagram. Diabatic Operation.

EXERGY LOSSES (kJe/h) To = 15°C	ADIABATIC COLUMN Irr heat transfer	mass transfer	Efl	Pex	DIABATIC COLUMN Irr heat transfer	mass transfer	Efl	Pex
CONDENSER	4287	-	3629	7916	622	-	-	622
1st plate	-	17,4	69,9	87,3	992,3	4,2	44,8	1041,3
2nd	-	16,0	70,0	86,0	553,5	12,0	45,0	610,5
3rd	-	114,7	70,8	185,5	249,0	35,8	45,3	330,1
4th	-	316,3	72,6	388,8	130,8	136,7	46,2	313,7
5th	-	504,7	75,5	580,2	82,3	404,9	48,3	535,5
6th	-	547,8	84,1	631,9	89,2	461,1	53,9	604,2
7th	-	173,7	91,1	264,8	44,4	146,9	57,8	249,1
8th	-	28,6	92,1	120,7	6,7	35,3	58,8	100,7
9th	-	30,0	93,2	123,2	2,2	25,6	59,5	87,3
10th	-	8,6	93,4	102,0	0,8	8,9	59,7	69,3
REBOILER	3926	-	-	3926	3749	-	-	3749
TOTAL	8213	1757	4442	14412	6522	1271	519	8312

Table I: Exergy Losses in Adiabatic and Diabatic Operation.

EXERGY AND PINCH ANALYSES OF KRAFT PULP MILL

G. LOMBARDO - F. GUILLET - E. MURATORE
(C.T.P. BP 251 - 38044 GRENOBLE CEDEX 9 - France

S. VIINIKAINEN
(V.T.T/P.L.T - P.O. Box 221 - SF-40101 JYVASKYLA - Finland)

ABSTRACT

To locate areas with inefficient use of energy in kraft pulp mill, specific methods using exergy analysis and pinch technology have been developed and utilized. A static and modular simulation model has been developed to perform exergy analysis of every area of a mill. The aim of the study is to help the French kraft mills in energy saving and in internal energy use optimization. A typical model mill has been set up. It is utilized to give recommendations to the mill for improving energy production and utilization.

INTRODUCTION

The kraft pulp mills are intensive thermal and electric energy consumers. The aim of the study is to help the mills in internal energy use optimization which leads to important purchased energy savings. This is being done by applying an exergy analysis tool to six French kraft mills.

For the analysis the French pulp and paper institute (CENTRE TECHNIQUE du PAPIER - CTP) has developed a modular "Flowsheet Simulator (1), (2), (3).

The studies carried out by CTP show that the exergetic analysis could be used to identify and locate energy degradations and to draw up a quantitative inventory of these degradations.

The modelling has been applied to the mills with the following approach :

- firstly, each individual study ,
- secondly, comparison of the results,
- finally, simulations from the "Model Mill" led to recommendations.

Moreover, in the case of one mill, pinch analysis has been done in parallel with exergy analysis to compare results from the two methodologies.

EXERGETIC ANALYSIS AND PINCH TECHNOLOGY PRINCIPLES

1. Exergetic analysis principles

Basing to the second law of thermodynamics, it could be seen that in the case of the irreversible cycle, the irreversibility produces an entropy variation ΔS with decrease ($T_o * \Delta S$) of the mechanical work compared to a reversible cycle

The exergy function of a fluid is defined by :

$$B = H - ToS \ (kJ.s^{-1})$$

with To : environment temperature (°K)
H : fluid enthalpy ($kJ.s^{-1}$)
S : fluid entropy ($kJ.°K^{-1}.s^{-1}$)

In the same way as the enthalpy balance, the exergy balance can be performed for a whole mill or for a mill department or for a specific operation. From the reference state, a specific method has been used to elaborate the necessary thermodynamic characteristics of all the substances circulating through the whole mill (4), (5).

2. Pinch Technology
Pinch technology is mainly designed for analysing the use of thermal energy in a specified process. The complete analyses for an existing process include the evaluation of theoretical minimum external energy, identification of the inefficient process parts (in terms of energy) and optimum design of the heat exchanger network including the necessary investments and lay-out etc (6).

KRAFT MILL MATHEMATICAL MODEL

- Model structure

The pulp making process can be defined as a complex network of unit operations (figure 1) where the operation of each unit depends not only on its own characteristics, but also on the interactions with the rest of the process.

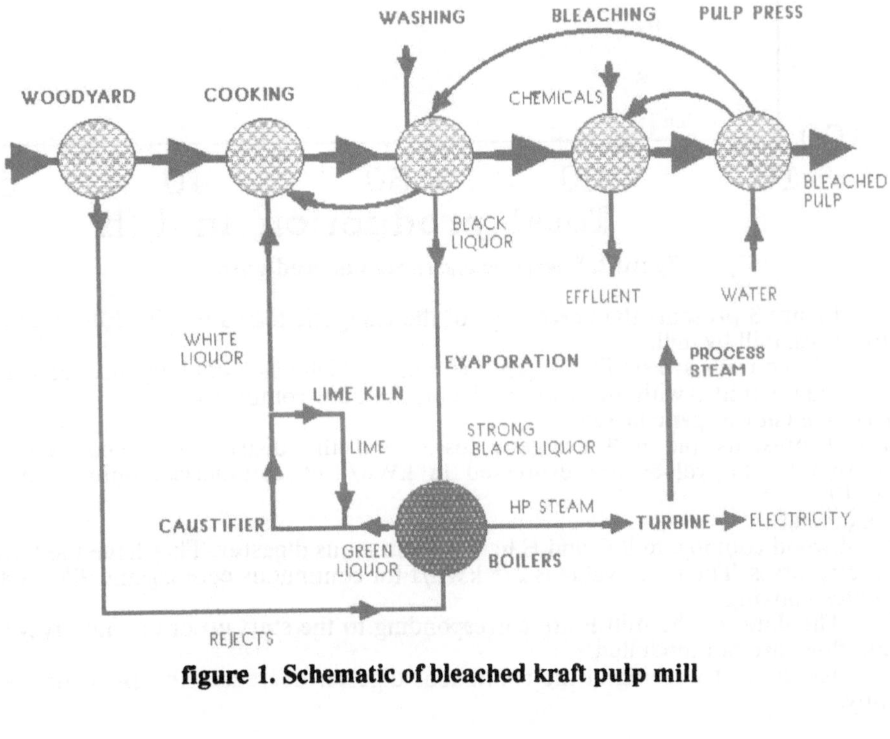

figure 1. Schematic of bleached kraft pulp mill

- Model results presentation.

The completed model presents as outputs :

. the general results of mass, enthalpy and entropy balances as components of all the branch vectors,

. exergetic yields and exergetic losses from each operation,

. thermal and exergetic losses for each outlet flow.

MASS, THERMAL AND EXERGY BALANCES DETERMINATION FOR SIX FRENCH PULP MILLS.

1. Exergetic losses by mill

Figure 2 presents the exergetic losses of the six mills, versus the production. The mills with high production have proportionally lower exergetic losses than the mills with low and medium production.

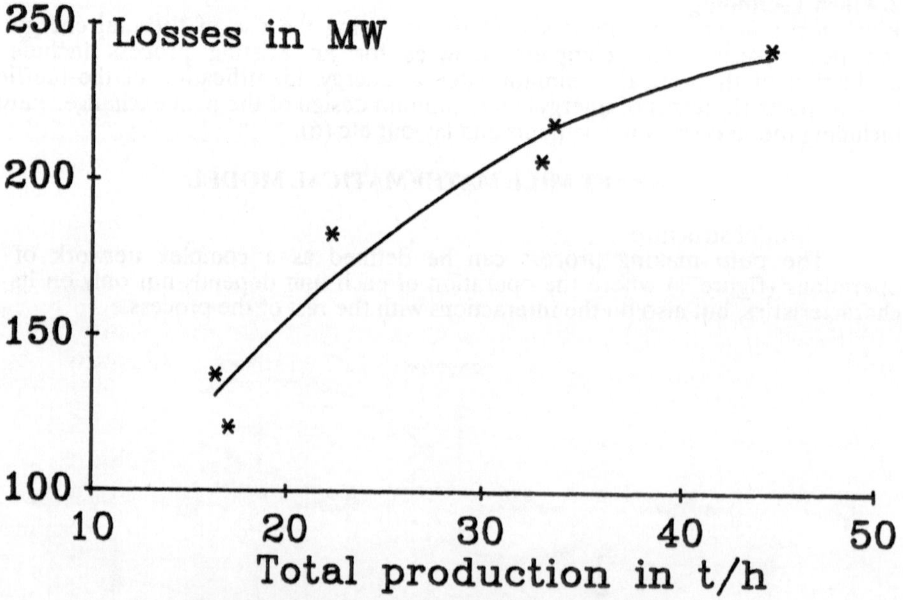

figure 2. Exergetic losses versus production

Figure 3 presents the percentage of the exergetic losses for the different parts of the mills, mill by mill.

All the mills are similar, except the mill F which has a cooking with important losses and the mill B with low losses in the black liquor combustion.

2. Department exergetic losses

Table 1 presents the mill exergetic losses and the distribution department by department. The values are expressed in kWh/t of commercial pulp or paper (A.D.T).

2.1. Cooking

For softwood cooking, mill C and E have a continuous digester. They have the lowest exergetic losses. The mean value is 275 kWh/t for continuous cooking and 370 kWh/t for batch cooking.

The data for the mill F are corresponding to the start up of the mill, recycled steam flows are not modelled.

The hi-heat washing of the continuous digester decreases the exergetic losses greatly.

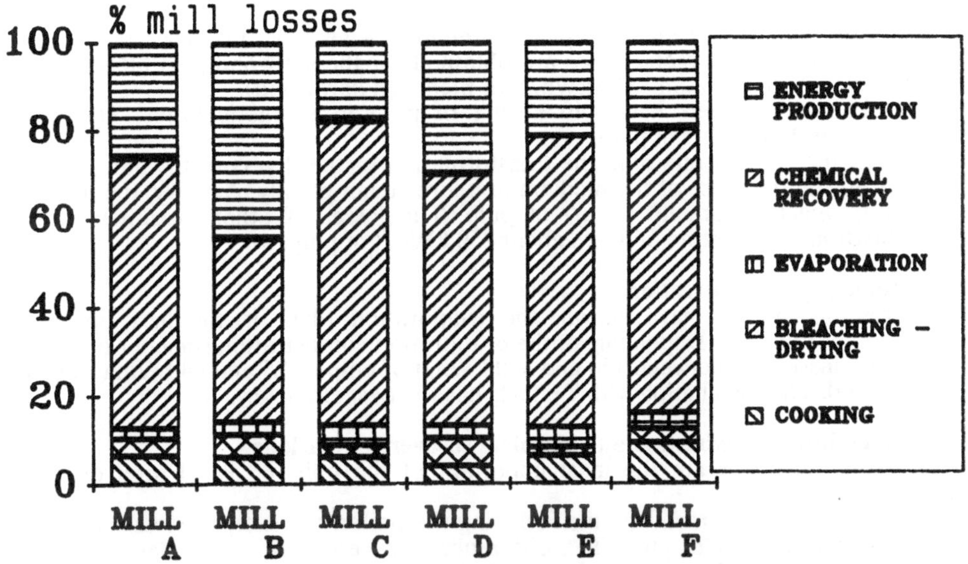

figure 3. Exergetic losses distribution by departments

	MILL A	MILL B	MILL C	MILL D	MILL E	MILL F
Pulp production (t/h)	17	16,3	33,5	22,3	44,5	32,9
Total exergetic losses						
kWh/pulp ADT	7086	8421	6521	8197	5453	6272
Exergetic losses						
Cooking						
. kWh/pulp ADT	464	510	397	338	348	588
. % total losses	6,6	6,1	6,1	4,1	6,4	9,3
Bleaching - Pulp drying						
. kWh/pulp ADT	268	435	173	515	106	192
. % total losses	3,8	5,1	2,7	6,3	1,9	3,1
Evaporation						
. kWh/pulp ADT	165	244	296	227	248	225
. % total losses	2,3	2,9	4,5	2,8	4,5	3,6
Combustion - Chemicals recovery						
. kWh/pulp ADT	4332	3491	4465	4650	3577	4014
. % total losses	61,1	41,4	68,5	56,7	65,7	64,0
Energy production						
. kWh/pulp ADT	1857	3741	1190	2467	1174	1253
. % total losses	26,2	44,5	18,1	30,1	21,5	20,0
. kWh/product ADT	1901	3053	1311	2620	1114	1334

Table 1. Exergetic losses by department (kWh/ADT

The exergetic losses are 20 kWh/t of pulp for the whole washing with hi-heat washing and from 30 to 50 kWh/t for the washing on filter drums.

The black liquor flash in a continuous cooking avoids some exergetic losses in comparison with the pulp blowing in a blow tank for the batch cooking.

2.2. Pulp bleaching and pulp drying
Four mills manufacture bleached pulp. Three mills have paper machines.

The lowest exergetic losses are obtained for the softwood pulp bleaching (70 kWh/t bleached softwood - 135 kWh/t for bleached hardwood).

The pulp dryers have 140 kWh/t of pulp in exergetic losses. The paper machine has 250 kWh/t of pulp.

In the case of one mill, the transformation of the dryers of the pulp mill has permitted the decrease of the medium pressure steam greatly. Actually, the losses of this pulp machine are 40 kWh/t of pulp. The losses of other machines without modification, in the same mill, are about 250 kWh/t of pulp.

2.3. Black liquor evaporation
As previous studies have shown, the evaporation plant is the best optimized department in a mill. Depending of the evaporation configuration (number of effects, etc...) the losses are between 170 and 300 kWh/t of pulp or 25 and 44 kWh/t of water evaporated. The installation with 7 effects has the best efficiency.

2.4. Black liquor combustion and chemical recovery operations
The exergetic losses are the most important of the mill (65 % of total exergetic losses).

For the whole recovery line including the boiler, the exergetic losses are from 3600 to 4600 kWh/t of pulp. The difference between these values depend on the black liquor dry solids content before the boiler and the pressure of the live steam.

The losses are from 2500 to 3000 kWh/t of dry solids. The boiler with the highest steam pressure (60 bar) has the lowest exergetic losses.

2.5. Energy production and utilization
The mill producing bleached pulp and paper, needs more consumption of steam. This entails greater exergetic losses, about 3000 kWh/t of paper.

The mills without paper machine have low exergetic losses.

For the mill with an efficient paper machine, the consumption of MP steam is very low. The exergetic losses are the lowest, 1100 kWh/t of paper.

SIMPLIFIED PINCH ANALYSIS FOR ONE PULP MILL

1. Initial data and consumption
Pinch analysis was carried out by VTT, Combustion and Thermal Engineering Laboratory. Initial data was taken from the simulation results of the exergy analysis. For the analysis a software called HEATNET (NEL, National Engineering Laboratory, UK) was used.

Only the process streams which are currently connected with a heat exchanger, or those outlet flows offering relativity easy heat recovery were included. Currently used secondary heat streams, other than heat exchanger connections, are kept unchanged. Also no economical evaluation concerning the heat transfer areas and respective capital costs were calculated.

2. Composite curves and pinch-point
Comparison with the current situation in steam use (37731 kW) reveals quite remarkable reduction potential. In this study dTmin value 15°C is used for further calculations.

The minimum heating requirements (alternative dTmin values in brackets) were 9910 kW (0°C), 14921 kW (10°C), 18642 kW (15°C) or 23313 kW (20°C).

In figure 4 the composite curves and in figure 5 the grand composite curve are presented.

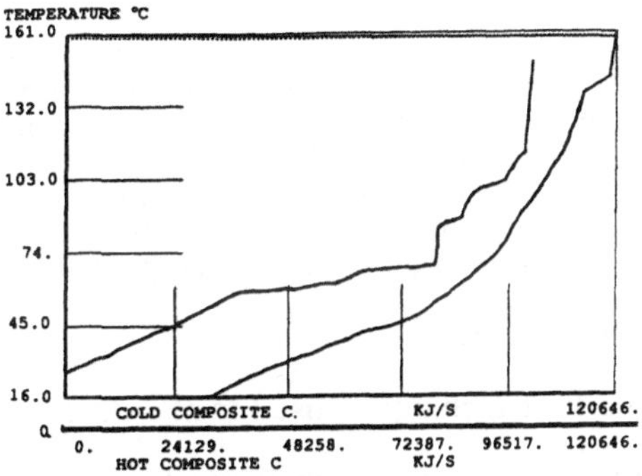

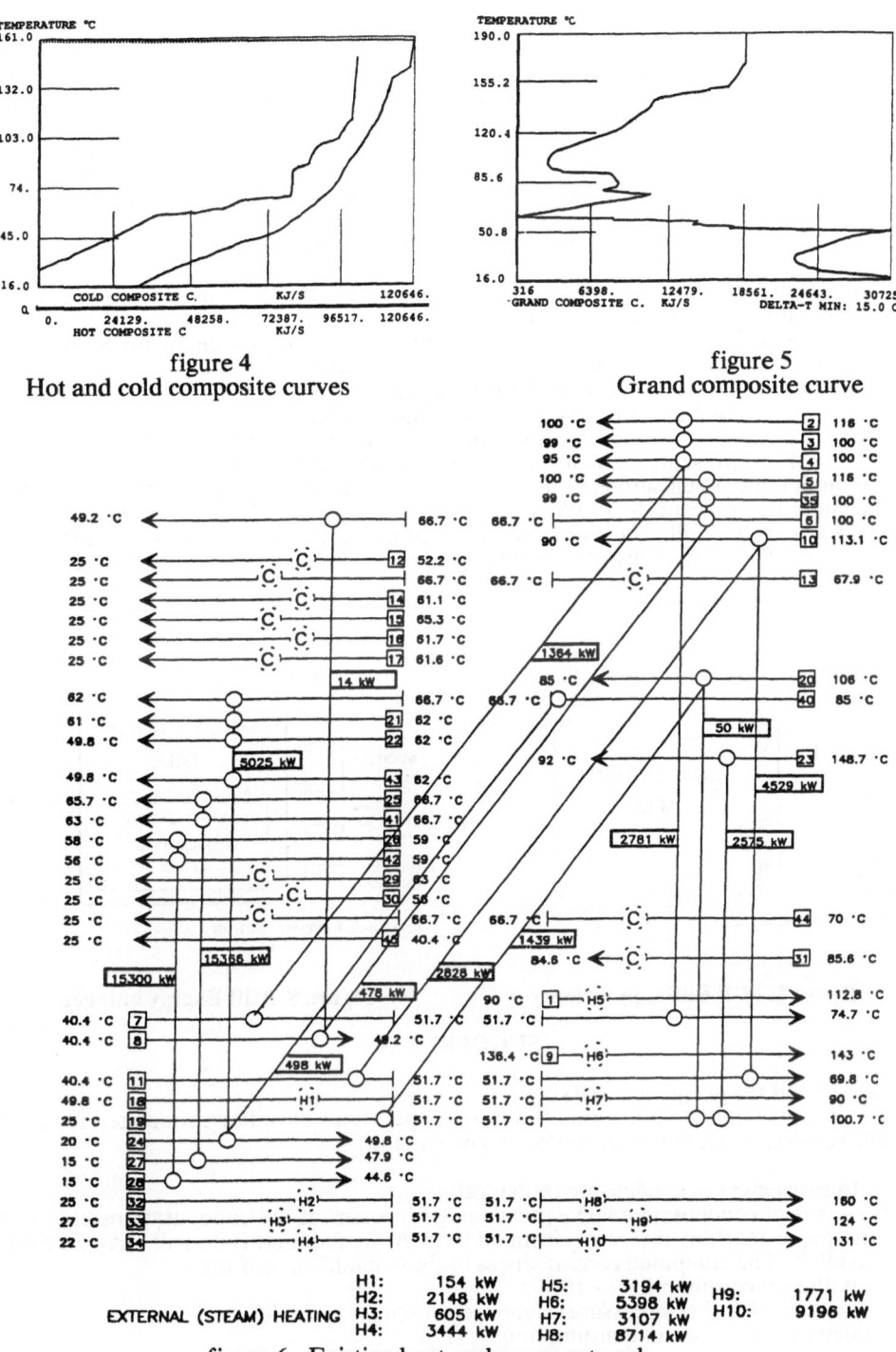

figure 4
Hot and cold composite curves

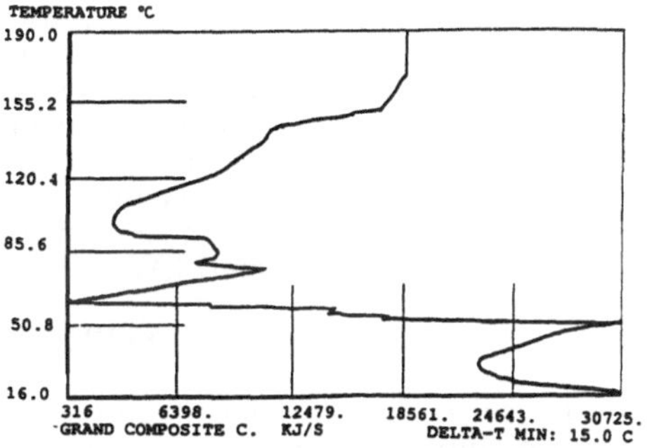

figure 5
Grand composite curve

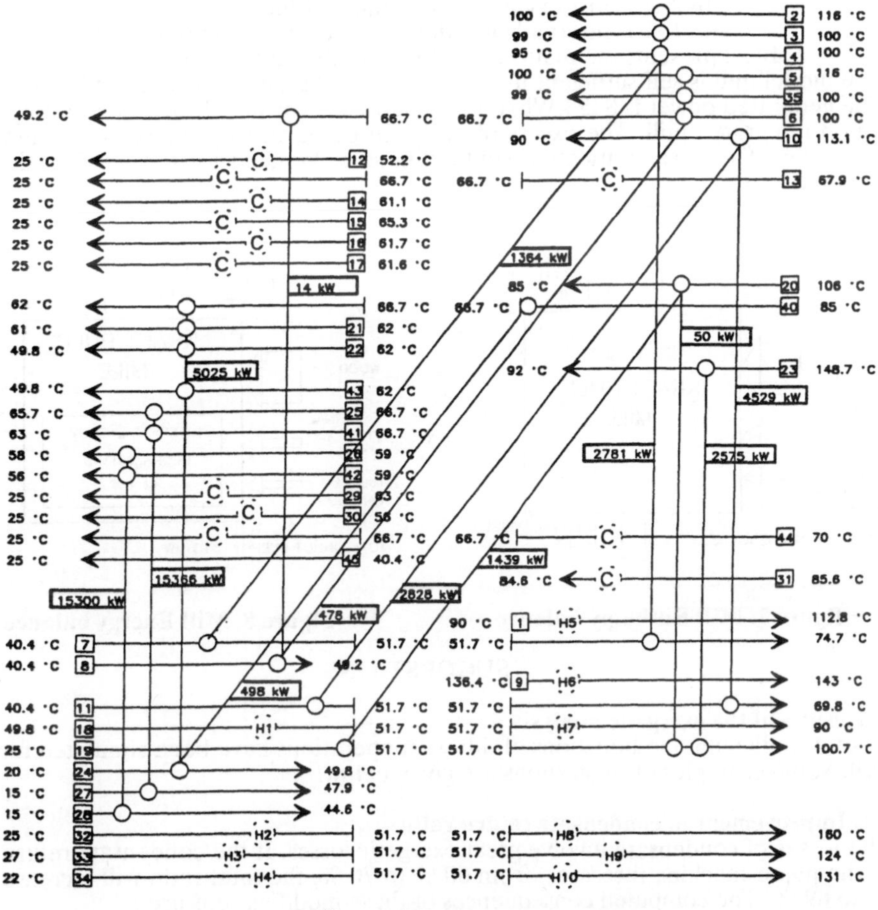

EXTERNAL (STEAM) HEATING

H1:	154 kW	H5:	3194 kW
H2:	2148 kW	H6:	5398 kW
H3:	605 kW	H7:	3107 kW
H4:	3444 kW	H8:	8714 kW
		H9:	1771 kW
		H10:	9196 kW

figure 6 - Existing heat exchanger network

3. The existing heat exchanger network
In figure 6 the grid diagram of the existing network is presented. A heat exchanger is shown with a circle in a hot stream and a circle in a cold steam connected with a line. In some cases one real heat exchanger is shown in the grid diagram using two lines above and under the pinch-point. A circle with a dashed line and "C" inside represent a fictitious cooler in order to cool the outlet steam to the ambient temperature (7).

The diagram is cut at the pinch-point. The external heat use below the pinch-point is 6351 kW. Also 6607 kW of heat is being transferred over the pinch-point.

TYPICAL MODEL MILL

A "typical model mill" has been based on the best techniques used in the different French pulp mills found in the exergetic analysis.

Figure 7 presents the whole mill enthalpy balance which figure 8 gives the corresponding exergy balance with regard to exergy input :

- 38.7 % is found with the produced,
- 18.2 % is lost in gaseous and liquid effluents,
- 43.1 % is internally lost due to the degradation of energy.

Table 2 presents the main exergy losses in the mill operations.
The model mill is self sufficient in energy. No purchased fuel is needed for the boiler. The generated power is 825 kWh/t, this is sufficient for the electricity consumption of a non integrated mill. The excess of bark can be burnt to produce steam and more electricity. The bark gasification could eliminate the need for the purchased fuel in the lime kiln.

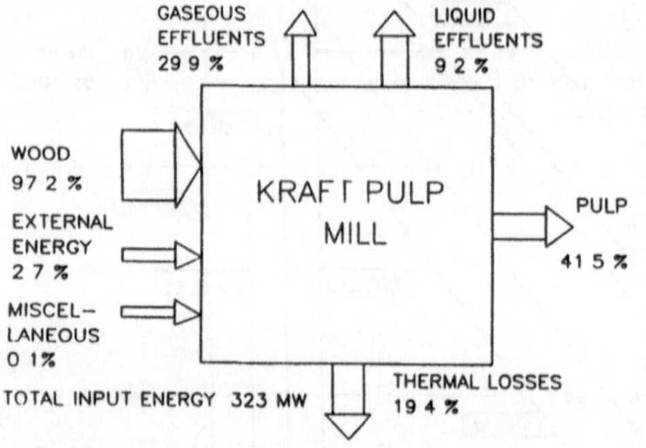

figure 7. Mill Enthalpy balance

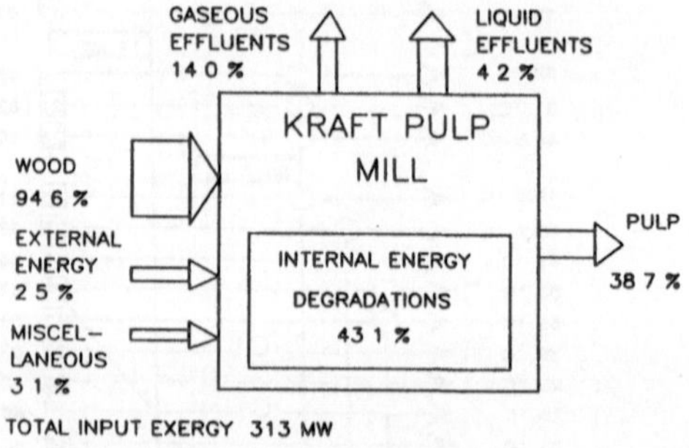

Figure 8. Mill Exergy balance

SUGGESTIONS

1. Results of the exergetic analysis
A lot of individual modifications of different operations have been simulated for each mill. Some examples of suggestions are given below.

1.1. Improvement in condensate return ratio
The losses of condensate involve great exergetic losses. If the condensate return ratio of the paper machine increases from 60 % to 70 %, the total ratio will increase from 63 to 69 %. The computed consequences of these modifications are :

- boiler oil consumption : - 10 %
- degazing BP (back pressure) steam consumption : - 25 %
- demineralized water consumption : - 17 %

OPERATION	EXERGETIC LOSSES in kW	% OF TOTAL LOSS
Recovery boiler	80500	45,7
Digester	6400	3,6
Lime kiln	3500	2,0
Dissolving tank	2800	1,6
Air heater	2700	1,5
Evaporator	2500	1,4
Caustifier	2200	1,3
Bark boiler	2200	1,3
Pulp machine	2100	1,2

Table 2. Main exergy losses in mill operations

1.2. New design of process
In one mill, the pulp washing operations are done using two different processes with recycling and mixing the two effluents. If the design is changed using the same process for the two washing operations, the efficiency increases and the dry solids content of the black liquor going to the pre-evaporation increases from 14,5 to 17,5 %. The computed consequence is BP steam consumption decreasing by 25 %. The gas consumption to the bark boiler is reduced to zero.

1.3. Simulation of a mill partly optimized
This simulation includes :

- suppression of water stream on the foam towers
- increase on the condensate return ratio (6 %)
- reduction of BP and MP steam pressure (1 bar for MP and 0,5 for BP)
- a new turbine.

The main results of computation of this simulation are presented in table 3.

OPERATIONS	VARIATION /REFERENCE CASE
Bark boiler oil consumption	+ 38 %
HP steam production	+ 3 %
Cogeneration power	+ 45 %
BP steam consumption	- 5 %
Purchased electricity	- 25 %

Table 3. Mill partly optimized

2. Results of the pinch analysis
Example suggestions concentrate on the cooking and bleaching. Due to relatively complicated modifications in the air ducts etc, the example suggestions do not include the pulp dryer area. The same goes for the recovery boiler and bark boiler combustion air preheating.

The following suggestions could be considered more precisely (existing and new heat exchangers) :

a) the heat from two bleaching effluents for mild water is 2505 kW and 1134 kW.

b) the heat from excess hot water to make-up feed water is 3840 kW.

c) the heat from the hot black liquor to white liquor preheating hot water production and mild water is 1133 kW, 3261 kW and 2957 kW.

d) the heat from the degazing to mild water is 3253 kW.

e) the heat from dissolving tank vent to make-up feed water is 1678 kW.

Altogether these suggestions reduce the external steam us by 9758 kW and pinch-point crossing by 4038 kW. External cooling over the pinch is reduced to the dissolving tank vent (after the heat recovery for make-up feed water).

CONCLUSIONS

The mean distribution between the exergetic losses of the different parts of the mill is as follows :

- cooking and washing	6 %
- bleaching and pulp dryer or paper machine	4 %
- black liquor evaporation	3 %
- chemical recovery operation	60 %
- energy production and utilization	27 %

For the cooking department, the use of a continuous digester decreases the exergetic losses. Some modifications on the paper machine can be realized to increase the exergetic yields.

The black liquor combustion is the area with the most important exergetic losses.

The differences between the mills allow the selection of the most energetic efficient installations. These selected installations were used to set up a typical model mill. A "typical model mill" has been set up and the software of this model is available for mill people training and for further studies to check consequences of mill modifications.

Exergy analysis determine the different points of energetic losses for all the mill operation. The pinch technology allows the determination of the energetic losses and to involve some proposal to modify some operations, but it can be applied only on the heat exchanger network of the mill. Exergy and pinch analyses performed simultaneously in one mill have been demonstrated to be complementary.

ACKNOWLEDGEMENTS

The authors are grateful to ADEME (Agence de l'Environnement et de la Maitrise de l'Energie - French Environment and Energy Agency -) for the financial support of this study. They thank also mill personnel for participation in data collection and results analysis.

REFERENCES

(1) Muratore E, Weisbuch H, Monzie P, Ramaz A, ATIP, vol 27, n°2, 1973
(2) Muratore E, Monzie P, ATIP, vol 28, n°1, 1974
(3) Muratore E, Gelus M, ATIP, vol 29, n°5, 1975
(4) Paccot C, INPG thesis, Grenoble, 02/02/1987
(5) Paccot C, Muratore E, Estebe J, ATIP, vol 43, n°1, 1989
(6) Linhoff B, Hindmarsh E. The pinch design method for heat exchanger networks. Chemical Engineering Science, vol 38, n°5, 1983
(7) Persson L, et al. Demonstration av pinchtehniken vid Varo bruk (in swedish) CIT 1990:2

INDEX OF CONTRIBUTORS